W9-DFJ-554

LINEAR OPTIMIZATION

W. Allen Spivey

Professor of Statistics

Graduate School of Business Administration
The University of Michigan

Robert M. Thrall

Professor and Chairman

Department of Mathematical Sciences
Rice University

HOLT, RINEHART AND WINSTON, INC.

New York Chicago San Francisco Atlanta Dallas
Montreal Toronto London Sydney

Library of Congress Catalog Card Number 70-125474

ISBN 0-03-084173-9

Printed in the United States of America

23456 038 98765432

Preface

Our major purpose in writing this book has been to combine a constructive development of the simplex algorithm with a presentation which is both mathematically sound and intended for a reader who is not primarily a mathematician. Our pedagogical philosophy is that mastery of mathematical programming is most rapidly achieved for most students by a multi-layered approach in which a topic may be first explored via a special case and then attacked in more and more generality until finally the research frontier is reached. This approach has in its favor the feature that a student who stops short of the most general theory still may have achieved a usable introductory understanding of many important topics.

In line with this philosophy we have treated the simplex algorithm at three levels. The early chapters contain as the first layer some geometrical background for linear programming and a variety of applications of linear programming to real world problems, and as the second layer a constructive development of the algorithm itself with many illustrations. Our third layer is a development of the schema approach of Tucker, Gomory, and Balinski. This is postponed to later chapters because we believe that despite its great mathematical elegance it is not as important to the attainment of our primary goal as is the material incorporated into the earlier chapters.

Our pedagogical position is illustrated again in our treatment of the assignment and transportation problems in Chapters 6 and 7 where only the lower layers are treated in this book. In the same amount of space or even less we could have presented a theory of network flows leading to the

Dennis-Minty-Fulkerson out-of-kilter algorithm. However, we believe that in a first course in linear optimization one should develop the assignment algorithm on its own and move on to the somewhat more complex Graves-Thrall algorithm for the capacitated transportation problem. This exposes the student to these important classes of linear programming problems and to special purpose algorithms for solving them; it also gives the student a background with which he can readily read and comprehend one of the several excellent treatments of network flows currently available. As for the Graves-Thrall algorithm itself, we believe that time may well show that there is a place for an efficient primal algorithm for capacitated transportation problems as a special purpose supplement to the more general out-of-kilter algorithm.

The book is concerned only with concepts in linear optimization and does not include topics in integer and nonlinear programming.

We have attempted to free the simplex algorithm from some of the notations and other features which seem to have lingered on, without any current justification, from the early work in linear programming. For example, we have emphasized the adjusted costs, using the symbol c_j^*, instead of the traditional symbolism $c_j - z_j$, although we have gone beyond most texts by providing an economic justification for the numbers z_j as well as an algebraic definition.

Our discussion of duality and postoptimality analysis in Chapter 5 provides another instance of our multilayer approach in that we first develop, illustrate, and apply duality for the symmetric case and then go on to the general case. The chapter may well be the most important part of the book for the reader who is interested primarily in the applications of linear programming.

A word of explanation may be needed for our extensive use of the concept of sequence of vectors. Traditionally there has been some confusion in books on both linear algebra and linear programming in the use of the phrase *set of vectors* when an ordered set (or sequence) of vectors is under discussion. This is more than a semantic quibble when one wishes one's exposition to be consistent with the needs of computer implementation, since a computer requires a precise specification of position. We have therefore placed much emphasis on developing a notation for enumerating components of vectors and identifying corresponding columns of matrices. This frequently takes the form of associating a basic sequence with each iteration of the simplex algorithm. We also believe that the explicit use of basic sequence enables us to give a more easily comprehended discussion of important features of special topics such as decomposition and upper bounding.

The reader whose college mathematics was limited to calculus will find in Appendixes A and B an exposition of those parts of finite mathematics which are assumed as background in the main text. Moreover, readers whose mathematics courses went beyond calculus may find the algorithmic flavor of Appendix B a useful supplement to an abstract course in linear algebra and the development of many topics in Appendix A is considerably deeper than in the conventional current underclass finite mathematics courses. Since our development involves a considerable amount of mathematical notation, we have included symbols in the index in alphabetical position when possible, and otherwise under the listing "Symbols."

The general structure of the book is as follows:

Chapters 1 and 2: Geometrical introduction and discussion of the modelling aspects of linear optimization.
Chapters 3, 4, 5, and 8: Development of the general theory.
Chapters 6, 7, 9, 10, and Appendix C: Development of special structures and applications.
Appendixes A and B: Background material on foundations and linear algebra.

The concluding sections of Chapters 3 and 4 and all of Chapter 8 contain the mathematically more advanced portions of the theory of linear programming. Except for Section 9.5, the chapter on game theory does not depend on the more advanced mathematical theory. We indicate two possible selections of material from this text for one semester courses.

A suggested sequence for the student with applied interests is: Chapters 1 and 2; Chapter 3 (omitting Section 3.10); Chapters 4 (omitting Section 4.4), 5, 6, 7, and 9 (omitting Section 9.5); Sections from the appendices.

For students with mathematical interests, the following sequence is appropriate: Chapter 3; Chapters 4, 5, and 8; Section 5 of Chapter 6 and Chapter 7; Chapters 9 and 10.

April 1970

W. A. S.
R. M. T.

Contents

LINEAR OPTIMIZATION

Basic Concepts in Linear Programming

1

1.1. A SIMPLE LINEAR PROGRAMMING PROBLEM.

We introduce the concept of linear programming by considering a miniature diet problem. Suppose in shopping for the family food supply the only foods available are potatoes and steak. The decision as to how much of each to buy is to be made entirely on the basis of dietary and economic considerations; tastiness is excluded as a decision factor. In this simplified problem the only nutrients considered are carbohydrates, vitamins, and proteins. We shall assume that dietitians have studied nutritional needs and have concluded that the minimum daily requirements for an acceptable diet are:

$$\begin{array}{c} \text{8 units of carbohydrates,} \\ \text{19 units of vitamins,} \\ \text{7 units of proteins.} \end{array} \tag{1.1}$$

For the purpose of our discussion, we shall assume that laboratory analyses show that the numbers of units of each nutrient contained in a unit of each food are as given in Table 1.1.

We define a *diet* to consist of a choice of x units of potatoes and y units of steak. A diet that satisfies the minimum requirements must have

$$\begin{aligned} 3x + y &\geq 8, \\ 4x + 3y &\geq 19, \\ x + 3y &\geq 7, \\ x &\geq 0, \\ y &\geq 0. \end{aligned} \tag{1.2}$$

The last two inequalities insure that only nonnegative amounts of steak and potatoes are admissible.

Table 1.1. DIETARY DATA.

	Per unit of potatoes	*Per unit of steak*	*Minimum requirements*
Units of carbohydrates	3	1	8
Units of vitamins	4	3	19
Units of proteins	1	3	7
Unit cost	25	50	

We can simplify the consideration of the problem by introducing some notation. Let

$$X = \begin{bmatrix} x \\ y \end{bmatrix}$$

denote a *diet vector*, and let

$$B = \begin{bmatrix} 8 \\ 19 \\ 7 \end{bmatrix}$$

be called a *requirements* vector. Any vector X which satisfies (1.2) is termed a *feasible diet* or a *feasible vector* or a *feasible program*. For example, if $X = \begin{bmatrix} 4 \\ 5 \end{bmatrix}$, the reader can verify that $x = 4$ and $y = 5$ satisfies the system (1.2) and so X is a feasible vector. Similarly, one can show that $\begin{bmatrix} 1 \\ 1 \end{bmatrix}$ and $\begin{bmatrix} 11 \\ -1 \end{bmatrix}$ are not feasible vectors. In this problem it is possible to characterize, in a rather simple fashion, all the feasible vectors. The inequalities

$$\begin{aligned} 3x + y &\geq 8, \\ 4x + 3y &\geq 19, \\ x + 3y &\geq 7, \end{aligned}$$

represent regions in the x, y plane lying, respectively, on or above the lines

$$\begin{aligned} 3x + y &= 8, \\ 4x + 3y &= 19, \\ x + 3y &= 7. \end{aligned}$$

The inequalities $x \geq 0$, $y \geq 0$ represent the first quadrant. Thus all the feasible vectors lie in the shaded region of the graph in Figure 1.1. The points

$$\begin{bmatrix} 7 \\ 0 \end{bmatrix}, \begin{bmatrix} 4 \\ 1 \end{bmatrix}, \begin{bmatrix} 1 \\ 5 \end{bmatrix}, \begin{bmatrix} 0 \\ 8 \end{bmatrix},$$

which are clearly seen to be corner points, are called *vertices* or *extreme points*. Similar points occur in linear programming problems in higher-dimensional spaces. We use the concept of a vertex and other related

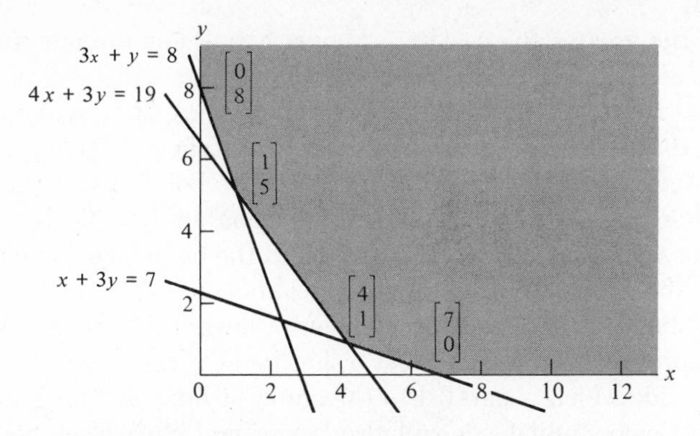

Figure 1.1 The feasibility region.

geometrical concepts intuitively for illustrative purposes only, in the early part of this book. A precise treatment relating these geometrical concepts to our general algebraic development of the simplex algorithm for solving linear programming problems is given in Section 3.10.

We suppose that the unit cost of potatoes is 25 and the unit cost of steak is 50 and that we cannot get price discounts for buying larger quantities. Clearly, then, the cost of a particular diet will be $z = 25x + 50y$. We introduce a cost vector

$$C = \begin{bmatrix} 25 \\ 50 \end{bmatrix};$$

the decision problem is to find, from among all the feasible diets, the one (or ones) which minimize the cost. A diet which minimizes z is called an *optimal feasible diet*.

The cost equation $z = 25x + 50y$ defines a family of parallel lines called *equicost lines*. All diets on any one line have the same cost; the higher

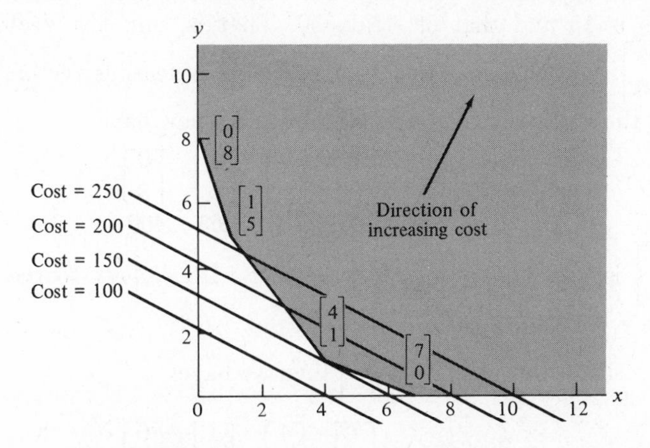

Figure 1.2 Equicost lines.

the line, the greater the cost (see Figure 1.2). Thus to determine optimal feasible vectors we look for those feasible vectors which lie on the lowest equicost line—one which intersects or touches the feasibility region. Suppose that X^* is an optimal feasible diet. Then consider the (unique) equicost line through X^*. Unless X^* is on the boundary of the feasibility region, there are lower equicost lines which contain feasible diets and hence X^* cannot be optimal. Thus X^* must be on the boundary. Now if X^* is not a vertex, the equicost line through X^* cannot cross the bounding side on which X^* lies lest there be a feasible diet of lower cost. Hence, we conclude that an optimal feasible diet must either be at a vertex or be a point on a bounding side which is parallel to the equicost lines. In this latter case, all points on this bounding side will also be optimal. Since each bounding side has at least one vertex, we conclude that *if there are any optimal feasible vectors at all there will be a vertex which is optimal*. This means that in practice we need only examine costs at vertices of the feasibility region. This consideration reduces the problem to finite terms and is of basic importance in the theory of linear programming.

In our problem Figure 1.2 indicates $\begin{bmatrix} 4 \\ 1 \end{bmatrix}$ to be an optimal diet. We can, however, arrive at this vector without recourse to graphs (the significance of nongraphical solution procedures becomes evident when we consider the difficulties entailed in drawing pictures in spaces of dimension three or higher). One way to proceed in this problem is to compute the costs at each of the vertices and then choose the vertex with the minimal cost. Thus we have:

$$\begin{array}{lcccc} \text{Vertex} & \begin{bmatrix} 7 \\ 0 \end{bmatrix} & \begin{bmatrix} 4 \\ 1 \end{bmatrix} & \begin{bmatrix} 1 \\ 5 \end{bmatrix} & \begin{bmatrix} 0 \\ 8 \end{bmatrix} \\ \text{Cost} & 175 & 150 & 275 & 400. \end{array} \tag{1.3}$$

Next, suppose the cost vector is changed so that the unit cost of potatoes is 15 and that of steak, 50. That is, our cost vector is now $C = \begin{bmatrix} 15 \\ 50 \end{bmatrix}$. This change does not affect the set of feasible vectors; only the costs at the vertices must be recalculated. We now have

$$\begin{array}{lcccc} \text{Vertex} & \begin{bmatrix} 7 \\ 0 \end{bmatrix} & \begin{bmatrix} 4 \\ 1 \end{bmatrix} & \begin{bmatrix} 1 \\ 5 \end{bmatrix} & \begin{bmatrix} 0 \\ 8 \end{bmatrix} \\ \text{New cost} & 105 & 110 & 265 & 400 \end{array} \tag{1.4}$$

and $\begin{bmatrix} 7 \\ 0 \end{bmatrix}$ is now an optimal feasible diet with respect to the new cost function $z = 15x + 30y$.

If, instead, we take $C = \begin{bmatrix} 15 \\ 45 \end{bmatrix}$ then we have

$$\begin{array}{lcccc} \text{Vertex} & \begin{bmatrix} 7 \\ 0 \end{bmatrix} & \begin{bmatrix} 4 \\ 1 \end{bmatrix} & \begin{bmatrix} 1 \\ 5 \end{bmatrix} & \begin{bmatrix} 0 \\ 8 \end{bmatrix} \\ \text{New cost} & 105 & 105 & 240 & 360. \end{array} \tag{1.5}$$

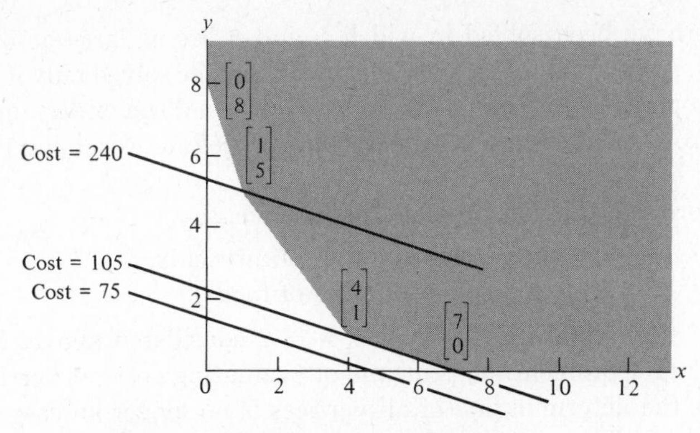

Figure 1.3 Multiple optimal programs.

There are now two optimal vertices, $\begin{bmatrix} 4 \\ 1 \end{bmatrix}$ and $\begin{bmatrix} 7 \\ 0 \end{bmatrix}$ (see Figure 1.3). Since the segment joining vertices is parallel to the equicost lines, every diet $\begin{bmatrix} x \\ y \end{bmatrix}$ on it is also optimal. Here the equicost lines and the segment all have slope $-1/3$.

We can now define a more general linear programming problem. A generalization of (1.2) is the system

$$
\begin{aligned}
a_{11}x_1 + a_{12}x_2 + \cdots + a_{1n}x_n &\geq b_1, \\
\vdots \qquad\qquad\qquad & \quad \vdots \\
a_{p1}x_1 + a_{p2}x_2 + \cdots + a_{pn}x_n &\geq b_p, \\
x_j \geq 0, \quad (j = 1, \cdots, n)&.
\end{aligned}
\tag{1.6}
$$

Given the pn coefficients a_{ij} and the p by 1 requirements vector

$$
B = \begin{bmatrix} b_1 \\ \cdot \\ \cdot \\ \cdot \\ b_p \end{bmatrix},
$$

the problem is to find an n by 1 program vector X satisfying (1.6) and such that

$$
z = c_1x_1 + c_2x_2 + \cdots + c_nx_n \text{ is a minimum,} \tag{1.7}
$$

where the c_j are also given. A program vector X which satisfies (1.6) is called *feasible* and if it also *satisfies* (1.7) it is called *optimal feasible.* The conditions or constraints (1.6) are called *feasibility conditions.*

Problems in which the number p of constraints is as large as 1,000 to 2,000 and the number of variables n as large as 3,000 to 4,000 are fairly common in applications today and can be solved on a large-scale computer.

Problems have been solved in which p and n are as large as 30,000 and 3,000,000, respectively. Problems of this size can be solved only if they have special properties which enable them to be broken down or decomposed into smaller problems that can be solved (some procedures for doing this will be presented in Chapter 10).

A more realistic diet problem might have, say,

$$p = 9 = \text{number of nutrients},$$
$$n = 77 = \text{number of foods},$$

instead of the miniature case $p = 3$, $n = 2$ considered above. In solving this or a larger problem the principle of evaluating costs at vertices is still valid, but the determination of all vertices is no longer an easy or even a practical task. The number of potential vertices in this diet problem, for example, is of the same order as the number of possible bridge hands! For still larger problems the number of vertices is finite but can be extremely large. Practical methods for solving such problems exist, however, and will be developed in later chapters.

1.2. EXISTENCE OF OPTIMAL PROGRAMS.

Not every linear programming problem has an optimal program vector. For example, suppose that a fourth substance is found in potatoes and steak in the amounts of 3 and 4 units per unit, respectively, and suppose further that no more than 12 units of this fourth substance can be tolerated in the diet. Thus, we have the additional condition

$$3x + 4y \leq 12$$

or

$$-3x - 4y \geq -12.$$

The region defined by this inequality has no vectors or diets in common with the feasible region for the original problem. Hence there are no

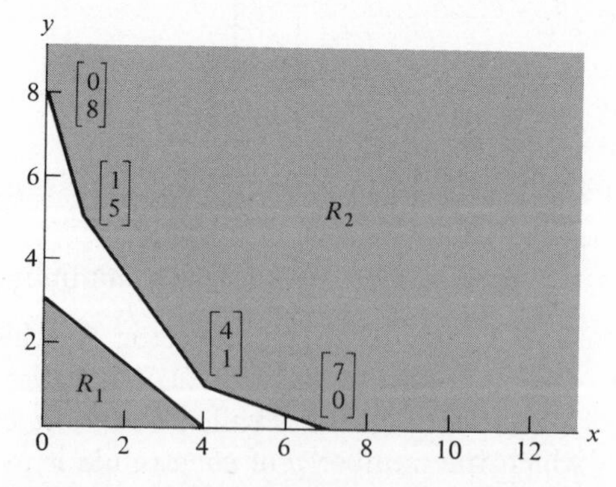

Figure 1.4 Infeasible problem; $R_1 \cap R_2$ is empty.

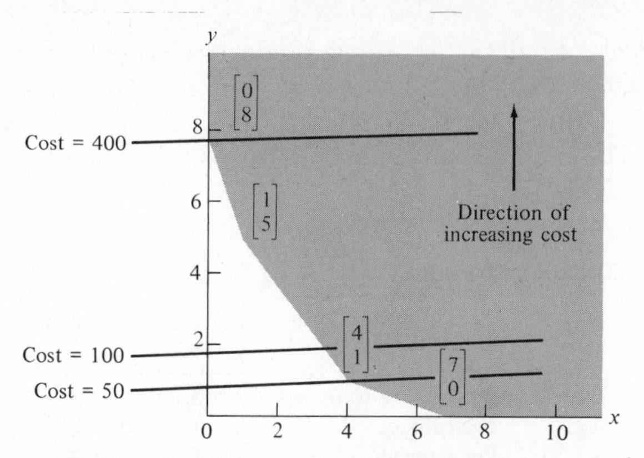

Figure 1.5 Feasible programs but no optimal program.

feasible vectors for the modified problem, and, of course, there is no optimal vector (see Figure 1.4).

For another example with no optimal program, suppose that for each unit of potatoes which a customer takes, the shopkeeper gives him 1 unit of money. The cost function is then $-x + 50y$. There is now no minimum cost in the region of feasibility and, hence, no optimal vector (see Figure 1.5). The lack of a minimum for a feasible problem is possible only when, as here, the feasibility region is unbounded (that is, there is no upper bound on the sum of the absolute values of the coordinates of feasible vectors). Although this outcome is perfectly valid mathematically it may violate one's intuitive feeling that no real world situation is infinitely profitable. A closer look at the diet problem leads to a realization that there are some additional restrictions that were not included in the initial formulation. For example there is an upper bound to the amount of food that any person can eat in a day, and to the amount of potatoes that can be transported in a day. An additional restriction such as $x + y \leq 100$ expressing some activity bound would replace the original feasibility region by one of finite size.

We shall see that the two examples above illustrate the only ways in which a linear programming problem can fail to have an optimal feasible vector: (1) it can have no feasible vector or (2) the objective function (cost) has no finite lower bound. As a practical matter, whenever case (2) occurs one should search for additional constraints which may have been overlooked in setting up the feasibility conditions.

1.3. A MINIATURE PRODUCTION PROBLEM.

In the examples given above our objective was to minimize a linear function relative to certain linear inequalities. We will now discuss an example where we wish to maximize a linear function relative to a set of linear inequalities.

Consider a small factory which produces picture frames, cabinets, and figurines. To produce those items only wood and labor are assumed to be required, but there are restrictions on the amounts of wood and labor available. The requirements, restrictions, and values of output are summarized in Table 1.2.

Table 1.2. TECHNOLOGY TABLE.

	Units of wood	*Units of labor*	*Unit value of output*
Per picture frame	3	1	8
Per cabinet	4	3	19
Per figurine	1	3	7
Available material	25	50	

Let x_1, x_2, and x_3, respectively, be the numbers of picture frames, cabinets, and figurines produced in a given time period. We call $X = \begin{bmatrix} x_1 \\ x_2 \\ x_3 \end{bmatrix}$ a *program vector*. To satisfy our technology limitations or feasibility requirements we must have

$$\begin{aligned} 3x_1 + 4x_2 + x_3 &\leq 25, \\ x_1 + 3x_2 + 3x_3 &\leq 50, \end{aligned} \tag{1.8}$$

and

$$x_i \geq 0, \quad (i = 1, 2, 3). \tag{1.9}$$

We also assume that management has contracted in advance for the stated amounts of wood and labor. Hence, their costs need not enter into our considerations. Thus we wish to maximize our return from sales

$$w = 8x_1 + 19x_2 + 7x_3, \tag{1.10}$$

subject to the inequalities (1.8) and (1.9). For each program vector X, (1.10) gives the *value* of the program.

The region of feasibility is given in Figure 1.6. It can be visualized as a truncated tetrahedron where the tetrahedron is the portion of the first octant cut off by the plane $3x_1 + 4x_2 + x_3 = 25$ and the truncating plane is $x_1 + 3x_2 + 3x_3 = 50$. There are six vertices and this time (in constrast with the diet problem) the feasibility region is bounded. As before, an optimal program will occur at a vertex. We can determine the program value (profit) at each vertex and find the following:

Vertex	*Program value*
P_1	$66\frac{2}{3}$
P_2	$118\frac{3}{4}$
P_3	$134\frac{3}{8}$
P_4	150
P_5	$116\frac{2}{3}$
P_6	0

Thus an optimal program vector is $P_4 = \begin{bmatrix} 0 \\ 2\frac{7}{9} \\ 13\frac{8}{9} \end{bmatrix}$, and for it $w = 150$.

Note that in the program which yields a maximum return no picture frames are produced and that we have used up all the wood and labor available. Interpretations of fractional values in a program may be of concern to the reader. We assume that partly finished products can be completed in the following period and therefore accept fractions of a product as meaningful. For example, when the production line in an automobile assembly plant is shut down at the end of a week, the partially built automobiles on the line are not destroyed but are completed in the next work period.

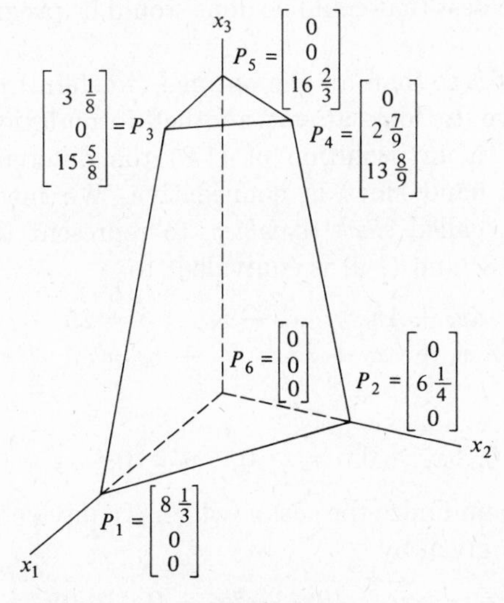

Figure 1.6 Feasibility region.

Suppose next that the price of figurines increases from 7 to 100. Then the program values at the vertices are:

Vertex	*Program value*
P_1	$66\frac{2}{3}$
P_2	$118\frac{3}{4}$
P_3	$1587\frac{1}{2}$
P_4	$1433\frac{3}{9}$
P_5	$1666\frac{2}{3}$
P_6	0

An optimal program is now $P_5 = \begin{bmatrix} 0 \\ 0 \\ 16\frac{2}{3} \end{bmatrix}$. Note that now we produce only figurines, and also, since

$$3 \cdot 0 + 4 \cdot 0 + 1 \cdot 50/3 = 50/3 < 25,$$

an optimal program does not use all the wood available; indeed under a corporate policy which required use of all available resources (that is, a no-waste policy) the best that could be done would be program $P_3 = \begin{bmatrix} 3\frac{1}{8} \\ 0 \\ 15\frac{5}{8} \end{bmatrix}$ with a return 79-1/6 less than for the optimal (wasteful) program.

It is instructive to look at yet another formulation of the wood-products problem. In any solution of (1.8) the difference between the right-hand and left-hand sides is nonnegative. We introduce two new variables x_4 and x_5, called *slack variables*, to represent these differences. Then the system (1.8) and (1.9) is equivalent to

$$\begin{aligned} 3x_1 + 4x_2 + x_3 + x_4 &= 25 \\ x_1 + 3x_2 + 3x_3 + x_5 &= 50 \end{aligned} \tag{1.11}$$

and

$$x_1 \geq 0, \quad x_2 \geq 0, \quad x_3 \geq 0, \quad x_4 \geq 0, \quad x_5 \geq 0. \tag{1.12}$$

We now wish to minimize the cost z (which is equivalent to maximizing the return $w = -z$) given by

$$z = -8x_1 - 19x_2 - 7x_3 + 0x_4 + 0x_5 \tag{1.13}$$

subject to (1.11) and (1.12) (the change to a minimization problem is made so that one standard computational form can be used throughout the book). Physically, x_4 represents the amount of unused wood and x_5 the amount of unused labor in any program. An optimal feasible program is then, of course, $x_2 = 25/9$, $x_3 = 125/9$, $x_1 = x_4 = x_5 = 0$. (In the case of the price change for figurines an optimal program has $x_2 = 0$ and $x_4 > 0$.)

Next, we solve (1.11) for x_2 and x_3 in terms of the remaining variables and use the result to eliminate x_2 and x_3 from (1.13). This gives

$$\begin{aligned} \frac{8}{9}x_1 + x_2 \qquad + \frac{1}{3}x_4 - \frac{1}{9}x_5 &= \frac{25}{9} \\ -\frac{5}{9}x_1 \qquad + x_3 - \frac{1}{3}x_4 + \frac{4}{9}x_5 &= \frac{125}{9} \end{aligned} \tag{1.14}$$

and

$$z = -150 + 5x_1 + 4x_4 + x_5. \tag{1.15}$$

A word of explanation here may help explain the form of (1.14). We say that these equations "represent the system (1.11) solved for x_2 and x_3." Actually from the instructions one might have expected to see equations written in the form $x_2 = \cdots$, $x_3 = \cdots$, but since this can be obtained from (1.14) by a few transpositions we accept (1.14) as the solved form. Our reason for preferring the form (1.14) will become clear when we introduce the simplex method.

We observe that the problem defined by (1.11), (1.12), and (1.13) is exactly the same as the one defined by (1.14), (1.12), and (1.15). The simplex method consists of systematically passing from one formulation of a problem to another until finally one is obtained in which either (1) an optimal program is attained or (2) it can be seen that no optimal program exists.

For the wood-products problem the formulation (1.14), (1.12), and (1.15) shows that the given program

$$x_1 = 0,\ x_2 = \frac{25}{9},\ x_3 = \frac{125}{9},\ x_4 = 0,\ x_5 = 0$$

is optimal, since any feasible solution with any of x_1, x_4, or x_5 greater than zero will increase z above -150. It also shows a great deal more. For example, (1.15) indicates that each unit of picture frames made increases the cost z by 5. Moreover, it tells us that the marginal values for wood and labor are 4 and 1, respectively. Thus, if additional wood or labor were offered to the manager of the factory at prices y_4 and y_5, respectively, he would be able to use them profitably provided $y_4 < 4$ and $y_5 < 1$. Alternatively, if another manufacturer offers to purchase his wood or contract for his labor it will be profitable to divert these resources provided that the prices y_4 and y_5 offered satisfy the conditions $y_4 > 4$ or $y_5 > 1$.

Equations (1.14) show how the production would have to be adjusted for each unit change in x_1, x_4, or x_5. For example, if 2 units of picture frames are to be made we have $x_1 = 2$ and hence $x_2 = 25/9 - (8/9)(2) = 1$, $x_3 = 125/9 + (5/9)(2) = 15$. Similarly, if the manager sells 3 units of wood, then $x_4 = 3$ and $x_2 = 25/9 - 1 = 16/9$, $x_3 = 125/9 + 1 = 134/9$. If he contracts for 6 units of labor then $x_5 = -6$ and $x_2 = 25/9 - 6/9 = 19/9$, $x_3 = 125/9 + 24/9 = 149/9$. This last shift is easy to understand; with more labor available some of the wood supply must be shifted from cabinets to figurines because the latter have a higher labor-to-wood utilization factor (3 as compared with 3/4).

There is sometimes confusion in the interpretation of positive and negative values for the slack variables. *A positive value for a slack variable represents a program in which not all of some available quantity is used.* For example $x_4 = 2$ means that two units of wood are not used. Thus if in some period we find that we have only 23 units of wood available instead of our usual 25, we set $x_4 = 2$. On the other hand, *a negative value of a slack variable represents an infeasible solution to the original problem* but can be used to describe a changed problem in which additional units of the item which it represents may be available. For example, a negative value of x_4, say $x_4 = -5$, represents an infeasible solution, that is, one in which 30 units of wood are used; thus $x_4 = -5$ can be used to determine the best program if 5 extra units of wood become available.

Equations (1.14) also show bounds on the size of any feasible changes in x_1, x_4, and x_5. We cannot have $x_1 > 25/8$ lest x_2 become negative. Similarly, $-125/3 \leq x_4 \leq 25/3$ and $-25 \leq x_5 \leq 125/4$. Note that manufacturing activities are rarely reversible (never so if labor is consumed in production); for an irreversible activity, negative values of the corresponding program variable are never considered. Thus (since picture frames do require labor) we do not consider negative values for x_1. On the other hand, both positive and negative values for slack variables are considered, since there are valid interpretations for them. If in (1.15) any of the coefficients of x_1, x_4, x_5 had been negative the given program would not have been optimal.

The simplex method introduced in Chapter 3 is based on the ideas just discussed and exploits them in a systematic way.

PROBLEMS

1.1. Plot the feasibility region defined by the following system of inequalities:

$$\begin{aligned} x_1 + x_2 &\leq 1, \\ x_1 \qquad &\leq 2, \\ 2x_1 - x_2 &\leq 5, \\ -x_1 + x_2 &\leq 2, \\ - x_2 &\leq 4. \end{aligned}$$

Find the maximum and minimum values of the linear function $2x_1 + 3x_2 + 2$ over this region.

1.2. Plot the feasibility region defined by

$$\begin{aligned} x_1 \qquad &\geq 0, \\ x_2 &\geq 0, \\ -x_1 - \ x_2 &\leq 1, \\ -x_1 + 2x_2 &\leq 4, \\ -x_1 + 2x_2 &\geq -5. \end{aligned}$$

Find maximum and minimum values of the function $2x_1 - x_2 + 5$ over this region.

1.3. Plot the feasibility region defined by

$$\begin{aligned} 2x_1 - x_2 &\leq 0, \\ x_1 + x_2 &\geq 140, \\ 3x_1 + x_2 &\leq 300, \\ x_1 &\geq 0, \\ x_2 &\geq 0, \end{aligned}$$

and find the points (vectors) in this region yielding a maximum for the function $z = x_1 + x_2$.

1.4. A manufacturer can use one or more production processes which are available to him. The first and second processes produce an output X and the third and fourth produce an output Y. There are three inputs to each of these processes: labor in man-weeks, pounds of raw material A, boxes of raw material B. Each process has different input requirements; therefore revenues yielded by the various processes differ even for processes producing the same output. In determining a week's production schedules, the manufacturer cannot use more than the available amounts of manpower and of the two raw materials. The relevant information is presented in the table below. Suppose that the revenue per unit of output sold for process 1 is

	One unit of output X		*One unit of output Y*		
	Process 1	*Process 2*	*Process 3*	*Process 4*	*Input availability*
Man-weeks	1	2	1	2	20
Pounds of A	6	5	3	2	100
Boxes of B	3	4	9	2	75
Production levels	x_1	x_2	x_3	x_4	. . .

\$6.00, and for the remaining processes is \$4.00, \$7.00, and \$5.00. Write out the linear programming version of this problem.

1.5. Find the values of x and y that maximize $z = 2x + 5y$ subject to the conditions

$$\begin{aligned} x + y &\leq 600, \\ 0 \leq x &\leq 300, \\ 0 \leq y &\leq 500. \end{aligned}$$

1.6. A convex polygon has the vertices

$$\begin{bmatrix}-1\\0\end{bmatrix}, \quad \begin{bmatrix}3\\4\end{bmatrix}, \quad \begin{bmatrix}0\\-3\end{bmatrix}, \quad \begin{bmatrix}1\\6\end{bmatrix}.$$

Can you find a system of linear inequalities whose solution set or feasibility region is this polygon?

1.7. Does the function $z = 3x + 2y$ have a maximum value subject to the conditions

$$\begin{aligned} -2x + 3y &\leq 9, \\ -x + 5y &\leq 20, \\ x &\geq 0, \\ y &\geq 0? \end{aligned}$$

Does the function have a minimum value subject to these conditions?

1.8. Find values of x and y that minimize $z = 20x + 15y$ subject to

$$\begin{aligned} 10x + 3y &\geq 20, \\ x + y &\geq 4, \\ 2x + 7y &\geq 14, \\ x &\geq 0, \\ y &\geq 0. \end{aligned}$$

1.9. A drill and a lathe are used exclusively in the manufacture of two products P_1 and P_2. The time units necessary for each step (drilling and turning) in the manufacture of both products are shown in the table below. If the lathe

	P_1	P_2
Lathe	1	3
Drill	1	1
Sales return Per unit of Product sold[1]	\$1.00	\$2.00

[1]Assume that all items manufactured are sold at these prices.

and drill are available for a maximum of 16 and 7 time units per week, respectively, find the maximum weekly profit in the manufacture of each item. Note: in this simple problem we assume that profit is a linear function of the number of items sold (if 10 units of P_2 are sold, the profit is \$20.00, if 30 units are sold the profit is \$60.00). Obviously, this is not always the case; profit per unit may decrease as a result of sales discounts or because of the cost of "tie-in" sales arrangements that must be made in order to increase the number of units sold.

1.10. Machines M_1, M_2, and M_3 are used in the manufacture of products F_1, F_2, F_3 and are available on a daily basis for maximums of 48, 24, and 24 hours, respectively. Profits on the sale of F_1, F_2, and F_3 are \$.50, \$6.00, and \$5.00 per unit, respectively. Using the following table, which gives the time required from each machine in the manufacture of each product, compute by the graphical methods of this chapter both an optimal feasible program (x_1, x_2, x_3) and the maximum profit.

	F_1	F_2	F_3	*Availabilities*
M_1	8	12	6	48
M_2	2	3	6	24
M_3	6	2	0	24
Profit per item (dollars)	.50	6.00	5.00	

Hint for quick algebraic solution: add slack variables and solve the resulting system for the program variables corresponding to F_2, F_3 and the slack variable corresponding to M_3.

Model Building With Linear Programming

2

2.1. MATHEMATICAL MODELS.

Linear programming, the subject of this book, provides mathematical models for use in dealing with decision problems in the real world. Successful application of a linear programming model—as of any mathematical model—requires knowledge of the model itself, knowledge of the real-world problems that are to be dealt with, and judgment in fitting these together. In order to discuss clearly these aspects of the model building process, we first turn our attention to the question: "What is a mathematical model?" This leads to a discussion of linear models and then to the subject of model building with linear programming.

A mathematical system consists of a set of assertions or axioms from which consequences are derived by mathematical (logical) argument. The axioms contain one or more primitive terms which are undefined and have no meaning in the mathematical system; they will usually consist of statements about the existence of a set of elements, relations and operations on the elements, and properties of these relations and operations (mathematical systems and examples of systems are discussed in Appendices A and B).

The term "model" is currently used in two ways. In the traditional development of mathematics a model for a set of axioms is a concrete mathematical system which satisfies these axioms. For example, the concept of field in mathematics is realized by the concrete model of the real numbers. In this context the role of a model is to demonstrate that the axioms are

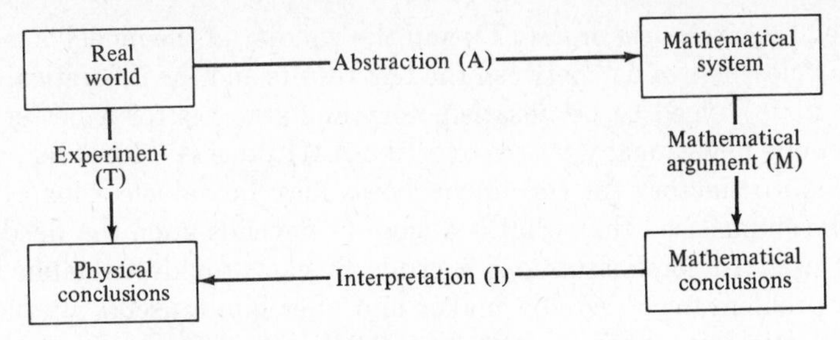

Figure 2.1 Mathematical model.

not self-contradictory and that they have some nonempty domain of application. More recently the term "mathematical model" has been applied to a mathematical system which abstracts some real-world situation. It is in this sense that the term is currently used in operations research and in the behavioral sciences, and we will follow this usage here. Figure 2.1 illustrates the role that mathematical models play in an application.

The analyst begins with some aspects of the real world and by means of a process called *abstraction* (A), maps his object system into some mathematical system called the mathematical model. Under this mapping the undefined concepts of the abstract system are images of features of the real world, and for the model to be valid the mathematical axioms must be the translations into the language of the model of valid properties of the real-world system being studied.

Following the abstraction the analyst uses *mathematical argument* (M) to arrive at theorems or mathematical conclusions. The conclusions are then given *interpretation* (I) into physical conclusions about the real world. This process is, in part, the reverse of the abstraction process but it is more than that, since the mathematical argument may introduce new defined concepts whose proper interpretation requires some care.

It should be clear that successful modeling requires judgment in making suitable abstractions (A) and in making interpretations (I). Moreover, the necessary judgment requires some knowledge of the real-world problem and some knowledge of the mathematical argument or model being used. Since few individuals possess an adequate store of these kinds of knowledge, a team approach is often used in operations research. Analysts develop the mathematical argument (M) and engage in (A) and (I) by means of consultation with persons intimately associated with the real-world problems. Successful execution of (A) and (I) then depends upon effective communication between the individuals concerned; poor communication can rarely be offset by the sophistication of the mathematical argument (M) and good communication can sometimes make a far more important contribution to a successful application than (M).

The physical conclusions arrived at through the AMI process may be checked by *experiment* or *test* (T), and the validity of the model is measured by the "closeness of fit" between the test results and the theoretical results. If the fit is judged to be unsatisfactory one searches for a better model, that is, one makes many passes over the AMI process. Moreover, a model which is satisfactory for certain purposes may be valueless for others. It must be emphasized that what is a close fit depends upon the needs of the model user. In some situations a crude fit can provide valuable insights into a problem for a decision maker and offer him a useful alternative to past practice; in others, such as controlling an astronaut's capsule in flight, a high degree of accuracy is required. Finally, a complex mathematical model may be made up of many blocks of the type illustrated in Figure 2.1. We will return to this point after considering some illustrative examples.

(a) *The Falling-body Model.* Suppose a baseball is dropped from observation level of the Washington Monument. How fast would it be traveling when it reached the ground (500 feet below) and how long would the fall take? If we neglect air resistance the Newtonian model

$$\text{Force} = \text{Mass times Acceleration}$$

is a relevant abstraction. Mathematical argument (M) leads to the differential equation

$$-g = d^2y/dt^2,$$

where $g = 32\ \text{ft/sec}^2$ is the gravitational constant, y measures height above the ground in feet, and t measures time in seconds. Integrating gives the velocity equation

$$-gt + v_0 = dy/dt,$$

where v_0 = initial velocity = 0 since the ball is dropped, not thrown. Integrating again we get

$$-gt^2/2 + y_0 = y,$$

where y_0 = initial displacement = 500. The ball strikes the ground when $y = 0$, that is, when

$$t = (2y_0/g)^{1/2} = 5.59.$$

Substituting this value of t in the velocity equation gives

$$dy/dt = -178.88.$$

The minus sign indicates that the ball is falling rather than rising; the terminal speed is 178.88 feet per second or approximately 122 miles per hour.

In this model step (I) was simple, consisting of interpretations of the equations in terms of distance traveled and time. Step (T) of Figure 2.1 might then consist of stopwatch measurements of actual experiments.

Suppose these reveal that the agreement between the mathematical conclusions and real-world results was not sufficiently close. One might then make another pass over the AMIT process by adding a term to the initial differential equation to account for air resistance. If we assume a resisting force proportional to the negative of the velocity, the new equation becomes

$$d^2y/dt^2 = -g - K\,dy/dt,$$

where the constant K depends on the nature of the falling object; K would clearly be larger for a parachute than for a baseball. Integrating this equation under the assumption of initial height y_0 used above and initial velocity zero we get

$$y = y_0 + g(-Kt - e^{-Kt} + 1)K^2,$$

and

$$dy/dt = -g(1 - e^{-Kt})/K.$$

According to this equation, as t becomes large, the velocity gradually increases in absolute value to an upper limit g/K called the terminal striking speed. For a man and open parachute this value is about 15 feet per second, whereas for a man in free fall it is about 120 feet per second.

The refined model contains the constant K which must be evaluated (usually experimentally) for each choice of a falling body. No single experiment will suffice both to define K and to test the model. Careful measurement of the time of fall from the top of the monument to the ground would give a value for t, call it t_f, from which K could be determined by solving the equation

$$0 = 500 + 32(-Kt_f - e^{-Kt_f} + 1)/K^2.$$

As is oftentimes the case, refinement of the original model requires additional effort, interpretations, and expense. Whether the refinement is desirable depends upon an assessment of the difference in usefulness of the two models and its relation to the inconvenience or cost of making the refinement. A trade off of this kind is usually involved in model refinement.

(b) *Simple Inventory Model.* The supply manager of a business is responsible for purchasing and storing a commodity H which is used at a constant rate which amounts to R pounds per year. We use a simple inventory model to provide the AMI parts of the analysis. The commodity is readily available for immediate delivery at a cost of C_1 dollars per pound and the cost of placing an order is C_2 dollars. Supplies on hand have a holding (or inventory) cost of C_3 dollars per pound per year. Company policy is to order a constant amount q each time the supply on hand is exhausted. It is the duty of the supply manager to determine the value of q which will minimize the average total annual costs C related to handling commodity H.

We have

$$C = G_1 + G_2 + G_3$$

where $G_1 = C_1R$ is the annual cost for R pounds of H, G_2 is the average annual cost of placing orders, and G_3 is the average annual holding cost. The average number of orders per year will be R/q, hence $G_2 = C_2R/q$.

Let r denote the amount of H on hand at any time t. Then the graph of r will have the form given in Figure 2.2, where t is measured in years. The linear behavior of r in each cycle is a consequence of our assumption of constant usage rate. A single order of size q will last q/R years, and the geometry of Figure 2.2 makes it clear that the average value of r is $q/2$ in each cycle (that is, r equals area divided by base). Hence, the average annual holding cost is $G_3 = C_3q/2$, and our total cost C becomes

$$C = C_1R + C_2R/q + C_3q/2.$$

Now if we assume that q can take any positive real value we can differentiate this expression giving

$$dC/dq = -C_2R/q^2 + C_3/2$$

and the critical value of q (the value which makes dC/dq zero)

$$q_0 = (2C_2R/C_3)^{1/2}$$

can be shown to be optimal.

Suppose, for example, that $R = 10{,}000$, $C_1 = 10$, $C_2 = 50$, $C_3 = 1$; then $q_0 = 1{,}000$ and there are ten orders placed per year. Also, $C_0 = 100{,}000 + 500 + 500 = 101{,}000$. In contrast to the optimal case, ordering 50 times per year would give $C = 102{,}600$ and ordering just once per year would give $C = 105{,}050$.

These two examples illustrate several general features of model building. The falling-body model is descriptive in that it describes a real-world situation without leading to any decision as to a course of action. In contrast the inventory model is a decision model. It mediates in a situation where there are two opposing factors (one, keep order size small so as to minimize holding costs and two, keep order size large so as to minimize ordering costs) and provides, relative to the model, an optimal balance.

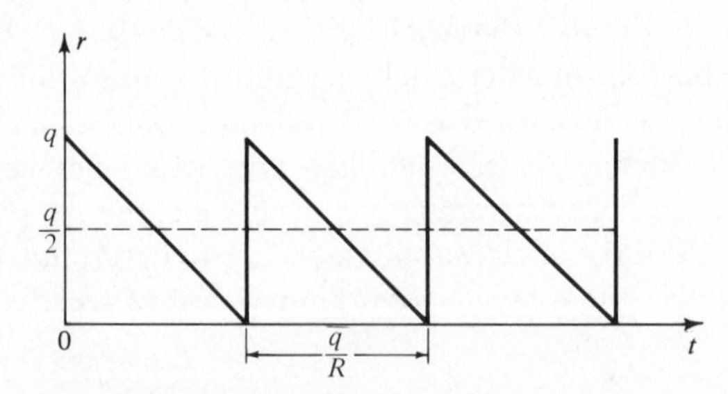

Figure 2.2 Inventory and average inventory.

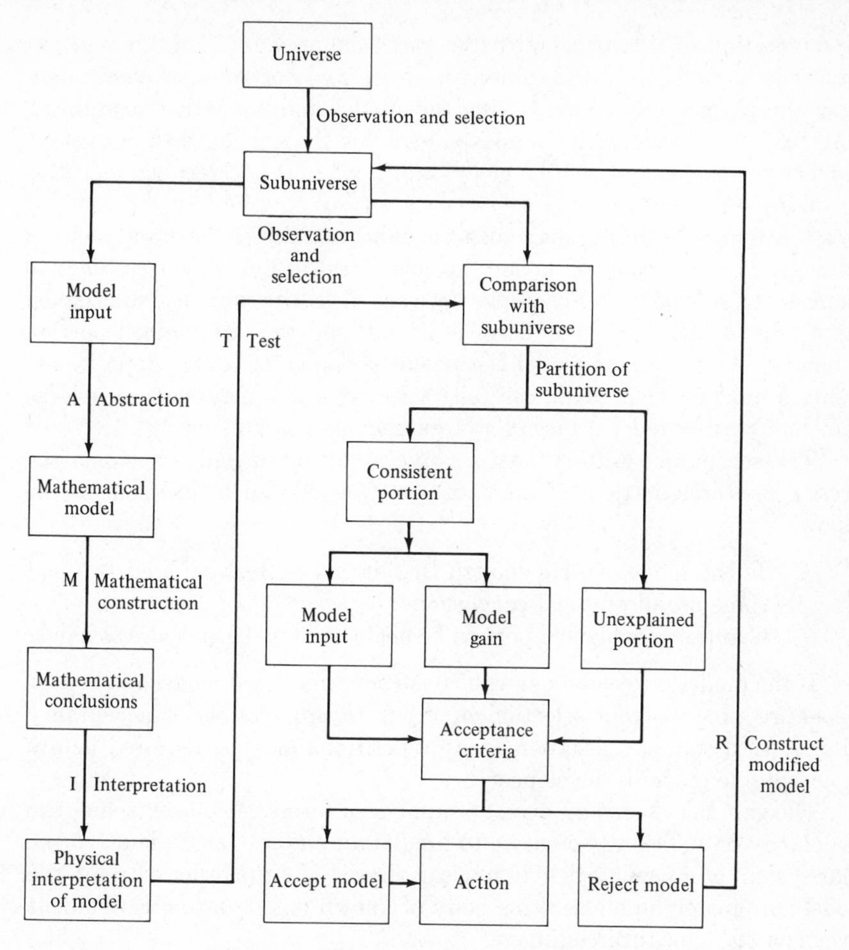

Figure 2.3 Role of mathematical model.

The first attempt to build a model is not likely to lead to a satisfactory formulation; indeed, it frequently initiates a sequence of (hopefully) better and better models until a suitable one is developed or a point is reached when what is available must be combined with the experience, intuition, and judgment of the decision maker in selecting a course of action.

Refinement of a model can be discussed more fully by means of an expanded version of Figure 2.1. The boxes of that figure appear on the left side of Figure 2.3 and the remainder illustrates the process of model acceptance or rejection.

The model builder begins by selecting the subuniverse he wishes to study and selects as model input that portion of the subuniverse from which he hopes to explain what remains. Following through the A, M, I stages as before we come to the T (test) stage in which we compare the physical interpretation of the model with the given subuniverse. This comparison results in a partition of the subuniverse into two portions, one containing everything that is consistent with the model and the other containing what is inconsistent with or unexplained by the model. The consistent portion, of course, contains the model input; what is left over can be called the *model gain*.

Ideally one wishes to maximize the gain (relative to the input) and to minimize the unexplained portion; however, realization of either objective alone is not sufficient. For example, if we use the entire subuniverse as model input, the A, M, I, T steps become trivial, and the unexplained portion is empty. We reject the model because it has no gain. At the opposite extreme, a model with a small, carefully selected input may lend to a large gain but be rejected because of a large unexplained portion.

The acceptance criteria must involve a suitable balance between the sizes of the three portions of the partition. One asks such questions as the following:

1. Is the model simple enough that it can be manipulated to yield some useful physical conclusions?
2. Can the unexplained portion be neglected for the purpose at hand?

If the model is rejected one can construct a modified model and repeat the entire process from selection of input to application of acceptance criteria. In a complex situation many repetitions may be required before an acceptable model is developed.

The gain may take any one of a number of forms. At one extreme, the model may be sufficiently creative to predict an undiscovered planet in the solar system or a new particle in nuclear physics. At the other extreme the model may merely organize some body of known results into a convenient form as a basis for future study.

The unexplained portion may include minor effects that can be neglected as unimportant for a given investigation, for example, air resistance in a short flight or minor variations in an assumed constant-demand inventory system; or it may include items *irrelevant* to the purpose at hand, such as questions about life on Mars when one is constructing a dynamic model of the solar system. This irrelevant class can sometimes be eliminated or reduced in size by more careful delineation of the initial subuniverse.

However, factors which are unimportant or irrelevant for one study may become sufficiently significant for a different study to require a revision of the model. For example, the model of Euclidean plane geometry is reasonably satisfactory for giving directions in a city but must be rejected for navigation across an ocean because here the curvature of the surface of the earth becomes significant and a different geometry must be invoked.

2.2. LINEAR MODELS.

An activity can be defined broadly as a process which transforms an input into an output.[1] Suppose that an activity operates on some input quantity or vector X, yielding an output quantity or vector Y. This can be described by writing

$$X \xrightarrow{A} Y = AX, \tag{2.1}$$

where AX represents the result of applying the activity A to the input X. If for all inputs X, X_1, X_2 and any real number or scalar k

$$X_1 + X_2 \xrightarrow{A} Y_1 + Y_2 \qquad \text{(additivity)} \tag{2.2}$$

where $Y_1 = AX_1$ and $Y_2 = AX_2$, and

$$kX \xrightarrow{A} kY, \qquad \text{(constant returns to scale)} \tag{2.3}$$

where $Y = AX$, then we say that A is a *linear activity* and any mathematical model having properties (2.2) and (2.3) is called a linear model.

Typical models in operations research involve the optimization of some function, called an objective function, of the input variables subject to various restrictions or constraints. The illustrative models in Chapter 1 are of this type. In linear optimization models, with which this book is concerned, both the objective function and the constraints must satisfy the linearity conditions (2.2) and (2.3).

The additivity property (2.2) means that in order for any optimization model to be linear, the output and total returns (value of the objective function) associated with the sum of inputs are equal to the sum of the outputs and returns associated with the individual inputs. Property (2.3) essentially means that the activity is characterized by *constant returns to scale;* the total amounts of each output and the associated return are strictly proportional to the amount of input. For example, the direct effect of doubling the input to any activity is the doubling of both output and returns.

The mathematical model of the diet problem introduced in Chapter 1 is one of linear optimization (it is easy to verify that it satisfies (2.2) and (2.3)). In using this model to choose a diet we were implicitly assuming that a linear model is a suitable representation of the corresponding diet problem, or in the language of Figure 2.1, that the relevant mathematical system (M) is that of cost minimization subject to the inequalities (1.2). For example, we assumed that there are 3 units of carbohydrates, 4 units of vitamins and

[1]The concept of a linear activity will be developed more fully in Chapter 5.

1 unit of protein in a unit of potatoes. Two additional units of potatoes in the diet would then add 6 units of carbohydrates, 8 units of vitamins, and 2 units of protein to the diet. Strict proportionality would prevail and this seems a reasonable assumption to make in this problem. Moreover, the diet is assumed to consist of x units of potatoes and y units of steak and we assume that the units of carbohydrate, vitamin, and protein in potatoes are additive with respect to the units of corresponding constituents in steak so that the total number of units of carbohydrate in the diet is the sum of the number of units of this in potatoes and in steak (again a reasonable assumption in this problem).

Elsewhere in this chapter and throughout this book activities are discussed that have been widely (and usefully) assumed to be linear. Strictly speaking there are few truly linear activities but linear models can provide useful approximations nevertheless. Departures from linearity can be unimportant and the advantages of using a linear model can compensate for the error involved. More importantly, although an activity is not linear throughout its entire range of operation, it can be approximately linear over a part of the range and economic factors can make it very likely that the activity is operated in this part much of the time. For example, in the diet problem assumptions of linearity are reasonable only for the relatively small values of x and y that appear to be relevant in view of the objective (cost) function.

Some activities are, however, inherently nonlinear (that is, either (2.2) or (2.3) or both are not satisfied at all). Examples of situations that may not have constant returns to scale are provided by activities in which a discounting factor is present. If larger quantities of output can be sold only at lower prices, then it may not be appropriate to use a linear model in which it is required that returns and outputs be strictly proportional to inputs. In production problems the process may be in a stage of increasing returns; in this case the technology is such that for a given increase in inputs the outputs increase more than proportionately. In a stage of decreasing returns, on the other hand, a given increase in inputs generates a less than proportional increase in output. Moreover, activities can be such that inputs are not additive; some inputs when used in a given combination may produce more (or less) output than when employed singly or in combination with other inputs.

When is a linear model appropriate? A general, operational answer to this question is not possible. Judgment and experience are necessary and these are fortified by an understanding of the model-building process, the nature and assumptions of linear models, and a knowledge of the real-world problem itself.

The remaining sections of this chapter deal with the building of linear models. In Sections 2.4 and 2.9 an additional complication, namely restriction of decision variables to integer values, changes the problem from linear programming to integer programming. In the other sections, linear-programming models result. These can all be regarded as introductory exposures to the problem-formulation experience and the last of these, somewhat more lengthy than the others, moves one closer still to the kind of thinking that is necessary for successful applications of linear programming to real-world problems. It is also hoped that these examples will provide motivation for the study of the theory of linear programming that begins in Chapter 3.

2.3. A GASOLINE-BLENDING PROBLEM.

One of the most significant applications of linear programming is to petroleum refinery operations. Here one finds a process which has been usefully approximated by a linear model. We consider a miniature problem in the blending of gasoline, where three types of raw gasoline are to be blended to make two types of gasoline for use in automobiles, premium and regular gasoline. The problem is to maximize the profit using the available amounts (or less) of raw gasoline to make these two products.

Table 2.1. BLENDING DATA.

	Performance rating	*Vapor pressure*	*Capacity*	*Cost/price*
	$\geq$	$\leq$		
Raw gasoline type 1	108	4	32,000	10
Raw gasoline type 2	90	10	20,000	6
Raw gasoline type 3	73	5	38,000	5
Premium gasoline	100	6		9
Regular gasoline	80	9		7

A blend is defined by six numbers, which are obtained as values of six variables, called the *decision variables:*

x_1 = amount of raw gasoline of type 1 to be used in premium gasoline,
x_2 = amount of raw gasoline of type 2 to be used in premium gasoline,
x_3 = amount of raw gasoline of type 3 to be used in premium gasoline,
x_4 = amount of raw gasoline of type 1 to be used in regular gasoline,
x_5 = amount of raw gasoline of type 2 to be used in regular gasoline,
x_6 = amount of raw gasoline of type 3 to be used in regular gasoline.

The profit made in using a unit of raw gasoline of type i for premium gasoline is

$$p_i = \text{sales value of premium gasoline} - \text{cost of gasoline } i;$$

for regular gasoline it is

$$p_{i+3} = \text{sales value of regular gasoline} - \text{cost of gasoline } i.$$

Therefore,

$$p_1 = 9 - 10 = -1, \qquad p_4 = 7 - 10 = -3,$$
$$p_2 = 9 - 6 = 3, \qquad p_5 = 7 - 6 = 1,$$
$$p_3 = 9 - 5 = 4, \qquad p_6 = 7 - 5 = 2.$$

The total profit to be maximized is

$$w = \sum_{j=1}^{6} p_j x_j = -x_1 + 3x_2 + 4x_3 - 3x_4 + x_5 + 2x_6. \tag{2.4}$$

The capacity restrictions are

$$\begin{aligned} x_1 + x_4 &\leq 32{,}000, \\ x_2 + x_5 &\leq 20{,}000, \\ x_3 + x_6 &\leq 38{,}000. \end{aligned} \tag{2.5}$$

The technological requirements for premium gasoline are

$$\begin{aligned} 108x_1 + 90x_2 + 73x_3 &\geq 100(x_1 + x_2 + x_3), \\ 4x_1 + 10x_2 + 5x_3 &\leq 6(x_1 + x_2 + x_3). \end{aligned} \tag{2.6}$$

For regular gasoline they are

$$\begin{aligned} 108x_4 + 90x_5 + 73x_6 &\geq 80(x_4 + x_5 + x_6), \\ 4x_4 + 10x_5 + 5x_6 &\leq 9(x_4 + x_5 + x_6). \end{aligned} \tag{2.7}$$

We must also have

$$X \geq 0. \tag{2.8}$$

No discussion seems necessary for the capacity restrictions expressed in (2.5). The determination of the technological requirements for the blends is more complex. To obtain (2.6) and (2.7) we first assume that the total volume of each blend is equal to the sum of the volumes of the inputs of raw gasoline from which it is made and secondly that the composite "performance rating times volume" for each blend is the sum of the corresponding composites for the inputs and similarly for "vapor pressure times volume." Thus the performance rating of the premium blend is given by the quotient

$$q = (108x_1 + 90x_2 + 73x_3)/(x_1 + x_2 + x_3).$$

The requirement that $q \geq 100$ is converted into the first relation of (2.6) when we clear fractions. Note that performance rating itself is not linear.

The inequalities (2.6) and (2.7) present one feature which requires special handling in the simplex algorithm. This feature is the absence of a constant term, that is, a term free of the unknowns; a linear relation with constant term equal to zero leads to the phenomenon known as degeneracy.

A solution of this problem is given in Section 3.4, where degeneracy is treated.

2.4. ADVERTISING MEDIA SELECTION PROBLEM.[2]

The management of an electric-products company has decided to spend up to $1,000,000.00 on the advertising of women's electric razors that it manufactures. The advertising budget is to be spent in 12 consumer magazines in which full-page, four-color advertisements can be purchased. Let x_i be the number of dollars spent on advertising in magazine i, x_1 being the number of dollars to be spent on advertising in *Cosmopolitan*, x_2 in *Mademoiselle*, x_3 in *Family Circle*, x_4 in *Good Housekeeping*, x_5 in *McCall's*, x_6 in *Modern Romances*, x_7 in *Modern Screen*, x_8 in *Motion Picture*, x_9 in *True Confessions*, x_{10} in *Woman's Day*, x_{11} in *Seventeen*, and x_{12} in *Ladies Home Journal*.

Management seeks advice from an advertising agency as to how to reach a desired audience most effectively. The agency indicates that an objective function should be developed first so that the goals of the advertising program can be incorporated into an optimization model. Even though there is not a perfect relationship between the number of reader exposures to advertising and the sales response to advertising, management is advised that an appropriate goal is to maximize the number of "effective" exposures given the advertising budget. In other words, the objective is not to maximize the number of exposures of all readers to the advertising, but rather to maximize the number of potential buyers exposed to the advertising. Clearly, some advertising media are better than others for reaching desired audiences, so the first step is to develop an "effectiveness rating" for each magazine so that management can be reasonably sure that the media chosen "match" the audience that should be reached as specified by the promotional objectives.

Suppose that market-research studies have shown that the desired market for shavers is largely composed of women who are from 18 to 44 years of age, have incomes of $7,000.00 or more, live in metropolitan areas, and who are married. Moreover, suppose these findings indicated that the desired market components have a relative importance represented by the weights given in Table 2.2.

Table 2.2. CONSUMER CHARACTERISTICS.

Characteristics	*Weights*
Age (18–44)	.50
Income ($7,000.00 or over)	.25
Metropolitan location	.15
Married	.10
Total	1.00

[2]Adapted from Reference 17.

Now suppose for the magazine *Good Housekeeping* the data in Table 2.3 are available on readership characteristics.

Table 2.3. READERSHIP CHARACTERISTICS FOR GOOD HOUSEKEEPING.

Category	*Percent of readers in category*
Age (18–44)	60
Income ($7,000 or more)	33
Metropolitan location	70
Married	95

In order to determine effective readership for this magazine the agency multiplies each of the desired weights by the corresponding percent incidence in the readership of the magazine and divides by 100,

$$\frac{(60)(.50) + (33)(.25) + (70)(.15) + (95)(.10)}{100} = .583.$$

If the magazine reaches 5,075,814 women the effective readership of the magazine is (.583)(5,075,814) = 2,959,200 readers. Finally, suppose the four-color, full-page rate for the magazine is $27,400.00; the quotient 2,959,200/27,400 gives an "effective readings per dollar spent" of 108. This number will then appear as the coefficient of the variable x_4 in the objective function.

Suppose that an analysis similar to that above was made for each of the twelve magazines and the results are in Table 2.4.

Table 2.4. EFFECTIVENESS RATINGS FOR MEDIA.

Media	*Effective readings per dollar spent in media*	*Cost of one full-page, four-color advertisement*
Cosmopolitan	158	$ 5,500
Mademoiselle	263	5,950
Family Circle	106	33,600
Good Housekeeping	108	27,400
McCall's	65	42,840
Modern Romances	176	3,275
Modern Screen	285	3,415
Motion Picture	86	2,248
True Confessions	120	25,253
Woman's Day	51	32,550
Seventeen	190	8,850
Ladies Home Journal	101	35,000

The objective then is to maximize

$$f = 158x_1 + 263x_2 + 106x_3 + 108x_4 + 65x_5 + 176x_6 + 285x_7 + 86x_8 + 120x_9 + 51x_{10} + 190x_{11} + 101x_{12}. \tag{2.9}$$

Since management has decided to spend up to \$1,000,000.00 on advertising, one of the constraints must be

$$\sum_{i=1}^{12} x_i \leq 1{,}000{,}000. \tag{2.10}$$

In addition, management wants to assure that no more than 12 insertions are made in any one magazine and that it wishes the number of insertions in *Mademoiselle* and *Ladies Home Journal* to be less than or equal to 7 and 2, respectively. Multiplying the per page cost in Table 2.4 by the appropriate number of insertions gives

$$\begin{array}{ll} x_1 \leq 66{,}000 & x_7 \leq 40{,}980 \\ x_2 \leq 41{,}650 & x_8 \leq 26{,}976 \\ x_3 \leq 403{,}200 & x_9 \leq 303{,}026 \\ x_4 \leq 339{,}280 & x_{10} \leq 390{,}600 \\ x_5 \leq 514{,}080 & x_{11} \leq 106{,}200 \\ x_6 \leq 39{,}300 & x_{12} \leq 70{,}000. \end{array} \tag{2.11}$$

Suppose also that management wishes to specify minimum expenditures in certain of the magazines, say

$$\begin{aligned} x_2 &\geq 17{,}850 \\ x_3 &\geq 67{,}200 \\ x_5 &\geq 42{,}840 \\ x_{10} &\geq 32{,}550. \end{aligned} \tag{2.12}$$

Finally, management desires an expenditure of no more than \$320,000.00 in four of the magazines

$$x_3 + x_9 + x_{10} + x_{12} \leq 320{,}000, \tag{2.13}$$

and, of course

$$x_j \geq 0 \qquad (j = 1, \cdots, 12). \tag{2.14}$$

The linear programming model of the media-selection problem is then to maximize (2.9) subject to (2.10) through (2.14). This model does not provide a complete solution since there is no guarantee that an optimal solution will result in an allocation for each journal which is some integer multiple of the full-page cost. For example, if in an optimal solution $x_1 = \$10{,}000.00$, where the cost per page is \$5,500.00, it is unclear how one would utilize this information for a decision. It may be useful to round such components of an optimal solution up or down to an integer value; however, there is no assurance that such a rounding process will result in a truly optimal solution to the media-selection problem or, indeed, even a feasible solution to the linear programming model. Actually, this problem should be formulated by means of the model of integer programming, in which the decision variables are the number of pages to purchase. General methods for dealing with integer programming problems are beyond the scope of this book.

2.5. PORTFOLIO SELECTION PROBLEM.

An investor seeks advice from an employee of a stock-brokerage firm. He does not wish to invest in a mutual fund but prefers to diversify his investments by allocating his funds among the following: bonds, deposits in savings and loan associations, preferred stock, blue-chip common stock, and growth stocks. As a first step he wants to decide how to allocate his funds among these classes of investments optimally (he will defer to a later date the problem of making optimal selections within each investment class).

The investor, together with his investment advisor, estimates an expected annual yield for each class of investment. The broker helps the investor develop a measure of confidence in each of these estimates in the form of a risk factor representing the probability that the estimated annual yield would be deficient. Finally, the research staff of the firm provides the investor with forecasts, for each class of investment, of the average number of years that the expected yield would be realized. The relevant data appear in Table 2.5.

Table 2.5. FINANCIAL DATA.

Class of investment	*Expected annual yield (percent)*	*Risk factor*	*Average term of investment (years)*
Bonds	5.75	.05	10
Savings and loan deposits	5.00	.01	3
Preferred stock	6.25	.15	5
Blue-chip common stock	7.50	.28	3
Growth stock	14.50	.60	2

Let

x_1 = percent of portfolio invested in bonds,
x_2 = percent of portfolio invested in savings and loan deposits,
x_3 = percent of portfolio invested in preferred stocks,
x_4 = percent of portfolio invested in blue-chip common stocks,
x_5 = percent of portfolio invested in growth stocks.

The first requirement is that all his funds be invested (the investor is assumed to have set aside an appropriate sum in the form of cash),

$$\sum_{i=1}^{5} x_i = 100. \tag{2.15}$$

The investor specifies that he does not wish to have a probability of more than .20 of failing to obtain the expected earnings from his portfolio. This is a nonlinear constraint; however, using the risk factors we can formulate a linear approximation to this constraint as follows,

$$.05x_1 + .01x_2 + .15x_3 + .28x_4 + .60x_5 \leq 20. \qquad (2.16)$$

He also wants a weighted average investment period of at least four years,

$$10x_1 + 3x_2 + 5x_3 + 3x_4 + 3x_5 \geq 400 \qquad (2.17)$$

and, of course,

$$x_j \geq 0 \qquad (j = 1, \cdots, 5). \qquad (2.18)$$

The investor then desires to select a portfolio in such a way as to maximize his expected yield,

$$y = 5.75x_1 + 5x_2 + 6.25x_3 + 7.5x_4 + 14.5x_5, \qquad (2.19)$$

subject to the requirements (2.15) through (2.18).

2.6. THE TRANSPORTATION AND ASSIGNMENT PROBLEMS.

We next consider a problem of distributing wheat from three elevators to four flour mills. In Figure 2.4 the initial column gives the number of carloads of wheat available at each elevator and the initial row gives the number of carloads needed at each mill. Note that

$$5 + 9 + 4 = 10 + 3 + 2 + 3 = 18,$$

that is, the total amount available is equal to the total amount needed. The body of the table gives the shipping costs from each elevator to each mill. Let x_{ij} denote the number of carloads to be shipped from the ith elevator to the jth mill. We assume that shipment of fractional carloads is not permitted. Our problem is to supply all of the mills at a minimum total transportation cost. It can be shown that an optimal program here is

$$x_{13} = 2, \quad x_{14} = 3, \quad x_{21} = 6, \quad x_{22} = 3, \quad x_{31} = 4,$$
$$\text{all other } x_{ij} = 0, \qquad (2.20)$$

and the minimum cost is 64 (the method of solution will be discussed later).

	Mill	1	2	3	4	
	Carloads needed	10	3	2	3	
Elevator	Carloads available					
1	5	6	4	3	2	
2	9	5	2	4	4	Shipping
3	4	4	2	2	4	costs

Figure 2.4 The wheat distribution problem.

More generally we consider the problem of distributing any commodity from a group of producing centers, called *origins*, to a group of receiving centers, called *destinations*. This is called the *transportation* or *distribution problem*. Let

$$\begin{aligned}
p &= \text{number of origins,}\\
n &= \text{number of destinations,}\\
a_{ij} &= \text{cost of shipping from origin } i \text{ to destination } j,\\
b_i &= \text{amount of stock at origin } i,\\
c_j &= \text{amount of stock delivered to destination } j,\\
x_{ij} &= \text{amount of stock shipped from origin } i \text{ to destination } j.
\end{aligned}$$

We assume that total demand and total supply are in balance, that is,

$$\sum_{i=1}^{p} b_i = \sum_{j=1}^{n} c_j, \tag{2.21}$$

and since there are no negative shipments,

$$x_{ij} \geq 0 \qquad \text{for all } i \text{ and } j. \tag{2.22}$$

The following equations indicate that the proper total amount is shipped from each origin:

$$\begin{aligned}
x_{11} + x_{12} + \cdots + x_{1n} &= b_1\\
x_{21} + x_{22} + \cdots + x_{2n} &= b_2\\
\vdots \qquad\qquad & \quad \vdots\\
x_{p1} + x_{p2} + \cdots + x_{pn} &= b_p.
\end{aligned} \tag{2.23}$$

Similarly, the following equations indicate the proper total amount is received at each destination:

$$\begin{aligned}
x_{11} + x_{21} + \cdots + x_{p1} &= c_1\\
x_{12} + x_{22} + \cdots + x_{p2} &= c_2\\
\vdots \qquad\qquad & \quad \vdots\\
x_{1n} + x_{2n} + \cdots + x_{pn} &= c_n.
\end{aligned} \tag{2.24}$$

The above equations define a feasible shipping program. The cost function to be minimized, subject to these conditions, is

$$z = \sum_{i,j} a_{ij} x_{ij}. \tag{2.25}$$

Note that here the feasibility requirements do not include a condition that the x_{ij} be integers. This omission is made possible by an important theorem which states that in the transportation problem all of the vertices of the feasibility region have integer components if each b_i and c_j is an integer (see Section 7.3).

The restraints (2.23) and (2.24) display an interesting pattern when corresponding terms are aligned vertically in the respective equations. We illustrate with the case of the wheat distribution problem which has $p = 3$, $n = 4$ (in Chapter 7 we will comment further on the structural properties of the transportation problem):

$$\begin{array}{lllllllllllll}
x_{11} + x_{12} + x_{13} + x_{14} & & & & & & & & & & & & = 5 \\
& & & & x_{21} + x_{22} + x_{23} + x_{24} & & & & & & & & = 9 \\
& & & & & & & & x_{31} + x_{32} + x_{33} + x_{34} & & & & = 4 \\
x_{11} & & & + x_{21} & & & & + x_{31} & & & & & = 10 \\
& x_{12} & & & + x_{22} & & & & + x_{32} & & & & = 3 \\
& & x_{13} & & & + x_{23} & & & & + x_{33} & & & = 2 \\
& & & x_{14} & & & + x_{24} & & & & + x_{34} & & = 4.
\end{array}$$

The special case of the transportation problem in which $p = n$ and $b_1 = \cdots = b_n = c_1 = \cdots = c_n = 1$ is called the *assignment problem*. The number a_{ij} then represents the cost of assigning man i to job j. One wishes the one-to-one assignment of men to jobs which minimizes the total cost. If $n = 20$ there are 20! possible assignments. How long would it take a computer to solve such a problem by examining each assignment? Assuming that a computer could examine one assignment per microsecond (one millionth of a second), it has been calculated that more than 50,000 years would be required. However, solving such a problem by means of a linear programming algorithm would require less than one minute.

Transportation and assignment problems arise in various guises of which the following example illustrates one. A certain firm classifies persons applying for jobs under one of eight categories according to such factors as education, sex, age, background experience, and so forth as shown in Table 2.6. The categories are labeled $P_1, \cdots, P_8$. It happens that the firm is short of personnel in the list of jobs given below beside a table filled with 0's and 1's; a 1 in the ith row under the column J_j indicates

Table 2.6. A TRANSPORTATION PROBLEM INVOLVING ASSIGNMENT OF PEOPLE.

	Job category							
Personnel category	*Janitor* J_1	*Typist* J_2	*Office Boy* J_3	*Secretary* J_4	*Manager* J_5	*Foreman* J_6	*Assembly woman* J_7	*Receptionist* J_8
P_1	1	0	0	0	0	0	1	0
P_2	1	0	0	0	0	0	0	1
P_3	0	1	0	1	0	0	0	1
P_4	0	0	0	1	1	0	0	0
P_5	1	0	1	0	0	0	1	0
P_6	0	0	0	0	1	1	0	0
P_7	0	0	0	0	1	0	0	0
P_8	0	1	1	0	0	1	0	1

that anyone in category or classification P_j is suitable for placement in J_i. A zero indicates unsuitability. We assume that each person can be placed in at most one job.

The problem here is to fill a maximum number of jobs given that the plant needs (2, 5, 7, 4, 2, 3, 17, 2) persons in jobs $(J_1, J_2, J_3, J_4, J_5, J_6, J_7, J_8)$, respectively, while the 61 applicants are classified (4, 27, 11, 4, 5, 2, 7, 1), respectively, in classifications $(P_1, P_2, P_3, P_4, P_5, P_6, P_7, P_8)$.

Let x_{ij} persons in classification P_j be assigned to job J_i, let there be b_j applicants classified P_j while the demand is for c_i individuals in job J_i, and let e_{ij} be zero if the entry in row i and column j of the table is zero and let e_{ij} be ∞ otherwise. Then we wish to maximize

$$w = \sum_i \sum_j x_{ij} \tag{2.26}$$

subject to

$$\sum_j x_{ij} \leq b_i \qquad \text{for each } i, \tag{2.27}$$

$$\sum_i x_{ij} \leq c_j \qquad \text{for each } j, \tag{2.28}$$

$$0 \leq x_{ij} \leq e_{ij} \qquad \text{for each } i \text{ and } j. \tag{2.29}$$

Also the x_{ij} must be integers. Taking into account (2.29), the inequalities (2.27) take the form

$$\begin{aligned}
&x_{11} + x_{17} \leq 4\\
&x_{21} + x_{28} \leq 27\\
&x_{32} + x_{34} + x_{38} \leq 11\\
&x_{44} + x_{45} \leq 4\\
&x_{51} + x_{53} + x_{57} \leq 5\\
&x_{65} + x_{66} \leq 2\\
&x_{75} \leq 7\\
&x_{82} + x_{83} + x_{86} + x_{88} \leq 1.
\end{aligned}$$

We leave it as an exercise for the reader to derive a similar set of conditions from (2.28) and (2.29).

2.7. PRODUCTION SCHEDULING.

A manufacturer wishes to schedule production for four months in advance to meet known monthly demands. He can produce 1,200 items per month on regular shifts and by using overtime he can produce up to 400 additional items per month at a unit cost of \$7.00 per item in excess of unit cost for regular production. It costs \$3.00 per month to hold an item. His monthly sales will be, respectively, 900, 1,100, 1,700, and 1,400. He wishes to determine an optimal production schedule, that is, the one which minimizes total costs of production and storage but still meets sales demand.

A program can be formulated in terms of eight decision variables, namely, how much of each class of production to use each month. Let x_i, y_i, respectively denote the number of units made using regular and overtime production in month i ($i = 1, 2, 3, 4$). Let z_i represent the number of units held over from month i to month $i + 1$.

The equations below express the monthly balance between sales and storage:

$$\begin{aligned} x_1 + y_1 - z_1 \qquad &= 900 \\ x_2 + y_2 + z_1 - z_2 &= 1{,}100 \\ x_3 + y_3 + z_2 - z_3 &= 1{,}700 \\ x_4 + y_4 + z_3 - z_4 &= 1{,}400. \end{aligned} \tag{2.30}$$

The inequalities

$$\begin{aligned} x_i &\leq 1{,}200 \\ y_i &\leq 400 \end{aligned} \qquad (i = 1, 2, 3, 4) \tag{2.31}$$

express the production limitations, and we must also have

$$x_i \geq 0,\ y_i \geq 0,\ z_i \geq 0 \qquad (i = 1, 2, 3, 4). \tag{2.32}$$

Subject to the feasibility conditions we wish to minimize the cost

$$z = 7(y_1 + y_2 + y_3 + y_4) + 3(z_1 + z_2 + z_3 + z_4) + kz_4. \tag{2.33}$$

Here z_4 is, of course, the total excess of production over demand and k is the unit penalty for unsold items. The information given in the statement of the problem is insufficient to determine k but it turns out that whatever value k has (so long as $k > 0$) the optimal solution will have $z_4 = 0$; hence a precise evaluation of k is not needed.

That $z_4 = 0$ should be intuitively obvious since there is no advantage in having leftover stock, whereas there is a holding cost for unsold items.

Although the formulation given above is a natural one it turns out that a second formulation, seemingly more complicated, leads to a simpler numerical solution. In the second formulation we introduce 40 decision variables. For $i = 1, 2, 3, 4$, let x_{ij} represent the amount of regular shift production in period i used to satisfy demand in period j ($j = 1, 2, 3, 4$)and let x_{i5} represent the unused regular shift production capacity. Let y_{ij} ($i = 1, 2, 3, 4$; $j = 1, 2, 3, 4, 5$) be defined similarly for overtime production. For $i \leq j < 5$ the cost coefficient for x_{ij} is $3(j - i)$, and for y_{ij} is $7 + 3(j - i)$. For x_{i5} and y_{i5} the cost coefficients are zero. Since an item cannot be used before it is made we must have $x_{ij} = y_{ij} = 0$ for $i > j$; we achieve this by assigning infinite (or very large finite) cost coefficients to these variables. Now the problem becomes a transportation problem with 8 origins and 5 destinations. The resulting cost matrix is given in Figure 2.5. The border numbers indicate the feasibility conditions. The regular and overtime production capacities are indicated by the row border numbers; the first four column border numbers indicate the monthly

	900	1,100	1,700	1,400	1,300
1,200	0	3	6	9	0
400	7	10	13	16	0
1,200	∞	0	3	6	0
400	∞	7	10	13	0
1,200	∞	∞	0	3	0
400	∞	∞	7	10	0
1,200	∞	∞	∞	0	0
400	∞	∞	∞	7	0

Figure 2.5 Cost matrix for scheduling problem.

demands and the fifth column border number 1,300 represents the total excess of capacity over demand. For solution by hand computation the ∞ symbols indicate cells where no assignment can be made; for computer solution these symbols could be replaced by some finite cost M large enough to guarantee that the corresponding activity will not appear in the final solution. For example, choosing $M = 100$ would be satisfactory in this problem.

The advantage of this apparently more complex formulation is that the solution algorithms for transportation problems are much simpler numerically than those for linear programming problems in general (see, in particular, the algorithm of Chapter 7 for the capacitated transportation problem).

2.8. MAKE-OR-BUY DECISIONS.

A "make-or-buy" decision of the following sort must be made in an automobile plant: there are five engine parts E_1, E_2, E_3, E_4, E_5, which can either be manufactured in the plant or bought from other concerns. Six machines M_1, M_2, M_3, M_4, M_5, M_6 are used to produce the five parts with the following machine times per part:

	M_1	M_2	M_3	M_4	M_5	M_6
E_1	0.04	0.02	0.02	0.00	0.03	0.06
E_2	0.00	0.01	0.05	0.15	0.09	0.06
E_3	0.02	0.06	0.00	0.06	0.20	0.20
E_4	0.06	0.04	0.15	0.00	0.00	0.05
E_5	0.13	0.02	0.00	0.00	0.01	0.05

No more than 50 hours are available on each machine while 1,000 of each component (engine part) are required. Denoting by $x_1, \cdots, x_5$, respectively, the numbers of parts $E_1, \cdots, E_5$ manufactured and by $x_6, \cdots, x_{10}$, respectively, the number of parts $E_1, \cdots, E_5$ bought, we are given the following costs:

Part	*Price/unit*	*Part*	*Price/unit*
x_1	2.55	x_6	3.10
x_2	2.47	x_7	2.60
x_3	4.40	x_8	4.50
x_4	1.90	x_9	2.25
x_5	2.50	x_{10}	2.80

The problem faced by the automobile manufacturer is to reduce his cost of all five parts (1,000 of each) to a minimum, that is, to minimize

$$\begin{aligned} z = 2.55x_1 + 2.47x_2 + 4.40x_3 + 1.90x_4 + 2.50x_5 \\ + 3.10x_6 + 2.60x_7 + 4.50x_8 + 2.25x_9 + 2.80x_{10} \end{aligned} \tag{2.34}$$

subject to the following constraints arising from machine capacity limitations

$$\begin{aligned}
0.04x_1 \qquad\qquad + 0.02x_3 + 0.06x_4 + 0.13x_5 &\leq 50, \\
0.02x_1 + 0.01x_2 + 0.06x_3 + 0.04x_4 + 0.02x_5 &\leq 50, \\
0.02x_1 + 0.05x_2 + \qquad\qquad + 0.15x_4 \qquad\qquad &\leq 50, \\
0.15x_2 + 0.06x_3 \qquad\qquad\qquad\qquad &\leq 50, \\
0.03x_1 + 0.09x_2 + 0.20x_3 \qquad\qquad + 0.01x_5 &\leq 50, \\
0.06x_1 + 0.06x_2 + 0.20x_3 + 0.05x_4 + 0.05x_5 &\leq 50,
\end{aligned}$$

the further constraints arising from the procurement requirements

$$\begin{aligned}
x_1 + x_6 &= 1{,}000, \\
x_2 + x_7 &= 1{,}000, \\
x_3 + x_8 &= 1{,}000, \\
x_4 + x_9 &= 1{,}000, \\
x_5 + x_{10} &= 1{,}000,
\end{aligned}$$

and the nonnegativity constraints

$$x_i \geq 0 \qquad (i = 1, \cdots, 10).$$

2.9. THE TRAVELING SALESMAN PROBLEM.

One of the best known linear programming problems is that of the traveling salesman who wishes to visit a number n of cities and return home at minimum cost. Let

$$a_{ij} = \text{the cost of going from city } i \text{ to city } j,$$

and consider a given tour which includes each city exactly once. Define

$$x_{ij} = \begin{cases} 0 \text{ if the tour does not lead directly from city } i \text{ to city } j, \\ 1 \text{ if the tour does lead directly from city } i \text{ to city } j. \end{cases} \tag{2.35}$$

For the tour indicated in Figure 2.6: $x_{12} = 1$, $x_{27} = 1$, $x_{76} = 1$, $x_{65} = 1$,

$x_{53} = 1$, $x_{34} = 1$, $x_{41} = 1$ and all other $x_{ij} = 0$. Since the tour from any one city cannot lead directly back to that city again, it is evident that

$$x_{11} = x_{22} = \cdots = x_{ii} = \cdots = x_{nn} = 0. \tag{2.36}$$

Also the path from one city can lead to exactly one other city, hence,

$$\begin{aligned} x_{11} + x_{12} + \cdots + x_{1n} &= 1 \\ x_{21} + x_{22} + \cdots + x_{2n} &= 1 \\ \vdots \quad\quad \vdots \quad\quad\quad & \vdots \\ x_{n1} + x_{n2} + \cdots + x_{nn} &= 1. \end{aligned} \tag{2.37}$$

Furthermore, each city is visited exactly once; therefore,

$$\begin{aligned} x_{11} + x_{21} + \cdots + x_{n1} &= 1 \\ x_{12} + x_{22} + \cdots + x_{n2} &= 1 \\ \vdots \quad\quad \vdots \quad\quad\quad & \vdots \\ x_{1n} + x_{2n} + \cdots + x_{nn} &= 1. \end{aligned} \tag{2.38}$$

We wish to rule out a noncyclic route; the conditions above do not exclude the case where there are two or more disconnected tours. This obviously is not what is wanted. Thus we must introduce the following additional restrictions:

$$x_{i_1 i_2} + x_{i_2 i_3} + \cdots + x_{i_{r-1} i_r} + x_{i_r i_1} \leq r - 1 \tag{2.39}$$

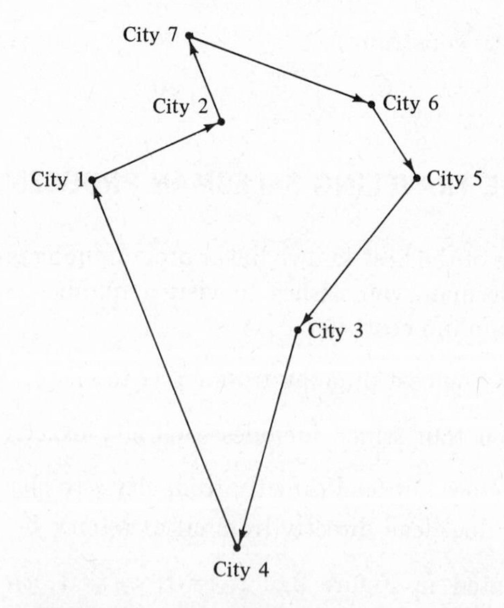

Figure 2.6 The salesman's trip.

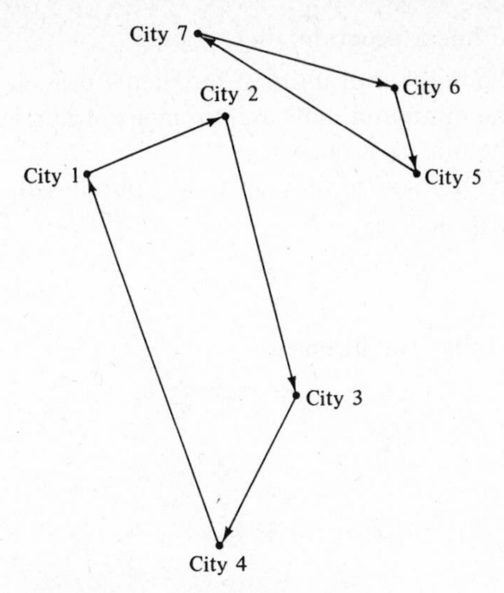

Figure 2.7 An improper trip.

for each $r < n$ and for all choices of i_j ($j = 1, \cdots, r$). For example in Figure 2.7 condition (2.39) is not satisfied since

$$x_{12} + x_{23} + x_{34} + x_{41} = 1 + 1 + 1 + 1 = 4 \not\leq 4 - 1.$$

The total cost of the trip is

$$z = \sum_{i,j} a_{ij}x_{ij}. \tag{2.40}$$

If the conditions above were the only factors to include in a model we would have a linear program. However, another constraint is that the x_{ij} be permitted to take on only the values 0 and 1. This constraint changes the model from one of linear programming to one of integer programming—a very significant complication which takes it beyond the scope of this book.

In the sense of pure mathematics the integer programming problem is trivial since there are only finitely many paths and one of these must minimize the cost. But for $n = 13$ there are more than a billion feasible paths and one sees that even modern computing machines would be hard put to handle problems of this magnitude directly. Actually, there is as yet no effective general method of solution for the traveling salesman problem, although a number of large problems have been solved by particular methods. An especially interesting case has $n = 48$ and involved the visiting a specified city in each of the then 48 states (see Reference 14).

2.10. THE GENERAL DIET PROBLEM.

We will consider a diet problem which involves n foods and p different nutrients, where, say, $n = 77$ and $p = 9$. (This was the size of a diet problem studied by G. J. Stigler, Reference 37, in 1945, prior to the development of linear programming). Let

$$\begin{aligned} a_{ij} &= \text{the number of units of nutrient } i \text{ in food } j,\\ b_i &= \text{the minimum daily requirement of nutrient } i,\\ c_j &= \text{the unit cost of food } j,\\ x_j &= \text{the number of units of food } j \text{ purchased.} \end{aligned}$$

We wish to minimize the cost

$$z = \sum_{j=1}^{n} c_j x_j \tag{2.41}$$

subject to the feasibility conditions

$$\begin{array}{c} a_{11}x_1 + \cdots + a_{1n}x_n \geq b_1 \\ \cdot \qquad\qquad \cdot \\ \cdot \qquad\qquad \cdot \\ \cdot \qquad\qquad \cdot \\ a_{p1}x_1 + \cdots + a_{pn}x_n \geq b_p \end{array} \tag{2.42}$$

$$x_j \geq 0. \tag{2.43}$$

The inequalities (2.42) assure that all nutritional requirements shall be met and (2.43) expresses the fact that we cannot reverse the operation of eating and thereby reconstruct foods from nutrients.

It is interesting to note that Stigler proposed a solution to this problem based on an examination of some but not all of the possible choices. He did not claim that his procedure was optimal but believed that the total cost, expressed on an annual basis, could not be reduced much below that associated with his solution. Subsequent consideration of this problem, when the efficiency of the simplex method was first examined, revealed that Stigler's solution on a cost-per-year basis was only 24 cents higher than the true minimum cost. Such a saving could nonetheless be significant however, depending upon the volume of operation. For example, a minimum cost diet provided for thousands of poverty stricken individuals in a nation of low per capita income could be an enormous social advantage if it could reduce costs by 24 cents per person per year.

2.11. THE CONTRACT AWARDS PROBLEM.

Consider the case in which contracts are to be awarded for the manufacture of m items. There are n contractors or sources available. The awards are to be made in such a way as to minimize the cost of purchase.

To simplify the problem we will require that each contractor bid on all the items. (If a contractor does not want to bid, he will make his bid so high that it will surely be rejected.) Let

$$\begin{aligned} c_{ij} &= \text{unit price for item } i \text{ as manufactured by contractor } j,\\ x_{ij} &= \text{the number of units of } i \text{ awarded to contractor } j,\\ a_i &= \text{minimum requirement for the } i\text{th item.} \end{aligned}$$

We are to minimize the buyer's total cost

$$z = \sum_{i,j} c_{ij} x_{ij}$$

subject to the following limitations

$$x_{11} + \cdots + x_{1n} \geq a_1, \cdots, x_{m1} + \cdots + x_{mn} \geq a_m.$$

A contractor can impose certain other limitations. He can refuse to take awards less than a given value or greater than another given value. That is, either

$$L_j \leq c_{1j}x_{1j} + \cdots + c_{mj}x_{mj} \leq U_j$$

or

$$x_{1j} = x_{2j} = \cdots = x_{mj} = 0,$$

where

$$L_j = \text{minimum total dollar award acceptable to } j$$

and

$$U_j = \text{maximum total dollar award acceptable to } j.$$

Furthermore, there may be limits on the number of units of i accepted by contractor j. That is, either $q_{ij} \leq x_{ij} \leq Q_{ij}$, or $x_{ij} = 0$, where q_{ij} and Q_{ij}, respectively, are the minimum and maximum numbers of units i accepted by contractor j. In any event,

$$x_{ij} \geq 0.$$

The objective function of this problem can be changed from linear to nonlinear form by dropping the assumption that c_{ij} is independent of the number of items ordered. For example, a discount might be given if a very large number of the items were ordered. In such a case it still might be possible to consider the new nonlinear objective as being piecewise linear.

2.12. MAINTAINING A PROFITABLE ECOLOGICAL BALANCE.[3]

The following model is not only an interesting application of linear programming to a bioeconomic situation, it also illustrates how information can be sifted and interpreted to formulate a useful mathematical model.

In the Edwards Plateau country of west Texas the vegetation is easily modified as a result of grazing by animals from a mixed vegetation to a dominance of grasses, or forbs, or browse, or various combinations of

[3]Used by permission of Professor G. M. Van Dyne, Colorado State University.

these. It is common to see in pastures in this area herds of cattle, bands of sheep, and flocks of mohair goats. Whitetail deer and wild turkey are common if there is sufficient browse and mixed vegetation. Catfish will thrive in ponds if there is suitable vegetation cover to prevent silting in the water. Ranchers sell beef, wool, mutton, mohair and they lease deer and rights to hunt turkey and to fish. The relative monetary income values per animal are: cattle, 10; goats, 1; sheep, 1 (wool and mutton combined); deer, 0.5; turkey, 0.05; and fish, 0.001. A rancher owns 10 sections of such land which has a maximum carrying capacity of 10 animal units per section per year. The animal-unit equivalents for the various species per animal are: cattle, 1; sheep, 0.2; goats, 0.25; deer, 0.3; and essentially zero for turkeys and fish. To properly organize his operation for livestock production he has to have at least 20 cattle and at least 20 goats on his ranch. The rancher wants at least some sheep and some deer on his ranch. He can maintain the desired vegetation cover for turkey and fish if he has (a) cattle, sheep, goats, and deer, (b) cattle, sheep, and goats, or (c) cattle, goats, and deer, but no more than 75 percent of the grazing load (measured in animal units) may be due to cattle and goats combined. Of course, he wants his total stocking rate of all organisms combined to be equal to or less than the carrying capacity of the range. Furthermore, he can put in no more than one pond per section, each of which will support no more than 500 fish and can harvest no more than 25 percent of the catfish per pond per year. The requirements and habits of the wild turkey are such that he cannot maintain more than 2 flocks of 10 birds per flock per section and he cannot harvest more than 20 percent of the population per year. The rancher keeps only castrated male goats for mohair and shears them once each year. His cattle, sheep, and deer harvests which will maintain a given population are respectively about 25, 35, and 15 percent of the population per year (the 35 percent for sheep includes both wool and mutton).

The first step is to isolate the salient points from this description. They are:

1. The relative monetary income values per animal are:

cattle	10.
goats	1.
sheep	1.
deer	.5
turkey	.05
fish	.001.

2. There are 10 sections of land, each with a maximum carrying capacity of 10 animal units per year.
3. Animal-unit equivalents for the species are:

cattle	1.
goats	.25
sheep	.2
deer	.3
turkey	0.
fish	0.

4. The ranch must have at least 20 cattle and 20 goats.
5. There must be some sheep and some deer.
6. If there are to be any turkey and fish, there must be

   (a) cattle, sheep, and goats

   or

   (b) cattle, goats, and deer.

7. Cattle and goats can comprise no more than 75 percent of the animal units.
8. Total animal units cannot exceed the carrying capacity.
9. A maximum of 500 fish per section, with a 25 percent annual harvest.
10. A maximum of 20 birds per section, with a 20 percent annual harvest.
11. Harvest percentages:

cattle	25
goats	100
sheep	35
deer	15.

12. How many individuals of each species should the rancher have in order to maximize annual profit?

In the first place the rancher wants an answer that is something like: "Have 30 cattle, 22 sheep, and so forth." He will be understandably upset if he is told that to realize a maximum profit, he requires 33.25 cattle and 21.7 sheep—that is to say, the rancher is certainly anticipating the answer to be in integers. However, any programming problem that imposes integer constraints on the variables will be difficult to solve. Thus, the first simplification is to treat the problem as a programming problem without an integer restriction. It may even turn out that all or some of the optimal values of the variables of this problem will be integers. If not, the solution can be rounded to the nearest integer solution. It is important to realize that this rounded solution may *not* be the best integer solution (that is, it may not be a true optimal solution to the original problem). However, if

p_0 is the maximum profit for the continuous problem,
p_1 is the profit obtained by rounding the optimal solution,
p_2 is the maximum profit to the integer problem,

then clearly $p_1 \leq p_2 \leq p_0$. If $p_0 - p_1$ is very small, the rancher can be advised that the answer is "close enough."

The next step is to determine the variables. Since he has 10 parcels of land, each being able to support 6 different species, the first inclination is to use 60 variables x_{ij}, $i = 1, \cdots, 6$, $j = 1, \cdots, 10$. Then x_{ij} will represent the number of species i to be placed on parcel j. However, since the conditions for survival are the same in any parcel, it is much simpler to consider the entire ten parcels as a single unit. Then only 6 variables x_i, $i = 1, \cdots, 6$ are needed, where x_i represents the total number of species i. In the final solution, 1/10 of x_i can be placed in each parcel. (Or other adjustments can be made, at the rancher's preference. Note that we cannot maximize profit over a single parcel. The constraints would require *some* deer on *every* parcel, which is not required in the given problem.)

Thus we have the following 6 variables:

x_1: number of cattle,
x_2: number of goats,
x_3: number of sheep,
x_4: number of deer,
x_5: number of turkeys,
x_6: number of fish.

The conditions then give the following constraints: Condition 2 says the maximum carrying capacity available is 100 units. Condition 3 gives the animal units for each species. Then, Condition 7 gives the constraint,

$$1x_1 + .25x_2 \leq 75.$$

Condition 8 becomes

$$1x_1 + .25x_2 + .2x_3 + .3x_4 \leq 100,$$

and Condition 4 is

$$x_1 \geq 20,$$
$$x_2 \geq 20.$$

Condition 5 is

$$x_3 \geq \text{"some,"}$$
$$x_4 \geq \text{"some,"}$$

where "some" is the minimum number of sheep and deer the rancher wants. Since the rancher is vague about this, let us assume that "some" = 1. The lower bound constraints become

$$x_1 \geq 20,$$
$$x_2 \geq 20,$$
$$x_3 \geq 1,$$
$$x_4 \geq 1.$$

Now we see that Condition 6 is irrelevant. The lower bound constraints guarantee that Conditions 6 (a) and (b) will *always* be satisfied. Hence, it will always be possible to have turkey ($x_5 > 0$) and fish ($x_6 > 0$). Condition 9 gives

$$x_6 \leq 5000,$$
$$x_5 \leq 200.$$

Finally, we need only calculate the profit. Condition 1 gives profit per animal. Not all animals can be harvested, however. Harvest percentages are given in Conditions 9, 10, and 11. The annual profit from each species is

cattle	$(10)(.25x_1)$,
goat	$(1)(x_2)$,
sheep	$(1)(.35x_3)$,
deer	$(.5)(.15x_4)$,
turkey	$(.05)(.20x_5)$,
fish	$(.001)(.25x_6)$.

Thus the rancher's problem is the following: subject to the constraints,

$$\begin{aligned} &x_1 + .25x_2 \leq 75, \\ &x_1 + .25x_2 + .2x_3 + .3x_4 \leq 100, \\ &x_1 \geq 20, \\ &x_2 \geq 20, \\ &x_3 \geq 1, \\ &x_4 \geq 1, \\ &0 \leq x_5 \leq 200, \\ &0 \leq x_6 \leq 5000, \end{aligned}$$

maximize the objective function

$$z = 2.5x_1 + x_2 + .35x_3 + .075x_4 + .01x_5 + .00025x_6.$$

It now becomes clear that the only constraints affecting x_5 and x_6 (number of turkeys and fish) are their upper bound constraints. So to maximize profit, the rancher should have as many turkeys and as many fish as possible, regardless of the other species. Thus, each parcel of land should have a pond with 500 fish and should support 2 flocks of 10 birds each.

The annual profit from birds and fish is thus

$$(.01)(200) + (.00025)(5000) = 2 + 1.25 = 3.25.$$

Next, in order to have only nonnegativity constraints, introduce four new variables:

$$\begin{aligned} y_1 &= x_1 - 20, \\ y_2 &= x_2 - 20, \\ y_3 &= x_3 - 1, \\ y_4 &= x_4 - 1, \end{aligned} \qquad y_i \geq 0, \qquad (i = 1, \cdots, 4).$$

So y_i is the amount by which species i exceeds its lower bound. Note that

$$\begin{aligned} x_1 &= y_1 + 20, \\ x_2 &= y_2 + 20, \\ x_3 &= y_3 + 1, \\ x_4 &= y_4 + 1. \end{aligned}$$

So the original problem is equivalent to the following: subject to

$$\begin{aligned} &(y_1 + 20) + .25(y_2 + 20) \leq 75, \\ &(y_1 + 20) + .25(y_2 + 20) + .2(y_3 + 1) + .3(y_4 + 1) \leq 100, \\ &y_1 \geq 0, \quad y_2 \geq 0, \quad y_3 \geq 0, \quad y_4 \geq 0, \end{aligned}$$

maximize

$$z_0 = 2.5(y_1 + 20) + (y_2 + 20) + .35(y_3 + 1) + .075(y_4 + 1) + 3.25.$$

Simplification gives: subject to

$$\begin{aligned} &y_1 + .25y_2 \leq 50, \\ &y_1 + .25y_2 + .2y_3 + .3y_4 \leq 74.5, \\ &y_i \geq 0, \qquad (i = 1, \cdots, 4), \end{aligned}$$

maximize

$$z_0 = 2.5y_1 + y_2 + .35y_3 + .075y_4 + 73.675.$$

This problem is easily solved by the simplex algorithm (the numerical solution is assigned as an exercise in the next chapter). The optimal solution turns out to be:

$$\begin{aligned} y_1^* &= 0, \\ y_2^* &= 200, \\ y_3^* &= 122.5, \\ y_4^* &= 0. \end{aligned}$$

Thus for maximum annual profit, the rancher's selection should be:

20	cattle for a profit of	50.
220	goats for a profit of	220.
123.5	sheep for a profit of	43.225
1	deer for a profit of	.075
200	turkeys for a profit of	2.
5000	fish for a profit of	1.25,

giving a profit of 316.55 units.

Of course, the rancher will have to have only 123 sheep, reducing his profit by .175. He will not be able to realize any profit on his single deer, which he insisted on having. So his profit is further reduced by .075. Thus, his total profit, with an integral number of species, is 316.30.

Since the *best* integral solution can give no more than 316.55 profit, it is probably not worth finding it. If the rancher now decides that "some" deer means more than one, then the problem can be reformulated and solved again.

If it is required that each parcel have an integral number of species to itself, then we must still specify the distribution of the species over the 10 parcels. Each parcel then will have

2 cattle,
22 goats,
2 flocks of 10 birds each,
1 pond with 500 fish.

Putting an integral number of sheep on each parcel means that only 12 sheep can be on any parcel. This further reduces the profit by 1.05, since only 120 sheep are present. (*Now*, the total profit is only 315.25 but the *best* integral solution can give a profit no greater than 316.55).

Note that the 2 cattle, 22 goats, and 12 sheep on each parcel use a total of 9.9 land units. Thus, no single parcel can support the deer. However, the deer can presumably roam over all 10 parcels, in which case there are enough land units to support three deer, which are still not enough for a deer harvest.

PROBLEMS

In each of the following problems formulate a linear programming model by setting up appropriate constraints and objective function (solving the resulting problems will form exercises for later chapters).

2.1. A small refinery blends five raw gasoline types to produce two grades of motor fuel, regular and premium. The number of barrels per day of each raw gasoline type available, the performance rating and cost per barrel are given in the following table.

	Performance rating	*Barrels/day*	*Cost/barrel*
Raw gasoline type 1	70	2000	.80
Raw gasoline type 2	80	4000	.90
Raw gasoline type 3	85	4000	.95
Raw gasoline type 4	90	5000	1.15
Raw gasoline type 5	99	3000	2.00

Regular motor fuel must have a performance rating of at least 85 and premium at least 95. The refinery's contract requires that 8,000 barrels/day of premium be produced; at the same time, the refinery can sell its entire output of both premium and regular for \$3.75/barrel and \$2.85/barrel, respectively.

Any raw gasoline types (not blended into motor fuel) having a performance rating of 90 or more are sold at \$2.75/barrel and types with performance rating of 85 or less are sold at \$1.25/barrel.

2.2. A manufacturer of products F_1 and F_2 wishes to schedule his production two months in advance. The costs of production over a monthly period for each product are given in the table below for both regular and overtime work. The costs of keeping either F_1 and F_2 in inventory are included for the first month (it being assumed they will be sold the second month).

	F_1			F_2		
	Cost per hour	*Inventory cost per unit*	*Demand (in units of output)*	*Cost per hour*	*Inventory cost per unit*	*Demand (in units of output)*
Month 1						
regular	4	1	10	6	2	8
overtime	8	1		11	2	
Month 2						
regular	4		15	6		12
overtime	8			11		

Each unit of F_1 requires 8 hours of production time and each unit of F_2 requires 12 hours of production time. We assume: (1) that 176 hours of regular production time are available each month and (2) that 88 hours of overtime are available each month.

2.3. A producer wishes to maximize profits arising from the production of two products X and Y. Both products are manufactured in a two-stage process in which all initial operations are performed in machine center 1 and all final operations can be performed in either machine center 2A or 2B. Centers 2A and 2B differ from each other in that for any given product they yield different time-unit rates and different unit profits. We also assume that a certain amount of overtime has been made available in machine center 2A for the manufacture of X and Y. Since the use of overtime results in changes (decreases) in unit profits (but not in unit rates), we denote separately, by machine center 2AA, the overtime use of 2A. The unit time required to manufacture products X and Y, the hours available at each center, and the unit profits are given in the table below. To simplify discussion, let X_1, X_2, and X_3 in the table denote the three different combinations for producing product X and Y_1, Y_2, and Y_3, denote the three different combinations of producing Y. The problem is to determine how much of each product should be made through the use of each possible combination of machine centers so as to maximize profits. Let x_1, x_2, x_3 and y_1, y_2, y_3 represent the level of operation of the activities X_1, X_2, X_3 and Y_1, Y_2, Y_3, respectively.

Discuss carefully the meaning of the linearity assumptions in the context of this problem.

Operation	Machine center	Product X x_1	x_2	x_3	Product Y y_1	y_2	y_3	Hours available
No. 1	1	.01	.01	.01	.03	.04	.03	800
	2A	.03			.05			750
No. 2	2AA		.02			.05		125
	2B			.03			.06	950
Unit profit		.25	.18	.37	.82	.58	.73	—

2.4. Suppose the costs of shipping a product from origins O_1, O_2, and O_3 to destinations D_1, D_2, and D_3 are shown in the table.

	D_1	D_2	D_3
O_1	73	40	9
O_2	62	93	96
O_3	95	65	80

Further, suppose the amount of stock available at the origins (the b_i in the notation of Section 2.6) is $b_1 = 8$, $b_2 = 7$, and $b_3 = 9$ and the amount to be delivered to the destinations (the c_j) is $c_1 = 6$, $c_2 = 8$, and $c_3 = 10$. We require that all stock available at origins is to be shipped, and we wish to meet all demands at minimum shipping cost.

2.5. A metal products company produces waste cans, filing cabinets, file boxes for correspondence, and lunch boxes. Its inputs are sheet metal of two different thicknesses, called A and B, and labor. Input-output relationships for the company are shown in the following table. The sales revenue per

	Waste cans	*Filing cabinets*	*Correspondence boxes*	*Lunch boxes*
Sheet metal A	6	0	2	3
Sheet metal B	0	10	0	0
Labor	4	8	2	3

unit of waste cans, filing cabinets, correspondence boxes, and lunch boxes are \$1.98, \$44.95, \$9.95, and \$2.98, respectively. There are 225 units of sheet metal A available in the company's inventory, 300 of sheet metal B, and 190 units of labor. What is the company's optimal sales revenue?

2.6. A patient on a special diet is allowed to eat only steak and rice. The following table gives nutritional values, costs, and minimum dietary requirements.

	Rice	*Steak*	*Minimum requirement*
Units of carbohydrates	2	1	3
Units of proteins	1	4	4
Unit cost	3	7	-

2.7. A farmer can raise sheep, hogs, and cattle. He has space for 40 sheep, or 60 hogs, or 30 cattle, or any pro rata combination of these (that is, 4 sheep, 6 hogs, or 3 cows use the same amount of space). The expected profits per animal are $5.00 for sheep, $4.00 for hogs, and $10.00 for cattle. He must by law raise at least as many hogs as sheep and cattle combined.

2.8. The Cubans have two types of ballistic missiles: type 1 (New York range) weighs 12 tons, type 2 (Washington D.C. range) weighs 10 tons. They have an inventory of 10 of type 1 and 6 of type 2, some of which are to be shipped back to Russia via two Czech destroyers. The volume of the missiles poses no problem, but ship C has a weight capacity of 100 tons, and ship D a weight capacity of 60 tons. If the Czechs charge the Cubans $500.00 per ton for material on ship C and $600.00 per ton on ship D, how should they load their ships to capitalize on the plight of the unfortunate Cubans?

2.9. An electronics defense products company is producing three lines of products for sale to the government: transistors, micromodules, and circuit assemblies. It has four process areas: transistor production, circuit printing and assembly, transistor and module quality control, and circuit assembly test and packing.

Production of one transistor requires 0.1 standard hour of transistor production area capacity, and 0.5 standard hours of transistor quality control area capacity and $.70 in direct costs.

Production of one micromodule requires 0.4 standard hours of the circuit printing and assembly area capacity, 0.5 standard hours of the quality control area capacity, 3 transistors and $.50 in direct costs.

Production of one circuit assembly requires 0.1 standard hour as the capacity of the circuit printing area, 0.5 standard hour of the test and packing area, 1 transistor, 3 micromodules, and $2.00 in direct costs.

Suppose that the three products may be sold in unlimited quantities at prices of $2.00, $8.00, and $25.00 each, respectively. There are 200 hours of production time open in each of the four process areas in the coming month.

2.10. Among the products of the Fuji Electronics Company of Japan is a line of transistor radios. All of the components for these radios are produced in Japan except the transistors which are imported from the United States. Fuji Electronics learns that because of a dock strike in New York City they will not receive their orders of transistors until at least a month after the date planned on. An inventory shows 4,000 TR1, 3,000 TR2, and 1,000 TR3 transistors in stock. Their line of radios requires transistors in the amounts shown below.

	Radio	*TR1*	*TR2*	*TR3*	*Unit profit (dollars)*
x_1	9 TR Model	6	2	1	20
x_2	7 TR Model	3	3	1	12
x_3	5 TR Model	3	1	1	8
x_4	3 TR Model	1	1	1	3

2.11. Jon and Sam are part of a boy-scout troop going on a camping trip and have been assigned the task of carrying all or part of the following items by their scoutmaster Scrooge McDuck: 9 cast iron frying pans, 12 pup tents, and 6 long-handled axes. These items have the following characteristics:

	Volume (ft^3)	*Weight* (*lbs*)
Frying Pans	2	4
Tents	3	2
Axes	1	5

Jon is tall and thin and is good at carrying bulky items but the weight tires him out. Sam is short, lifts weights, is proud of his strength but has trouble wrapping his arms around a large box of camping equipment. Their capacities are as follows:

	Volume (ft^3)	*Weight* (*lbs*)
Jon	40	30
Sam	20	60

To offer the boys an incentive Mr. McDuck has kindly offered them the following handsome sums:

1 pup tent 8 cents,
1 frying pan 6 cents,
1 axe 10 cents.

2.12. The Schocter and Scramble Company produces bars of soap in six different sizes: large, medium, personal, 3 ounces, 2 ounces, and 1 ounce. The production scheduling system program indicates that next month's schedule should be

18,800 containers of large size,
12,500 containers of medium size,
16,700 containers of personal size,
1,950 containers of 3 ounce size,
1,750 containers of 2 ounce size,
700 containers of 1 ounce size.

Actual monthly productions are supposed to vary from the predicted schedule by no more than 10 percent per miniature brand size (3, 2, and 1 ounce) and 5 percent for the other sizes.

Production requirements: A given amount of operating time is required to make one container of each type of soap. The relevant data are:

.013 unit operating hours/container of large soap,
.012 unit operating hours/container of medium soap,
.011 unit operating hours/container of personal soap,
.037 unit operating hours/container of 3 ounce soap,
.037 unit operating hours/container of 2 ounce soap,
.038 unit operating hours/container of 1 ounce soap.

Evaluation of unit maintenance requirements indicates that there is a maximum of 980 unit operating hours available for the given month.

Packing department requirements: The following unit packing hours are required to impress, wrap, and package each type of soap:

.015 unit packing hours for one container of large soap,
.015 unit packing hours for one container of medium soap,
.017 unit packing hours for one container of personal soap,
.031 unit packing hours for one container of 3 ounce soap,
.038 unit packing hours for one container of 2 ounce soap,
.076 unit packing hours for one container of 1 ounce soap.

Packing can be done on three units. Unit 1 can impress and wrap only large and 3 ounce brand sizes. Unit 2 can only work on medium and 2 ounce sizes. Unit 3 can only work on personal and 1 ounce sizes. Setup costs are negligible and only one size can be impressed at a time on its designated unit.

Evaluation of the packing unit maintenance requirements indicates there are 330 packing unit hours available on unit 1, 269 on unit 2 and 325 on unit 3.

Costs: Approximate operating cost/container are:

1.44 per container for large size,
1.36 per container for medium size,
1.54 per container for personal size,
2.30 per container for 3 ounce size,
3.76 per container for 2 ounce size,
4.23 per container for 1 ounce size.

These costs are based on the full utilization of the packing department (that is, unit 1 used for 330 hours, unit 2 used for 269 hours and unit 3 used for 325 hours). Find an optimal actual production schedule.

2.13. At Big River Air Force Base the engine buildup shop is manned by three civil service employees. Because of current budget requirements the Director of Maintenance requires the officer in charge not to work his men overtime but to limit their work week to forty hours. There are three types of aircraft on the base: T-33's, B-47's and, F-106's and these use J-33, J-47, and J-75 engines, respectively. The maintenace officer wishes to program engines from the factory into the buildup shop in such a way as to maximize the total number of flying hours for the base. The engine buildup requirements are as follows.

	J-33	*J-47*	*J-75*	
Engine mechanic	3	1	4	Man-hours per engine
Engine electrician	2	2	3	
General mechanic	4	1	2	
Flying hours per engine before overhaul	100	75	80	

Because of predicted engine loss it is required that the number of J-33 engines in buildup per week be at least equal to the sum of the numbers of the other engines in build up.

2.14. Three grain elevators E_1, E_2, and E_3 contain respectively 10, 20, and 22 units of grain. All the grain is to be shipped to two flour mills requiring 25 and 27 units, respectively. The shipping costs per unit are:

Mill / *Elevator*	M_1	M_2
E_1	20	10
E_2	5	8
E_3	6	19

Can you determine an optimal solution and minimum cost by trial-and-error experimentation?

2.15. Shown below is a map of a traveling salesman's territory involving four cities, each of which he must visit exactly once in a trip. If the cost of going from one city to another is the same in either direction, what is the maximum number of cost-distinct trips possible? The costs of traveling from city i to city j are shown in the following table:

j / i	*1*	*2*	*3*	*4*
1	0	2	6	2.5
2	2	0	2	1
3	6	2	0	3.5
4	2.5	1	3.5	0

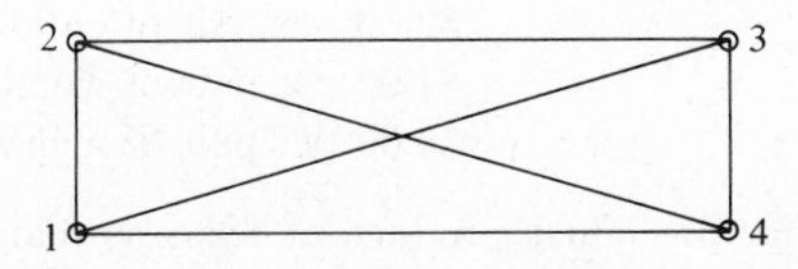

Can you trace out the salesman's most economical trip?

2.16. International Silverplate Company produces cups, trophies, and mugs at a unit profit of \$20.00, \$30.00, and \$10.00 respectively. From the following technology table, can you extract a linear programming problem? If the data are incomplete supply some additional (hypothetical) data.

	Labor hours	*Pounds of silver*
Cups	3	1
Trophies	5	4
Mugs	1	2

2.17. Formulate the following as a transportation problem:

							Origin supply
	5	3	7	3	8	5	3
	5	6	12	5	7	11	4
	2	8	3	4	8	2	2
	9	6	10	5	10	9	8
Destination requirement	3	3	6	2	1	2	

Numbers in the box denote shipping cost per unit.

2.18. A resort merchant has short-term credit available to him of up to $10,000 at a cost of two percent for the three months of the skiing season. He has assembled the following data on his operations.

	Cost	*Sales price*	*Maximum number sold*
Skis	30	50	200
Ski poles	10	20	400
Boots	12	30	300

Assume that he sells at most two-thirds of his stock of boots at retail and that the remainder are disposed of at $8.00 per pair. All his salesmen work on a commission, so his sales cost consists of a commission of 10 percent of retail price for items sold by salesmen. Rent for his shop is $600.00 per season. What is the merchant's maximum profit and how does he achieve it?

2.19. The Neet Shoe Company can produce as many as 200 pairs of men's shoes per day, 400 pairs of women's shoes per day, and 800 pairs of children's shoes per day. The profits on the shoes are as follows:

$2.00 per pair of children's shoes,
$4.00 per pair of women's shoes,
$6.00 per pair of men's shoes.

The factory must produce at least as many pairs of men's and women's shoes as the number of children's shoes less one hundred.

2.20. As a toy manufacturer, employing 50 laborers and running on a 40-hour work week, you find that your best-selling item is a line of stuffed St. Bernard dogs. These dogs come in three sizes: miniature, regular, and large. Stuffing, covering, and labor is required in the manufacturing of these dogs. Your distributor informs you that a retailer is having a grand opening sale in one week. The retailer has agreed to buy as many dogs as you can supply in that time provided the number of miniatures is twice the number of regulars and large combined. Your inventory records show that you have 800 pounds of stuffing and 1,000 square yards of covering on hand. The table indicates the amount of raw materials used for each size dog and the selling price per unit.

	Pounds of stuffing	*Square yards of covering*	*Labor*	*Selling price (dollars)*
Miniature	2	1	5	2.00
Regular	4	2	8	4.00
Large	8	4	10	7.00

2.21. Solid rocket fuels A, B, and C are to be blended to provide a satisfactory fuel at minimum cost. Three factors are measured for each fuel type as follows:

	Thrust work per 10 cubic inches	*Corrosion factor per 10 cubic inches*	*Weight in pounds per 10 cubic inches*
Fuel A	10	10	6
Fuel B	5	4	10
Fuel C	2	6	8

The fuels must be blended so that the average thrust work per 10 cubic inches is at least 4.2. The corrosion factor for the mixture cannot exceed 6.4 per 10 cubic inches and the weight must be less than or equal to 8.0 pounds per 10 cubic inches. If fuel A costs $1,000.00 per 10 cubic inches, fuel B costs $500.00 and fuel C costs $800.00, what is the minimum cost for 1,000 pounds of fuel which are required to put a man in orbit?

2.22. The Soft Suds Brewing and Bottling Company, because of faulty planning, was not prepared for a long weekend coming up. There was to be a big party at the University of Michigan and Gus Guzzler, the manager, knew that Soft Suds would be called upon to supply the refreshments. However, the raw materials required had not been ordered and could not be obtained before the weekend. Gus took an inventory of the available supplies and found the following:

Malt	500 units
Hops	250 units
Yeast	250 units.

Soft Suds produces three types of pick-me-ups: light beer, dark beer, and malt liquor, with the following specifications:

	Requirements per gallon		
	Malt	*Hops*	*Yeast*
Light beer	2	1	2
Dark beer	4	3	2
Malt liquor	7	2	2

The light beer brings $1.00/gallon profit, the dark beer 2.00/gallon profit, and the malt liquor 3.00/gallon profit. Knowing the students will buy whatever is made available, how should Gus plan production to maximize profits?

2.23. Rattan Incorporated makes living-room sets, dining-room sets, and bedroom sets. The production process has three stages: carpentry, upholstery, and finishing. Based on past sales, the company forecasts that demand for living-room sets will be at least equal to the demand for dining-room sets and bedroom sets combined. Raw materials used are rattan, wood, and padding. The supply of raw material is assumed to be unlimited for the planning period. The following are known and the company wants to know what product mix will maximize profit.

Product	*Operation*	*Time (hours/set)*	*Profit (dollars/set)*
Living-room set	Carpentry	9	
	Upholstery	5	60.00
	Finishing	3	
Dining-room set	Carpentry	8	
	Upholstery	2	50.00
	Finishing	5	
Bedroom set	Carpentry	10	
	Upholstery	7	100.00
	Finishing	6	

Labor available for each operation:

Carpentry	440 hours/week
Upholstery	260 hours/week
Finishing	205 hours/week.

2.24. A charter flying company in Florida contracts to deliver in any quantity the following items to a client.

Item	*Density (pounds/cubic foot)*	*Profit (dollars/pound)*
Ammunition	80	0.20
Guns	40	0.30
Drugs	20	0.10

The company has two air planes. Airplane A can carry 1500 pounds or 100 cubic feet of cargo, whichever is less. Airplane B can carry 2500 pounds of cargo or 200 cubic feet, whichever is less. One restriction imposed: no more than 100 pounds of drugs may be delivered with each delivery. (One flight each by both A and B is defined as a delivery.)

2.25. A company manufactures drill rod from coils of wire. The manufacturing process has three stages which we denote as S_1, S_2, and S_3. It can sell the rod in any quantity in the following sizes: 0.5000 inch, 0.3750 inch, 0.2500 inch. The following data are known and the company wants to know what product mix will maximize profit.

Size	*Operation*	*Time (hours/100 pounds)*	*Profit (dollars/100 pounds)*
0.5000	S_1	.10	2.15
	S_2	.05	
	S_3	.07	
0.3750	S_1	.23	4.98
	S_2	.13	
	S_3	.09	
0.2500	S_1	.26	6.05
	S_2	.22	
	S_3	.11	

Departmental capacities	
S_1	800 hours/week
S_2	600 hours/week
S_3	500 hours/week

2.26. Set up the following problem by using linear programming. A ship has three cargo holds, forward, aft, and center. The capacity limits are:

Forward	2,000 tons	100,000 cubic feet
Center	3,000 tons	135,000 cubic feet
Aft	1,500 tons	30,000 cubic feet.

The following cargoes are offered and the shipowners may accept all or any part of each commodity:

Commodity	*Amount (tons)*	*Volume (cubic feet/ton)*	*Profit (dollars/ton)*
A	6,000	60	6
B	4,000	50	8
C	2,000	25	5

In order to preserve the trim of the ship, the weight in each hold must be proportional to the capacity in tons. How should the cargo be distributed so as to maximize profits?

2.27. The Spanish company *LANZ IBERICA, S. A.*, manufactures three models of tractors: 28 HP, 38 HP, and 60 HP. The market conditions are such that they can sell as many tractors as they produce, at least during the planning period. At the end of a certain unusual run, after having completed the normal production, the firm has all the parts to assemble four 28 HP tractors, seven 38 HP tractors, and two 60 HP tractors, except one of the gears of the gearshift, which is different for each of the three models. It is interested in finishing as many of these 13 tractors as soon as possible in order to send them to a certain customer.

To manufacture these gears one must use two different machines plus the corresponding labor. The factory has the opportunity to use these machines during the following two weeks, because of the beginning of a new run. The available hours of the two machines and labor are the following:

Machine 1	40 hours
Machine 2	43 hours
Labor	56 hours.

The three types of gears for the three tractor models require the following amount of time (number of hours) to be produced:

	Machine 1	*Machine 2*	*Labor*
1 gear tractor 28 HP	1	5	2
1 gear tractor 38 HP	3	4	5
1 gear tractor 60 HP	2	1	6

A 28 HP tractor is known to contribute $500.00 to the firm's profit, and contributions to profit per 38 HP tractor and per 60 HP tractor are $1,000.00, and $200.00, respectively.

Determine the optimal profit that can be expected after the finished tractors are sold.

2.28. Formulate the following as a transportation problem.

							Destination requirements
	2	0	4	0	5	3	4
	1	6	6	0	1	6	3
	0	7	0	2	4	0	2
	2	6	5	4	1	3	4
	4	1	4	0	3	4	2
Supplies at origins	5	2	3	1	2	2	

2.29. Formulate the following as a transportation problem.

						Destination requirements
	7	3	5	16	7	25
	6	8	4	12	10	30
	10	5	2	11	4	35
	13	9	6	14	5	80
	9	8	4	9	3	60
	11	7	9	10	5	25
	4	8	3	13	7	15
Supplies at origins	30	60	90	35	55	

2.30. Three types, I, II, and III of steel scrap are to be shipped to the Crushem Steel Company for remelting. The shipper can mix the three kinds in the available shipping car, which has a 100 cubic foot capacity and a 7,000 pound net weight limit. He wishes to maximize his profit from the sale of this scrap. The steel company requires that at least 60 cubic feet of type II be sent.

The availabilities, densities, and profit to the shipper are shown below:

	Cubic feet available	*Pounds per cubic foot*	*Dollars per cubic foot profit*
Scrap I	180	100	5
Scrap II	100	50	4
Scrap III	50	25	2

What should the shipper's program be in order to maximize profit?

2.31. A foreman of a production line has seven jobs to fill and seven persons available for the jobs. On the basis of tests and other data the manager has a numerical productivity rating available for each person on each job. These data are:

		Job						
		1	2	3	4	5	6	7
	1	6	8	12	4	1	9	6
	2	3	4	10	7	3	2	3
	3	5	3	8	9	6	4	8
Person	4	1	4	1	12	10	2	5
	5	7	5	3	3	9	3	2
	6	8	8	9	9	6	7	4
	7	2	6	10	8	9	1	9

Productivity ratings

(a) Assuming that a person can perform one job only if he is assigned to it, what is the best or optimal assignment?

(b) Is the best assignment one in which each person is assigned to the job he can do best (the job for which his personal productivity is greatest)?

2.32. The IBUM Corporation manufactures and assembles low-cost computers on a job-shop basis. The company has fallen behind in its monthly production schedule and finds that at least 20 finished computers must be completed within the next week. All the necessary parts have been fabricated but two subassembly operations and the final assembly operations remain to be performed. The assembly sequence is:

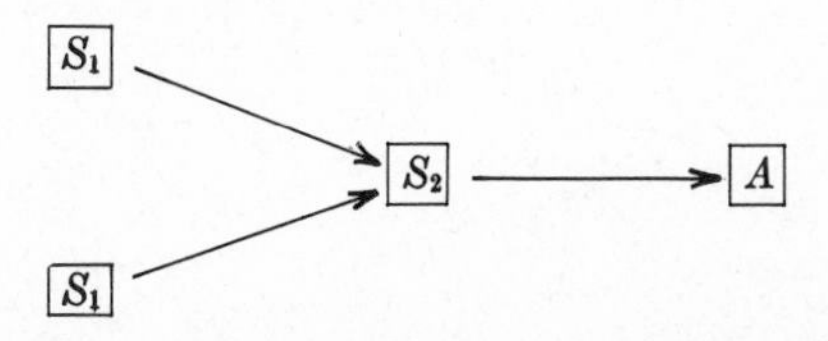

S_2, denoting subassembly 2, requires two subassemblies S_1; final assembly A utilizes the one subassembly S_2. A total of 200 man-hours is authorized for the next week in order to meet the contract. If the contract is met it is desired to use the remaining man-hours available to build up stock. Engineering standards show that 10 man-hours/S_1, 10 man-hours/S_2 and 40 man-hours/A are required. Standard cost rates for the three operations to be performed are \$6.00/S_1 hour, \$4.00/S_2 hour, and \$5.00/A hour.

Is the contract to be met?

The Simplex Algorithm

3

3.1. PREVIEW OF THE SIMPLEX ALGORITHM.

In the previous chapters we have seen that a wide variety of processes can be described mathematically by linear programming models. These models may involve either linear inequalities or linear equations (or both) in which the variables are most frequently nonnegative real numbers. We have also seen that by introducing a new nonnegative variable, called a slack variable, we can make any linear inequality into a linear equation. In the present introductory section we give a preview of the simplex algorithm using simple problems involving equations in nonnegative variables. In later sections of the chapter we will develop the algorithm in a form which covers the general case and in Section 3.10 we give a formal treatment of the geometry underlying the simplex algorithm. We have leaned heavily on the work of the early founders of the subject, Dantzig (Reference 13) and Charnes and Cooper (Reference 10), in the development of this chapter, and we have adapted freely from the concepts of Tucker (Reference 44) and Balinski and Gomory (Reference 2) in the reduction from standard to canonical form. In particular, Sections 3.5 through 3.8 below were inspired by the work of Balinski and Gomory (Reference 2) and G. Graves (Reference 21). A full discussion of the history of the subject is to be found in Dantzig (Reference 13) and Charnes and Cooper (Reference 10).

The diet problem (introduced in Chapter 1) illustrated the fact that the key to linear programming is the determination of vertices of the feasibility region. The simplex algorithm is a step-by-step process for successive determination of vertices each of which is better (less costly) than the preceding one. From the algebraic point of view a vertex is determined by solving a system of equations for certain variables (called basic) in terms of the remaining ones (nonbasic).

We begin with a simple illustrative example in which a starting vertex is immediately available. Consider the problem: given the feasibility conditions

$$x_i \geq 0 \qquad (i = 1, \cdots, 5), \tag{3.1}$$

$$\begin{aligned} x_1 \quad + 3x_3 - 2x_4 + 7x_5 &= 9, \\ x_2 + 3x_3 - 3x_4 - 2x_5 &= 2, \end{aligned} \tag{3.2}$$

find a vector X which minimizes

$$z = 2x_1 + x_2 + 12x_3 - 3x_4 + 17x_5$$

or, equivalently,

$$z = \qquad 3x_3 + 4x_4 + 5x_5 + 20. \tag{3.3}$$

The system (3.2) is in solved form with x_1, x_2 determined in terms of x_3, x_4, x_5. The reduction from the first to the second expression for z is made by using (3.2) to eliminate x_1 and x_2 from the first expression for z. [To do this, begin with a system of three equations the first two of which are (3.2) and the third is the first expression for z. Eliminating x_1 and x_2 from the third equation in this system results in a system the first two equations of which are again (3.2) and the third of which is (3.3)].

We say that (3.2) determines x_1, x_2 as *basic* (or dependent) variables and x_3, x_4, x_5 as *nonbasic* (or independent) variables. A system of equations is said to be in *canonical form* relative to a given sequence of basic variables if (1) each basic variable appears in exactly one equation and has coefficient one in that equation and (2) each equation contains exactly one basic variable (with nonzero coefficient). The *cost function* is said to be in *canonical form* if in it each basic variable has coefficient zero. Thus here we say that the system (3.2) and (3.3) is in canonical form relative to the basic variables x_1, x_2.

If a system is in canonical form relative to a sequence of basic variables and we set each nonbasic variable equal to zero, each basic variable is then equal to the constant on the right-hand side in its equation. The resulting *solution* is said to be associated with the given canonical form and is called *basic*. For example, the initial basic solution associated with (3.2) and (3.3) is

$$x_1 = 9, x_2 = 2, x_3 = x_4 = x_5 = 0, z = 20. \tag{3.4}$$

Since the right-hand sides in (3.2) are positive this solution is feasible [that is, it satisfies (3.1)]. Clearly it is also optimal since no nonnegative values of the nonbasic variables x_3, x_4, x_5 can decrease the right-hand side of (3.3) below 20.

This inference is made possible by two properties: (a) there are no basic variables in the expression (3.3) for z, and (b) the coefficients of the nonbasic variables in (3.3) are nonnegative.

We next consider a slightly altered problem where we preserve property (a) but not (b). Suppose we retain the feasibility conditions (3.1) and (3.2) and seek a vector X which minimizes

$$z = 2x_1 + x_2 + 12x_3 - 11x_4 + 17x_5$$

or, equivalently,

$$z = \qquad 3x_3 - 4x_4 + 5x_5 + 20. \tag{3.5}$$

The associated basic solution is still given by (3.4) but this time it is obvious that introducing positive values of the nonbasic variable x_4 into a solution will reduce z. Indeed, for every $x_4 \geq 0$ (leaving the other nonbasic variables unchanged)

$$x_1 = 9 + 2x_4, \qquad x_2 = 2 + 3x_4, \qquad x_3 = 0, \qquad x_4 = x_4, \qquad x_5 = 0 \tag{3.6}$$

is a feasible vector with corresponding cost

$$z = 20 - 4x_4. \tag{3.7}$$

The situation is illustrated in Figure 3.1.

Clearly z can be made as small (large negative) as we wish by choosing x_4 sufficiently large. Hence z has no lower bound and there is no optimal solution.

Finally, consider another choice of objective function

$$z = 2x_1 + x_2 + 6x_3 - 3x_4 + 17x_5$$

or, equivalently,

$$z = -3x_3 + 4x_4 + 5x_5 + 20. \tag{3.8}$$

The associated basic solution is still given by (3.4); we can reduce z below 20 by introducing positive values of the nonbasic variable x_3. The improved solutions have the form

$$x_1 = 9 - 3x_3, \qquad x_2 = 2 - 3x_3, \qquad x_3 = x_3, \qquad x_4 = x_5 = 0 \tag{3.9}$$

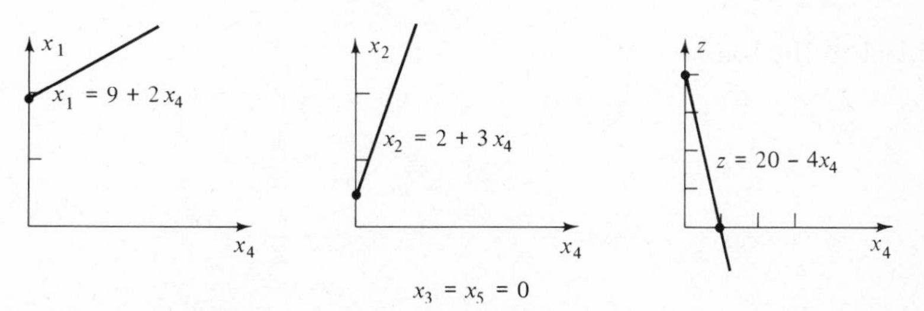

Figure 3.1 Effect of change in x_4.

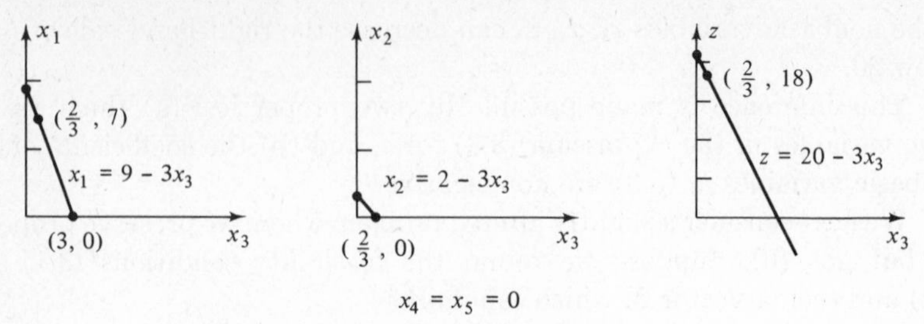

Figure 3.2 Effect of change in x_3.

with corresponding cost

$$z = 20 - 3x_3. \tag{3.10}$$

We see algebraically or from Figure 3.2 that if we choose x_3 too large we violate (3.1); indeed $x_3 = 2/3$ reduces x_2 to 0 and hence no larger value of x_3 than 2/3 is admissible. For $x_3 = 2/3$ we have

$$x_1 = 7, \quad x_2 = 0, \quad x_3 = \frac{2}{3}, \quad x_4 = x_5 = 0, \quad z = 18. \tag{3.11}$$

Comparing (3.4) with (3.11) we see that x_2 and x_3 have apparently changed roles in their classification as basic and nonbasic variables. Accordingly, we solve (3.2) for x_1 and x_3 and obtain

$$\begin{aligned} x_1 - x_2 \qquad\quad + x_4 + 9x_5 &= 7, \\ \frac{1}{3}x_2 + x_3 - x_4 - \frac{2}{3}x_5 &= \frac{2}{3}, \end{aligned} \tag{3.12}$$

and using these equations to eliminate x_3 from (3.8) we get

$$z = x_2 \qquad + x_4 + 3x_5 + 18. \tag{3.13}$$

This system (3.12), (3.13) is in canonical form relative to the basic variables x_1 and x_3 and the associated basic solution is given by (3.11) and is feasible. Since in addition the canonical cost function (3.13) now has properties (a) and (b) we know that the solution (3.11) is optimal.

As another example, consider the following problem:

$$\text{Minimize } z = -\frac{1325}{2} + \frac{25}{2}x_1 - \frac{5}{2}x_4 + \frac{5}{4}x_5 + 15x_7$$

subject to the constraints

$$\begin{aligned} & x_i \geq 0 \qquad (i = 1, \cdots, 7) \\ & \frac{1}{4}x_1 + x_2 \qquad - \frac{1}{4}x_4 + \frac{5}{8}x_5 \qquad - \frac{1}{2}x_7 = \frac{45}{4}, \\ & \frac{1}{4}x_1 \qquad + x_3 + \frac{3}{4}x_4 - \frac{3}{8}x_5 \qquad + \frac{1}{2}x_7 = \frac{5}{4}, \\ & \qquad\qquad\qquad x_4 + \frac{4}{8}x_5 + x_6 - \ x_7 = 10. \end{aligned} \tag{3.14}$$

The problem is at once seen to identify x_2, x_3, and x_6 as basic variables and the associated basic solution is

$$x_1 = 0, \quad x_2 = \frac{45}{4}, \quad x_3 = \frac{5}{4}, \quad x_4 = x_5 = 0,$$

$$x_6 = 10, \quad x_7 = 0, \quad z = -\frac{1325}{2}.$$

Notice, however, that in the expression for z the coefficient of x_4 is negative so that by allowing x_4 to have a positive value, we discover that

$$\begin{aligned} x_2 &= \frac{45}{4} + \frac{1}{4}x_4, \\ x_3 &= \frac{5}{4} - \frac{3}{4}x_4, \\ x_6 &= 10 - x_4. \end{aligned}$$

The largest value of x_4 consistent with the second equation of (3.14) is 5/3 (see Figure 3.3). We then have

$$x_1 = 0, \quad x_2 = \frac{35}{3}, \quad x_3 = 0, \quad x_4 = \frac{5}{3}, \quad x_5 = 0,$$

$$x_6 = \frac{25}{3}, \quad x_7 = 0, \quad z = -\frac{2000}{3}. \tag{3.15}$$

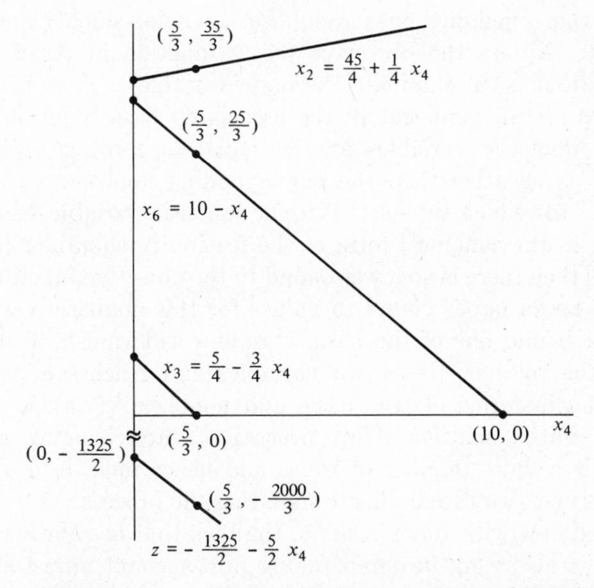

Figure 3.3 Consolidated display of effects of introducing a new activity.

Again, two variables have changed roles, namely x_3 and x_4. Solving (3.14) for x_2, x_4, and x_6 gives

$$\begin{aligned}
\frac{1}{3}x_1 + x_2 + \frac{1}{3}x_3 \qquad + \frac{1}{2}x_5 \qquad - \frac{1}{3}x_7 &= \frac{35}{3}, \\
\frac{1}{3}x_1 \qquad + \frac{4}{3}x_3 + x_4 - \frac{1}{2}x_5 \qquad + \frac{2}{3}x_7 &= \frac{5}{3}, \\
-\frac{1}{3}x_1 \qquad - \frac{4}{3}x_3 \qquad + x_5 + x_6 + \frac{1}{3}x_7 &= \frac{25}{3},
\end{aligned}$$

and for these new basic variables the corresponding canonical cost function is

$$z = -666\frac{2}{3} + \frac{40}{3}x_1 + \frac{10}{3}x_3 + \frac{50}{3}x_7.$$

That we now have an optimal feasible solution can be seen by the nonnegativity of the coefficients of the x_i in our expression for z. The absence of an x_5 term in this expression indicates the existence of other optimal vectors X; this topic is treated in Section 3.9. (This problem comes from the situation described in Table 3.1 (Section 3.2). A full solution in simplex form is given at the end of Section 3.3.)

These four simple examples give almost the full background for the simplex algorithm. The first two illustrate what turn out to be the two possible terminal states and the third and fourth illustrate the main step of the simplex algorithm. In essence, the simplex algorithm rests on solving the feasibility equations for certain of the variables (basic or dependent) in terms of the remaining ones (nonbasic or independent) and using this solution to express the objective (cost) function in terms of nonbasic variables (that is, to eliminate the basic variables). Then if the resulting coefficients are all nonnegative the associated basic solution (the one in which all nonbasic variables are set equal to zero) is optimal. If any coefficient is negative then the corresponding nonbasic variable can be introduced to reduce the cost. If that nonbasic variable has no positive coefficient in the canonical form of the feasibility equations [for example, x_4 in (3.2)] then there is no lower bound to the objective function; otherwise there will be an upper bound to values for this nonbasic variable and at this upper bound one of the basic variables will vanish. Then one interchanges the roles of these two variables and reaches a new canonical formulation in terms of new basic and nonbasic variables and with an improved initial solution. This process of interchanging can only be carried out a finite number of times and hence must lead eventually to one of the first two cases, thus terminating the process.

Armed now with these insights, the task that lies ahead is to develop the simplex algorithm in detail taking into account numerous structural properties such as mixtures of equations and inequalities, degeneracy, multiple optimal solutions, and variables which can be of either sign, restricted, or free.

An important feature of this development is the ability to transform a problem into a new one which is considered easier to handle. Two problems are said to be *equivalent* if they are so related that the solution (or solutions) of one can be readily obtained from that (or those) of the other.[1]

As a simple example of equivalent problems consider first the diet problem of Chapter 1 described by inequalities (1.2) and cost function $z = 25x + 50y$ and second the problem defined by inequalities (1.2) but with the objective (cost) function z replaced by

$$w = d + 25x + 50y$$

obtained by setting $w = z + d$. Here d might represent the cost of a trip to the grocery store to purchase and bring home the foods. Clearly w gives a more complete description of total cost.

Since the feasibility conditions are the same and since for every vector X we have $w - z = d$, it is clear that an optimal diet vector X^* for the first problem will also be optimal for the second problem and vice versa. Thus the two problems are seen to be equivalent.

This argument is sometimes used to justify setting $d = 0$ in the initial cost function $z = d + c_1x_1 + \cdots + c_nx_n$ for any linear programming problem, as is done in many discussions of linear programming models. However, in our formal development we retain the d term for two reasons. First, this procedure provides for a description of "absolute" as contrasted with merely "relative" cost. Second, and more important, it states the problem initially in a form which fits later stages as well; for even though $d = 0$ initially, after simplex transformations there will be a nonzero constant term in the cost function.

In the development of the simplex algorithm which follows we will make frequent transformations to equivalent problems which will in some cases involve introduction of additional variables or even of additional constraints. But in each case optimal solutions of the original problem will be obtainable from those of the transformed "equivalent" problems.

3.2. MATRIX FORMULATION; STANDARD AND CANONICAL FORMS.

We begin our discussion of linear programming in this section. Key concepts are those of standard and canonical form. Their introduction brings us to a stage where further discussion and development are greatly facilitated by the use of matrix and vector notation and language. Readers who are not familiar with these concepts are referred to

[1]In some cases the equivalence is quite formally defined so that it is a mathematical equivalence, that is, a reflexive, symmetric, and transitive relation on a set. In other cases the term "equivalence" is used in a more general sense; this is the case in the class of production planning models introduced in Section 3.2 [compare 3.32 and 3.33]. See also Section A.11, Appendix A.

Appendix B which contains a concise summary of matrix and vector operations and other parts of linear algebra which are needed for our development of linear programming (Sections B.1 through B.4 are particularly relevant at this stage). The present section also includes the introduction of some special notational conventions for vectors and matrices which are useful in our treatment of linear programming.

For our purposes the term "vector" will mean a column of real numbers. Thus we can regard a vector as being a matrix having only one column, that is, a column vector. For example,

$$\begin{bmatrix} 3 \\ 2 \\ -1 \end{bmatrix}, \quad \begin{bmatrix} 1 \\ 0 \\ 0 \\ 0 \end{bmatrix}, \quad \begin{bmatrix} x_1 \\ x_2 \\ x_3 \\ x_4 \\ x_5 \end{bmatrix} \tag{3.16}$$

are vectors. The number of rows in a vector is called its *degree*. Thus the three vectors displayed above have degrees 3, 4, 5, respectively. We use the term *row vector* to denote a matrix having only one row. We will ordinarily encounter row vectors as transposes of column vectors. For example,

$$[3 \quad 2 \quad -1] = \begin{bmatrix} 3 \\ 2 \\ -1 \end{bmatrix}^T, \qquad [x_1 \quad x_2 \quad x_3 \quad x_4 \quad x_5] = \begin{bmatrix} x_1 \\ x_2 \\ x_3 \\ x_4 \\ x_5 \end{bmatrix}^T. \tag{3.17}$$

We denote matrices and vectors by capital letters.

Our approach to the simplex algorithm is to begin with a class of linear programming problems which is characterized by the presence of a number of special features which facilitate the solution process. We proceed to adapt the basic algorithm successively to more general classes of problems until, finally, we reach the most general class.

A general linear programming problem involves optimization of a linear objective function subject to both inequality and equality constraints in both nonnegative and free variables. A linear programming problem is said to be in *standard form* if it is stated as a minimization problem and if its constraints, other than those on the signs of the variables, are equations in nonnegative variables, that is, if it can be written in the form

$$\text{Minimize} \qquad z = d + C^T X \tag{3.18}$$

subject to the constraints

$$AX = B \tag{3.19}$$

and

$$X \geq 0, \tag{3.20}$$

where (1) A is a p by n matrix called the *coefficient matrix*, (2) B is a p by 1 vector called the *constraint vector*, (3) C is an n by 1 vector called the *cost vector*, (4) d is a scalar called the *constant cost term*, and (5) X is an n by 1 variable vector called the *program vector*.

We illustrate matrix formulation with some examples from Chapters 1 and 2. The wood products problem as expressed by Equations (1.11) and (1.13) can be put in the form above by choosing

$$A = \begin{bmatrix} 3 & 4 & 1 & 1 & 0 \\ 1 & 3 & 3 & 0 & 1 \end{bmatrix}, \quad B = \begin{bmatrix} 25 \\ 50 \end{bmatrix}, \quad C = \begin{bmatrix} -8 \\ -19 \\ -7 \\ 0 \\ 0 \end{bmatrix}, \quad d = 0. \tag{3.21}$$

If we express the wood products problem by the equivalent Equations (1.14) and (1.15), then we have instead of (3.21)

$$A = \begin{bmatrix} \frac{8}{9} & 1 & 0 & \frac{1}{3} & -\frac{1}{9} \\ -\frac{5}{9} & 0 & 1 & -\frac{1}{3} & \frac{4}{9} \end{bmatrix}, \quad B = \begin{bmatrix} \frac{25}{9} \\ \frac{125}{9} \end{bmatrix}, \quad C = \begin{bmatrix} 5 \\ 0 \\ 0 \\ 4 \\ 1 \end{bmatrix}, \quad d = -150. \tag{3.22}$$

The wheat distribution problem of Section 2.6 is in standard form with

$$A = \begin{bmatrix} 1 & 1 & 1 & 1 & & & & & & & & \\ & & & & 1 & 1 & 1 & 1 & & & & \\ & & & & & & & & 1 & 1 & 1 & 1 \\ 1 & & & & 1 & & & & 1 & & & \\ & 1 & & & & 1 & & & & 1 & & \\ & & 1 & & & & 1 & & & & 1 & \\ & & & 1 & & & & 1 & & & & 1 \end{bmatrix}, \qquad B = \begin{bmatrix} 5 \\ 9 \\ 4 \\ 10 \\ 3 \\ 2 \\ 4 \end{bmatrix}, \tag{3.23}$$

$$C^T = [6\ 4\ 3\ 2\ 5\ 2\ 4\ 4\ 4\ 2\ 2\ 4], \qquad d = 0.$$

Here the blank positions in A represent zero coefficients.

The gasoline blending problem of Section 2.3 is not in standard form. Each of the seven relations in (2.5.), (2.6), and (2.7) must first be reduced to equation form by introduction of a slack variable. The resulting system of constraints can then be written in the form:

$$\begin{aligned} x_1 + x_4 + x_7 &= 32{,}000, \\ x_2 + x_5 + x_8 &= 20{,}000, \\ x_3 + x_6 + x_9 &= 38{,}000, \\ -8x_1 + 10x_2 + 27x_3 + x_{10} &= 0, \\ -2x_1 + 4x_2 - x_3 + x_{11} &= 0, \\ -28x_1 - 10x_5 + 7x_6 + x_{12} &= 0, \\ -5x_1 + x_5 - 4x_6 + x_{13} &= 0, \\ x_j &\geq 0, \qquad (j = 1, \cdots, 13). \end{aligned} \tag{3.24}$$

The slack activities enter the profit equation with zero coefficients giving

$$w = -x_1 + 3x_2 + 4x_3 - 3x_4 + x_5 + 2x_6 + 0x_7 + \cdots + 0x_{13}. \quad (3.25)$$

Now maximizing w is the same as minimizing $z = -w$ and with z thus defined the blending problem is in standard form with

$$A = \begin{bmatrix} 1 & & & 1 & & & 1 & & & & & & \\ & 1 & & & 1 & & & 1 & & & & & \\ & & 1 & & & 1 & & & 1 & & & & \\ -8 & 10 & 27 & & & & & & & 1 & & & \\ -2 & 4 & -1 & & & & & & & & 1 & & \\ & & & -28 & -10 & 7 & & & & & & 1 & \\ & & & -5 & 1 & -4 & & & & & & & 1 \end{bmatrix}, \quad B = \begin{bmatrix} 32{,}000 \\ 20{,}000 \\ 38{,}000 \\ 0 \\ 0 \\ 0 \\ 0 \end{bmatrix},$$

$$C^T = [\ 1\ -3\ -4\quad 3\ -1\ -2\ 0\ 0\ 0\ 0\ 0\ 0\ 0], \qquad d = 0. \quad (3.26)$$

We leave to the reader verification that introduction of slack activities will put the diet problem of Section 1.1 in standard form, where

$$A = \begin{bmatrix} 3 & 1 & -1 & 0 & 0 \\ 4 & 3 & 0 & -1 & 0 \\ 1 & 3 & 0 & 0 & -1 \end{bmatrix}, \quad B = \begin{bmatrix} 8 \\ 19 \\ 7 \end{bmatrix}, \quad (3.27)$$

$$C^T = [25 \quad 50 \quad 0 \quad 0 \quad 0], \qquad d = 0.$$

We observe that any linear programming problem whose variables are restricted to be nonnegative can be expressed in an equivalent standard form as follows:

1. Make the goal one of minimization. This requires the changing of signs in the objective function in case the original goal was maximization.
2. Introduce a slack variable into each inequality. The slack variables are assigned zero coefficients in the objective function.

For the simplex algorithm standard form is not enough. First, we need the equations in solved form; next, we need the cost function in a special form; and, finally, we need an initial feasible solution. For a precise statement of what is required we consider a problem in standard form with p equations and n variables and begin by selecting a sequence $S = (s_1, s_2, \cdots, s_p)$ of p distinct integers from the set $J_n = \{1, 2, \cdots, n\}$, and let G be the sequence consisting of the integers $g_1, \cdots, g_{n-p}$ that are in J_n but not in S and arranged in natural order ($g_1 < g_2 < \cdots < g_{n-p}$). Let A_j denote the jth column of A. Then a problem in standard form is said to be in *canonical form relative to the basic sequence* S if the following three conditions are satisfied:

C1. Canonical coefficients, $\quad [A_{s_1} \cdots A_{s_p}] = I_p$;
C2. Canonical costs, $\quad c_{s_i} = 0, \quad (i = 1, \cdots, p)$;
C3. Canonical constants, $\quad B \geq 0$.

We call S the *basic sequence*, its elements $s_1, \cdots, s_p$ are called *basic indices*, the variables $x_{s_1}, \cdots, x_{s_p}$ are called *basic variables*, the columns $A_{s_1}, \cdots, A_{s_p}$ of A are called *basic columns*. Correspondingly, G is the *nonbasic sequence* and we speak of *nonbasic indices, variables*, and *columns*.

The problem defined by

$$A = \begin{bmatrix} 1 & 0 & -2 & 1 & 1 & 0 & 2 \\ 2 & 1 & 1 & 0 & -2 & 0 & 2 \\ -3 & 0 & 2 & 0 & 1 & 1 & 3 \end{bmatrix}, \qquad B = \begin{bmatrix} 3 \\ 7 \\ 2 \end{bmatrix},$$

$$C = \begin{bmatrix} 3 \\ 0 \\ -2 \\ 0 \\ -4 \\ 0 \\ 7 \end{bmatrix}, \qquad d = 14,$$

is in canonical form relative to the sequence $S = (4, 2, 6)$. Here $G = (1, 3, 5, 7)$. Note that we use parentheses with S and G to indicate that the order in which the numbers occur is significant. The fact that the 4 comes first in S indicates that A_4 is the first unit vector $U_1 = \begin{bmatrix} 1 \\ 0 \\ 0 \end{bmatrix}$; similarly, A_2 is the second unit vector $U_2 = \begin{bmatrix} 0 \\ 1 \\ 0 \end{bmatrix}$, and A_6 is the third unit vector $U_3 = \begin{bmatrix} 0 \\ 0 \\ 1 \end{bmatrix}$.

We introduce a notation for submatrices which will be useful in many contexts and which in particular will afford a simple formulation of conditions C1 and C2.

Let A be a p by n matrix, let C be an n by 1 column vector, and let $R = (r_1, \cdots, r_q)$ be any sequence of integers with $1 \leq r_i \leq n$. Then we denote by A_R the p by q matrix whose columns in order (from left to right) are $A_{r_1}, \cdots, A_{r_q}$, and we denote by C_R the q by 1 column vector whose components in order are $c_{r_1}, \cdots, c_{r_q}$, that is,

$$A_R = [A_{r_1} \cdots A_{r_q}], \qquad C_R = \begin{bmatrix} c_{r_1} \\ \cdot \\ \cdot \\ \cdot \\ c_{r_q} \end{bmatrix}. \tag{3.28}$$

Thus for the example above if $R = (2, 7, 3, 1)$ then

$$A_R = \begin{bmatrix} 0 & 2 & -2 & 1 \\ 1 & 2 & 1 & 2 \\ 0 & 3 & 2 & -3 \end{bmatrix}, \qquad C_R = \begin{bmatrix} 0 \\ 7 \\ -2 \\ 3 \end{bmatrix},$$

and for $S = (4, 2, 6)$,

$$A_S = \begin{bmatrix} 1 & 0 & 0 \\ 0 & 1 & 0 \\ 0 & 0 & 1 \end{bmatrix}, \qquad C_S = \begin{bmatrix} 0 \\ 0 \\ 0 \end{bmatrix}$$

so that conditions $C1$ and $C2$ are satisfied.

In general, the conditions C1 and C2 can be written in matrix notation as

C1. Canonical coefficients, $A_S = I_p$;
C2. Canonical costs, $C_S = 0$.

When dealing with a p by n matrix A, we denote by $U_1, U_2, \cdots, U_p$ the unit vectors of degree p, that is,

$$U_1 = \begin{bmatrix} 1 \\ 0 \\ \cdot \\ \cdot \\ \cdot \\ 0 \end{bmatrix}, \qquad U_2 = \begin{bmatrix} 0 \\ 1 \\ \cdot \\ \cdot \\ \cdot \\ 0 \end{bmatrix}, \qquad \cdots, \qquad U_p = \begin{bmatrix} 0 \\ 0 \\ \cdot \\ \cdot \\ \cdot \\ 1 \end{bmatrix}.$$

For example, if $p = 5$ we have the five unit vectors,

$$U_1 = \begin{bmatrix} 1 \\ 0 \\ 0 \\ 0 \\ 0 \end{bmatrix}, \quad U_2 = \begin{bmatrix} 0 \\ 1 \\ 0 \\ 0 \\ 0 \end{bmatrix}, \quad U_3 = \begin{bmatrix} 0 \\ 0 \\ 1 \\ 0 \\ 0 \end{bmatrix}, \quad U_4 = \begin{bmatrix} 0 \\ 0 \\ 0 \\ 1 \\ 0 \end{bmatrix}, \quad U_5 = \begin{bmatrix} 0 \\ 0 \\ 0 \\ 0 \\ 1 \end{bmatrix}.$$

We observe that U_j is the jth column of I_p $(j = 1, \cdots, p)$. Therefore condition C1 means that all of the unit vectors appear as columns of the coefficient matrix; more specifically $A_{s_1} = U_1$, $A_{s_2} = U_2, \cdots, A_{s_j} = U_j$, $\cdots, A_{s_p} = U_p$. If there is some doubt as to what degree is under discussion, we can use the more complete notation $U_1^p, \cdots, U_j^p, \cdots, U_p^p$ to designate the unit vectors of degree p. For example,

$$U_2^3 = \begin{bmatrix} 0 \\ 1 \\ 0 \end{bmatrix}, \qquad \text{and} \qquad U_2^5 = \begin{bmatrix} 0 \\ 1 \\ 0 \\ 0 \\ 0 \end{bmatrix}.$$

This more complete notation may be desirable when vectors of several different degrees are being considered simultaneously.

If C1 and C2 hold then the pair X^S, z^S given by

$$\begin{aligned} x_{s_i}^S &= b_i, && (i = 1, \cdots, p), \\ x_g^S &= 0 && \text{for all } g \in G, \\ z^S &= d, \end{aligned} \tag{3.29}$$

is a solution of (3.19) and is called the *basic solution associated with* S. The vector X^S is called the *basic program* associated with S. The reason for the word "basic" will be given later. This solution is feasible if and only if C3 holds. Thus in the illustrative example above we have the basic solution

$$X^S = \begin{bmatrix} 0 \\ 7 \\ 0 \\ 3 \\ 0 \\ 2 \\ 0 \end{bmatrix}, \qquad z^S = 14.$$

To say that the coefficients are canonical (that C1 holds) means essentially that the system (3.19) has been solved for the basic variables $x_{s_1}, \cdots, x_{s_p}$ in terms of the nonbasic variables $x_{g_1}, \cdots, x_{g_{n-p}}$. When we set each nonbasic variable x_{g_j} equal to zero then the equations reduce to $x_{s_i} = b_i$ $(i = 1, \cdots, p)$. Hence the vector X^S defined in (3.29) is indeed a solution to (3.19). The property C2 states that the equations have been used to eliminate the basic variables from the cost function, and C3 states that in the solved form of the equations the constant terms b_i are non-negative.

Thus again using our illustrative example the system $AX = B$, when written out in full, takes the form

$$\begin{aligned} x_1 \qquad - 2x_3 + x_4 + x_5 \qquad\quad + 2x_7 &= 3, \\ 2x_1 + x_2 + x_3 \qquad - 2x_5 \qquad\quad + 2x_7 &= 7, \\ -3x_1 \qquad + 2x_3 \qquad + x_5 + x_6 + 3x_7 &= 2. \end{aligned}$$

Here we clearly regard the system as having been solved for the basic variables x_4, x_2, x_6 in terms of the nonbasic variables x_1, x_3, x_5, x_7. What may seem a bit strange at first is the fact that we have not transposed the nonbasic variables to the right-hand side (as is done in school algebra). The reason for this is that the simplex algorithm requires continual changes in role from basic to nonbasic for variables and hence it is convenient to keep all of the variables together on the left-hand side.

We have defined the concept "in canonical form with respect to a basic sequence S" without having fully explored the concept of basic sequence by itself. Let A be a p by n matrix. A sequence $S = (s_1, \cdots, s_p)$ is said to be *basic for* A if and only if the system (3.19) can be solved for $x_{s_1}, \cdots, x_{s_p}$ in terms of the remaining variables. Thus we see that in a solved system the terms "basic variable" and "dependent variable" are synonomous as are the terms "nonbasic variable" and "independent variable."

In the nomenclature of linear algebra a sequence S is basic for A if and only if the p columns of A_S are linearly independent, that is, if and only if the matrix A_S has an inverse (is nonsingular). Indeed, let S be a basic sequence; then

$$X_S = A_S^{-1}B - A_S^{-1}A_G X_G \tag{3.30}$$

is an explicit expression for the basic variables in terms of the nonbasic variables.

Again in the nomenclature of linear algebra a sequence $S = (s_1, \cdots, s_p)$ is basic if and only if the columns $A_{s_1}, \cdots, A_{s_p}$ of A form a basis for the space V_p of all vectors of degree p. It is this property that gives rise to the term "basic."

For our purposes the most important of these equivalent characterizations is the first, namely, that a basic sequence is one such that the corresponding basic variables can be solved for uniquely in terms of the remaining (nonbasic) variables. This elementary characterization is sufficient for presenting the simplex algorithm, whereas the more sophisticated characterizations are useful in treating the theory of linear programming.

Note that the concept of basic sequence has nothing to do with the nonnegativity of the basic program for (3.19) obtained from (3.30) by taking $X_G = 0$ and $X_S = A_S^{-1}B$. However, if S is basic for A and if $B' = A_S^{-1}B$ and $A' = A_S^{-1}A$, then the problem defined by (3.18),

$$A'X = B' \tag{3.31}$$

and (3.20) satisfies C1 relative to S and is equivalent to the initial problem.

Although we are not yet ready to discuss the problem of reducing a general linear programming problem to standard or canonical form it is worth mentioning one important special case where this reduction is especially simple. Consider a linear programming problem

$$\text{minimize} \qquad z = f_0 + F^T Y$$

subject to the constraints

$$\begin{aligned} DY &\leq B, \\ Y &\geq 0, \end{aligned} \tag{3.32}$$

where the system has p constraints (not counting $Y \geq 0$) and m variables, and where $B \geq 0$. Let $U = [u_i]$ be a p by 1 vector. Then (3.32) is equivalent to the problem

$$\text{minimize} \qquad z = f_0 + F^T Y + 0^T U$$

subject to the constraints

$$\begin{aligned} DY + I_p U &= B, \\ Y &\geq 0, \\ U &\geq 0. \end{aligned} \tag{3.33}$$

This "equivalence" follows from the fact that if z^*, Y^*, U^* is an optimal solution of (3.33) then z^*, Y^* is an optimal solution of (3.32) ,and conversely if z^*, Y^* is an optimal solution of (3.32) and we set $U^* = B - DY^*$ then z^*, Y^*, U^* is an optimal solution of (3.33).

Next we let

$$A = [D \quad I_p], \qquad X = \begin{bmatrix} Y \\ U \end{bmatrix}, \qquad C = \begin{bmatrix} F \\ 0 \end{bmatrix}, \qquad d = f_0 \tag{3.34}$$

and (3.33) takes on the standard form of (3.18), (3.19), (3.20), with $n = m + p$. Moreover, this system is in canonical form relative to the basic sequence $S = (m + 1, \cdots, n)$ and has the nonbasic sequence $G = (1, \cdots, m)$. The associated basic (feasible) solution is

$$X^S = \begin{bmatrix} 0 \\ B \end{bmatrix}, \qquad z^S = d. \tag{3.35}$$

The problem (3.32) is a mathematical model for an important class of production planning problems. Consider, for example, a factory which has technological capacity for m products and suppose that these products use, in all, p raw materials (including labor). Suppose further that top management has contracted for a fixed amount of each of these raw materials, and that each product has a ready market at known price. Then the decision facing the factory manager is to decide which product mix will maximize his total profit.

The manager obtains from his industrial engineering staff a p by m technology matrix $D = [d_{ij}]$ where d_{ij} is the number of units of the ith raw material used in producing one unit of the jth product. His decision takes the form of determining an m by 1 vector Y, called his *program vector*, where y_j is the number of units of the jth product to be made. Let $B = [b_i]$ be a p by 1 vector where b_i denotes the number of units of the ith raw material that has been allocated to his factory. Then the relations $DY \leq B$ and $Y \geq 0$ express the constraints bounding management's decision. Let $P = [p_j]$ where p_j denotes the known market price for the jth product. Let d represent his total fixed costs including costs of the raw materials (which he must pay for whether or not he uses them and which we assume have no salvage value). Then he wishes to maximize the profit $P^TY - d$. This is equivalent to minimizing $z = d - P^TY$. If we let $f_0 = d$ and $F = -P$ the manager's problem takes the form (3.33) with $B \geq 0$.

Each variable u_i is called a *slack variable*, since u_i represents the amount of the ith raw material that is left over when a given program vector Y is selected. Let Y, U be any feasible vector. If u_i is positive we say that the ith constraint is *slack* and if u_i is zero we say that the ith constraint is *tight*. Our technical use of the term "slack" resembles its everyday connotations of unused or leftover capacity.

As an illustrative numerical example, we consider a small factory whose product line consists of four items: chairs, tables, desks, and bookcases.

Table 3.1. TECHNOLOGY TABLE FOR WOOD PRODUCTS FACTORY.

	Chair	*Table*	*Desk*	*Bookcase*	*Units available*
Lumber grade 1	2	4	4	2	50
Lumber grade 2	1	1	3	3	25
Labor	2	3	5	3	40
Price per unit	20	50	80	50	

Each column of the body of Table 3.1 gives the number of units of the three inputs required to produce one unit of the corresponding output, the numbers at the right give the number of units of each input that have been contracted for by management, and the numbers at the bottom give the prevailing market prices per unit of each output. This problem can be transformed into one of minimization by change of signs in the objective function. The problem can be put into the form (3.32) if we let

$$D = \begin{bmatrix} 2 & 4 & 4 & 2 \\ 1 & 1 & 3 & 3 \\ 2 & 3 & 5 & 3 \end{bmatrix}, \quad B = \begin{bmatrix} 50 \\ 25 \\ 40 \end{bmatrix}, \quad F = \begin{bmatrix} -20 \\ -50 \\ -80 \\ -50 \end{bmatrix},$$

$$f_0 = 0, \quad Y = \begin{bmatrix} y_1 \\ y_2 \\ y_3 \\ y_4 \end{bmatrix}. \tag{3.36}$$

Here y_j represents the number of units to be made of the product described in the jth column of Table 3.1, $j = 1, 2, 3, 4$.

Then following the development of (3.33) and (3.34) we put the problem in the form (3.18), (3.19), (3.20) with

$$A = \begin{bmatrix} 2 & 4 & 4 & 2 & 1 & 0 & 0 \\ 1 & 1 & 3 & 3 & 0 & 1 & 0 \\ 2 & 3 & 5 & 3 & 0 & 0 & 1 \end{bmatrix}, \quad B = \begin{bmatrix} 50 \\ 25 \\ 40 \end{bmatrix},$$

$$C = \begin{bmatrix} -20 \\ -50 \\ -80 \\ -50 \\ 0 \\ 0 \\ 0 \end{bmatrix}, \quad d = 0, \quad X = \begin{bmatrix} x_1 \\ x_2 \\ x_3 \\ x_4 \\ x_5 \\ x_6 \\ x_7 \end{bmatrix} = \begin{bmatrix} y_1 \\ y_2 \\ y_3 \\ y_4 \\ u_1 \\ u_2 \\ u_3 \end{bmatrix}. \tag{3.37}$$

This problem is in canonical form with respect to the basic sequence $S = (5, 6, 7)$. The corresponding basic solution is

$$X^S = \begin{bmatrix} 0 \\ 0 \\ 0 \\ 0 \\ 50 \\ 25 \\ 40 \end{bmatrix},$$

$$\begin{aligned} z^S &= d + C^T X^S \\ &= 0 - 20\cdot 0 - 50\cdot 0 - 80\cdot 0 - 50\cdot 0 + 0\cdot 50 + 0\cdot 25 + 0\cdot 40 \\ &= 0. \end{aligned}$$

The cost

$$\begin{aligned} z &= d + C^T X \\ &= -20x_1 - 50x_2 - 80x_3 - 50x_4 \end{aligned}$$

can be reduced below z^S by introducing any one of the first four activities.

Suppose then, that x_1 is assigned some positive value, say t, but that x_2, x_3, x_4 remain zero. Then $x_1 = t$, $x_2 = x_3 = x_4 = 0$, $x_5 = 50 - 2t$, $x_6 = 25 - t$, $x_7 = 40 - 2t$ and $z = -20t$. Now $t \leq 20$, lest x_7 become negative thus violating the condition $X \geq 0$. For $t = 20$ we obtain the improved solution

$$X = \begin{bmatrix} 20 \\ 0 \\ 0 \\ 0 \\ 10 \\ 5 \\ 0 \end{bmatrix}, \qquad z = -20 \cdot 20 = -400. \tag{3.38}$$

The roles of x_1 and x_7 have been interchanged. If we solve the system for x_1, x_5, x_6 we obtain

$$\begin{aligned} x_2 - x_3 - x_4 + x_5 \qquad\qquad - x_7 &= 10, \\ -\frac{1}{2}x_2 + \frac{1}{2}x_3 + \frac{3}{2}x_4 \qquad + x_6 - \frac{1}{2}x_7 &= 5, \\ x_1 + \frac{3}{2}x_2 + \frac{5}{2}x_3 + \frac{3}{2}x_4 \qquad\qquad + \frac{1}{2}x_7 &= 20, \end{aligned}$$

and z becomes

$$z = -400 - 20x_2 - 30x_3 - 20x_4 + 10x_7.$$

The new basic solution is given by (3.38). Note the improvement in z from 0 to -400. Of course, cost of -400 can be interpreted as a profit of 400.

One might ask if slack activities need be assigned zero value. This is not necessary and indeed if leftover raw materials have salvage values it is desirable to assign nonzero values to slack activities. Thus, if we change the production-planning problem by introducing salvage values, say h_i for each leftover unit of the ith raw material, then our model will fit the form (3.33) if we retain the same constraints but replace the cost function $z = d - P^T Y$ by

$$z = d - P^T Y - H^T U, \tag{3.39}$$

where H is the p by 1 vector with components h_i.

We note although that this modified model still satisfies C1 and C3 relative to the basic sequence $S = (m + 1, \cdots, n)$ it no longer satisfies C2. This situation is easily remedied for we can use the equation $DY + I_p U = B$ to eliminate U from the cost function. Thus we obtain

$$\begin{aligned} z &= d - P^T Y - H^T U \\ &= d - P^T Y - H^T(B - DY) \\ &= (d - H^T B) + (-P^T + H^T D)Y \end{aligned} \tag{3.40}$$

which does satisfy C2.

Thus, in principle, we see that it is easy to obtain C2 if we have C1 satisfied. Later we will use this idea to provide a computationally convenient method for obtaining C2 given C1.

For example, if we modify the numerical example by assigning salvage values h_1, h_2, h_3, respectively, to grade 1 lumber, grade 2 lumber, and labor and carry out the calculations of (3.40) we get the adjusted cost function

$$z = -50h_1 - 25h_2 - 40h_3 + (2h_1 + h_2 + 2h_3 - 20)x_1 + (4h_1 + h_2 + 3h_3 - 50)x_2 + (4h_1 + 3h_2 + 5h_3 - 80)x_3 + (2h_1 + 3h_2 + 3h_3 - 50)x_4, \qquad (3.41)$$

which is in canonical form relative to S. In particular, if $h_1 = 3$, $h_2 = 2$ $h_3 = 0$ (leftover labor is not worth much) we get

$$z = -200 - 12x_1 - 36x_2 - 62x_3 - 38x_4.$$

3.3. THE SIMPLEX ALGORITHM, NONDEGENERATE CASE.

The simplex algorithm applies to a problem in canonical form. We develop the algorithm under the assumption that conditions C1, C2, and C3 all hold.

In this section we also assume that (a) every basic program X^S has exactly p nonzero components (that is, X_S^S has no zero components), and that (b) every sequence of p distinct columns from A is independent (in other words, every sequence S of p distinct integers is basic for A). This is called the *nondegenerate case*.[2] We will discuss the *degenerate case* later. The assumption of nondegeneracy implies, in particular, that $B > 0$ and hence that (3.29) gives a basic program X^S with p nonzero components. Each basic program involves a sequence of p columns which form a basis for V_p. Each cycle of the algorithm involves replacing one basis by another obtained by discarding one old basic column and introducing a new one. There can be four major phases in each cycle.

D1. Decision as to whether
Case I: X^S is optimal.
Case II: There is no optimal solution.
Case III: An improved basic solution is available.

If either Case I or II occurs we stop. In Case III the cycle continues with the remaining three phases.

D2. Determination of the pivot column, that is, which new column k to introduce, or which nonbasic variable should become basic.

[2]In what follows we actually use only (a). Some authors use (a) as the definition of the nondegenerate case.

D3. Determination of the pivot row h or of which old column s_h to discard, or which basic variable should become nonbasic.

D4. Determination of the canonical form corresponding to the new basic sequence S'.

We then enter a new cycle with basic sequence S', with corresponding basic program $X^{S'}$, and corresponding value of the cost function $z^{S'}$.

Any second solution X of $AX = B$ must by virtue of C1 have its p basic coordinates determined in terms of the nonbasic coordinates by

$$x_{s_i} = b_i - \sum_{g \in G} a_{ig} x_g \qquad (i = 1, \cdots, p). \tag{3.42}$$

For any choice of the x_g the vector X defined by (3.42) will be a solution of $AX = B$ and it will be feasible if, in addition, it is nonnegative. Thus we may construct new feasible vectors by choosing any nonnegative scalars x_g $(g \in G)$ for which each of $x_{s_1}, \cdots, x_{s_p}$ is nonnegative.

For any feasible X we have (using C2)

$$z = d + C^T X = d + \sum_{i=1}^{p} c_{s_i} x_{s_i} + \sum_{g \in G} c_g x_g = d + \sum_{g \in G} c_g x_g. \tag{3.43}$$

Since $X \geq 0$ we conclude that

Case I: If $c_j \geq 0$ for $j = 1, \cdots, n$ then X^S is an optimal program.

For, under the hypothesis of Case I and using (3.43), we see that $z - d$ is a sum of nonnegative terms $c_j x_j$ and hence $z \geq d$ for all feasible vectors X. (It should be noted that since $c_j = 0$ for $j \in S$ we use the full range $j = 1, \cdots, n$ in stating the hypothesis for Case I.)

Case II: If for some k, $c_k < 0$ and for this k also $A_k \leq 0$ ($a_{ik} \leq 0$ for $i = 1, \cdots, p$) then there is no lower bound to z and hence no optimal solution.

To see this consider (3.42) with $x_k \geq 0$ but keeping $x_g = 0$ for $g \neq k$, $g \in G$. Note that $k \in G$ since $c_j = 0$ for $j \in S$. Then, for all $x_k \geq 0$, $X \geq 0$ since all x_g for $g \in G$ were selected as nonnegative and since from (3.42)

$$x_{s_i} = b_i - a_{ik} x_k \geq b_i \geq 0 \qquad (i = 1, \cdots, p). \tag{3.44}$$

Here the first inequality is a consequence of the hypothesis

$$A_k = \begin{bmatrix} a_{1k} \\ \cdot \\ \cdot \\ \cdot \\ a_{ik} \\ \cdot \\ \cdot \\ \cdot \\ a_{pk} \end{bmatrix} \leq 0.$$

Now, for this X

$$z = d + \sum_{i=1}^{p} c_{s_i}(b_i - a_{ik}x_k) + \sum_{\substack{g \in G \\ g \neq k}} c_g x_g + c_k x_k = d + c_k x_k \qquad (3.45)$$

since each $c_{s_i} = 0$ and each $x_g = 0$ for $g \neq k$. By taking x_k large enough, z can be made as small (that is, large negative) as we please; hence, there is no lower bound for $z = d + C^T X$.

This case is easily illustrated geometrically by considering the system of linear equations

$$\begin{aligned} x_{s_i} &= b_i - a_{ik}x_k, \qquad (i = 1, \cdots, p), \\ z &= d + c_k x_k. \end{aligned}$$

The first of these equations can be graphed for any i, since it is then an equation in the two variables x_i and x_k. We have $a_{ik} \leq 0$ for each i; if $a_{ik} < 0$, then as x_k becomes larger and larger, x_{s_i} also assumes larger and larger positive values. If $a_{ik} = 0$, then x_{s_i} is a constant (positive) value as x_k becomes larger and larger. Both of these possibilities are illustrated in

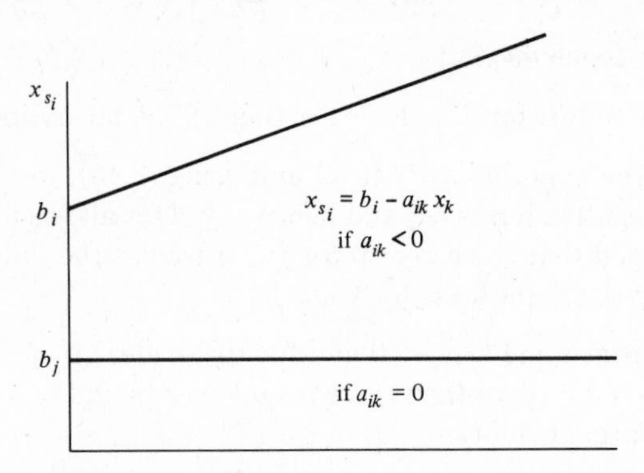

Figure 3.4 Case II.

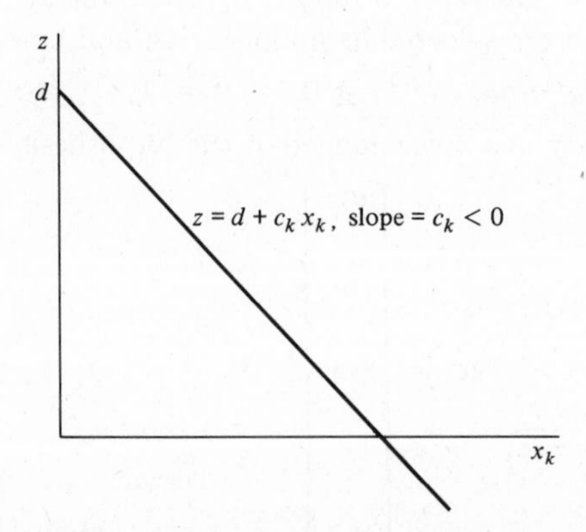

Figure 3.5 Cost function in Case II.

Figure 3.4. Also, from the graph of the cost equation $z = d + c_k x_k$, in Figure 3.5 we see that $z \to -\infty$ as $x_k \to \infty$, so that there is no lower bound for z when $a_{ik} \leq 0$.

Case III: If for some k, $c_k < 0$ but for this k there is at least one i for which $a_{ik} > 0$ $(A_k \nleq 0)$, then there is a better basic program, that is, a basic program $X^{S'}$ for which $z^{S'} < z^S = d$.

To see this, choose X as in Case II, except that now we must choose x_k small enough (but still positive) so that

$$x_{s_i} = b_i - a_{ik}x_k \geq 0 \qquad (i = 1, \cdots, p). \tag{3.46}$$

If $a_{ik} \leq 0$, $x_i \geq 0$ for all $x_k \geq 0$. However, if $a_{ik} > 0$, to have $x_i \geq 0$ we must have

$$x_k \leq b_i/a_{ik}.$$

This bound on x_k is easily illustrated; we have $a_{ik} > 0$, and we see in Figure 3.6 that as x_k becomes larger x_{s_i} must get smaller. Also, when $x_{s_i} = 0$ we have $b_i - a_{ik}x_k = 0$, or

$$x_k = \frac{b_i}{a_{ik}}.$$

Then it is clear that x_k can be chosen to be any positive number such that

$$x_k \leq \frac{b_i}{a_{ik}}.$$

If more than one a_{ik} is positive, choose

$$x_k^* = \min\left\{\frac{b_i}{a_{ik}} \mid \text{for all } i \text{ for which } a_{ik} > 0\right\} = \frac{b_h}{a_{hk}}. \tag{3.47}$$

This choice is also clear geometrically; let a_{rk}, a_{qk}, a_{jk}, and a_{hk} each be positive and suppose that the graphs of the corresponding linear equations

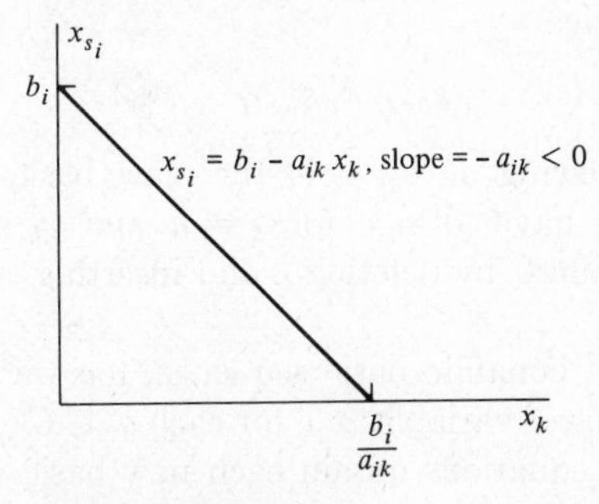

Figure 3.6 Case III.

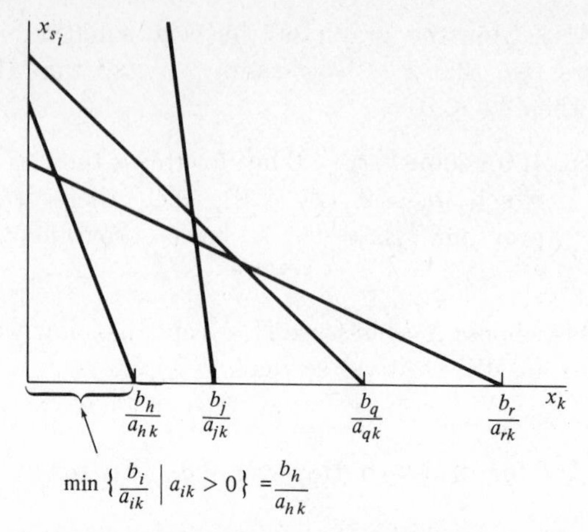

Figure 3.7 Illustration of choice rule, Equation (3.47).

are as shown in Figure 3.7. The choice rule (3.47) leads to the selection that is illustrated.

Now denote by X^* and z^* the vector and value of the objective function given by (3.46) and (3.45) for this value x_k^* of x_k. We conclude that

(1) $X^* \geq 0$,
(2) $z^* < d = z^S$,
(3) X^* is again a basic program.

Conclusion (1) follows at once from (3.47). Conclusion (2) is established as follows. Given the choice of h in (3.47) we have

$$z^* = z^S + c_k x_k^* = z^S + c_k b_h / a_{hk}.$$

From the nondegeneracy assumption we have $b_h > 0$. Hence $x_k^* > 0$ and the cost z^S is reduced by adding the *negative* amount $c_k b_h / a_{hk}$; it follows that

$$z^* < z^S.$$

To establish (3) we observe that our choice of x_k^* gives $x_h^* = 0$ to make up for $x_k^* > 0$. The sequence

$$S' = (s_1, \cdots, s_{h-1}, k, s_{h+1}, \cdots, s_p)$$

obtained from S by replacing s_h by k is the new basic sequence and $X^* = X^{S'}$, $z^* = z^{S'}$. We have $s_i' = s_i$ for $i \neq h$ and $s_h' = k$. The new nonbasic G' is obtained from G by deleting k and inserting s_h in the proper place.

Since $a_{hk} \neq 0$, S' is a bonafide basic sequence, for we can solve the hth equation for x_k in terms of variables $x_{g'}$, for each $g' \in G'$, and substituting this in the remaining equations obtain each new basic variable $x_{s'_i}$ in terms of the new nonbasic variables $x_{g'_i}$.

By our nondegeneracy assumption $x^*_{s_i} > 0$ for all $i \neq h$ in the range 1 to p so that $X^* = X^{S'}$ has exactly p nonzero coefficients, namely those with subscripts in S' (the new basic sequence).

The foregoing discussion settles not only D1 but also D2 and D3, namely;

D2. The new (incoming) basic index can be any k (pivot column) for which $c_k < 0$ and for which A_k has at least one positive component.

D3. Once k has been selected as the incoming basic index, the outgoing basic index s_h is (uniquely) determined by the h (pivot row) for which

$$x_k^* = b_h/a_{hk} = \min \{b_i/a_{ik} \text{ for } a_{ik} > 0\}.$$

(The uniqueness of h follows from the fact that otherwise $X^*_{S'}$ would have a zero component contrary to our nondegeneracy assumption.)

A convenient way to summarize the three cases above is the following. If one cannot find a pivot column, then an optimal solution has been attained (Case I); if one can find a pivot column for which there is no pivot row, then the objective function is unbounded over the feasible region (Case II); if one can find a pivot row and a pivot column, one executes a pivot operation and obtains an improved solution (Case III).

As we have observed, the passage from old to new basic solution gives the nonzero improvement

$$z^{S'} - z^S = z^{S'} - d = c_k x_k^* < 0.$$

This improvement, which accompanies each pivot transformation, guarantees that the process will not *circle*; that is, it will not return at any stage to a basic sequence already used. Since the number of possible basic sequences is finite (being at most $n!/(n-p)!$) this absence of circling guarantees that after a finite number of pivots the problem must leave Case III and therefore that the process will terminate eventually either in Case I or in Case II.

In practice the number of pivotal transformations needed turns out to be surprisingly small so that the simplex algorithm is efficient enough to be practical for modern digital computation, even though there is as yet no mathematical proof of any theorem to assure results as good as those observed in practice.

Although there may be several possible choices for the incoming index it is customary to choose the k for which c_k has the most negative value. The reason for this is that in the absence of any information about x^*_k, maximizing $-c_k$ tends to maximize the improvement $z - d$. This method of choice is known as the "gradient rule."

We recall that the pivot column k has been determined by D2 and the pivot row h has been determined by D3. We call (h, k) the *pivot position* and a_{hk} the *pivot element*. We turn now to D4, called the *pivot process*, which determines the new canonical form. This is the heart of the simplex algorithm and involves essentially the same Gaussian pivotal operations used in reducing a matrix to echelon form (see Section 11 of Appendix B for a discussion of pivot operations in terms of matrices). As a consequence of C1 we have

$$A_j = \sum_{i=1}^{p} a_{ij} A_{s_i} \qquad (j = 1, \cdots, n) \qquad \text{and}$$
$$B = \sum_{i=1}^{p} b_i A_{s_i}. \tag{3.48}$$

The coefficients a'_{ij} in the new canonical form will have the property

$$A_j = \sum_{\substack{i=1 \\ i \neq h}}^{p} a'_{ij} A_{s_i} + a'_{hj} A_k \tag{3.49}$$

and since S' is a basic sequence the a'_{ij} are uniquely defined by (3.49).

Clearly, we can obtain (3.49) by solving the kth equation of (3.48) for A_{s_h} and substituting the result in the remaining equations of (3.48). Thus from

$$A_k = \sum_{i=1}^{p} a_{ik} A_{s_i}$$

we obtain first

$$A_{s_h} = \sum_{\substack{i=1 \\ i \neq h}}^{p} (-a_{ik}/a_{hk}) A_{s_i} + (1/a_{hk})\, A_k. \tag{3.50}$$

Then, for $j = 1, \cdots, n$, we have

$$A_j = \sum_{\substack{i=1 \\ i \neq h}}^{p} a_{ij} A_{s_i} + a_{hj} A_{s_h}$$
$$= \sum_{\substack{i=1 \\ i \neq h}}^{p} a_{ij} A_{s_i} + a_{hj} [\sum_{\substack{i=1 \\ i \neq h}}^{p} (-a_{ik}/a_{hk}) A_{s_i} + (1/a_{hk}) A_k]$$
$$= \sum_{\substack{i=1 \\ i \neq h}}^{p} (a_{ij} - a_{ik} a_{hj}/a_{hk}) A_{s_i} + a_{hj}/a_{hk} A_k. \tag{3.51}$$

Comparison of (3.51) and (3.49) gives the new coefficient matrix $A' = [a'_{ij}]$ where

$$a'_{ij} = a_{ij} - a_{ik} a_{hj}/a_{hk}, \qquad (i \neq h, j = 1, \cdots, n)$$
$$a'_{hj} = a_{hj}/a_{hk}, \qquad (j = 1, \cdots, n). \tag{3.52}$$

Similarly, we get for the new right-hand side B' the vector $B' = [b'_i]$, where

$$\begin{aligned} b'_i &= x^{S'}_{s'i} = b_i - a_{ik}b_h/a_{hk}, \qquad i \neq h \\ b'_h &= x^{S'}_{s'h} = b_h/a_{hk}. \end{aligned} \tag{3.53}$$

Finally, to achieve C2, the canonical cost for the new basic sequence, we add a suitable multiple of the old pivot row to the old cost row. The multiplying factor $-c_k/a_{hk}$ is chosen so as to obtain $c'_k = 0$ as required by C2. Thus we have

$$\begin{aligned} c'_j &= c_j - c_k a_{hj}/a_{hk}, \qquad (j = 1, \cdots, n) \\ -d' &= -z^{S'} = -d - c_k b_h/a_{hk}. \end{aligned} \tag{3.54}$$

The subscripts tend to obscure the true simplicity of the pivot operation; Figure 3.8 gives a presentation without the use of subscripts.

We can represent the initial form of the problem by the $(p + 1)$ by $(n + 1)$ matrix

$$M = \begin{bmatrix} -d & C^T \\ B & A \end{bmatrix} \tag{3.55}$$

and the derived form by

$$M' = \begin{bmatrix} -d' & C'^T \\ B' & A' \end{bmatrix};$$

M is called the *initial tableau matrix*, and M' the *derived tableau matrix*. We call $[-d \quad C^T]$ row zero of M and call $\begin{bmatrix} -d \\ B \end{bmatrix}$ column zero of M. Thus if i and j are both positive integers, $m_{ij} = a_{ij}$, $m_{00} = -d$, $m_{0j} = c_j$, $m_{i0} = b_i$. Suppose that α is in the pivot position (h, k) of M, that δ is in any position (i, j) neither in the pivot row nor in the pivot column, that β is the element in position (h, j), that is, in the pivot row and in the same column as δ, and that γ is the element in position (i, k), that is, in the pivot column and in the same row as δ. Thus in Figure 3.8 we show the pivot row h, an arbitrary row i ($i \neq h$ but including the possibility $i = 0$, the cost row), the pivot column k, the outgoing basic column s_h, and an arbitrary column j ($j \neq k$ but including the possibility $j = 0$, the constant column), both for the initial tableau and for the derived tableau.

	k	j	s_h
h	α	β	1
i	γ	δ	0

initial tableau

	k	j	s_h
h	1	$\beta' = \beta/\alpha$	$1/\alpha$
i	0	$\delta' = \delta - \gamma\beta/\alpha$	$-\gamma/\alpha$

derived tableau

Figure 3.8 Illustration of pivot algebra.

The fact that the equation $\delta' = (\alpha\delta - \gamma\beta)/\alpha$ has the determinant of the two-by-two matrix $\begin{bmatrix} \alpha & \beta \\ \gamma & \delta \end{bmatrix}$ as its numerator is no accident and indeed expressions of this type appear in classical linear algebra and in statistics.

In Chapters 4 and 5 below we will study more closely the connection between the tableau matrices M and M'.

The foregoing formulas relating to the pivot process solve D4 in theory; however, for computational purposes the simplex tableau provides a convenient device for handling D1, D2, D3, and D4. The passage from M to M' can be described in words as follows:

D4. To replace s_h by k in the basic sequence (that is, to replace the column A_{s_h} by the column A_k in the basis),

(1) *Normalize:* divide row h of the initial tableau M by the pivot element a_{hk}, this will make the entry in position (h, k) equal to 1.

(2) *Sweep out:* by adding suitable multiples of this new hth row to all other rows sweep out the kth column, that is, make the elements in position (i, k) equal to zero for all $i \neq h$. Thus the ith row of M' is $-a_{ik}/a_{hk}$ times the hth row of M plus the ith row of M, $i \neq h$.

These two operations together are described as *pivoting on* position (h, k).

The resulting M' will be the derived tableau and will be in canonical form relative to the new basic sequence

$$S' = (s_1, \cdots, s_{h-1}, k, s_{h+1}, \cdots, s_p).$$

Note that all of the descriptions D1 through D4 are applicable to any canonical form, in other words, they apply regardless of the location of the basic columns (those in which columns of I_p appear).

Starting with the initial tableau M (see Table 3.2) we apply D1. If either Case I or Case II obtains we are done. In Case I the initial basic program, whose nonzero entries are the elements of the B column, is optimal. If Case III obtains we then choose the incoming index k according to D2. Once k is determined, the appended column is used to calculate the quotients b_i/a_{ik} for all i for which $a_{ik} > 0$, as needed for applying D3. The smallest of these, b_h/a_{hk}, determines the outgoing index s_h. The resulting derived tableau M' is copied directly under M beginning with the cost row. One then starts a new cycle with M' and continues cycle by cycle until ultimately Case I or Case II is reached and the computation is complete.

We now illustrate these concepts with the wood products problem given in canonical form by (3.21). The full set of simplex tableaus is given in Table 3.3. In the second step 2 is the incoming index since c_2 is the smallest of the c_j. The outgoing index is $s_1 = 4$ since $25/4 \leq 50/3$. The changes of basis are made according to the pivot process D4 and the process continues until at some stage all the c_j are nonnegative. Here the optimal solution is $x_2 = 25/9$, $x_3 = 125/9$, $x_1 = x_4 = x_5 = 0$, $z = -150$. The minimum cost is -150 (that is, the maximum return in the initial problem is 150).

Table 3.2. THE INITIAL SIMPLEX TABLEAU FOR S = (1, 2, . . . , p).

	x_1	$\cdots$	x_p	$\cdots$	x_j	$\cdots$	x_n	b_i/a_{ik}
$-d$	0	$\cdots$	0	$\cdots$	c_j	$\cdots$	c_n	
b_1	1	$\cdots$	0	$\cdots$	a_{1j}	$\cdots$	a_{1n}	
$\cdot$	$\cdot$		$\cdot$		$\cdot$		$\cdot$	
$\cdot$	$\cdot$		$\cdot$		$\cdot$		$\cdot$	
$\cdot$	$\cdot$		$\cdot$		$\cdot$		$\cdot$	
b_p	0	$\cdots$	1	$\cdots$	a_{pj}	$\cdots$	a_{pn}	

Table 3.3. THE WOOD PRODUCTS PROBLEM.

	x_1	x_2	x_3	x_4	x_5	b_i/a_{ik}	
0	−8	−19	−7	0	0		
25	3	4	1	1	0	25/4 $h = 1$	
						$s_h = 4$	$S = (4, 5)$
50	1	3	3	0	1	50/3 $k = 2$	
475/4	25/4	0	−9/4	19/4	0		
25/4	3/4	1	1/4	1/4	0	25 $k = 3$	
						$h = 2$	$S' = (2, 5)$
125/4	−5/4	0	9/4	−3/4	1	125/9 $s_h = 5$	
150	5	0	0	4	1		
25/9	8/9	1	0	1/3	−1/9		$S'' = (2, 3)$
125/9	−5/9	0	1	−1/3	4/9		

The number 150 in the last row is $-d$. The last row represents the equation $z - (-150) = 5x_1 + 4x_4 + 1x_5$ and in general represents $z - d = \Sigma_{j=1}^{n} c_j x_j$ and hence it is $-d$ which appears rather than d.

For an interpretation of the symbolism of Figure 3.8 in terms of the problem solved in Table 3.3 we have, from the first tableau of the latter, $h = 1$, $s_h = 4$, $k = 2$. Then corresponding to Figure 3.8 we have the data in Figure 3.9. For another example, see Table 3.4. Here we wish to find numbers x_i such that $x_1A_1 + x_2A_2 + \cdots + x_8A_8 = B$ and $x_i \geq 0$ and $x_1 + x_2 + x_3 + x_4 + x_5$ is maximized, that is, $-x_1 - x_2 - x_3 - x_4 - x_5$ is minimized.

In cycle 1 we have a tie for the smallest c_j. Column 1 is selected arbitrarily as the pivot column; ties among the c_j have no significance.

	$k = 2$	$j = 3$	$s_h = 4$
$h = 1$	$\alpha = 4$	$\beta = 1$	1
$i = 2$	$\gamma = 3$	$\delta = 3$	0

initial tableau

	$k = 2$	$j = 3$	$s_h = 4$
$h = 1$	1	$\beta' = \frac{\beta}{\alpha} = \frac{1}{4}$	$\frac{1}{\alpha} = \frac{1}{4}$
$i = 2$	0	$\delta' = \delta - \frac{\gamma\beta}{\alpha} = 3 - \frac{3}{4} = \frac{9}{4}$	$-\frac{\gamma}{\alpha} = -\frac{3}{4}$

derived tableau

Figure 3.9 Illustration of pivot algebra.

Although we chose the smallest value of c_j at each cycle, this is not necessary; as we observed earlier, the process will work successfully if any negative value is chosen.

In cycle 4 the c_j for $j = 1, 2, 3$ are expected to be zero. The other two zeros in this row indicate that there are additional basic solutions with the same minimal cost. These solutions can be obtained by additional simplex transformations, see Table 3.19 below. The optimal solution given by the fourth tableau is

$$X^{S'''} = \begin{bmatrix} 7/39 \\ 4/39 \\ 10/39 \\ 0 \\ 0 \\ 0 \\ 0 \end{bmatrix}, \qquad z^{S'''} = -7/13.$$

As a further illustration of our algorithm, we apply the simplex algorithm in tabular form to the problem given by (3.37).

The first tableau is the following:

	x_1	x_2	x_3	x_4	x_5	x_6	x_7	b_i/a_{ik}
0	−20	−50	−80	−50	0	0	0	
50	2	4	4	2	1	0	0	50/4
25	1	1	3	3	0	1	0	25/1
40	2	3	5	3	0	0	1	40/3

, $S = (5, 6, 7)$.

Table 3.4. EXAMPLE OF SIMPLEX SOLUTION.

	x_1	x_2	x_3	x_4	x_5	x_6	x_7	x_8	b_i/a_{ik}	
0	−1	−1	−1	−1	−1	0	0	0		
1	1	3	2	13/5	10/3	1	0	0	1 $h = 3$	
1	2	0	5/2	1	1/2	0	1	0	1/2 $s_h = 8$	$S = (6, 7, 8)$
1	3	2	1	8/5	1	0	0	1	1/3 $k = 1$	
1/3	0	−1/3	−2/3	−7/15	−2/3	0	0	1/3	b_i/a_{i3}	
2/3	0	7/3	5/3	31/15	3	1	0	−1/3	2/5 $h = 2$	
1/3	0	−4/3	11/6	−1/15	−1/6	0	1	−2/3	2/11 $s_h = 7$	$S' = (6, 7, 1)$
1/3	1	2/3	1/3	8/15	1/3	0	0	1/3	1 $k = 3$	
5/11	0	−9/11	0	−27/55	−8/11	0	4/11	1/11	b_i/a_{i2}	
4/11	0	39/11	0	117/55	104/33	1	−10/11	3/11	4/39 $h = 1$	
2/11	0	−8/11	1	−2/55	−1/11	0	6/11	−4/11	— $s_h = 6$	$S'' = (6, 3, 1)$
3/11	1	10/11	0	6/11	4/11	0	−2/11	5/11	3/10 $k = 2$	
7/13	0	0	0	0	0	3/13	2/13	2/13		
4/39	0	1	0	3/5	8/9	11/39	−10/39	1/13		
10/39	0	0	1	2/5	5/9	8/39	14/39	−4/13		$S''' = (2, 3, 1)$
7/39	1	0	0	0	−4/9	−10/39	2/39	5/13		

Instead of using the gradient rule which is only useful "most of the time," we pivot in the second column and in accordance with D3 choose the first row for pivoting. By use of D4 we achieve the next tableau as follows. Once

								b_i/a_{ik}
625	5	0	−30	−25	25/2	0	0	
25/2	1/2	1	1	1/2	1/2	0	0	25
25/2	1/2	0	2	5/2	−1/4	1	0	5
5/2	1/2	0	2	3/2	−3/4	0	1	5/3

$, S' = (2, 6, 7).$

again we ignore the gradient rule and pivot on column 4, choose the third row as our pivot in accordance with D3, since $5/3 = \min \{25, 5, 5/3\}$. Our next and final tableau is:

666 2/3	40/3	0	10/3	0	0	0	50/3
35/3	1/3	1	1/3	0	1/2	0	−1/3
25/3	−1/3	0	−4/3	0	1	1	1/3
5/3	1/3	0	4/3	1	−1/2	0	2/3

$, S'' = (2, 6, 4).$

Here we recognize an optimal program,

$$X^{S''} = \begin{bmatrix} 0 \\ 35/3 \\ 0 \\ 5/3 \\ 0 \\ 25/3 \\ 0 \end{bmatrix}$$

and a minimum cost $z^{S''} = -666\ 2/3$. Of course, in terms of the original problem as stated in Section 3.2 (Table 3.1) our maximum return is 666 2/3.

3.4. SIMPLEX ALGORITHM: HANDLING OF DEGENERACY.

We recall that in a nondegenerate problem no set of p columns in the matrix $[B \quad A]$ is dependent. This guarantees that every basic program has exactly p nonzero coordinates and hence that there is a definite reduction in the cost z following each simplex transformation. Because of this reduction in cost at each stage we can never return to a basis previously used, and thus we are guaranteed a termination of the process in a finite number of steps (compare Section 3.3 above).

Degeneracy is indicated in a linear programming problem in two ways:

at least one component of the vector B is zero, or at some cycle in the simplex algorithm there are ties for h in the application of the choice rule (3.47). If a tie is resolved in the latter case and an iteration is executed, then at least one zero will appear in the B column of the next tableau. This is the major complication of degeneracy for, suppose that hth component of B is zero initially and that $a_{hk} > 0$; then $x_k^* = b_h/a_{hk} = 0$ in the application of D3. There is no improvement in the value of z. This raises the possibility that in solving a degenerate linear programming problem basic sequences may be repeated again and again, thus resulting in endless *circling* with no exit from Case III and no determination of an optimal solution.

A simple example of a degenerate problem which circles is due to Beale (Reference 4),

$$\text{minimize } z = -\frac{3}{4}x_4 + 20x_5 - \frac{1}{2}x_6 + 6x_7$$

subject to

$$\begin{aligned}
x_1 + \frac{1}{4}x_4 - 8x_5 - \phantom{\frac{1}{2}}x_6 + 9x_7 &= 0,\\
x_2 + \frac{1}{2}x_4 - 12x_5 - \frac{1}{2}x_6 + 3x_7 &= 0,\\
x_3 \qquad\qquad\qquad + \phantom{\frac{1}{2}}x_6 \qquad\quad &= 1,\\
x_j &\geq 0, \qquad (j = 1, \cdots, 7).
\end{aligned}$$

The iterations for this problem are shown in Table 3.5. At each stage or cycle ties result in use of the choice rule (3.47); the vector having the smallest subscript is chosen to go out of the basis at each cycle. The basic sequence for the first cycle is $S = (1, 2, 3)$ and is the same as that for the last cycle shown; that is, given this problem and the tie-breaking rule stated above, we have circling after six iterations (note there is no improvement in the value of the cost function at any iteration and the last tableau is the same as the first).

Circling is not a serious computational hazard in practice, however, and many procedures are available for avoiding it. One of the simplest is the following tie-breaking rule: suppose that

$$\begin{aligned}
x_k^* &= \min\left\{\frac{b_i}{a_{ik}} \mid a_{ik} > 0\right\}\\
&= \frac{b_{h_s}}{a_{h_s k}}, \qquad (s = 1, \cdots, r),
\end{aligned} \tag{3.56}$$

where $r > 1$. Then consider all ratios

$$t_{sj} = \frac{a_{h_s j}}{a_{h_s k}}, \qquad (s = 1, \cdots, r), \tag{3.57}$$

and for each $j = 1, \cdots, n$ in turn eliminate all s from future consideration except the one or more which give t_{sj} algebraically minimal until finally a j

is reached for which there is a unique minimal t_{sj}. Then select h_s as the pivot row (of course, k is the pivot column).

We prove below that such a j always exists and that using this rule circling will never occur. One may have several cycles with zero improvement. However, ultimately one either reaches D1, Case I or Case II, or comes to a cycle where each zero in the B column has in the same row in the kth column a nonpositive entry so that there is definite improvement in z with the next transformation.

Three degenerate problems are illustrated in Tables 3.6, 3.7, and 3.8. In the first cycle of Table 3.6 we have a three-way tie for h: $r = 3$, $h_1 = 1$,

Table 3.5. EXAMPLE OF CIRCLING.

								b_i/a_{ik}
0	0	0	0	−3/4	20	−1/2	6	
0	1	0	0	1/4	−8	−1	9	0
0	0	1	0	1/2	−12	−1/2	3	0
1	0	0	1	0	0	1	0	–
0	3	0	0	0	−4	−7/2	33	
0	4	0	0	1	−32	−4	36	–
0	−2	1	0	0	4	3/2	−15	0
1	0	0	1	0	0	1	0	–
0	1	1	0	0	0	−2	18	
0	−12	8	0	1	0	8	−84	0
0	−1/2	1/4	0	0	1	3/8	−15/4	0
1	0	0	1	0	0	1	0	1
0	−2	3	0	1/4	0	0	−3	
0	−3/2	1	0	1/8	0	1	−21/2	–
0	1/16	−1/8	0	−3/64	1	0	3/16	0
1	3/2	−1	1	−1/8	0	0	21/2	2/21
0	−1	1	0	−1/2	16	0	0	
0	2	−6	0	−5/2	56	1	0	0
0	1/3	−2/3	0	−1/4	16/3	0	1	0
1	−2	6	1	5/2	−56	0	0	–
0	0	−2	0	−7/4	44	1/2	0	
0	1	−3	0	−5/4	28	1/2	0	–
0	0	1/3	0	1/6	−4	−1/6	1	0
1	0	0	1	0	0	1	0	–
0	0	0	0	−3/4	20	−1/2	6	
0	1	0	0	1/4	−8	−1	9	
0	0	1	0	1/2	−12	−1/2	3	
1	0	0	1	0	0	1	0	

$h_2 = 3$, $h_3 = 4$. This tie is resolved by considering the first column; for we have

$$\frac{a_{11}}{a_{13}} = -3, \qquad \frac{a_{31}}{a_{33}} = 2, \qquad \frac{a_{41}}{a_{43}} = -2,$$

and hence select $h = 1$.

In cycle two there is a two-way tie for h. Now $a_{31}/a_{31} = a_{41}/a_{41} = 1$, but $a_{32}/a_{31} = -3/10$, $a_{42}/a_{41} = -3$; hence $h = 4$. We have not shown the calculations for cycle three, but pivoting at the indicated position $(h,k) = (3, 4)$ will yield a tableau in optimal form.

Table 3.7 illustrates the case in which a component of B is zero initially.

The gasoline-blending problem (see Sections 2.3 and 3.2) provides an interesting example of degeneracy. Table 3.8 shows a complete solution of the problem by means of the simplex algorithm.

The maximum total profit is \$88,928.57; four cycles are required in the solution procedure. Note that three iterations or cycles are required in order to get an increase in value of the objective function; at the end of cycles one and two the value of the objective function remains at 0 and at the end of cycle three it increases to \$47,500.00. The simplex algorithm might therefore at first be considered to be relatively inefficient in the solution of this particular problem, since it might appear that we would be able to solve the problem more quickly if activity A_5 were brought in (x_5 given a positive value) earlier. However, since no variable is introduced and later excluded from any basic sequence, no process could reach the final tableau in less than four cycles. Sometimes shortcuts can be taken in hand computation by scanning the entire tableau and eliminating cycles that the simplex algorithm would have included; this is rarely feasible with machine calculations, since computers usually can perform calculations much more quickly than they scan a set of numbers and identify one or more which satisfy a given property.

A tie-breaking rule that works well for machine calculation is to make a random choice of h from the set of tied values. This rule is especially convenient if the library of subroutines available on the computer being used includes one for generating random numbers.

If there is exactly one zero in the B column, say $b_i = 0$, and if there are several negative c_j the following rule for selecting k may be useful. If for some j we have $a_{ij} \leq 0$ and $c_j < 0$ let $k = j$. If, however, $a_{ij} > 0$ for all j for which $c_j < 0$ we choose k so that

$$\frac{c_k}{a_{ik}} = \min \left\{ \frac{c_j}{a_{ij}} \mid c_j < 0 \right\}. \tag{3.58}$$

This choice of k gives pivot position (i, k) and guarantees either termination of the calculation in one cycle or a definite improvement in z in the next cycle with pivot position having $h \neq i$. The proof of this statement is left as an exercise for the reader. (This was the rule used in cycle three of the example of Table 3.6).

Table 3.6. A DEGENERATE PROBLEM.

									b_i/a_{ik}	
14	−1	2	−3	1	0	0	0	0		
2	−3	2	1	5	1	0	0	0	2	$k = 3$
5	2	0	2	1	0	1	0	0	5/2	$h = 1$
4	4	1	2	−4	0	0	1	0	2	$s_h = 5$
6	−6	−3	3	1	0	0	0	1	2	
20	−10	8	0	16	3	0	0	0		
2	−3	2	1	5	1	0	0	0		$k = 1$
1	8	−4	0	−9	−2	1	0	0	1/8	$h = 4$
0	10	−3	0	−14	−2	0	1	0	0	$s_h = 8$
0	3	−9	0	−14	−3	0	0	1	0	
20	0	−22	0	−92/3	−7	0	0	10/3		
2	0	−7	1	−9	−2	0	0	1		$k = 4$
1	0	20	0	85/3	6	1	0	−8/3	3/85	$h = 3$
0	0	27	0	98/3	8	0	1	−10/3	0	$s_h = 7$
0	1	−3	0	−14/3	−1	0	0	1/3		

Table 3.7. ANOTHER DEGENERATE PROBLEM.

							b_i/a_{ik}	
−3	0	0	0	2	−7/2	0		
0	1	0	0	−2	5/2	0	0	$k = 5$
2	0	1	0	−1	−3/2	1		$h = 1$
1	0	0	1	1	2	−1	1/2	$s_h = 1$
−3	7/5	0	0	−4/5	0	0		
0	2/5	0	0	−4/5	1	0		$k = 4$
2	3/5	1	0	−11/5	0	1	10/11	$h = 3$
1	−4/5	0	1	13/5	0	−1	5/13	$s_h = 3$
−35/13	15/13	0	4/13	0	0	−4/13		
4/13	2/13	0	4/13	0	1	−4/13		$k = 6$
37/13	−1/13	1	11/13	0	0	2/13	37/2	$h = 2$
5/13	−4/13	0	5/13	1	0	−5/13		$s_h = 2$
3	1	2	2	0	0	0		
6	0	2	2	0	1	0		
37/2	−1/2	13/2	11/2	0	0	1		
15/2	−1/2	5/2	5/2	1	0	0		

Because of the theoretical interest in and the importance of degeneracy we give a proof that the tie-breaking rule of (3.56) and (3.57) does indeed avoid circling and that with it the simplex algorithm will terminate after a finite number of cycles in either Case I or Case II of Phase D1.

A vector $D = \begin{bmatrix} d_1 \\ \cdot \\ \cdot \\ \cdot \\ d_m \end{bmatrix}$ in V_m is said to be *lexicographically positive*, written $D \succ 0$, if for some h with $1 \leq h \leq m$ we have $d_h > 0$ but $d_i = 0$ for $i = 1, \cdots, h - 1$. For two vectors D', D'' we say that D' is *lexicographically greater* than D'', written $D' \succ D''$, if $D = D' - D'' \succ 0$, that is, if for some h, $d'_h > d''_h$ but $d'_i = d''_i$ for $i = 1, \cdots, h - 1$. We write $D' \succeq D''$ if either $D' = D''$ or $D' \succ D''$. For any two vectors D' and D'' exactly one of the alternatives $D' \succ D''$, $D' = D''$, $D'' \succ D'$ is true. The relation $\succeq$ defines a chain order on V_m (the reader may wish to read Appendix A.13 on order relations before proceeding).

Suppose that we have a linear programming problem (3.18), (3.19), (3.20) in canonical form with respect to some basic sequence S and, moreover, satisfying the condition C3′:

C3′. Each row of $[BA]$ is lexicographically positive.

Note that C3′ is satisfied if $B > 0$ or if $B \geq 0$ and $S = (1, \cdots, p)$. Hence, if C1, C2, C3 hold then by permuting columns we can also satisfy C3′.

Now choose a pivot column k according to D2. If $a_{ik} \neq 0$, denote by R_{ik} the ith row of $[BA]$ divided by a_{ik},

$$R_{ik} = \frac{1}{a_{ik}}[b_i a_{i1} \cdots a_{in}];$$

then select h so that

$$\text{D3}'. \quad R_{hk} \prec R_{ik} \qquad \text{for all } i \neq h \text{ with } a_{ik} > 0,$$

or, alternatively,

$$R_{hk} = \frac{1}{a_{hk}}[b_h a_{h1} \cdots a_{hn}] = \text{lex min} \left\{ \frac{1}{a_{ik}}[b_i a_{i1} \cdots a_{in}] \mid a_{ik} > 0 \right\}.$$

This h is uniquely defined by the properties of $\prec$ and the fact that no two rows of a matrix A which contains all the columns of the identity matrix (that is, satisfies C1) can be proportional. Hence we cannot have $R_{ik} = R_{i'k}$ unless $i = i'$.

We first show that after pivoting at (h, k) we obtain a new matrix $[B^* \quad A^*]$ which satisfies $C3'$. First, we observe that row h of $[B^* \quad A^*]$ is just

Table 3.8. A BLENDING PROBLEM.

0	1	−3	−4	3	−1	−2	0	0	0	0	0	0	0
32,000	1	0	0	1	0	0	1	0	0	0	0	0	0
20,000	0	1	0	0	1	0	0	1	0	0	0	0	0
38,000	0	0	1	0	0	1	0	0	1	0	0	0	0
0	−8	10	[27]	0	0	0	0	0	0	1	0	0	0
0	−2	4	−1	0	0	0	0	0	0	0	1	0	0
0	0	0	0	−28	−10	7	0	0	0	0	0	1	0
0	0	0	0	−5	1	−4	0	0	0	0	0	0	1
0	−5/27	−41/27	0	3	−1	−2	0	0	0	4/27	0	0	0
32,000	1	0	0	1	0	0	1	0	0	0	0	0	0
20,000	0	1	0	0	1	0	0	1	0	0	0	0	0
38,000	8/27	−10/27	0	0	0	1	0	0	1	−1/27	0	0	0
0	−8/27	10/27	1	0	0	0	0	0	0	1/27	0	0	0
0	−62/27	118/27	0	0	0	0	0	0	0	1/27	1	0	0
0	0	0	0	−28	−10	[7]	0	0	0	0	0	1	0
0	0	0	0	−5	1	−4	0	0	0	0	0	0	1
0	−5/27	−41/27	0	−5	−27/7	0	0	0	0	4/27	0	2/7	0
32,000	1	0	0	1	0	0	1	0	0	0	0	0	0
20,000	0	1	0	0	1	0	0	1	0	0	0	0	0
38,000	8/27	−10/27	0	[4]	10/7	0	0	0	1	−1/27	0	−1/7	0
0	−8/27	10/27	1	0	0	0	0	0	0	1/27	0	0	0
0	−62/27	118/27	0	0	0	0	0	0	0	1/27	1	0	0
0	0	0	0	−4	−10/7	1	0	0	0	0	0	1/7	0
0	0	0	0	−21	−33/7	0	0	0	0	0	0	4/7	1

47,500	20/108	−214/108	0	0	−29/14	0	0	0	5/4	11/108	0	3/28	0
22,500	100/108	10/108	0	0	−10/28	0	1	0	−1/4	1/108	0	1/28	0
20,000	0	1	0	0	$\boxed{1}$	0	0	1	0	0	0	0	0
9,500	2/27	−10/108	0	1	10/28	0	0	0	1/4	−1/108	0	−1/28	0
0	−8/27	10/27	1	0	0	0	0	0	0	1/27	0	0	0
0	−62/27	118/27	0	0	0	0	0	0	0	1/27	1	0	0
38,000	8/27	−10/27	0	0	0	1	0	0	1	−1/27	0	0	0
199,500	168/108	−210/108	0	0	39/14	0	0	0	21/4	−21/108	0	−5/28	1
88,928.57	20/108	136/1512	0	0	0	0	0	29/14	5/4	11/108	0	3/28	0
29,642.86	100/108	85/189	0	0	0	0	1	10/28	−1/4	1/108	0	1/28	0
20,000	0	1	0	0	1	0	0	1	0	0	0	0	0
2,357.14	2/27	−170/278	0	1	0	0	0	−10/28	1/4	−1/108	0	−1/28	0
0	−8/27	10/27	1	0	0	0	0	0	0	1/27	0	0	0
0	−62/27	118/27	0	0	0	0	0	0	0	1/27	1	0	0
38,000	8/27	−10/27	0	0	0	1	0	0	1	−1/27	0	0	0
143,785.71	168/108	−3576/756	0	0	0	0	0	−39/14	21/4	−21/108	0	−5/28	1

R_{hk}, which is lexicographically positive. Next, we observe that for $i \neq h$ row i of $[B^* \quad A^*]$ is

$$[b_i a_{i1} \cdots a_{in}] - \frac{a_{ik}}{a_{hk}}[b_h a_{h1} \cdots a_{hn}] = [b_i a_{i1} \cdots a_{in}] - a_{ik} R_{hk}$$
$$= a_{ik}(R_{ik} - R_{hk}).$$

Now suppose $a_{ik} > 0$; we have $R_{ik} - R_{hk} \succ 0$ by the choice of h. Hence $a_{ik}(R_{ik} - R_{hk}) \succ 0$. On the other hand, if $a_{ik} < 0$, the ith row of $[B^* \quad A^*]$ can be written

$$[b_i a_{i1} \cdots a_{in}] + (-a_{ik}) R_{hk}.$$

The ith row of $[B \quad A]$ is lexicographically positive, $-a_{ik} > 0$, $R_{hk} \succ 0$, so the sum of the vectors is lexicographically positive and $[B^* \quad A^*]$ satisfies C3′.

Next we consider the payoff or cost row of the new tableau. For it we have

$$[-d^* \quad C^{*T}] = [-d \quad C^T] + (-c_k) R_{hk} \succ [-d \quad C^T].$$

Therefore each cycle results in a lexicographic increase in the payoff row and hence no tableau can ever recur at a later position (after one or more cycles). Circling is therefore impossible and since there are only finitely many basic sequences the algorithm must eventually reach one of the terminal Cases I or II.

It may be helpful to observe that the comparisons needed to determine h in D3′ can be made in precisely the same manner as they are made in determining h in (3.47). Starting with the first two candidates, discard the larger and compare the remaining element with the next candidate. This process continues until all but the smallest candidate have been eliminated. To determine which of

$$R_{qk} = \frac{1}{a_{qk}}[b_q a_{q1} \cdots a_{qn}]$$

and

$$R_{ik} = \frac{1}{a_{ik}}[b_i a_{i1} \cdots a_{in}]$$

is lexicographically the smaller, we compare in turn their corresponding components which are the quotients

$$\frac{b_q}{a_{qk}}, \quad \frac{b_i}{a_{ik}}$$
$$\frac{a_{q1}}{a_{qk}}, \quad \frac{a_{i1}}{a_{ik}}$$
$$\vdots$$

until the first unequal pair is found. The sense of the inequality between the quotients in that pair determines which vector is smaller in the lexicographic sense. For example, in the top tableau of Table 3.6 we have for $k = 3$,

$$R_{13} = [2 \quad -3 \quad 2 \quad 1 \quad 5 \quad 1 \quad 0 \quad 0 \quad 0]$$

and

$$R_{43} = [2 \quad -2 \quad -1 \quad 1 \quad 1/3 \quad 0 \quad 0 \quad 0 \quad 1/3].$$

Then $R_{13} \prec R_{43}$ since in the first unequal position, namely the second, we have $-3 < -2$.

It is this feature of the comparison between vectors that is sytematized in the discussion relating to (3.57) above which calls for making comparisons column by column rather than row by row. The columnwise comparisons can be made more efficiently.

We note that we have now proved that any linear programming problem in canonical form either has a basic optimal feasible solution or there is no lower bound to the cost for feasible solutions. In the following four sections we show how to obtain canonical form. More precisely, we establish a procedure which either (1) demonstrates that a linear programming problem has no feasible solution or (2) leads to an equivalent problem in canonical form.

Our method of treatment deviates from the form of the simplex algorithm which is still found in most texts. Two major differences are, first, avoidance of artificial variables (see Section 4.2 for a definition of artificial variable and of an alternate formulation of the simplex algorithm which does employ artificial variables) and, second, a method of handling free variables which results in an equivalent problem in the standard form (3.18), (3.19), (3.20) having both fewer equations and fewer variables as contrasted with the increased number of variables required in the traditional treatment.

Our discussion builds up gradually, step by step from problems in canonical form to those in the most general form and shows how to handle each added complexity as it is encountered.

3.5. REDUCTION TO CANONICAL FORM, I: CANONICAL COSTS.

In Sections 3.3 and 3.4 we presented the simplex algorithm for a problem in canonical form, that is, one which initially satisfies conditions C1, C2, C3 of Section 3.2. In this and the three sections which follow we show how to achieve canonical form even for the most general class of linear programming problem. In the present section we assume that we have a problem in the standard form (3.18), (3.19), (3.20) for which C1 holds and we show how to achieve C2.

Suppose that $S = (s_1, \cdots, s_p)$ is the basic sequence, and suppose that for some h, $c_{s_h} \neq 0$. Then, if we subtract c_{s_h} times row h from the cost row we get a new cost row $[-d' \; c'_1 \cdots c'_n]$ with the following properties:

$$\begin{aligned} -d' &= -d - c_{s_h} b_h, \\ c'_j &= c_j - c_{s_h} a_{hj}, \qquad (j = 1, \cdots, n), \\ c'_{s_i} &= c_{s_i}, \qquad i \neq h, \\ c'_{s_h} &= 0. \end{aligned}$$

If we carry out this operation, consecutively, for $h = 1, \cdots, p$ we obtain a new cost row $[-d^* \; c_1^* \cdots c_n^*]$ with

$$\begin{aligned} -d^* &= -d - c_{s_1} b_1 - \cdots - c_{s_p} b_p \\ c_j^* &= c_j - c_{s_1} a_{1j} - \cdots - c_{s_p} a_{pj}, \qquad (j = 1, \cdots, n) \end{aligned} \tag{3.59}$$

and, in particular, with $c^*_{s_h} = 0$ $(h = 1, \cdots, p)$, that is, the new cost satisfies condition C2.

What this process really amounts to is carrying out p consecutive pivot operations on the respective positions $(1, s_1), \cdots, (p, s_p)$. Since C1 holds the only effect of these pivots is to modify the cost row as described by (3.59); the constant column B and the coefficient matrix A are unchanged.

As an aid to computation the following procedure for calculating the c_j^* is given. Let

$$\begin{aligned} z_0 &= c_{s_1} b_1 + \cdots + c_{s_p} b_p, \\ z_j &= c_{s_1} a_{1j} + \cdots + c_{s_p} a_{pj} \qquad (j = 1, \cdots, n). \end{aligned} \tag{3.60}$$

Then (3.59) becomes

$$\begin{aligned} -d^* &= -d - z_0, \\ c_j^* &= c_j - z_j \qquad (j = 1, \cdots, n). \end{aligned} \tag{3.61}$$

To calculate the z_j it is convenient to append an auxiliary initial column C_S (compare (3.28) above) to the left of the tableau with c_{s_h} entered in row h. Then we note that

$$z_0 = C_S^T B, \qquad z_j = C_S^T A_j, \qquad (j = 1, \cdots, n).$$

We enter a new auxiliary row $[z_0 \; z_1 \cdots z_n] = [z_0 \; Z^T]$ just below the original tableau and then subtract this auxiliary row from the original cost row to obtain the new canonical cost row $[-d^* \; c_1^* \cdots c_n^*]$. See Tables 3.9 and 3.10 for the general format and an illustrative example. If the original tableau is

$-d$	C^T
B	A

then the modified tableau which satisfies both C1 and C2 is

$-d^*$	C^{*T}
B	A

.

Note that B is unchanged in this operation; hence if C3 holds initially we now have reached a canonical form. In the next section we consider what to do when C3 does not hold.

Table 3.9. CALCULATION OF CANONICAL COSTS.

	$-d$	C^T
C_S	B	A
	z_0	Z^T
	$-d^*$	C^{*T}

$$z_0 = C_S^T B$$
$$z_j = C_S^T A_j \qquad (j = 1, \cdots, n)$$
$$-d^* = -d - z_0$$
$$c_j^* = c_j - z_j \qquad (j = 1, \cdots, n)$$

In the illustrative example of Table 3.10 we give three tableaus. The first (a) shows the problem with its noncanonical costs; the second (b) shows the calculation of the canonical costs; and the third (c) shows the resulting tableau, which in this case is canonical since $B \geq 0$. In practice one uses only the second of these; then in applying the simplex algorithm one ignores the C_S column, the c_j row, and the z_j row. To sample the actual numerical calculation we observe that

$$z_1 = C_S^T A_1 = 4 \cdot 2 + (-2) \cdot (-1) + 5 \cdot 3 = 25,$$
$$z_3 = C_S^T A_3 = 4 \cdot 1 + (-2) \cdot 0 + 5 \cdot 0 = 4,$$
$$z_6 = C_S^T A_6 = 4 \cdot (-5) + (-2) \cdot 2 + 5 \cdot 2 = -14.$$

Table 3.10. CALCULATION OF CANONICAL COSTS, ILLUSTRATIVE EXAMPLE.

3	2	5	4	−3	−2	1
3	2	0	1	4	0	−5
4	−1	0	0	2	1	2
1	3	1	0	−4	0	2

(a) Initial tableau, $S = (3, 5, 2)$.

c_j	3	2	5	4	−3	−2	1
4	3	2	0	1	4	0	−5
−2	4	−1	0	0	2	1	2
5	1	3	1	0	−4	0	2
z_j	9	25	5	4	−8	−2	−14
c_j^*	−6	−23	0	0	5	0	15

(b) Calculation of canonical costs c_j^*.

−6	−23	0	0	5	0	15
3	2	0	1	4	0	−5
4	−1	0	0	2	1	2
1	3	1	0	−4	0	2

(c) Canonical tableau.

In addition to providing a way to obtain canonical costs this procedure can also be used to provide a partial check on the numerical accuracy of the simplex calculations. Suppose for example that the tableau

$$\boxed{\begin{matrix} -d & C^T \\ B & A \end{matrix}}$$

occurs at some cycle in the calculations and that the initial tableau (not necessarily canonical) was

$$\boxed{\begin{matrix} -d^0 & C^{0T} \\ B^0 & A^0 \end{matrix}}.$$

Then we could apply the process to

$$\boxed{\begin{matrix} -d^0 & C^{0T} \\ B & A \end{matrix}}$$

and obtain a canonical cost row $[-d^* \quad C^{*T}]$. Now, unless $d^* = d$ and $C^* = C$ we know that there has been a numerical error. We will provide a more comprehensive check for numerical accuracy in Section 4.1.

Historically, this process, sometimes called the $(c_j - z_j)$ process, was used by some writers at each cycle of the simplex algorithm to obtain the new cost row. This practice still prevails in some current presentations of the simplex algorithm; however the $(c_j - z_j)$ process as applied to a problem already in canonical form is inefficient both computationally and pedagogically as contrasted with the pivot procedure as presented in Section 3.3.

The z_j are *cost adjustment numbers* and will be given an economic interpretation in Section 5.5.

3.6. REDUCTION TO CANONICAL FORM, II: CANONICAL CONSTANTS.

Suppose next that the problem in standard form satisfies C1 and C2 but not C3. We use the number q of nonnegative elements of B as a measure of feasibility and suppose that the rows have been permuted (if necessary) so that the last q elements of B are nonnegative and the first $p - q$ elements of B are negative. We then consider an auxiliary problem whose initial tableau consists of the last $q + 1$ rows of the tableau for the main problem with the $(p - q)$th row of the initial tableau as the auxiliary cost row. This auxiliary problem is in canonical form with respect to the sequence $S' = (s_{p-q+1}, \cdots, s_p)$; we use the auxiliary problem to determine a pivot column, pivot row, and pivot element, but we carry out the pivot operations on the full tableau of the main problem. The main problem remains in a form that satisfies C1 and C2.

If at any cycle the $(p - q)$th element in the constant column of the main problem becomes nonnegative we choose a new auxiliary problem with a number $q' > q$ of nonnegative elements in the constant column. (We recall in the simplex algorithm the upper left-hand corner element $-d$ either increases or remains the same, and the same holds for b_{p-q} in the auxiliary problem.)

If the auxiliary problem reaches the terminal state of Case I with b_{p-q} still negative we conclude that the main problem is not feasible, for in this case the $(p - q)$th equation, which is playing the role of the cost equation for the auxiliary problem, requires that

$$b_{p-q} = a_{p-q,1}x_1 + \cdots + a_{p-q,n}x_n \tag{3.62}$$

with the left-hand side negative and every term on the right-hand side nonnegative.

If the auxiliary problem reaches the terminal state of Case II with $b_{p-q} < 0$, $a_{p-q,k} < 0$, and with $a_{ik} \leq 0$ for $i = p - q + 1, \cdots, p$, then we pivot in position $(p - q, k)$, that is, we choose as pivot row the cost row of the auxiliary problem and obtain

$$b'_{p-q} = \frac{b_{p-q}}{a_{p-q,k}}, \tag{3.63}$$

$$b'_i = b_i - b_{p-q}a_{ik}/a_{p-q,k} \qquad (i = p - q + 1, \cdots, p), \tag{3.64}$$

where the pivot operation is performed on the main problem and the primes denote corresponding elements in the new tableau of the main problem. Moreover, we see that

$$b'_{p-q} > 0$$

and that

$$b'_i \geq b_i \geq 0,$$

since all elements in the fractions in both (3.63) and (3.64) are negative. We then have $q' > q$; thus, the number of nonnegative elements of B in the main problem is larger than before. We can now construct a new and larger auxiliary problem and continue the process. A slight improvement on this rule is to pivot in position (h, k), where h is determined so that

$$\frac{b_h}{a_{hk}} = \max\left\{\frac{b_i}{a_{ik}} \mid b_i a_{ik} > 0\right\}. \tag{3.65}$$

Hence, in every case, after a finite number of pivots selected either by the simplex rule or by the special rule for Case II, we either increase the number of nonnegative constants or demonstrate infeasibility. We must eventually either reach a main tableau in full canonical form or arrive at the conclusion that the problem is infeasible.

In hand computation one might have the good fortune to obtain one or more nonnegative b_i for $i < p - q$ before b_{p-q} becomes positive; in this case it is, of course, desirable to permute rows at once and choose a new larger auxiliary problem. Indeed, it is not really necessary to permute rows at all provided a careful record is kept of which rows belong to the auxiliary problem.

For machine computation it is questionable whether searching for nonnegativity in the constant column is worthwhile, especially if one is using some variant of the revised simplex method (to be described below).

Table 3.11 gives an illustrative example with $p = 4$, $n = 7$, $q = 2$, and in which the two negative elements of B appear first, followed by the positive elements. The auxiliary problem (b) consists of the rows $q + 1, \cdots, p$ of tableau (a)—the rows 3 and 4—and it has row $p - q = 2$ of (a) as the cost row. The pivot column in (b) is that having a corresponding cost that is most negative, column 5, and the pivot position is (2, 5) in (b) or the position (4, 5) in the main problem (a). We pivot on the position (4, 5) in (a) getting Table 3.12 as the result. In this tableau the number of nonnegative elements in B is $q' = 3$.

We carry the calculations one cycle further, this time using the first row as the auxiliary cost row. Two elements in the auxiliary cost row are negative; since we can choose k by means of either of these, we let $k = 1$ and select (2, 1) as pivot position. We obtain the new first row,

$$\left[-\frac{14}{3} \quad 0 \quad 1 \quad \frac{17}{3} \quad 0 \quad 0 \quad \frac{4}{3} \quad \frac{1}{3}\right]$$

or, in equation form,

$$-\frac{14}{3} = x_2 + \frac{17}{3}x_3 + \frac{4}{3}x_6 + \frac{1}{3}x_7.$$

We are then in Case I with $b_{p-q} = b_1 < 0$ and have demonstrated infeasibility of the main problem.

Table 3.11. EXAMPLE ILLUSTRATING MAIN AND AUXILIARY PROBLEMS.

4	0	0	3	0	2	0	4
−2	0	1	3	0	4	0	−1
−2	0	0	2	0	−3	1	1
3	0	0	3	1	2	0	3
2	1	0	−1	0	2	0	2

(a) Initial tableau of main problem.

−2	0	0	2	0	−3	1	1
3	0	0	3	1	2	0	3
2	1	0	−1	0	2	0	2

(b) Auxiliary tableau.

Another example is provided by Table 3.13, in which the number of nonnegative elements in B is $q = 0$. We can then pivot at any position in row 3 where the entry is negative, say in position (3, 1). Pivoting there produces the problem in Table 3.13 (b) which is in canonical form with respect to the sequence $S = (3, 4, 1)$.

Table 3.12. EXAMPLE AFTER PIVOT AT (2, 5) IN AUXILIARY TABLEAU [AT (4, 5) IN MAIN TABLEAU].

2	−1	0	4	0	0	0	0
−6	−2	1	5	0	0	0	−5
1	3/2	0	1/2	0	0	1	4
1	−1	0	2	1	0	0	1
1	1/2	0	−1/2	0	1	0	1

Table 3.13. PROBLEM IN WHICH $q = 0$.

0	25	50	0	0	0
−8	−3	−1	1	0	0
−19	−4	−3	0	1	0
−7	−1	−3	0	0	1

(a) Main problem.

−175	0	−25	0	0	25
13	0	8	1	0	−3
9	0	9	0	1	−4
7	1	3	0	0	−1

(b) After pivot at (3, 1).

3.7. REDUCTION TO CANONICAL FORM, III: CANONICAL COEFFICIENTS.

In any linear programming problem involving inequality constraints such as (3.32) we can introduce slack variables (defined to be the difference between the greater side and lesser side) and get a basis with which to initiate the simplex algorithm. However, suppose that in the statement of the original problem both equality and inequality constraints appear. Specifically, let d be a scalar, let A_1, A_2, C_1, B_1, B_2 be real matrices with respective degrees p_1 by n_1, p_2 by n_1, n_1 by 1, p_1 by 1, p_2 by 1, and let X_1 be an n_1 by 1 vector. We consider the problem

$$\text{minimize } z = d + C_1^T X_1 \tag{3.66}$$

subject to the feasibility constraints

$$A_1 X_1 \leq B_1, \qquad A_2 X_1 = B_2, \qquad X_1 \geq 0. \tag{3.67}$$

We introduce a p_1 by 1 slack vector X_2 and obtain an equivalent problem in the standard form of (3.18), (3.19), (3.20) by taking $p = p_1 + p_2$, $n = n_1 + p_1$ and defining C, A, B, X by

$$C = \begin{bmatrix} C_1 \\ 0 \end{bmatrix}, \quad A = \begin{bmatrix} A_1 & I_{p_1} \\ A_2 & 0 \end{bmatrix}, \quad B = \begin{bmatrix} B_1 \\ B_2 \end{bmatrix}, \quad X = \begin{bmatrix} X_1 \\ X_2 \end{bmatrix}. \tag{3.68}$$

Unlike the pure inequality case when the slack columns provided a full basis, here we have initially only a partial basis. Rather than limit ourselves to the special case represented in (3.68), let us suppose that we have a problem in standard form and that we have a sequence of $q - 1$ distinct integers $S = (s_1, \cdots, s_{q-1})$, where $q \leq p$, and such that

$$A_{s_1} = U_1, \cdots, A_{s_{q-1}} = U_{q-1}, \tag{3.69}$$

where as before U_j is the natural basis vector of p coordinates, all of which are zero except for the jth coordinate which is 1. Our general plan is to pivot in nonbasic columns to obtain a full basis of unit vectors; we will then be in the case treated in Section 3.6. We examine the rows in order, beginning with row q; for row q there are three possible cases:

Case (a) $b_q = a_{q1} = \cdots = a_{qn} = 0$,
Case (b) $b_q \neq 0, \quad a_{q1} = \cdots = a_{qn} = 0$,
Case (c) for some k, $a_{qk} \neq 0$.

In Case (a) we delete row q and proceed with an equivalent problem in which $p' = p - 1$. This is the case of redundant or dependent equations. In Case (b) the equations are inconsistent and the original problem is infeasible. In Case (c) we pivot on position (q, k) and let $s_q = k$. Now $s_q \notin S$ because

Table 3.14. OBTAINING A BASIS.

3	−2	−1	4	0	0
3	1	3	2	1	0
1	0	−2	1	0	1
2	[1]	−1	1	0	0
0	2	2	3	0	0
7	0	−3	6	0	0
1	0	4	1	1	0
1	0	−2	1	0	1
2	1	−1	1	0	0
−4	0	4	[1]	0	0
31	0	−27	0	0	0
5	0	0	0	1	0
5	0	−6	0	0	1
6	1	−5	0	0	0
−4	0	4	1	0	0

of (3.69), so we obtain a new basic or canonical column. Moreover, the pivot operation leaves the columns $A_{s_1}, \cdots, A_{s_{q-1}}$ undisturbed, and we obtain a new sequence $S' = (s_1, \cdots, s_{q-1}, s_q)$. The process is inductive starting with $q \geq 1$ and ending either (1) with a full basis and a tableau satisfying C1 and C2 or (2) with the conclusion that the problem is infeasible. Note that we do not require $q - 1 > 0$; in other words, there need be no unit basis vectors initially.

We give an example in Table 3.14 in which $n_1 = 3$, $p_1 = p_2 = 2$. The pivot elements are indicated. The first is $(h, k) = (3, 1)$; the second tableau has column 1 basic and it shows that columns 4 and 5 remain canonical. After two pivots the partial basic sequence $s_1 = 4$, $s_2 = 5$ is expanded to the full sequence (4, 5, 1, 3). Note also that row four of the final tableau has a negative constant term and nonnegative coefficients and thus demonstrates infeasibility of the original problem.

3.8. REDUCTION TO CANONICAL FORM, IV: HANDLING FREE VARIABLES AND MIXTURES OF CONSTRAINTS.

We have by now considered how to deal with almost all linear programming problems encountered in practice (most involve constraints that are all inequalities or all equalities). We now complete our proof of convergence of the simplex algorithm by considering the most general form of a linear programming problem, that is, one in which both nonnegative and unrestricted (or free) variables and both equalities and inequalities appear. To simplify notation we first consider the case of equality constraints involving both nonnegative and free variables; the section is concluded by showing how the most general case can be reduced to this form.

Suppose we wish to

$$\text{minimize } z = d + C_1^T X_1 + C_2^T X_2 \tag{3.70}$$

subject to

$$A_1 X_1 + A_2 X_2 = B, \tag{3.71}$$
$$X_1 \geq 0, \tag{3.72}$$
$$X_2 \text{ free}, \tag{3.73}$$

where A_1 is p by n_1, A_2 is p by n_2, and the vectors in (3.72) and (3.73) are n_1 by 1 and n_2 by 1, respectively. This problem would be in standard form were it not for the free variables. A little reflection leads to the observation that basic free variables present no difficulty but that nonbasic (independent) free variables upset the logic of the simplex method. This suggests that we look for pivot operations which will make as many free variables as possible become basic variables. We will then be able to lay the corresponding equations aside, temporarily, and deal with a smaller problem in standard form.

We proceed inductively and assume, for some q with $0 \leq q \leq n_2$, that there are q basic free variables and that by permuting equations in (3.71) if necessary these basic free variables are in equations $p - q + 1, \cdots, p$ with x_{s_i} appearing in equation i, $i = p - q + 1, \cdots, p$. If $q < n_2$ there exists another free column k and for it there are three cases to consider:

$$\begin{aligned}
&\text{Case (a)}\ a_{1k} = \cdots = a_{p-q,k} = c_k = 0; \\
&\text{Case (b)}\ a_{1k} = \cdots = a_{p-q,k} = 0, \qquad c_k \neq 0; \\
&\text{Case (c)}\ \text{for some } h \text{ with } 1 \leq h \leq p - q, \qquad a_{hk} \neq 0.
\end{aligned}$$

If Case (a) holds for every nonbasic free column or if $q = n_2$ we consider the subproblem which remains when we delete all of the free columns and the last q rows. This subproblem is in standard form and therefore can be handled by the methods already developed. If the subproblem has a basic optimal feasible program X_1^0 we obtain a basic optimal feasible program X_1^0, X_2^0 for the main problem by setting each nonbasic free variable equal to zero and by setting

$$x_{s_i} = b_i - (a_{i1}x_1^0 + \cdots + a_{in_1}x_{n_1}^0), \qquad (i = p - q + 1, \cdots, p). \tag{3.74}$$

Moreover, the minimum value of z is the same for the main problem and for the subproblem.

If Case (b) holds then either there is no lower bound for the objective function or the problem is infeasible. For, suppose X^0 is any feasible vector with corresponding objective value z^0; then define a vector Y as follows

$$\begin{aligned}
y_{s_i} &= -a_{ik} \qquad (i = p - q + 1, \cdots, p), \\
y_k &= 1, \\
y_i &= 0, \text{ otherwise.}
\end{aligned} \tag{3.75}$$

The vector Y makes the left-hand sides of (3.71) zero so that the vector

$$X = X^0 + tY \tag{3.76}$$

(1) is feasible for all real numbers t, and (2) has the property that the objective function assumes the value $z = z^0 + tc_k$. It is clear that the constraint equations (3.71) are satisfied by X. Moreover, for all t, $X_1 = X_1^0 \geq 0$ so that X satisfies all of the feasibility requirements. Next, $z = d + C^TX = (d + C^TX^0) + tC^TY = z^0 + tc_k$ (since $c_{s_i} = 0$, $i = p - q + 1, \cdots, p$). Now by giving t the sign opposite that of c_k and by taking the absolute value of t sufficiently large we can make z assume negative values with arbitrarily large absolute value; thus, the objective function has no lower bound. Thus if Case (b) occurs we stop, assured that the problem has no solution.

Finally, if Case (c) holds we (1) pivot at position (h, k), (2) permute row h and row $p - q$, (3) set $s_{p-q} = k$ and thereby have completed an inductive step. We then iterate the process until eventually we reach Case (a), Case (b), or $q = n_2$.

We illustrate Case (a) by means of the following problem:

$$\text{minimize } z = -6 + 4x_2 \tag{3.77}$$

subject to

$$\begin{aligned} 3x_2 + x_3 \qquad\qquad &= 2, \\ x_1 - 2x_2 \qquad\qquad\qquad &= 4, \\ 2x_1 + 2x_2 + x_3 + x_4 + 2x_5 &= 3, \\ x_1,\ x_2,\ x_3 &\geq 0, \\ x_4,\ x_5 &\text{ free}, \end{aligned} \tag{3.78}$$

for which, in the notation of (3.71), (3.72), and (3.73), we have

$$A_1 = \begin{bmatrix} 0 & 3 & 1 \\ 1 & -2 & 0 \\ 2 & 2 & 1 \end{bmatrix}, \qquad A_2 = \begin{bmatrix} 0 & 0 \\ 0 & 0 \\ 1 & 2 \end{bmatrix}, \qquad C = \begin{bmatrix} 0 \\ 4 \\ 0 \\ 0 \\ 0 \end{bmatrix},$$

$$X_1 = \begin{bmatrix} x_1 \\ x_2 \\ x_3 \end{bmatrix}, \qquad X_2 = \begin{bmatrix} x_4 \\ x_5 \end{bmatrix}.$$

The corresponding tableau appears as Table 3.15. We see that $p = 3$, $n_1 = 3$, $n_2 = 2$. Column 4 is a basic column; the corresponding free variable is basic so we have $s_3 = 4$ and this variable appears with coefficient 1 in the last equation of (3.78). Also, $q = 1$ and there is another free column $k = 5$.

Table 3.15. TABLEAU FOR ORIGINAL PROBLEM.

6	0	4	0	0	0
2	0	3	1	0	0
4	1	−2	0	0	0
3	2	2	1	1	2
				Free columns	

We are in Case (a) because

$$\begin{aligned} a_{1k} &= a_{15} = 0, \\ a_{2k} &= a_{25} = 0, \\ c_k &= c_5 \ = 0. \end{aligned}$$

The corresponding subproblem is obtained by deleting columns 4 and 5 and the last row from the tableau in Table 3.15; this gives the tableau in Table 3.16, which is in canonical form. From this we get

$$X_1^0 = \begin{bmatrix} 4 \\ 0 \\ 2 \end{bmatrix},$$

which happens to be an optimal program for the subproblem.

Table 3.16. TABLEAU FOR SUBPROBLEM, CASE (a).

6	0	4	0
2	0	3	1
4	1	−2	0

Also, we set the nonbasic free variable x_5 equal to zero and from (3.74) we get

$$\begin{aligned} x_{s_3}^0 = x_4^0 &= b_3 - (a_{31}x_1^0 + a_{32}x_2^0 + a_{33}x_3^0) \\ &= 3 - [2(4) + 2(0) + 1(2)] = -7. \end{aligned}$$

Therefore

$$X_2^0 = \begin{bmatrix} -7 \\ 0 \end{bmatrix}$$

and

$$X^0 = \begin{bmatrix} X_1^0 \\ X_2^0 \end{bmatrix} = \begin{bmatrix} 4 \\ 0 \\ 2 \\ -7 \\ 0 \end{bmatrix}$$

Table 3.17. CASE (b), UNBOUNDED OBJECTIVE FUNCTION.

0	1	3	4	−2	0
2	−1	3	−3	0	0
−4	2	−6	3	0	0
5	1	2	1	3	1
				Free columns	

$$p = 3, \quad n_1 = 3, \quad n_2 = 2, \quad q = 1, \quad s_3 = 5, \quad k = 4,$$

$$X^0 = \begin{bmatrix} 1 \\ 1 \\ 0 \\ 0 \\ 2 \end{bmatrix}, \quad z^0 = 4, \quad Y = \begin{bmatrix} 0 \\ 0 \\ 0 \\ 1 \\ -3 \end{bmatrix},$$

$$z = 4 - 2t \to -\infty \text{ as } t \to +\infty.$$

is an optimal program for the original problem for which minimum $z = z^0 = -6$.

Table 3.17 illustrates Case (b), that of an unbounded objective function, and Table 3.18 gives an example of Case (c). In both examples the variables x_1, x_2, and x_3 are nonnegative and x_4 and x_5 are free. The subproblem obtained from the second tableau in the third example (Table 3.18) is

4	-1	9	2
2	-1	3	6

Unboundedness is indicated here by the column $\begin{bmatrix} -1 \\ -1 \end{bmatrix}$, so the original problem has no optimal solution.

Table 3.18. CASE (c), INCREASE IN q.

0	1	3	4	-2	0
2	-1	3	-3	1	0
-4	2	-6	3	-3	0
3	1	2	1	3	1

Free columns (last two columns)

$p = 3, \quad n_1 = 3, \quad n_2 = 2, \quad q = 1, \quad s_3 = 5, \quad k = 4.$

Pivot in row 1 and then interchange rows 1 and 2.

4	-1	9	-2	0	0
2	-1	3	-6	0	0
2	-1	3	-3	1	0
-3	4	-7	10	0	1

Now $q = 2$ for this tableau.

These three illustrations correspond to the possibilities that can occur in the simplex method. Case (a) displays optimality, Case (b) unboundedness, and the first tableau of Case (c) is one in which an improvement step can be made.

A free variable x can also be handled by replacing it by the difference of two nonnegative variables $x' - x''$. The earlier form of the simplex algorithm can then be used, and since the column vectors corresponding to the variables x' and x'' are negatives of each other, they form a dependent sequence and hence cannot both be in an optimal basis. However, this procedure increases the size of the problem and is not as satisfactory as the procedure developed above since the latter uses free variables to reduce the size of the problem.

We observed earlier that the most general form of a linear programming problem contains both equalities and inequalities and nonnegative and free variables:

$$\text{minimize } z = d + D_1^T Y_1 + D_2^T Y_2 \tag{3.79}$$

subject to

$$A_{11}Y_1 + A_{12}Y_2 \geq B_1 \quad (p_1 \text{ constraints}), \tag{3.80}$$
$$A_{21}Y_1 + A_{22}Y_2 = B_2 \quad (p_2 \text{ constraints}), \tag{3.81}$$
$$Y_1 \geq 0 \quad (n' \text{ variables}), \tag{3.82}$$
$$Y_2 \text{ free} \quad (n_2 \text{ variables}). \tag{3.83}$$

We introduce a slack vector Y_1' with p_1 coordinates, set $D_1' = 0$ and obtain the problem

$$\text{minimize } z = d + D_1^T Y_1 + D_1'^T Y_1' + D_2^T Y_2 \tag{3.79'}$$

subject to

$$A_{11}Y_1 - I_{p_1}Y_1' + A_{12}Y_2 = B_1, \tag{3.80'}$$
$$A_{21}Y_1 + 0Y_1' + A_{22}Y_2 = B_2, \tag{3.81'}$$
$$Y_1 \geq 0, \tag{3.82}$$
$$Y_1' \geq 0, \tag{3.84}$$
$$Y_2 \text{ free}. \tag{3.83}$$

This is now in the form of the problem defined by (3.70) through (3.73), where

$$A_1 = \begin{bmatrix} A_{11} & -I_{p_1} \\ A_{21} & 0 \end{bmatrix}, \quad A_2 = \begin{bmatrix} A_{12} \\ A_{22} \end{bmatrix}, \quad X_1 = \begin{bmatrix} Y_1 \\ Y_1' \end{bmatrix}, \quad X_2 = Y_2,$$

$$B = \begin{bmatrix} B_1 \\ B_2 \end{bmatrix}, \quad p = p_1 + p_2, \quad n_1 = n' + p_1, \quad C_1 = \begin{bmatrix} D_1 \\ D_1' \end{bmatrix}, \quad C_2 = D_2$$

3.9. MULTIPLE OPTIMAL SOLUTIONS.

When an optimal solution is reached, that is, when the final cost row is nonnegative for $j = 1, \cdots, n$ we always have $c_{s_i} = 0$, $i = 1, \cdots, p$, that is, the basic costs are all zero. If $c_k = 0$ for any nonbasic column k (and if the problem is nondegenerate) then we obtain a new basic optimal solution by applying a simplex transformation to introduce this k as a basic index. The basic index s_h to be discarded is determined by the usual decision rule.

If the problem is degenerate there may be several bases corresponding to the same solution. (For example, this always occurs in the personnel assignment problem for $n > 1$.) In this case an additional zero cost does not necessarily indicate an additional basic solution. We do not treat this situation.

In most practical problems, if there is more than one optimal basic solution it is not difficult to find all of them; however, in theory, it is not an easy problem to give a direct-search procedure for efficient determination of all such solutions.

The first tableau of Table 3.19 is an optimal tableau from Table 3.4. We have $c_j \geq 0$ and a nonbasic activity (activity 4) with zero cost. Let $k = 4$ so that 4 is the index of the incoming vector, and select s_h as usual. One then pivots at row 1, column 4; the next tableau gives an alternate optimal solution and activity 4 is now a basic activity. Again, a nonbasic activity has a zero cost; pivoting in column 5 as above results in still another basic optimal solution. The last tableau shows a third optimal solution.

The case of a nonbasic free variable with cost zero requires more attention. Since such a variable occurs in rows in which the associated basic variables are also free there is no restriction on the range; a nonbasic free variable can take any value between minus infinity and plus infinity. If any of these values results in a zero value for some basic free variable, then a new basic solution is obtained. This case arises rarely in practice and so we do not attempt to treat it in full detail.

Table 3.19. MULTIPLE OPTIMAL BASIC SOLUTIONS (CONTINUATION OF TABLE 3.4).

									b_i/a_{ik}	
7/13	0	0	0	0	0	3/13	2/13	2/13		
4/39	0	1	0	3/5	8/9	11/39	−10/39	1/13	20/117	$k = 4$
10/39	0	0	1	2/5	5/9	8/39	14/39	−4/13	50/78	$h = 1$
7/39	1	0	0	0	−4/9	−10/39	2/39	5/13		$s_h = 2$
7/13	0	0	0	0	0	3/13	2/13	2/13		
20/117	0	5/3	0	1	40/27	55/117	−50/117	5/39	3/26	$k = 5$
22/117	0	−2/3	1	0	−1/27	2/117	62/117	−14/39		$h = 1$
21/117	1	0	0	0	−4/9	−10/39	2/39	5/13		$s_h = 5$
7/13	0	0	0	0	0	3/13	2/13	2/13		
3/26	0	9/8	0	27/40	1	33/104	−30/104	9/104		
4/26	0	−5/8	1	1/40	0	3/104	54/104	−37/104		
7/26	1	1/2	0	3/10	0	−3/26	−1/13	11/26		

3.10. BASIC PROGRAMS AND EXTREME VECTORS.

The optimization process of linear programming was illustrated in Chapter 1 in terms of equicost lines, feasible regions, and vertices of feasible regions. It was stated that if there is an optimal feasible program, then there is a vertex which is optimal. The simplex algorithm developed in this chapter is an algebraic process dealing with basic programs. We wish now to examine relationships between the geometric concept of a vertex (or extreme vector as we will call it) and the algebraic concept of basic program.

Let T be the set of all feasible programs X of the linear programming problem,

$$\text{minimize } z = d + C^T X$$

subject to

$$AX = B, \qquad X \geq 0.$$

If $X_1, \cdots, X_r$ are any vectors in T, then for any scalars $k_1, \cdots, k_r$

$$X = k_1 X_1 + \cdots + k_r X_r \tag{3.85}$$

is also feasible provided that $k_1 + \cdots + k_r = 1$ and each $k_i \geq 0$. We recall from Appendix B.16 that X is called a convex combination or probability combination of the vectors $X_1, \cdots, X_r$. To verify that X is feasible we observe that

$$\begin{aligned} AX &= A(k_1 X_1 + \cdots + k_r X_r) = k_1 A X_1 + \cdots + k_r A X_r \\ &= (k_1 + \cdots + k_r) B = B; \end{aligned}$$

the nonnegativity of X follows from that of the k_i and X_i.

A set S is said to be *convex* if, given any vectors X_1 and X_2 in S, the line segment joining X_1 and X_2 is in S, that is, if $kX_1 + (1 - k)X_2 \in S$ for $0 \leq k \leq 1$. The above result for feasible vectors can be summarized by saying that T is a convex set.

A vector $X \in T$ is said to be *extreme* or a *vertex* if

$$X = kX_1 + (1 - k)X_2$$

for $0 < k < 1$ and $X_1, X_2 \in T$ requires $X_1 = X_2 = X$. It is easily verified that this condition is equivalent to requiring that it be impossible to express X as a convex combination of vectors in T all of which are different from X. In particular, X cannot be on the interior of any line segment which is in T.

Let $X = \begin{bmatrix} x_1 \\ \cdot \\ \cdot \\ \cdot \\ x_n \end{bmatrix}$ be any vector of degree n; if $x_i \neq 0$ we call i an *active* index for X and we denote the sequence of all active indices for X by $I(X)$, or $I(X) = (i_1, \cdots, i_r)$, $i_1 < i_2 < \cdots < i_r$.

Theorem 1. A feasible program X is extreme if and only if the sequence of columns $(A_{i_1}, \cdots, A_{i_r})$ of A is independent.

Proof: Suppose that $X = kX_1 + (1 - k)X_2$, where X_1, X_2 are in T. Then clearly $I(X_j) \subset I(X)$, $j = 1, 2$. Now let $Y = X_2 - X_1$; then

$$AY = AX_2 - AX_1 = B - B = 0.$$

Since $I(Y) \subset I(X_1) \cup I(X_2) = I(X)$, we have

$$A_{i_1}y_{i_1} + \cdots + A_{i_r}y_{i_r} = 0. \tag{3.86}$$

Now if $(A_{i_1}, \cdots, A_{i_r})$ is independent we must have $y_{i_1} = \cdots = y_{i_r} = 0$ and therefore, since $I(Y) \subset I(X)$, $Y = 0$. It follows that $X_1 = X_2 = X$. Conversely, if $(A_{i_1}, \cdots, A_{i_r})$ is dependent there exists a vector $Y \neq 0$ such that (3.86) holds and $I(Y) \subset I(X)$, that is, $y_j = 0$ for $j \notin I(X)$. Now since $I(Y) \subset I(X)$ and $X \geq 0$, we can choose X_1 and X_2 in T and a t sufficiently small so that $X_1 = (X - tY) \geq 0$ and $X_2 = (X + tY) \geq 0$; then the equation $X = (1/2)X_1 + (1/2)X_2$ shows that X is not extreme.

Corollary. There exists a finite sequence $K = (X^1, \cdots, X^k)$ of distinct vectors which contains all the extreme vectors in T.

Proof: By the theorem, each extreme vector is associated with an independent sequence of columns of A; hence the number of extreme vectors is finite.

Theorem 2. Let S be a basic sequence; then the vector X^S is extreme. Conversely, if the rows of A are independent and if X is extreme, then there exists a basic sequence S for which $X = X^S$.

Proof: Since $x_i^S = 0$ for every $i \notin S$, each index $j \in I(X^S)$ is a member of S. Since A_S is nonsingular by the definition of basic sequence, it follows that every sequence of distinct columns of A_S is independent and hence, in particular, $(A_{i_1}, \cdots, A_{i_r})$ is independent so that X is extreme.

Conversely, let p be the number of rows of A and suppose that X^0 is extreme. Choose $s_1 = i_1, \cdots, s_r = i_r$; if $r = p$ we have $X^0 = X^S$. If $r < p$ we use the fact that A has rank p and select additional indices $s_{r+1}, \cdots, s_p$ such that for $S = (s_1, \cdots, s_p)$, A_S is nonsingular. The equation $A_SY = B_S$ has a unique solution; but since $I(X^0) \subset S$ we must have $A_SX_S^0 = B_S$. It follows that $X^0 = X^S$ since $X_S^0 = X_S^S$ and $X_G^0 = X_G^S = 0$.

If the rows of the matrix A are dependent there can be no basic sequence and hence no basic programs. Therefore we may have extreme vectors which are not basic. However, if the system $AX = B$ is consistent we may always discard redundant equations to obtain a residual system $A'X = B'$ with the same solutions and for which the rows of the coefficient matrix A' are independent. For this derived system every extreme vector will be a basic program.

Now suppose that the rows of A are independent. Then we have $r = p$ unless the problem is degenerate. If $r < p$ there may be many choices for the indices $s_{r+1}, \cdots, s_p$ and hence in degenerate problems there may be many basic sequences corresponding to one extreme vector. For example, consider the second and third tableaus of Table 3.6. There is one extreme program for each of these, the vector

$$X = \begin{bmatrix} 0 \\ 0 \\ 2 \\ 0 \\ 0 \\ 1 \\ 0 \\ 0 \end{bmatrix};$$

corresponding to this there are two basic sequences: for the second tableau the basic sequence is $S_1 = (3, 6, 7, 8)$ for the third it is $S_2 = (3, 6, 7, 1)$.

Of course, if two sequences S_1 and S_2 differ only in the order of their elements but are equal as sets, then $X^{S_1} = X^{S_2}$. In the case of degeneracy we can have $X^{S_1} = X^{S_2}$ where the sequences S_1 and S_2 are not equal as sets as the example above indicates.

A subset C of V_n is said to be a *convex cone* if

$$Y = k_1 Y_1 + \cdots + k_r Y_r \tag{3.87}$$

is in C whenever each $Y_j \in C$ and each $k_j \geq 0$. The vector Y in (3.87) is said to be a *nonnegative combination* of $(Y_1, \cdots, Y_r)$. If Y is any nonzero vector in a convex cone C, then $Y = (1/2)(Y_1 + Y_2)$, where $Y_1 = (1/2)Y$ and $Y_2 = (3/2)Y$; hence *the origin is the only possible vertex in a convex cone.* If C contains any vector subspace of V_n, then the origin is not a vertex; in all other cases the origin is a vertex and C is said to be *pointed.*

If $H = [H_1 \cdots H_q]$ is an n by q matrix then we denote by $C(H)$ the set of all nonnegative combinations of columns of H. Clearly $C(H)$ is a convex cone and is called the cone generated by H (or by the columns of H). A convex cone is said to be *polyhedral* if it is generated by some matrix H.

A cone generated by a single nonzero vector is said to be a *ray.* Let C be a cone and let Y be a nonzero vector in C; the ray $C(Y)$ is said to be *extreme in* C if $Y = Y_1 + Y_2$ for $Y_j \in C, j = 1, 2$, requires that $Y_j \in C(Y)$ for $j = 1, 2$.

Let A be a p by n matrix and let U be the set of all solutions of the system

$$AY = 0, \qquad Y \geq 0. \tag{3.88}$$

It is easy to show that U is a pointed convex cone. Our next objective will be to show that U is polyhedral; the latter is the essential content of Theorem 4.

Theorem 3. A nonzero vector $Y \in U$ generates an extreme ray if and only if the columns of A_R are independent for each proper subsequence R of $I(Y)$.

Proof: Suppose that each A_R has independent columns and that $Y = Y_1 + Y_2$ with $Y_1, Y_2 \in U$. Then since all the vectors in U are nonnegative we must have $I(Y_j) \subset I(Y)$ for $j = 1, 2$. Let $I(Y) = (i_1, \cdots, i_r)$; then the independence hypothesis implies that rank $A_{I(Y)}$ is $r - 1$. Hence all solutions of $AZ = 0$ with $I(Z) \subset I(Y)$ are proportional to Y. It follows that Y_1 and Y_2 are proportional to Y so that $C(Y)$ is an extreme ray.

Conversely, suppose for some proper subsequence R of $I(Y)$ that the columns of A_R are dependent. Then there is a nonzero vector Z with $I(Z) \subset R$ and $AZ = 0$. Since $I(Z) \subset R \langle I(Y)$, Z is not proportional to Y. For t sufficiently small both $Y_1 = (1/2)Y - tZ$ and $Y_2 = (1/2)Y + tZ$ are nonnegative and hence in U, and neither can be proportional to Y, since Z is not proportional to Y. The equation $Y = Y_1 + Y_2$ then shows that Y is not an extreme ray.

Theorem 4. (A double-description theorem for pointed cones). There exists a finite sequence $H = (Y^1, \cdots, Y^h)$ of generators of extreme rays with the following properties:

1. no two vectors in H are proportional;
2. every vector $Y \in U$ can be written as a nonnegative combination of vectors in H, that is, $U = C(H)$.

Proof: Consider a sequence $Q = (q_1, \cdots, q_r)$ with $1 \leq q_1 < q_2 < \cdots < q_r \leq n$; if the rank of A_Q is $r - 1$ but every proper subsequence of columns of A_Q is independent, we call Q *admissible*. For a vector Y to generate an extreme ray it is necessary (but not sufficient) for $I(Y)$ to be admissible. With each admissible sequence Q we associate a unique vector $Z(Q)$ by stipulating the following requirements:

1. $Z(Q)$ has 1 for its first nonzero component,
2. $A_Q Z(Q) = 0$,
3. $I(Z(Q)) = Q$.

A nonzero vector satisfying (2) exists because of the hypothesis that A_Q has a column kernel of dimension 1 (see Appendix B.13). After normalization so as to achieve (1) we have a unique vector $Z(Q)$ satisfying (1), (2), and having $I(Z(Q)) \subset Q$. However, $I(Z(Q)) \langle Q$ contradicts the admissibility of Q so that we have (3) as well.

We now take for the sequence of vectors H any ordering of the set of all vectors $Z(Q)$ which lie in U, that is, vectors for which $Z(Q) \geq 0$. Since there can be only finitely many admissible sequences, H is finite. Moreover, it is clear that every extreme ray is generated by a member of H.

Suppose that $Y \in U$ and that $C(Y)$ is not an extreme ray; then proceed as in the converse part of the proof of Theorem 3 except for normalizing Z so that it has its first nonzero component positive. Set $t_1 = \min \{y_i/z_i \mid z_i > 0\}$ and let $Z_1 = Y - t_1 Z$. The construction guarantees that $Z_1 \in U$ and also that $I(Z_1) \langle I(Y)$. Now if $Z \geq 0$ let $Z_2 = t_1 Z$ and we have

$Y = Z_1 + Z_2$. On the other hand, if Z has negative components we set $t_2 = \min \{-y_i/z_i \mid z_i < 0\}$ and $Z_2 = Y + t_2Z$. Then $Y = t_2/(t_1 + t_2) Z_1 + t_1/(t_1 + t_2)Z_2$. In each case Y is expressed as a sum of vectors each with fewer nonzero components than has Y.

We can iterate this process until we obtain an expression for Y as a nonnegative combination of generators of extreme rays and hence of vectors in H.

We turn now to an examination of the structure of the feasible region T. Suppose X is not basic; then there exists a vector $Y \neq 0$ such that $I(Y) \subset I(X)$ and $AY = 0$. We can assume without loss of generality that the first nonzero component of Y is positive (that Y is lexicographically positive). We consider two cases: Case 1, $Y \geq 0$; Case 2, Y has a negative component. In either case there exists a unique t_1 for which $X_1 = X - t_1Y$ is an element of T with $I(X_1) \langle I(X)$. Indeed we have

$$t_1 = \min \{x_i/y_i \mid i \in I(X) \quad \text{and} \quad y_i > 0\}.$$

In Case 1 we can write

$$X = X_1 + t_1Y$$

where $X_1 \in T$, $Y \in U$, and $I(X_1) \langle I(X)$. In Case 2 we let $X_2 = X + t_2Y$, where

$$t_2 = \min \{-x_i/y_i \mid i \in I(X) \quad \text{and} \quad y_i < 0\},$$

and let $k = t_2/(t_1 + t_2)$. Then

$$X = kX_1 + (1 - k)X_2$$

where $0 < k < 1$ and $I(X_1) \langle I(X)$, $I(X_2) \langle I(X)$.

In any case we can write any nonextreme vector $X \in T$ in the form

$$X = k_1W_1 + \cdots + k_sW_s + Y, \tag{3.89}$$

where $W_j \in T$, $I(W_j) \langle I(X)$, $k_j \geq 0$, $\sum_{j=1}^{s} k_j = 1$ and $Y \in U$ (Y may be the zero vector). Now if any W_j is not basic we have a similar expression

$$W_j = k_{1j}W_{1j} + \cdots + k_{qj}W_{qj} + Y_j$$

and then

$$X = k_1W_1 + \cdots + k_{j-1}W_{j-1} + k_jk_{1j}W_{1j} + \cdots + k_jk_{qj}W_{qj} + k_{j+1}W_{j+1} + \cdots + (Y + k_jY_j), \tag{3.90}$$

which is again of the form (3.89), that is, X is a convex combination of vectors in T plus a vector in U. Moreover, the shrinking of the index sets $I(X) \rangle I(W_j) \rangle I(W_{ji})$ guarantees that we will ultimately reach an expression of the form (3.89) in which each vector W_j is extreme.

We summarize the preceding results in the following theorem, which is a part of the main theorem of double description for convex polyhedra.

Theorem 5. Let A be a p by n matrix, let B be a p by 1 vector, let T be the set of all solutions of the system $AX = B$, $X \geq 0$, let U be the set of all solutions of the system $AY = 0$, $Y \geq 0$; then there exists a sequence $H = (Y^1, \cdots, Y^h)$ of generators of extreme rays in U such that U consists of all nonnegative combinations Y of H and there exists a sequence $K = (X^1, \cdots, X^k)$ of extreme vectors in T such that T consists of all vectors X which can be expressed as a convex combination of vectors in K plus a nonnegative combination of vectors in H.

The discussion in this section is limited to a consideration of special cases of double description theorems for convex polyhedral cones and sets useful for our development of linear programming. A more complete treatment can be found in papers 2 and 3 of Reference 29.

PROBLEMS

3.1. Consider the following problem:

$$\text{maximize} \quad 2x_1 + 5x_2$$

subject to

$$\begin{aligned} x_1 + 4x_2 &\leq 24, \\ 3x_1 + x_2 &\leq 21, \\ x_1 + x_2 &\leq 9, \\ x_1 &\geq 0, \\ x_2 &\geq 0. \end{aligned}$$

(a) Incorporate slack variables and identify the initial basic sequence S, the initial basic program X^S; what is the value of the objective function for this solution?

(b) Solve the problem by graphical means. Then use the simplex algorithm to solve the problem and to verify your graphical solution. What is the final (optimal) basic solution? What is the basic sequence S for the final or optimal basic program?

(c) Are there any slack variables with positive value in an optimal program? If so, can you interpret this result?

3.2. Show that the simplex algorithm fails to provide an optimal solution to the problem of maximizing $2x_1 + x_2$ subject to

$$\begin{aligned} -x_1 + x_2 &\leq 1, \\ x_1 - 2x_2 &\leq 2, \\ x_1 &\geq 0, \\ x_2 &\geq 0. \end{aligned}$$

Show graphically why there is no optimal solution.

3.3. Solve the following problem:

$$\text{minimize } z = 2x_1 + x_2 - 3x_3 + 5x_4 + 8x_5$$

subject to

$$
\begin{aligned}
3x_1 - 5x_2 - 9x_3 + x_4 - 2x_5 &\le 6, \\
2x_1 + 3x_2 + x_3 + 5x_4 + x_5 &\le 9, \\
x_1 + x_2 - 5x_3 - 7x_4 + 11x_5 &\le 10, \\
x_j &\ge 0, \qquad (j = 1, \ldots, 5).
\end{aligned}
$$

Let S and X^S, z^S be the basic sequence and basic feasible solution for the first tableau, respectively, and S' and $X^{S'}$, $z^{S'}$ denote corresponding aspects of the final or optimal cycle. What are S, S', X^S, z^S, $X^{S'}$, $z^{S'}$?

3.4. Solve the following problem:

$$\text{minimize } z = 2x_1 + 3x_2 + x_3 + 5x_4$$

subject to

$$
\begin{aligned}
5x_1 - 3x_2 + 2x_3 - x_4 &\ge 3, \\
4x_1 + 2x_2 - 5x_3 - 3x_4 &\ge 7, \\
-x_1 + 3x_2 + 7x_3 + 2x_4 &\ge -4, \\
x_j &\ge 0, \qquad (j = 1, \ldots, 4).
\end{aligned}
$$

3.5. Solve the following problem:

$$\text{minimize } g = x_1 - 3x_2 + 2x_3$$

subject to

$$
\begin{aligned}
x_1 - 2x_2 + 5x_3 &\ge 3, \\
2x_1 + x_2 + x_3 &\ge 1, \\
x_j &\ge 0, \qquad (j = 1, 2, 3).
\end{aligned}
$$

3.6. Solve the following problem:

$$\text{maximize } z = x_1 - x_2 + x_3 - 3x_4 + x_5$$

subject to

$$
\begin{aligned}
3x_3 - x_4 + 2x_5 &\le 5, \\
x_1 + 2x_2 + x_4 + 5x_5 &\le 9, \\
-x_1 + x_4 &\le 2, \\
-x_2 + 3x_4 &\le 3, \\
x_j &\ge 0, \qquad (j = 1, \ldots, 5).
\end{aligned}
$$

(a) Review the notation of Section 3.3 and let S be the basic sequence for the first tableau, X^S, z^S the basic solution, and let S^* be the basic sequence for the optimal tableau or cycle and X^{S^*}, z^{S^*} be the basic solution for this cycle. What are S, S^*, and X^S, z^S, X^{S^*}, z^{S^*}? For the first tableau, what is G?

(b) Now consider the first and second cycles of your solution and the matrix M in Equation (3.55). If M' is the derived tableau for the second cycle, write out the matrices M and M'.

3.7. Solve the following problem:

$$\text{minimize } z = -10x_1 - 13x_2 - x_3 - x_4$$

subject to

$$\begin{aligned} 5x_1 + 2x_2 &\leq 20, \\ 3x_1 + 4x_2 &\leq 24, \\ x_1 &\geq 0, \\ x_2 &\geq 0. \end{aligned}$$

3.8. Solve Problem 2.5 using the simplex algorithm.

3.9. Solve Problem 2.6 using the simplex algorithm.

3.10. Solve Problem 2.9.

3.11. (a) Solve Problem 2.10.
(b) If it is decided to produce 50 of the 3TR model and 50 of the 5TR model (these are the expected demands for the two models for the coming month) how many 7TR and 9TR models should be made? Will the profit be reduced from that associated with an original optimal solution?

3.12. Solve Problem 2.11.

3.13. Discuss the use of lexicographic ordering to resolve degeneracy and illustrate your discussion by defining a suitable pivot position in the following tableau.

2	3	4	0	0	−5	0	0
1	3	−2	0	0	4	0	1
0	3	−2	0	0	4	1	0
0	2	2	0	1	2	0	0
0	5	3	1	0	1	0	0

3.14. Consider the degenerate problem shown in Table 3.5. Solve this problem by using the tie breaking rule introduced in Section 3.4.

3.15. Reduce each of the following tableaus to canonical form:

3	2	5	4	−3	−2	1
3	2	0	1	4	0	−5
4	−1	0	0	2	1	2
1	3	1	0	−4	0	2

,

0	−4	2	−2	0	0	0
−1	6	3	−7	1	0	0
1	0	2	4	0	1	0
1	1	5	2	0	0	1

.

3.16. Construct a linear programming problem which satisfies C1 and C2 but not C3. Also, let the problem have the property that Case II of the simplex algorithm occurs in the auxiliary problem at some stage, that is,

$$b_{p-q} < 0, \qquad a_{p-q,k} \leq 0, \qquad a_{ik} \leq 0, \qquad (i = p - q + 1, \ldots, p).$$

Show that after executing a pivot operation on position $(p - q, k)$ you decrease the number of negative constants in the main problem.

3.17. Given the following tableau

0	−3	−1	5	0	0
2	1	3	2	0	1
3	3	−2	−1	1	0
5	2	0	3	0	0

expand the partial basic sequence to a full basic sequence and find an optimal solution if one exists.

3.18. Given the tableau

0	2	3	1	−1	0	0
3	1	2	2	4	1	0
−2	−3	−7	1	−2	0	1
4	2	−3	−1	−1	0	0
			Free columns		Slack columns	

write out the corresponding linear programming problem. Does the problem have an optimal solution? If so, solve the problem.

3.19. Solve the following problem,

$$\text{minimize } z = -2 + 2x_1 + 3x_2 + x_3$$

subject to

$$\begin{aligned} 4x_1 - 6x_2 + 2x_3 &= 8, \\ 3x_1 - 4x_2 - x_3 &\geq 1, \\ x_1,\ x_2 &\geq 0, \\ x_3 &\text{ free.} \end{aligned}$$

3.20. Maximize $z = 4x_1 - x_2 + 2x_3$ subject to

$$\begin{aligned} 3x_1 - 4x_2 + 2x_3 &= 0, \\ 3x_2 + 2x_3 &\leq 500, \\ 7x_1 + x_2 &\geq 700, \\ x_1,\ x_2 &\geq 0, \\ x_3 &\text{ free.} \end{aligned}$$

3.21. Maximize $z = 2x_1 + x_2 - x_3$ subject to

$$\begin{aligned} x_1 &\geq 17, \\ x_1 + 2x_2 - x_3 &= 12, \\ x_1 &\leq 27, \\ x_2 &\geq 0, \\ x_3 &\text{ free.} \end{aligned}$$

3.22. Minimize $z = 2x_1 + 3x_2 - x_3 + x_4$
subject to

$$\begin{aligned} 2x_1 - x_2 + 7x_3 + x_4 &= 9, \\ x_1 + x_2 + 4x_3 &\geq 17, \\ x_2 + 7x_3 &\leq 21, \\ x_j &\geq 0, \qquad (j = 1, \ldots, 4). \end{aligned}$$

3.23. Maximize $f = x_1 + 3x_2 + 2x_3$
subject to

$$\begin{aligned} x_1 + 2x_2 + 3x_3 &\leq 13, \\ 2x_1 + x_2 + 5x_3 &= 20, \\ x_1 + x_2 - x_3 &\geq 8, \\ x_j &\geq 0, \qquad (j = 1, 2, 3). \end{aligned}$$

3.24. Minimize $z = x_1 - 2x_2 + 5x_3$
subject to

$$\begin{aligned} 2x_1 - x_2 - 3x_3 &= -3, \\ x_1 + 3x_2 + 4x_3 &\geq 1, \\ x_j &\geq 0, \qquad (j = 1, 2, 3). \end{aligned}$$

3.25. Minimize $f = -2x_1 + x_2 + 3x_3 - x_4$
subject to

$$\begin{aligned} x_1 + 3x_2 - x_3 + x_4 &= 1, \\ 2x_1 - x_2 + 2x_3 + 5x_4 &= 8, \\ x_1 - 2x_2 + 2x_3 + 3x_4 &= 7, \\ x_j &\geq 0, \qquad (j = 1, \ldots, 4). \end{aligned}$$

3.26. Minimize $z = x_1 + 6x_2 - 8x_3 + x_4 + 5x_5$
subject to

$$\begin{aligned} 5x_1 - 4x_2 + 12x_3 - 2x_4 + x_5 &= 21, \\ x_1 - x_2 + 5x_3 - x_4 + x_5 &= 8, \\ x_j &\geq 0, \qquad (j = 1, \ldots, 5). \end{aligned}$$

3.27. Solve Problem 2.4.

3.28. Solve Problem 1.4.

3.29. Solve Problem 2.3.

3.30. Solve Problem 2.13.

3.31. Solve Problem 2.7.

3.32. Solve Problem 2.22.

3.33. Solve Problem 2.25.

3.34. Solve Problem 2.26.

3.35. Find an alternate optimal solution from the following tableau:

2000/3	40/3	0	10/3	0	0	0	50/3
35/3	1/3	1	1/3	0	1/2	0	−1/3
25/3	−1/3	0	−4/3	0	1	1	1/3
5/3	1/3	0	4/3	1	−1/2	0	2/3

3.36. Discuss the simplex algorithm for solving a linear programming problem in the form

$$\text{minimize } z = d + C^T X$$

subject to

$$A_1 X \geq B_1, \quad (p_1 \text{ inequalities}),$$
$$p = p_1 + p_2,$$
$$A_2 X = B_2, \quad (p_2 \text{ equalities}),$$
$$X \geq 0.$$

Include the following topics in your discussion:

(a) definition of standard and canonical form;
(b) choice of pivot when in canonical form;
(c) reduction to standard form;
(d) obtaining canonical costs given canonical coefficients;
(e) obtaining canonical constants given canonical coefficients;
(f) obtaining canonical coefficients.

Also give a brief discussion of the ways in which free variables can be dealt with in solving a linear programming problem.

3.37. Solve Problem 2.18. How much of an increase in the rate of interest could occur before a change is necessary in the merchant's optimal program?

3.38. A gas station sells gasoline in eight different grades at the prices and octane ratings shown in the table below.

Grade	*Price*	*Octane Rating*
190	30.9¢	93.5
200	34.9	95
210	35.9	96.5
220	36.9	98
230	37.9	99.5
240	38.9	101
250	39.9	102.5
260	40.9	104

Suppose a customer desires to purchase 20 gallons of grade 220; since this grade can be obtained in a variety of ways (for example, 10 gallons of grade 210 and 10 of 230), use a linear programming model to find the cheapest way to provide what the customer wants. Can you prove—by reformulating the model and without using the simplex algorithm—that an optimal program will have positive amounts only for grades 190 and 260?

3.39. Solve the following problem.[3] The word serendipity was coined by the English writer Horace Walpole after a fairy tale entitled *The Three Princes of Serendip*.

[3]From *Management Science*, Vol. 11, No. 4., Feb. 1965, p. C-45.

SERENDIPITY

The three princes of Serendip
Went on a little trip.
They could not carry too much weight;
More than 300 pounds made them hesitate.
They planned to the ounce. When they returned to Ceylon
They discovered that their supplies were just about gone
When, what to their joy, Prince William found
A pile of coconuts on the ground.
"Each will bring 60 rupees," said Prince Richard with a grin
As he almost tripped over a lion skin.
"Look out!" cried Prince Robert with glee
As he spied some more lion skins under a tree.
"These are worth even more—300 rupees each
If we can just carry them all down to the beach."
Each skin weighed fifteen pounds and each coconut, five,
But they carried them all and made it alive.
The boat back to the island was very small
15 cubic feet baggage capacity—that was all.
Each lion skin took up one cubic foot
While eight coconuts the same space took.
With everything stowed they headed to sea
And on the way calculated what their new wealth might be.
"Eureka!" cried Prince Robert. "Our worth is so great
That there's no other way we could return in this state.
Any other skins or nut which we might have brought
Would now have us poorer. And now I know what—
I'll write my friend Horace in England, for surely
Only he can appreciate our serendipity."

3.40. Solve Problem 2.20.

3.41. Solve Problem 2.16.

3.42. Is the following problem infeasible, unbounded, or does it have an optimal solution? If it has an optimal solution calculate it, illustrating main and auxiliary problems. If it is infeasible or unbounded, demonstrate why in terms of the procedures of Section 3.6.

$$\text{Minimize } z = x_1 - 3x_3 + 2x_5 + 3x_7$$

subject to

$$\begin{aligned}
2x_3 + x_4 + 2x_5 \qquad + 3x_7 &= 3,\\
x_1 + 2x_3 \qquad - 4x_5 + x_6 + x_7 &= -2,\\
x_1 - x_3 \qquad + x_5 \qquad + 2x_7 &= 2,\\
2x_2 + x_3 \qquad + 5x_5 \qquad - 3x_7 &= -1,\\
x_j &\geq 0, \qquad (j = 1, \ldots, 7).
\end{aligned}$$

3.43. Solve Problem 2.17.

3.44. Solve Problem 2.28.

3.45. Solve Problem 2.29.

3.46. Solve Problem 2.30.

3.47. Minimize $z = 2x_1 - x_2 + x_3$
subject to

$$\begin{aligned} 3x_1 + 5x_2 - x_3 &= 7, \\ x_1 - x_2 + 2x_3 + x_4 &= 6, \\ x_2 + x_5 &= 9, \\ 4x_1 + 3x_2 + 5x_3 &= 10, \\ x_j &\geq 0, \qquad (j = 1, \ldots, 5). \end{aligned}$$

3.48. Consider the diet problem of Sections 1.1 and 1.2 and the additional inequality restriction discussed in Section 1.2; show by means of the simplex method that this leads to an inconsistent system of constraints.

3.49. Solve the advertising media selection problem of Section 2.4.

3.50. Minimize $z = x_1 + x_2 + x_3$
subject to

$$\begin{aligned} x_1 + 2x_2 - x_3 &\leq 6, \\ 2x_1 - x_2 - x_3 &\leq 7, \\ x_2 + x_3 &\leq 11, \\ -x_1 - 2x_2 + x_3 &\leq -2, \\ x_j &\geq 0, \qquad (j = 1, 2, 3). \end{aligned}$$

3.51. Does the following problem have an optimal feasible solution? If so, calculate it using the gradient rule for selecting incoming columns. Is degeneracy encountered?

$$\text{Maximize } z = x_1 + 5x_2$$

subject to

$$\begin{aligned} 5x_1 + 6x_2 + x_3 &= 30, \\ 3x_1 + 2x_2 + x_4 &= 12, \\ -x_1 + 2x_2 + x_5 &= 10, \\ x_j &\geq 0, \qquad (j = 1, \ldots, 5). \end{aligned}$$

3.52. Construct a linear programming problem which illustrates Case (b) of Section 3.8.

3.53. Prove that the set of all optimal solutions of a linear programming problem is a convex set.

3.54. Prove that the set of all extreme rays of the solution set of $AY = 0$, $Y \geq 0$ is in one-to-one correspondence with the set of extreme vectors in the set of $AY = 0$, $\Sigma y_i = 1$, $Y \geq 0$.

3.55. Use the result of Problem 3.54 to prove Theorem 4 of Section 3.10.

3.56. (a) Formulate the following case as a linear programming problem.
(b) Solve this problem.

Oleum Company, Ltd. (A)[4]

[4]This case, Oleum Company, Limited (A), IM 1820, was prepared by S. W. Yost under the direction of J. L. McKenney of the Harvard University Graduate School of Business Administration as a basis for classroom discussion rather than to illustrate either effective or ineffective handling of administrative situations.

The Oleum Company is a holding corporation for a group of oil-producing, transporting, refining, and marketing companies located in various countries throughout the world. These affiliates, while maintaining operating autonomy, are given long-range direction by the Head Office of the Oleum Company. The Head Office provides planning guide lines for its operating affiliates on a yearly basis, reviewed semiannually, the purpose of which is to assure management that the overall operations of the company are coordinated in a manner which will yield the greatest total benefit to the parent company.

The planning group in the Head Office, located in London, is charged with the responsibility for developing long-range plans that will bring about these results and avoid suboptimization among and by the various affiliates. A clear distinction is drawn between planning and local operations. The Head Office plans are intended to offer broad guidelines, but make no attempt to dictate the day-to-day operational decisions made by affiliate management. Constant variability in process conditions, feed stock, and demand precludes rigid control. An illustration of the operating problems, for instance, is that each batch of crude has variations in it, even when from the same field, so that its yields, while similar to past performance, will not be identical. Similarly, equipment will run within a range of temperatures and pressures and, while output specifications will average out over the long run for a specific feed stock, on a given day for a given crude the results may be significantly different. Another important consideration is that a variety of crudes are being run on a continuous basis. If plans call for 40 percent crude Type A and 60 percent crude Type B, operating management will attempt to strike this proportion over the planning period, but certainly not on a daily basis. Such constraints as crude and product tankage, tanker arrival and departures, seasonal, secular, and random shifts in demand and product specifications all cause short-term variations from long-range plans.

In August 1964, Charles White, Manager of the Head Office planning group, was faced with the problem of developing the planning guidelines for Oleum's Far Eastern Area operations during 1965. In the past, operating plans for the coming year had been developed by the various affiliates, based on their knowledge of expected operations in their own market, crude availabilities and estimates of the most economical manner in which to program the refining equipment. These affiliate plans were then passed up to the Oleum Head Office in London for approval. Mr. White's job was to try and balance the various plans in such a manner as to be of greatest benefit to the whole company while taking into account all the local circumstances that had resulted in the affiliates' proposed plans.

Mr. White had met with Raymond Jackson and Stanley Cramer, two of his planning analysts, in March 1964. He had begun the meeting as follows:

White: "You both know of the problems that we have had in the past in coming up with the plans that our affiliates in the Far Eastern Area can live with and at the same time providing the most profitable overall operation for Oleum in the area. There always seems to be a hassle that continues through the year, even after the plans are approved, to modify this quantity or that specification because the plan does violence to local operating conditions. It seems to me that the approach we have used in the past just isn't doing the kind of job we need, and I would like to hear what you think the real problems are and what the possible solutions might be."

Jackson: "Well, we always seem to run into problems when we are trying to balance affiliates' plans because we unknowingly violate some obscure processing limitation, a crude availability, or something of that nature."

Cramer: "Yes, and even when we do get an apparent technically feasible balance, as derived from their plans, we find that the costs incurred may throw the overall economics out of bed. The results seem to be that the local managements, while cooperative, don't feel that our plans are too meaningful or realistic and are inclined to work around the plans to optimize their own operations."

White: "I know, Stan, it is just this sort of thing that weakens Oleum's overall operations. But when you confront the local boys with divergences from plan, they always have a dozen apparently valid reasons for explaining the necessity of their actions. They are very hesitant to commit themselves to any revisions in their plans that we suggest, apparently feeling that our grasp of the local situation is too limited to be of real value in planning."

Jackson: "Well, if you recall, a few years ago they were pretty skeptical of solving their refinery blending and other operating problems using linear programming techniques, but now most of the refinery managers are very enthusiastic about it. In fact, a lot of our present problems are the result of their taking divergent action on the basis of the economic analyses provided by the LP solutions."

White: "I think you're on the right track, Ray. The problem largely has been in not being able to integrate the detailed information we need to achieve a balanced overall plan. If we were working with the same LP operating models that they are using, for instance, we would be able to 'talk their language' when it comes to planning. If we could combine this kind of information with the other factors involved such as market demands, transportation costs for crude and product, crude oil types, availabilities, and so forth, we could start making plans that can stick, that will really represent operational guidelines and that will provide a defensible economic basis for making our area plans real working documents.

Of course, in order to gain our affiliates' support it will be necessary to convince them that our plans are based on their actual operations as represented in part by their own LP operative models. I think that it would be wise for both of you to go out there and work with them in developing your inputs for our overall planning model. However, I believe that a big part of your job will be in establishing rapport with the area's local management and staff so that we really will be 'talking their language' when we put everything together back here."

Involved in the area's planning were the refineries of two of Oleum's affiliates, their homeland marketing areas (called tributary areas) and three other marketing areas in nearby countries that could be supplied either by the Far Eastern refining affiliates or by products imported from distant Oleum affiliates located in the United States and Europe. Crude supplies for the area in 1965 were available from two sources, a relatively heavy crude (24° API) from Iran in the Middle East and a lighter crude (36° API) from the Brunei fields in Borneo. Under the terms of an agreement with the Netherlands Petroleum Company (N.P.M.) in Borneo, a fixed quantity of 40,000 B/D of Brunei crude, at a cost of 2.60$/B at the crude loading terminal was to be supplied during 1965 to the Far Eastern Area affiliates designated by the Oleum Company. On the other hand, the Oleum affiliate in Iran was prepared to supply any amount of crude up to 100,000 B/D at a cost of 2.00$/B at Abadan, the crude terminal on the Persian Gulf. (See Exhibit 1 for a map of the area.)

Late in July, Ray Jackson returned from his field assignment to Aussi Petroleums, Ltd., (APL) in Australia, where he had been working with local operating people in determining the various characteristics and constraints in their operations, which, except for crude oil production, are fully integrated. APL's refinery, located in Sydney, is designed for a 50,000 B/D throughput. Process equipment includes a catalytic cracking unit and sufficient auxiliary equipment to produce a specification quality product from the crude normally available for its use. A large part of Jackson's time had been spent on familiarizing himself with the linear programming model that APL uses for solving day-to-day operating plans at the refinery. This model totals 110 rows and 250 columns which define the operating characteristics of all the refinery's equipment using various modes of possible operation and the different crudes normally available.

Although the model had been designed to provide APL with the ability to solve operational problems, Jackson believed that he could use it to derive cost parameters and refining coefficients for planning purposes. With this goal in mind, and working with the local staff to assure himself that the results were realistic, Jackson used the refinery model to establish the yields and costs for the crudes available in 1965. To accomplish this, he assumed that the quality specifications for the three principal end products would remain constant during the planning period. In addition, although the refinery had processing flexibility that permitted a wide range of yields, he decided for planning purposes that the use of the values at highest and lowest conversion (process intensity) for the crudes available would suffice. The yields so derived are as follows:

Product	APL refinery yields			
	Brunei crude process intensity		Iranian crude process intensity	
	Low	High	Low	High
Motor Gasoline (Mogas)	.259	.365	.186	.312
Distillate (Dist.)	.371	.379	.229	.230
Fuel Oil (Fuel)	.317	.194	.503	.378
Total[5]	.947	.938	.918	.920

[5]The total for product output does not equal 1.000 because of the use of a portion of the throughput as fuel to operate the processing equipment as well as the production of minor byproducts such as coke, tar, and fuel gas. The latter are accounted for as credits against the manufacturing costs.

In using this data he realized that a certain amount of nonlinearity existed in the refining yields of products going from low process intensity to high process intensity; however, he felt that the error introduced by assuming a linear relationship would not be significant in the planning model he hoped to construct, and the linearity assumption would certainly reduce the size of the matrix. Jackson planned to test the significance of nonlinearity in the refining coefficients on the planning model's results after it was formulated by the Head Office planning staff.

Using the high and low conversion limits as the refinery's operating modes, Jackson was able to determine incremental costs for those manufacturing activities that vary as a function of the type of crude being processed and the level of process intensity being employed. These costs include special chemicals and catylsts used, utilities required, the tetraethyl lead (TEL) used to bring the product up to octane specification, and so on. Fixed charges (those not varying with changes in processing modes) are excluded. Deducted from the sum of the variable, or incremental manufacturing costs is the value of the byproducts produced by the particular process mode using the particular type crude. These so called byproducts are defined as minor products as contrasted to the three main products manufactured for Far Eastern Area consumption (namely; motor gasoline, distillate, and fuel oil), and consist of small quantities of specialty products and such things as coke, tar, and fuel gas. The net incremental costs Jackson found appropriate were:

Intensity level	Incremental manufacturing costs ($/B)	
	Brunei crude	Iranian crude
Low	0.12	0.15
High	0.28	0.30

APL management forecasted that fixed operating expenses at the refinery would total $10,038,000 during 1965.

Besides the refinery study, Jackson collected data on shipping costs to the Australian refinery. Normally, in this area crude is transported in tankers owned by a subsidiary of the Oleum Company; however, in cases where refinery shipping demands exceed the fleet capacity, spot charters of independent tankers are used. The owned fleet capacity is 6.3 47,000 DWT tanker equivalents.[6] The requirements for the transportation in terms of 47,000 DWT tanker equivalents per thousand barrels per day to the APL refinery is 0.05 for Brunei crude and 0.120 for Iranian crude.[7] The costs for the movement of crude, in the company's tanker fleet, from Borneo and the Middle East during 1965 to Australia are anticpated to be, 0.26$/B and 0.62$/B, respectively. These costs included port charges, tolls, bunker fuel, and other variable ship expenses. (For a detailed example of how these costs are determined, see Exhibit 2.) If a spot charter is required, an additional transportation expense of $3,200/47,000 DWT tanker equivalent per thousand barrels per day is incurred.

The last important area that Jackson investigated was the marketing and distribution of APL's output. APL's Australian market is its prime concern and this affiliate of Oleum's is solely responsible for meeting its demands. APL's market forecast for Australia in 1965 is 9,000 B/D of mogas, 10,000 B/D of distillate, and 11,000 B/D of fuel oil. In talking to the marketing manager, Jackson ascertained that the Australian market had established a fairly stable growth rate during the past ten years and that the market forecast was not expected to vary more than 10 percent in either direction. Jackson was also able to determine what the shipping costs for APL's products are to the other markets in the Far East; namely, Malaysia (M), Philippines (P), and New Zealand (N). As in the case of crude transportation, these costs are incremental including such variables as import duties, barge, tanker, and port costs.

	Distribution costs ($/B) from Australia to		
Product	*M*	*P*	*N*
Mogas	.40	.15	.10
Dist.	.40	.15	.10
Fuel	.60	.20	.15

[6]The standard-sized tanker operating in this area is 47,000 DWT (dead weight tons) and the fleet capacity is described by the company in terms of 47,000 DWT tanker equivalents.

[7]For example, if the APL refinery requires 10,000 B/D of Brunei crude this means that 0.5 47,000 DWT tanker equivalents of fleet capacity will be needed. If a 47,000 DWT tanker is used, whose capacity is 270,000 barrels of crude and whose round-trip time from Borneo to Sydney (including loading and unloading) is approximately, 13 1/2 days, then it can be seen that 135,000 barrels or roughly 50 percent of the tanker's capacity is necessary to meet refinery requirements. The tanker can be thought of as spending 50 percent of its time supplying APL with Brunei crude, leaving 270,000 barrels each trip it makes to Australia once every 27 days. Actually, the refinery is supplied by Oleum's tanker fleet operating in the area. This fleet is comprised of tankers of different capacities and serves the various area affiliates so that economical dispatching normally precludes the permanent assignment of a specific tanker to an affiliate.

Stanley Cramer, the planning analyst who had been sent to Japan to gather the same type of information as had Raymond Jackson in Australia, returned to London shortly after Jackson. Oleum's affiliate in Japan, the Nippon Oil Company (NOC), owns a refinery with a nominal daily capacity of 28,600 barrels. The process equipment includes a catalytic cracking unit; however, because of process flexibility, the crude throughput may be increased, if low-intensity operations are carried out. Thus while its nominal capacity at high intensity is 28,600 B/D, it can process as much product as 30,000 B/D at low intensity. However, treating facilities limit the volume of the type of Iranian crude that can be processed and still meet minimum product specification. With the present equipment the amount of Iranian crude processed cannot exceed 50 percent of the volume of Brunei crude. Based on these limitations the following yields for Brunei and Iranian crude were determined at the process extremes.

	NOC refinery yields			
	Brunei crude process intensity		*Iranian crude process intensity*	
Product	*Low*	*High*	*Low*	*High*
Mogas	.259	.350	.186	.300
Dist.	.371	.378	.229	.230
Fuel	.317	.210	.503	.390
Total	.947	.938	.918	.920

The incremental manufacturing costs are:

Process intensity	*Brunei crude ($/B)*	*Iranian crude ($/B)*
Low	.16	.20
High	.34	.39

Fixed refinery costs for 1965 were estimated by NOC staff personnel to be equivalent to $6,570,000.

Cramer had the same reservations in using this data as did Jackson, but decided the model would be complex enough this first year and that further refinements could be made later if proved necessary. It was his opinion that the construction of too detailed a model initially would unnecessarily increase the task of interpreting results and testing for sensitive parameters and coefficients.

The estimated demand in Japan for NOC products in 1965 was forecast as mogas 3,000 B/D, distillate 5,250 B/D, and fuel oil 6,750 B/D. The accuracy of the forecast was expected to be no worse than 10 percent on the low side and 20 percent on the high side.

Cramer found that the costs for transporting NOC products to other Far Eastern markets served by Oleum affiliates were as follows.

Product	Distribution costs ($/B) from Japan to M	P	N
Mogas	.20	.20	.10
Dist.	.20	.20	.10
Fuel	.40	.25	.10

NOC's requirements for tanker transportation in terms of 47,000 DWT equivalents per thousand barrels per day was .045 for Brunei crude and .110 for Iranian crude. Transportation costs for the oil was expected to be 0.24$/B and 0.59$/B from Borneo and Iran, respectively.

With the return of Jackson and Cramer to London from their field assignments, Mr. White called a meeting to discuss 1965 plans for the Far Eastern Area. Besides White, Jackson, and Cramer, also included in the meeting were Mr. F. T. H. Landsbert and Mr. M. Davies, Head Office representatives for Marketing and Operations, respectively.

Mr. White explained the studies that Jackson and Cramer had made in Australia and Japan and requested that Landsbert supply his group with a forecast of demand for 1965 in the other Oleum markets in the Far Eastern Area, namely, New Zealand, the Philippines, and Malaysia. Mr. Landsbert said that his group had already prepared a forecast of the market demands in the area and that the information would be available that afternoon.

Mr. White pointed out that to make their plans realistic and achieve overall economies for Oleum, the planning group needed information on all the feasible alternatives to the basic problem of supplying the Far Eastern Area markets with refined products. He said that attempting to consider the area as a separate entity might result in local refining and distributing operations that were more costly than importing products from affiliates outside of the area. Consequently, White requested that Mr. Davies support the planning group with information on the possibilities and economics of supplying area markets with products from Oleum affiliates located elsewhere around the world. Davies stated that an analysis of the operations in the United States and Europe would be necessary to assess product availabilities and costs. He estimated that he could obtain the necessary information within two weeks.

The next day Jackson and Cramer reviewed the market forecast that Landsbert had prepared for them. The report indicated that marketing operations in the Philippines in 1965 would require 2,500 B/D of motor gasoline, 3,500 B/D of distillate, and 4,000 B/D fuel oil. The prospects in New Zealand were for 3,360 B/D of motor gasoline, 4,200 B/D of distillate, and 4,440 B/D of fuel oil. The market conditions in Malaysia were expected to be unstable during 1965 but best estimates were that some 9,000 B/D, divided equally between motor gasoline and distillate, would be required, while 6,000 B/D was the expectation for fuel oil demand. To questions of forecast accuracy during the planning period, Landsbert's people felt that the Philippines' growth projection would be within 15 percent or 20 percent of the estimate. Malaysia's relatively unstable political and industrial environment caused them considerable concern and the estimate that their forecast could easily be off by 50 percent or more in either direction. New Zealand's market conditions exhibited the least growth and the greatest stability and the marketing forecasters confidently predicted sales to be within 10 percent to 15 percent of their estimate.

Some ten days later the report from Davies' operations group arrived. The report stated that because of the resurgent European economy, demand in that area is exceeding present refining capacity. Although additional units are in construction, a surplus of product could not be depended upon during 1965. On the other hand, United States operations on the West Coast expected a surplus of 10,000 B/D of distillate and 7,000 B/D of fuel oil during 1965. The price for distillate and fuel oil, at the loading port of Los Angeles, was quoted by the American affiliate at $2.10 and $1.25 per barrel, respectively. The burgeoning population growth in the western states and the high per capita auto ownership precluded any excess of motor gasoline for export. A politically sensitive situation in Malaysia had caused the U.S. State Department to request that petroleum products not be exported from the United States to Malaysia in the near future. The costs for transportation of the products to the Far Eastern markets as estimated by Davies are listed below and include all variable costs, such as duties, port costs, and tanker expenses:

	Product costs ($/B)	
Countries	*Dist.*	*Fuel*
Philippines	.75	.80
New Zealand	.60	.65

Following the receipt of this information, White asked Jackson and Cramer to develop a comprehensive plan for satisfying the Far Eastern market during 1965. He indicated that the plan should include the requirements for each type of crude available, the amount and proportion of crude to be processed by the Australian and Japanese refineries during 1965, and the distribution pattern for supplying the various markets, including import requirements, if necessary, as well as the local affiliate's share. Exhibit 3 is an outline of the type of information that White believed would be necessary to establish the planning documents he had in mind for Oleum's Far Eastern operations for 1965. He also requested that he be supplied with information regarding the marginal value of crude and product so that if further capital became available for investment in the area the relative economics of the various facets of the operations would be known and could be further evaluated. In this regard White was particularly interested in knowing what the value of additional capacity is at the two refineries. In determining these marginal values he told Jackson and Cramer that he was interested only in the variable costs and that fixed charges were not to be included in their calculations.

Oleum Company, Ltd. (a)
Exhibit 1

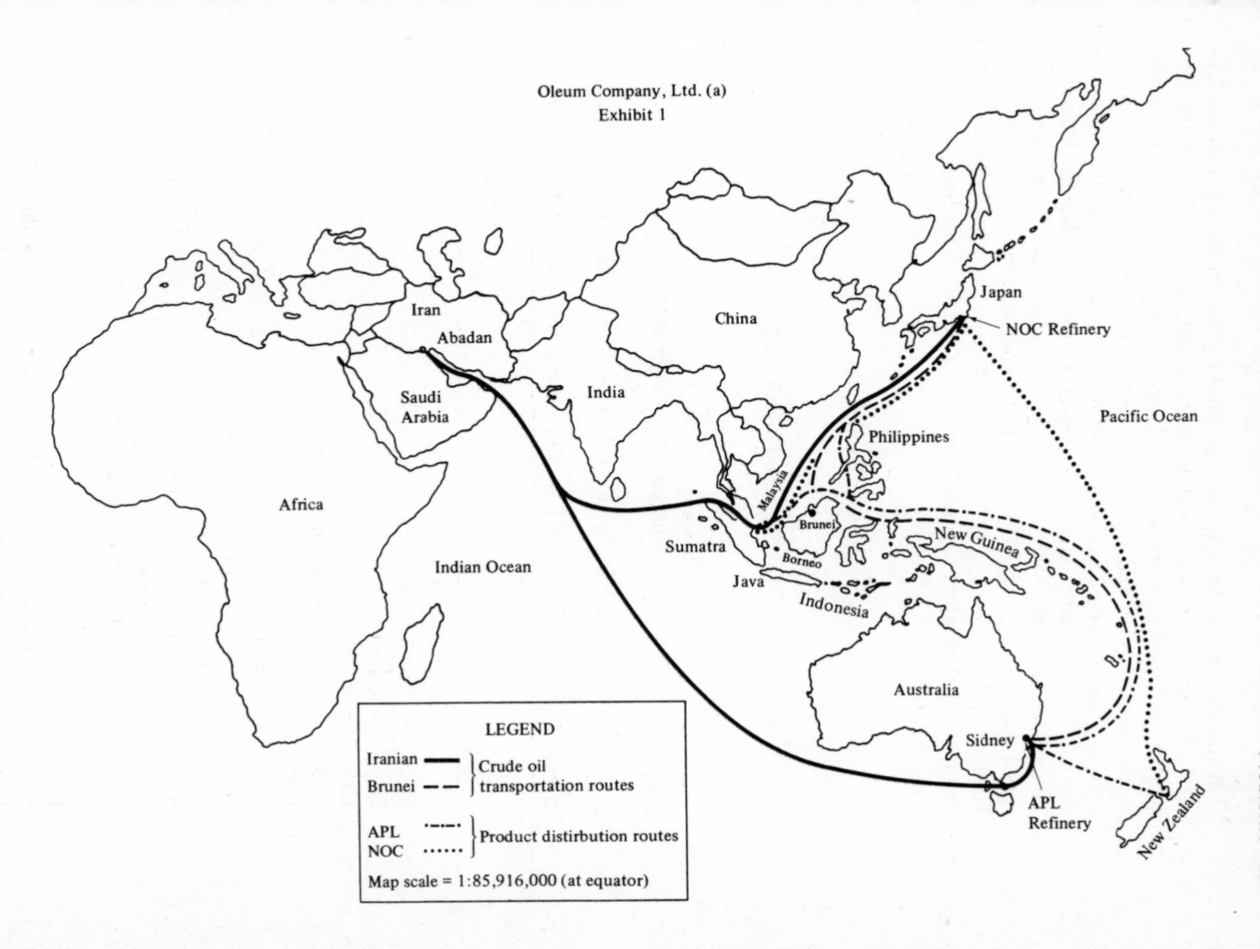

While discussing the approach to take for accomplishing their assignment, Jackson and Cramer decided that, in addition to obtaining the base solution, it would be worthwhile to determine the sensitivity of the model to the various costs and restrictions used in its construction. This would involve programming the computer to test how much each cost or restriction could be increased and/or decreased without changing the base solution, while holding the other costs or restrictions constant. These operations, called cost ranging and RHS (right hand side) ranging, respectively, were believed by both men to justify the extra time and expense in terms of the value of learning where the model was most sensitive to change.

Exhibit 2

OLEUM COMPANY, LTD. (A)

Shipping Data for Calculating Transportation Coefficients[8]

$$C = (D \cdot BST + BIP \cdot DIP) \cdot BC + LPC + DPC + R \cdot (DIP + DAS),$$

where $DAS = 2D/S$ and $DIP = TLP + FTDP \cdot N + VD/PR$.

	Description	*Typical cost factors for 47,000 DWT Tankers*
C	= voyage cost	
DAS	= days at sea	
D	= distance	
DIP	= days in port	
BST	= bunker requirements steaming	2.8 tons/hour
BIP	= bunker requirements in port	22 tons/day
TLP	= time loading	50 hours
N	= number of calls (normally equals 1 because of single porting)	
FTDP	= fixed time in discharging port	7 hours
VD	= total volume discharge (capacity minus requirements)	
PR	= pumping rate	940 tons/hour
BC	= bunker costs	$26/ton
LPC	= loading port cost	$100/call
DPC	= discharging port cost	$100/call
R	= operating costs	$925,000/year
S	= vessel speed	15 n. m./hour

[8]The formulation implies 100% utilization (assumes spot charter if not in continuous company service).

Exhibit 3

OLEUM COMPANY, LTD. (A)

Planning Data Required for Far Eastern Area Operations in 1965

1. *Summary*
   A. Crude transportation
      (1) Owned ships ________47,000 DWT EQ
      (2) Spot charter ________47,000 DWT EQ
      (3) Crude cost ________M$/D
   B. Refining operations
      (1) Crude run
          Brunei crude ________MB/D[9]
          Iranian crude ________MB/D
          Total ________MB/D
      (2) Cost
          Variable cost ________M$/D
          Total cost ________M$/D
          Cost/B crude ________$/B
   C. Distribution operations
      (1) Product import volumes (MB/D)

	Volume	*Max. vol.*	*Mar. value $/B*
Dist.	________	________	________
Fuel	________	________	________
Total	________		

      (2) Product import cost (total)________M$/D
      (3) Product transportation cost________M$/D
2. *Refinery operations*
   A. APL refinery (capacity________MB/D)
      (1) Crude summary

	Volume MB/D	*Mar. value $/B*
Brunei type	________	________
Iranian type	________	________

(value of additional capacity $/B/D = ____)

      (2) Product summary

	Volume MB/D	*Mar. value $/B*
Mogas	________	________
Dist.	________	________
Fuel	________	________

[9]Note: Brunei crude volume fixed by contract, incentive to process more (or less) = ________ $/B/D

(3) Manufacturing cost

Incremental ________M$/D

Total cost ________M$/D

Cost/B crude ________ $/B

B. NOC refinery (capacity________MB/D nominal)

(1) Crude summary

	Volume MB/D	*Mar. value $/B*
Brunei type	________	________
Iranian type	________	________
Total	________	

(value of additional capacity $/B/D = __)

(2) Product summary

	Volume MB/D	*Mar. value $/B*
Mogas	________	________
Dist.	________	________
Fuel	________	________

(3) Manufacturing cost

Incremental ________M$/D

Total cost ________M$/D

Cost ________ $/B

3. *Product distribution*

A. Distribution (MB/D)

Product	*Source*	*Market areas*
________		
________		
________		

B. Cost (cost of last barrel–$/B, no fixed costs included)

Product	*Market areas*
________	
________	
________	

Algebraic Interpretations of the Simplex Algorithm and Some Computational Considerations

4

4.1. A COMPUTATIONAL CHECK.

Let

$$M = \begin{bmatrix} -d & C^T \\ B & A \end{bmatrix} \tag{4.1}$$

denote the initial tableau matrix for a linear programming problem in standard form and let

$$M^* = \begin{bmatrix} -d^* & C^{*T} \\ B^* & A^* \end{bmatrix} \tag{4.2}$$

be a second tableau matrix for the same problem. We explore some relationships between M and M^*. Given the matrix M and any basic sequence $S = (s_1, s_2, \cdots, s_p)$ we define the matrix

$$M_S = \begin{bmatrix} 1 & C_S^T \\ 0 & A_S \end{bmatrix} \tag{4.3}$$

to be the square matrix of degree $p + 1$ consisting of the first unit vector of V_{p+1}, that is, the vector whose $p + 1$ components are

$$\begin{bmatrix} 1 \\ 0 \\ \cdot \\ \cdot \\ \cdot \\ 0 \end{bmatrix},$$

followed by columns $s_1, \cdots, s_p$ of M (as usual we consider the initial column

$$\begin{bmatrix} -d \\ B \end{bmatrix}$$

to be column 0 of the matrix M and consider $[-d \quad C^T]$ to be row 0 of M). The unit vector is included in the matrix M_S in order to account for the presence of z—the value of the objective function—in our calculations.

For example, the wood products problem is to minimize

$$z = -8x_1 - 19x_2 - 7x_3 + 0x_4 + 0x_5$$

subject to

$$\begin{aligned} 3x_1 + 4x_2 + x_3 + x_4 \qquad &= 25, \\ x_1 + 3x_2 + 3x_3 \qquad + x_5 &= 50, \\ x_j &\geq 0, \qquad (j = 1, \cdots, 5). \end{aligned}$$

The full tableau matrix M is the 3 by 6 matrix

$$M = \begin{bmatrix} 0 & -8 & -19 & -7 & 0 & 0 \\ 25 & 3 & 4 & 1 & 1 & 0 \\ 50 & 1 & 3 & 3 & 0 & 1 \end{bmatrix}, \tag{4.4}$$

in which the initial column M_0 is the vector

$$\begin{bmatrix} -d \\ B \end{bmatrix}$$

and the initial row (row 0) is the vector $[-d \quad C^T]$. Note that here, as in many problems, $d = 0$ in the initial tableau matrix. Now $S = (4, 5)$, $p = 2$, so $V_{p+1} = V_3$ and the initial column of the matrix M_S is the first unit vector of V_3 (more precisely, the first vector in the ordered unit basis of V_3) and

$$M_S = \left[\begin{array}{ccc|c} 1 & 0 & 0 & C_S^T \\ 0 & 1 & 0 & A_S \\ 0 & 0 & 1 & \end{array}\right].$$

We now return to the matrices M and M^* of (4.1) and (4.2). The pivot process on position (h, k) can be regarded as premultiplication of M by the pivot matrix

$$F = \begin{bmatrix} 1 & 0 & \cdots & -c_k/a_{hk} & \cdots & 0 \\ 0 & 1 & \cdots & -a_{1k}/a_{hk} & \cdots & 0 \\ \cdot & \cdot & \cdot & \cdot & \cdot & \cdot \\ \cdot & \cdot & \cdot & \cdot & \cdot & \cdot \\ \cdot & \cdot & \cdot & \cdot & \cdot & \cdot \\ 0 & 0 & \cdots & 1/a_{hk} & \cdots & 0 \\ \cdot & \cdot & \cdot & \cdot & \cdot & \cdot \\ \cdot & \cdot & \cdot & \cdot & \cdot & \cdot \\ \cdot & \cdot & \cdot & \cdot & \cdot & \cdot \\ 0 & 0 & \cdots & -a_{pk}/a_{hk} & \cdots & 1 \end{bmatrix} \begin{matrix} \text{Row number} \\ 0 \\ 1 \\ \cdot \\ \cdot \\ \cdot \\ h \\ \cdot \\ \cdot \\ \cdot \\ p \end{matrix} \tag{4.5}$$

$$\text{Column number} \quad 0 \quad 1 \quad \cdots \quad h \quad \cdots \quad p$$

(see also Appendices B.10 and B.11 for a discussion of pivot matrices).

For example, the first pivot operation in the solution of the wood-products problem (Table 3.3) was at position (1, 2). The pivot operation can be effected by premultiplication of the matrix M in (4.4) by the matrix

$$F_1 = \begin{bmatrix} 1 & 19/4 & 0 \\ 0 & 1/4 & 0 \\ 0 & -3/4 & 1 \end{bmatrix},$$

and for the second pivot the corresponding matrix is

$$F_2 = \begin{bmatrix} 1 & 0 & 1 \\ 0 & 1 & -1/9 \\ 0 & 0 & 4/9 \end{bmatrix}.$$

Similarly, for the first pivot in Table 3.4 the corresponding matrix is

$$F_3 = \begin{bmatrix} 1 & 0 & 0 & 1/3 \\ 0 & 1 & 0 & -1/3 \\ 0 & 0 & 1 & -2/3 \\ 0 & 0 & 0 & 1/3 \end{bmatrix}.$$

We observe that consecutive multiplications by matrices of the form (4.5) can be effected by means of a single matrix which is their product. From the wood products problem let $Q = F_2F_1$, where F_1 and F_2 are as given above; then

$$Q = \begin{bmatrix} 1 & 4 & 1 \\ 0 & 1/3 & -1/9 \\ 0 & -1/3 & 4/9 \end{bmatrix}$$

and the final tableau matrix of Table 3.3 can be obtained from the initial one by premultiplication of the latter by the matrix Q.

Moreover, if a sequence of pivot processes is effected by a single matrix multiplication, then the reverse sequence can be achieved by multiplying by the inverse of this same matrix. If the final tableau matrix of Table 3.3 is premultiplied by the matrix Q^{-1}, the initial tableau will result.

We wish to show that the matrix Q is independent of the particular sequence of pivot operations used to go from M to M^* (in other words, that the matrix Q is unique). Suppose now that Q results from a particular sequence of pivot operations and therefore that

$$M^* = QM \qquad \text{and} \qquad M = Q^{-1}M^*. \tag{4.6}$$

Since we never pivot in row 0, each pivot matrix F has as column 0 the unit vector

$$\begin{bmatrix} 1 \\ 0 \\ \cdot \\ \cdot \\ \cdot \\ 0 \end{bmatrix}$$

and hence so does their product Q and its inverse Q^{-1}. Now (by properties C1 and C2)

$$M^*_{S*} = \begin{bmatrix} 1 & & & \\ 0 & & & \\ \cdot & M^*_{s_1*} & \cdots & M^*_{s_p*} \\ \cdot & & & \\ \cdot & & & \\ 0 & & & \end{bmatrix} = I_{p+1}. \tag{4.7}$$

It follows from (4.6) that for each j, $j = 1, \cdots, n$, $M_j = Q^{-1}M^*_j$; also for $j = 0$ we have

$$M_o = Q^{-1}M^*_o = Q^{-1} \begin{bmatrix} 1 \\ 0 \\ \cdot \\ \cdot \\ \cdot \\ 0 \end{bmatrix} = \begin{bmatrix} 1 \\ 0 \\ \cdot \\ \cdot \\ \cdot \\ 0 \end{bmatrix},$$

so we have

$$M_{S*} = Q^{-1}M^*_{S*} = Q^{-1}, \tag{4.8}$$

$$M = M_{S*}M^*. \tag{4.9}$$

Equation (4.8) establishes the uniqueness of Q^{-1} and Q.

Equation (4.9) provides a numerical check of the accuracy of M^* under the hypothesis that M^*, but not necessarily M, is in canonical form. We call it a *reverse check* since it goes from M^* to M by premultiplication with a submatrix of M.

If M is in canonical form with respect to a basic sequence $S = (s_1, \cdots, s_p)$ then

$$M^* = M^*_S M, \qquad \text{that is} \qquad Q = M^*_S. \tag{4.10}$$

We call (4.10) a *direct check* since it multiplies M by a matrix all but one of whose columns are columns of M^* to obtain M^*. If a check (either direct or reverse) discloses an error one can then check in intermediate stages and soon locate the source of the error. For example, if M^* is the result of twelve cycles, one could first check at cycle 6. If this is correct check at cycle 9, if it is incorrect check at cycle 3, and so on.

We observe that when both M and M^* are in canonical form then

$$M_{S*}M^*_S = I_{p+1}; \tag{4.11}$$

thus we may regard the simplex method as being a systematic procedure for calculating inverses of the matrices M_{S*} of M and for specifying by selection of S^* which M_{S*} to form and to invert.

To help avoid confusion it should be observed that M_{S*} consists of the columns of the initial tableau matrix M determined by the later basic sequence S^*; on the other hand M_S^* consists of the columns of the later tableau matrix M^* determined by the initial basic sequence S. Symbolically,

$$Q^{-1} = M_{S^*} = \begin{bmatrix} 1 & & & \\ 0 & & & \\ \cdot & & & \\ \cdot & M_{s_1^*} & \cdots & M_{s_p^*} \\ \cdot & & & \\ 0 & & & \end{bmatrix} \tag{4.12}$$

and

$$Q = M_S^* = \begin{bmatrix} 1 & & & \\ 0 & & & \\ \cdot & & & \\ \cdot & M_{s_1}^* & \cdots & M_{s_p}^* \\ \cdot & & & \\ 0 & & & \end{bmatrix}. \tag{4.13}$$

The fact that these matrices are inverses of each other is no accident and is an important feature of the simplex algorithm. We note further that the relations $M_S = M_{S^*}^* = I_{p+1}$ are also important features of the simplex algorithm (compare Condition C1, Section 3.2). Thus each of the four symbols M_S, M_{S^*}, M_S^*, $M_{S^*}^*$ is meaningful; the second and third are the ones that are used in making computational checks.

The process introduced in Section 3.5 for obtaining canonical costs is a special case of the results of the present section. Let M^* be in canonical form with respect to S^* and let

$$M = \begin{bmatrix} -d & C^T \\ B^* & A^* \end{bmatrix}$$

be the same as M^* except that some different form of the cost function, usually the initial form, is substituted for the canonical costs. We have then from (4.8) that

$$Q^{-1} = M_{S^*} = \begin{bmatrix} 1 & C_{S^*}^T \\ 0 & I_p \end{bmatrix}, \tag{4.12'}$$

but now it is easy to invert Q^{-1} to obtain

$$Q = \begin{bmatrix} 1 & -C_{S^*}^T \\ 0 & I_p \end{bmatrix}$$

and hence

$$[-d^* \ C^{*T}] = [1 \ -C_{S^*}^T] \ M = [-d \ C^T] - [C_{S^*}^T B^* \ C_{S^*}^T A^*], \tag{4.14}$$

which coincides with (3.61).

Equation (4.14) is useful as a partial check on computational accuracy since the multiplications employed will in general involve all of the elements of M^*.

As an illustrative example of computational checks we use Table 3.4 and regard the initial tableau as M and the final tableau as M^*. These two tableaus are reproduced in Table 4.1 to facilitate the illustration.

Table 4.1. INITIAL AND OPTIMAL TABLEAUS.

$C_{S^*}^T$	x_1	x_2	x_3	x_4	x_5	x_6	x_7	x_8	
0	−1	−1	−1	−1	−1	0	0	0	Initial full tableau matrix M
1	1	3	2	13/5	10/3	1	0	0	
1	2	0	5/2	1	1/2	0	1	0	
1	3	2	1	8/5	1	0	0	1	
7/13	0	0	0	0	0	3/13	2/13	2/13	C_S^{*T} Final full tableau matrix M^*
4/39	0	1	0	3/5	8/9	11/39	−10/39	1/13	
10/39	0	0	1	2/5	5/9	8/39	14/39	−4/13	
7/39	1	0	0	0	−4/9	−10/39	2/39	5/13	

Here $S = (6, 7, 8)$ and $S^* = (2, 3, 1)$; M_S^* is the matrix made up of the bordering column

$$\begin{bmatrix} 1 \\ 0 \\ 0 \\ 0 \end{bmatrix},$$

and columns M_6^*, M_7^*, M_8^*, so that

$$M_S^* = \begin{bmatrix} 1 & & & \\ 0 & & & \\ \cdot & M_{s_1}^* & \cdots & M_{s_p}^* \\ \cdot & & & \\ \cdot & & & \\ 0 & & & \end{bmatrix}$$

becomes in this case

$$M_S^* = Q = \begin{bmatrix} 1 & 3/13 & 2/13 & 2/13 \\ 0 & 11/39 & -10/39 & 1/13 \\ 0 & 8/39 & 14/39 & -4/13 \\ 0 & -10/39 & 2/39 & 5/13 \end{bmatrix}.$$

The matrix M_{S^*} is formed from elements of the *initial* tableau and is given by

$$M_{S^*} = Q^{-1} = \begin{bmatrix} 1 & -1 & -1 & -1 \\ 0 & 3 & 2 & 1 \\ 0 & 0 & 5/2 & 2 \\ 0 & 2 & 1 & 3 \end{bmatrix}.$$

The reader can verify that

$$M_S^* M_{S^*} = I_4, \qquad M^* = M_S^* M, \qquad M = M_{S^*} M^*.$$

Formula (4.10) shows how to locate Q if M is in canonical form with respect to S. More generally from (4.9) we obtain

$$M^* = M_{S^*}^{-1} M \tag{4.15}$$

or $Q = M_{S^*}^{-1}$. This formulation emphasizes the fact that the simplex algorithm rests on successive calculations of inverses of submatrices of the initial tableau matrix M. Indeed, if we knew an optimal basic sequence S^* we could proceed directly from M to M^* merely by inverting M_{S^*}. In some situations one may have information which leads one to believe that some particular basic sequence S^* is either optimal or near optimal. If so, then one may save computational steps by calculating $M_{S^*}^{-1}$, proceeding directly to M^* and then carrying on with the simplex algorithm from M^* in case it is not optimal. For example, if S^* is such a sequence, then

$$Q^{-1} = M_{S^*} = \begin{bmatrix} 1 & C_{S^*}^T \\ 0 & A_{S^*} \end{bmatrix}; \tag{4.12''}$$

calculation of $M_{S^*}^{-1}$ in (4.15) is essentially equivalent to calculating $A_{S^*}^{-1}$ since

$$Q = M_{S^*}^{-1} = \begin{bmatrix} 1 & -C_{S^*}^T A_{S^*}^{-1} \\ 0 & A_{S^*}^{-1} \end{bmatrix} \tag{4.16}$$

(see Appendix B.14 for details). We emphasize that if the initial tableau matrix is in canonical form relative to S then $M_{S^*}^{-1}$ has the simpler form (4.13).

4.2. ARTIFICIAL VARIABLES.

In the early development of the simplex method artificial variables were introduced so that the simplex method could be used to determine feasibility by either constructing a basic feasible solution or demonstrating infeasibility. As we have seen in our presentation of the simplex method there is no need to introduce any auxiliary variables except for slack variables (which have a useful physical interpretation). Sections 3.6 and 3.7 above provide the key to our ability to avoid artificial variables.

However, from the point of view of computing efficiency and for certain other reasons which we will discuss later in this section we find it desirable to present a version of the simplex method which utilizes artificial variables. We begin by assuming (1) that the problem is in equation form, after slack variables have been adjoined, if necessary, and (2) that each constraint equation has a nonnegative constant term. Thus the original problem is assumed to have the form

$$\text{minimize } z = d + C_1^T X_1 + C_2^T X_2 \tag{4.17}$$

subject to the constraints

$$A_1 X_1 + A_2 X_2 = B, \qquad X_1 \geq 0, \qquad X_2 \text{ free}, \tag{4.18}$$

where

$$B \geq 0. \tag{4.19}$$

We assume that B, A_1, A_2 all have p rows; X_1 has n_1 rows; and X_2 has n_2 rows. We let $n = n_1 + n_2$ and have the initial tableau matrix

$$M = \begin{bmatrix} -d & C_1^T & C_2^T \\ B & A_1 & A_2 \end{bmatrix}. \tag{4.20}$$

Next we introduce an additional nonnegative vector X_3 with p rows, denote by E_p the column vector with p rows and each element 1, and consider the auxiliary problem

$$\text{minimize } w = E_p^T X_3 \tag{4.21}$$

subject to the constraints

$$B = A_1 X_1 + A_2 X_2 + I_p X_3, \qquad X_1 \geq 0, \qquad X_2 \text{ free}, \qquad X_3 \geq 0. \tag{4.22}$$

This problem satisfies C1 relative to the basic sequence $S = (n + 1, \cdots, n + p)$, it satisfies C3, and will also satisfy C2 if we replace (4.21) by the equivalent canonical cost formulation

$$w = E_p^T B - E_p^T A_1 X_1 - E_p^T A_2 X_2 = f + G_1^T X_1 + G_2^T X_2 \tag{4.21$'$}$$

which can be obtained by (1) solving (4.22) for X_3, (2) multiplying by E_p^T, and (3) substituting the resulting value of $E_p^T X_3$ in (4.21).

The form (4.21) of the cost function shows that $w \geq 0$ for all feasible programs of the auxiliary problem; hence the simplex algorithm must lead to an optimal basic program X_1^0, X_2^0, X_3^0 for this problem with corresponding cost $w^0 \geq 0$.

Now, if X_1', X_2' is a feasible program to the original problem then by taking $X_3' = 0$ we obtain a feasible program to the auxiliary problem with cost $w' = E_p^T 0 = 0$. This shows that *the original problem is feasible if and only if the auxiliary problem has optimal value* $w^0 = 0$, which in turn requires $X_3^0 = 0$ so that X_1^0, X_2^0 will be a feasible program to the original problem.

It is this connection between the original and auxiliary problems that motivates the use of artificial variables. The original problem is solved by a two-phase process: Phase (1) the artificial stage and Phase (2) the natural stage. In Phase 1, the auxiliary problem is solved by the simplex algorithm. If the resulting minimum artificial cost w^0 is positive the original problem is infeasible and we stop; otherwise, if $w^0 = 0$ then X_1^0, X_2^0 is a feasible program to the original problem and we proceed to the second (natural) stage. In this stage we return to the original problem and apply the simplex algorithm to it beginning with the basic sequence corresponding to the program X_1^0, X_2^0. At this stage we may drop the columns corresponding to the artificial variables.

We now discuss the formal mechanism used in the two stages and in the transition from the artificial stage to the natural stage. We begin with the augmented tableau

$$K = \begin{bmatrix} -E_p^TB & -E_p^TA_1 & -E_p^TA_2 & 0 \\ -d & C_1^T & C_2^T & 0 \\ B & A_1 & A_2 & I_p \end{bmatrix} = \begin{bmatrix} -f & G_1^T & G_2^T & 0 \\ -d & C_1^T & C_2^T & 0 \\ B & A_1 & A_2 & I_p \end{bmatrix} \tag{4.23}$$

which has $p + 2$ rows and $n + p + 1$ columns. The first two rows give, respectively, the artificial and the natural costs. In the artificial stage the natural cost row is transformed at each cycle so as to be always in canonical form but otherwise plays no role.

We treat the columns corresponding to the free variables just as described in Section 3.8. It is easy to show that in both Case (a) and Case (b), $g_k = 0$. The procedures in all three cases are as before, except that the continuing operations in Cases (a) and (c) refer to the auxiliary problem whereas the conclusion from Case (b) still applies to the original problem.

Thus, unless Case (b) occurs (which indicates that the original problem has no optimal feasible program) we conclude the artificial stage with an optimal program for the auxiliary problem and a final tableau of the form

$$K^* = \begin{bmatrix} -f^* & G_1^{*T} & G_2^{*T} & G_3^{*T} \\ -d^* & C_1^{*T} & C_2^{*T} & C_3^{*T} \\ B^* & A_1^* & A_2^* & A_3^* \end{bmatrix} \tag{4.24}$$

in canonical form relative to a basic sequence $S^* = (s_1^*, \cdots, s_p^*)$ where for some q with $0 \leq q \leq p$ we have $s_i^* \leq n_1$, $(i = 1, \cdots, p - q)$, $s_i^* > n_1$ $(i = p - q + 1, \cdots, p)$.

If $w^* = f^* \neq 0$ we stop, since the initial problem is infeasible. If $f^* = 0$ and if no s_j^* exceeds n then $G_1^* = 0$, $G_2^* = 0$, $G_3^* = E_p$ and

$$M^* = \begin{bmatrix} -d^* & C_1^{*T} & C_2^{*T} \\ B^* & A_1^* & A_2^* \end{bmatrix} \tag{4.25}$$

is a tableau matrix for the original problem from which we may [following Case (a) of Section 3.8] lay aside the last q rows and the last n_2 columns and obtain a submatrix which is in canonical form with respect to the basic sequence $S^{*\prime} = (s_1^*, \cdots, s_{p-q}^*)$ for a subproblem with no free variables. We then apply the simplex algorithm to this subproblem and if it has a basic optimal feasible program we use it as described in Section 3.8 to get a basic (optimal) feasible program for our original problem. For machine computation we may actually choose to continue pivoting in the large tableau of the auxiliary problem merely being careful to use pivot selection rules which avoid the last q rows and last $n_2 + p$ columns. No further changes will occur in the artificial cost row.

The case where $f^* = 0$ but $s_h^* > n$ for at least one h with $1 \leq h \leq p - q$ requires further discussion. We must have $b_h^* = 0$ lest f^* be positive [in accordance with (4.21)]. Now either (a) $a_{h1}^* = a_{h2}^* = \cdots = a_{hn_1}^* = 0$ or (b) there exists k with $1 \leq k \leq n_1$ such that $a_{hk}^* \neq 0$. In Case (b) we pivot at (h, k) and get a new basic sequence with one less artificial index which still yields a tableau in optimal form. If Case (a) occurs we conclude that there was a dependence relation among the original equality constraints; we then delete row h and proceed with the remaining tableau matrix in which p is replaced by $p - 1$. The resulting subproblem will be in canonical form and we must therefore eventually reach an equivalent problem with no artificial indices in the basic sequence.

For an illustration consider the problem,

$$\text{minimize } z = 15x_1 + 50x_2 + 0x_3 + 0x_4 + 0x_5 \tag{4.26}$$

subject to

$$\begin{aligned} 3x_1 + x_2 - x_3 &= 8, \\ 4x_1 + 3x_2 - x_4 &= 19, \\ x_1 + 3x_2 - x_5 &= 7, \\ x_j &\geq 0, \qquad (j = 1, \cdots, 5), \end{aligned} \tag{4.27}$$

which is in the form (4.17) through (4.19) with $d = 0$,

$$X_1 = \begin{bmatrix} x_1 \\ x_2 \\ x_3 \\ x_4 \\ x_5 \end{bmatrix},$$

and $X_2 = 0$. We can regard this as a diet problem (compare Section 1.1) and the variables x_3, x_4, x_5 as slack variables.

The auxiliary problem corresponding to (4.21) and (4.22) is

$$\text{minimize } w = 0x_1 + \cdots + 0x_5 + x_6 + x_7 + x_8 \tag{4.28}$$

subject to

$$\begin{aligned} 3x_1 + x_2 - x_3 + x_6 &= 8, \\ 4x_1 + 3x_2 - x_4 + x_7 &= 19, \\ x_1 + 3x_2 - x_5 + x_8 &= 7, \\ x_j &\geq 0, \qquad (j = 1, \cdots, 8), \end{aligned} \tag{4.29}$$

in which the artificial variables are x_6, x_7, and x_8. This problem would be in canonical form relative to the basic sequence (6, 7, 8) were it not for the artificial cost coefficients. Solving (4.29) for the artificial variables and substituting the result into (4.28) gives a cost function analagous to (4.21′). The auxiliary problem is then in canonical form.

The cost coefficients in (4.26) will be referred to as natural costs and those in (4.28) as artificial costs. The row labeled g_j' in the first tableau of Table 4.2 gives the result of solving (4.29) for the artificial variables and

Table 4.2. SOLUTION OF PHASE (1) OR AUXILIARY PROBLEM.

Artificial costs g_j							1	1	1		
Natural costs c_j		15	50	0	0	0					
g_j'	−34	−8	−7	1	1	1	0	0	0	b_i/a_{i1}	
c_j'	0	15	50	0	0	0	0	0	0		
	8	$\boxed{3}$	1	−1	0	0	1	0	0	$\boxed{8/3}$	$k = 1$
	19	4	3	0	−1	0	0	1	0	19/4	$h = 1$
	7	1	3	0	0	−1	0	0	1	7	$s_h = 6$
g_j'	−38/3	0	−13/3	−5/3	1	1	8/3	0	0	b_i/a_{i2}	
c_j'	−40	0	45	5	0	0	−5	0	0		
	8/3	1	1/3	−1/3	0	0	1/3	0	0	8	$k = 2$
	25/3	0	5/3	4/3	−1	0	−4/3	1	0	5	$h = 3$
	13/3	0	$\boxed{8/3}$	1/3	0	−1	−1/3	0	1	$\boxed{13/8}$	$s_h = 8$
g_j'	−45/8	0	0	−9/8	1	−5/8	17/8	0	13/8	b_i/a_{i3}	
c_j'	−905/8	0	0	−5/8	0	135/8	5/8	0	−135/8		
	17/8	1	0	−3/8	0	1/8	3/8	0	−1/8		$k = 3$
	45/8	0	0	$\boxed{9/8}$	−1	5/8	−9/8	1	−5/8	$\boxed{5}$	$h = 2$
	13/8	0	1	1/8	0	−3/8	−1/8	0	3/8	13	$s_h = 7$
g_j'	0	0	0	0	0	0	1	1	1		
c_j'	−110	0	0	0	−5/9	155/9	0	5/9	−155/9		
	4	1	0	0	−1/3	1/3	0	1/3	−1/3		
	5	0	0	1	−8/9	5/9	−1	8/9	−5/9		
	1	0	1	0	1/9	−4/9	0	−1/9	4/9		

substituting into (4.28)— it corresponds to the first row of the matrix K in (4.23). The last tableau corresponds to the matrix K^* in (4.24). In solving the Phase (1) problem we carry along the natural costs and transform them at each stage of the solution process but, as remarked earlier, they play no role with respect to the auxiliary problem.

An optimal basic sequence for the Phase (1) problem is (1, 3, 2); the artificial variables have been driven to zero in value in the final tableau of

Table 4.2 and the minimum value of (4.28) is zero. An optimal program for the auxiliary problem leads to a basic feasible program for the original problem (4.27),

$$X = \begin{bmatrix} 4 \\ 1 \\ 5 \\ 0 \\ 0 \end{bmatrix}.$$

Phase (2) is begun with this program and the corresponding natural costs. We could ignore the last three columns if we wished, since the artificial variables do not appear in the original problem; they are carried along here since they can be used for making computational checks by using the concepts developed in Section 4.1.

Table 4.3 displays a solution of the original problem. An optimal basic sequence is (1, 3, 4) and the minimum value of (4.26) is 105.

Table 4.3. SOLUTION OF PHASE (2) OR ORIGINAL PROBLEM.

Natural costs		15	50	0	0	0			
c_j'	−110	0	0	0	−5/9	155/9	0	5/9	−155/9
	4	1	0	0	−1/3	1/3	0	1/3	−1/3
	5	0	0	1	−8/9	5/9	−1	8/9	−5/9
	1	0	1	0	1/9	−4/9	0	−1/9	4/9
c_j'	−105	0	5	0	0	15	0	0	−15
	7	1	9/3	0	0	−1	0	0	1
	13	0	8	1	0	−1244/9	−1	0	1244/9
	9	0	9	0	1	−4	0	−1	−4

It is important to understand the difference between the roles of slack and artificial variables. A positive value for a slack variable indicates either that some capacity is not fully utilized or that some requirement is exceeded in an optimal solution. A positive value for an artificial variable indicates that some feasibility condition has been violated and hence no program involving a positive value for an artificial variable can lead to a feasible program for the original problem.

In the example above, each of the artificial variables has a value of zero in an optimal program for the auxiliary problem (indicating that the original problem is feasible). Recalling the interpretation of the original problem as a diet problem (see Section 1.1), an optimal program of Phase (2) with $x_1 = 7$, $x_2 = 0$, $x_3 = 13$, $x_4 = 9$, and $x_5 = 0$ means that an optimal diet calls for seven units of potatoes and zero units of steak (note the latter has a high price relative to the former). The variables x_3, x_4, and x_5 are slack variables and comparison of (4.27) with (1.2) indicates that the minimum requirements of carbohydrates and vitamins will be exceeded in this diet (x_3 and x_4 are positive), and the minimum requirement for proteins will be met exactly (x_5 is zero). In other words, given the costs and other data of the problem, a "potato economy" is optimal, like that of Ireland some years ago.

One may ask why we go to all of this trouble when the method developed in Chapter 3 will work satisfactorily. We give three reasons. First, the presence of a full initial basic sequence makes available the direct as well as the reverse check of Section 4.1. Second, the method introduced in Section 3.6 may be regarded as essentially introducing artificial variables one at a time; bringing them all in at once seems to be more efficient computationally, although there is as yet only scattered evidence for this statement. Third, programming the simplex algorithm for a computer may involve fewer subroutines when the two-phase process is used than for the method of Chapter 3.

For hand computation there are numerous shortcuts available in the use of artificial variables. If some of the slack columns can be used as part of a basic sequence then we may reduce the number of artificial variables employed. However, for hand calculations it is recommended that the reader use the algorithm as developed in Chapter 3.

4.3. THE REVISED SIMPLEX METHOD.

We have seen in Section 4.1 that if M is the initial full tableau matrix and if M^* is a later tableau matrix, then there exists a matrix Q for which $M^* = QM$. Actually, given M and Q there is no need to calculate all of M^*. The revised simplex method is based on this idea. We state the algorithm in inductive form by showing how, starting with M^* and Q, to obtain a corresponding matrix Q' for which $M' = Q'M$. In particular, the induction process begins with $M^* = M$ and $Q = I_p$. First we calculate the initial (cost) row of M^* and either, Case 1, find an index k for which $c^*_{hk} < 0$ or, Case 2, conclude that M^* is in optimal form.

In Case (1) we calculate column k of M^* and conclude either, Case 3, there is some $a_{ih} > 0$ or, Case 4, the problem has no optimal solution. In Case 3 we calculate column 0 (the B column) and apply the pivot selection rule to determine the next pivot position (h, k). (For this discussion we assume that ties for h are resolved by use of some random device.) Let M' be the matrix obtained from M^* after pivoting at (h, k); then

$$M' = F^*M^*, \tag{4.30}$$

where F^* is given by (4.5) (calculated from M^* rather than from M). Then, from (4.6) we get

$$M' = Q'M, \qquad \text{where} \qquad Q' = F^*Q. \tag{4.31}$$

This completes a full cycle of the revised simplex algorithm. Note that this form of the simplex algorithm requires (1) keeping a permanent record of the initial tableau matrix M, (2) keeping a temporary record of the transforming matrix Q which is discarded in favor of its successor Q' when it is calculated, and (3) calculating the initial row, the initial column, and one other column of M^*. These are used to decide the state of M^* and to determine F^* in case M^* is not in a terminal state and then are discarded. One might also wish to (4) keep a record of the basic sequence S'.

The revised simplex method is not recommended for hand calculation. For that matter, hand calculation in any form is hopelessly inefficient for all but very small problems. Thus the revised simplex method must be evaluated with machine calculation in mind.

It is not particularly difficult to devise a computer program which will put into effect the four parts of the revised simplex method. The method has several important advantages for digital calculation.

(a) All calculations are based on the initial data, and the accuracy of Q can be tested by reference to (4.8), in other words, Q is the inverse of M_S^*.

(b) If the original data are sparse (that is, have many zero coefficients) the calculations will be simpler and quicker at every cycle since this sparse matrix M is used throughout. In the ordinary algorithm sparseness is rapidly lost. Moreover, very large sparse problems can be handled with a computer memory which could not contain problems of the same size which are not sparse.

(c) The fact that not all of M^* need be calculated gives substantial computing economy.

In most current computer programs the method as described is modified in two ways. First, we begin with the matrix K of (4.23) instead of with M and introduce a computer routine for shifting from the artificial stage to the natural stage when the former is complete. Second, instead of storing Q we consider the factors F' and for each F' we store only one column, namely its single nonidentity column, together with a record of the pivot position. The algorithm in this form is called the *product form of the revised simplex method* and several computer algorithms for this method are in current use. We do not discuss the details here.

4.4. HOMOGENEOUS INEQUALITIES AND THE FARKAS LEMMA.

Although we will postpone our main discussion of inequalities until Chapter 8, it seems appropriate to give a brief introduction to the topic here. One of the major theorems in this branch of mathematics is the celebrated Farkas Lemma which in many developments of linear programming has been proved directly and then applied in the development of the theory itself (hence the name lemma). We reverse the usual process here and obtain the lemma as a corollary to our theorem on the convergence of the simplex algorithm.[1]

Farkas' Lemma. Let A be a p by n real matrix and let B be any vector in V_p; then either the system

$$AX = B, \qquad X \geq 0 \tag{4.32}$$

has a solution, or there exists a vector $Y \in V_p$ for which

$$Y^T A \leq 0, \qquad Y^T B > 0. \tag{4.33}$$

Moreover, only one of these alternatives is valid.

Proof: Suppose first that $Y^T A \leq 0$ and $Y^T B > 0$ and that $X \geq 0$. Then

$$Y^T(AX) = (Y^T A)X \leq 0$$

and hence we cannot have $AX = B$. This proves that at most one of the alternatives (4.32) and (4.33) can hold. To show that at least one of the alternatives is valid, let (4.32) be the feasibility conditions for a linear programming problem with any cost function $z = d + C^T X$ and define

$$M = \begin{bmatrix} -d & C^T \\ B & A \end{bmatrix}$$

as usual. If the problem has a feasible solution X, then clearly the first alternative of the lemma holds. If the problem is infeasible, then according to the results of Sections 3.6 and 3.7, after a sequence of pivots in M we reach a tableau

$$M^* = \begin{bmatrix} -d^* & C^{*T} \\ B^* & A^* \end{bmatrix}$$

which has a row

$$[b_h^* a_{h1}^* \cdots a_{hn}^*] \tag{4.34}$$

such that either

$$b_h^* < 0, \qquad a_{hj}^* \geq 0, \qquad (j = 1, \cdots, n), \text{ see (3.62)} \tag{4.35}$$

or

$$b_h^* \neq 0, \qquad a_{hj}^* = 0, \qquad (j = 1, \cdots, n), \text{ see (3.69 (b))}. \tag{4.36}$$

Now, according to (4.6), (4.12), and (4.13) we have

$$M^* = QM, \tag{4.37}$$

[1]This development is consistent with the long-term goals of Dantzig for using the simplex algorithm to prove the basic theorems of linear inequality theory. These goals have been realized in an elegant form by Balinski in paper 2 of Reference (15).

where Q has the form

$$Q = \begin{bmatrix} 1 & P \\ 0 & R \end{bmatrix}. \tag{4.38}$$

In particular, the row (4.34) of M^* will satisfy the equation

$$[b_h^* a_{h1}^* \cdots a_{hn}^*] = R^{(h)}[B\ A], \tag{4.39}$$

where $R^{(h)}$ is row h of R. Now let

$$Y = \frac{1}{b_h^*} R^{(h)T} \tag{4.40}$$

and we have from (4.35), (4.36), and (4.39) that

$$Y^T B = 1 > 0 \qquad \text{and} \qquad Y^T A = \frac{1}{b_h^*}[a_{h1}^* \cdots a_{hn}^*] \leq 0; \tag{4.41}$$

that is, Y satisfies (4.33).

We can paraphrase this proof by the statement that if a linear programming problem turns out to be infeasible, the sequence of pivots which establishes the infeasibility also serves to construct a solution Y for the second alternative, whereas if the problem is feasible, any feasible X gives a solution for the first alternative. Note also that the proof of Farkas' Lemma is completely constructive.

PROBLEMS

4.1. Consider your solution to Problem 3.3. Let M be the initial full tableau matrix and M^* be the final optimal tableau. Find M_{S*}, M_S^*, M_{S*}^*, and verify that $M_{S*}M_S^* = I_{p+1}$. Also verify that

$$M^* = M_S^* M, \qquad M = M_{S*}M^*.$$

4.2. Given the following sequence of tableaus and information that some entries in M^* are incorrect, indicate which columns of M^* contain errors.

$M =$

14	−1	2	−3	1	0	0	0	0
2	−3	2	1	5	1	0	0	0
4	2	0	2	1	0	1	0	0
4	4	1	2	−4	0	0	1	0
6	−6	−3	3	1	0	0	0	1

$M^* =$

20	3	5/2	0	0	0	1	1/2	0
0	−11/5	23/10	0	0	1	−7/5	9/10	0
0	3/5	−1/5	0	1	0	1/5	−1/5	0
2	6/5	1/10	1	0	0	2/5	1/10	0
0	−46/5	−21/10	0	0	0	−7/5	−1/10	1

4.3. Given the following tableaus and the information that M is correct but that errors exist in M^*, locate and correct the errors in M^*. After correcting errors, find matrices Q and Q^* such that $M^* = QM$ and $M = Q^*M^*$, respectively.

$M =$

0	2	1	−3	5	8	0	0	0
6	3	−5	−8	1	−2	1	0	0
9	2	2	1	5	2	0	1	0
10	1	1	−5	−7	11	0	0	1

$M^* =$

27	8	10	0	20	11	0	3	0
87	21	22	0	46	7	1	9	0
9	2	3	1	5	1	0	1	0
55	11	16	0	18	16	0	5	1

4.4. Return to your solution of Problem 3.30. Let M denote the initial tableau matrix and let M_1 and M_2 denote the second and third tableau matrices. Find a matrix Q_1 such that $M_1 = Q_1M$ and a matrix Q_2 such that $M_2 = Q_2M_2$. Now let Q_3 be a matrix such that $M_2 = Q_3M$ and verify that $Q_3 = Q_2Q_1$.

4.5. Consider the problem,

$$\text{minimize } z = -3x_1 - 2x_2 - x_3 + 5x_4$$

subject to

$$\begin{aligned} 3x_1 + 4x_2 + 5x_3 + 2x_4 &\leq 9, \\ x_1 + x_2 + 4x_3 + x_4 &= 5, \\ 2x_1 + 3x_2 + x_3 + 5x_4 &\geq 2, \\ x_j &\geq 0, \qquad (j = 1, \cdots, 4). \end{aligned}$$

(a) Introduce artificial variables, put the Phase (1) problem in canonical form and solve the Phase (1) problem.
(b) Continue with the Phase (2) problem, carrying along in your computations the columns corresponding to the artificial variables.
(c) What is an optimal solution to the Phase (1) problem?
(d) What is an optimal solution to the Phase (2) problem?
(e) Let M denote the initial tableau matrix of the Phase (1) problem and M^* an optimal tableau of the Phase (2) problem. Find a matrix Q such that $M^* = QM$. Verify that one can obtain M^* by premultiplying M by the matrix Q.

4.6. Solve the following problem by introducing artificial variables:

$$\text{minimize } z = 2x_1 + 3x_2 - x_3 + x_4$$

subject to

$$\begin{aligned} 2x_1 + x_2 - 7x_3 + x_4 &= 9, \\ x_1 + x_2 - 4x_3 \quad &\geq 17, \\ x_2 + 7x_3 \quad &\leq 21, \\ x_j &\geq 0, \qquad (j = 1, \cdots, 4). \end{aligned}$$

4.7. Solve Problem 3.19 using artificial variables. Is there more than one optimal solution to this problem? If so, determine an alternate optimal solution.

4.8. Solve Problem 3.20 using artificial variables.

4.9. Use artificial variables and a Phase (1) procedure to determine if the following problem has a feasible program:

$$\text{minimize } z = x_1 + x_2 + 2x_3$$

subject to

$$\begin{aligned} 2x_1 + x_2 + x_3 &\geq 12, \\ 7x_1 + 5x_2 + 2x_3 &\leq 35, \\ 8x_1 + 9x_2 - x_3 &\geq 72, \\ x_j &\geq 0, \qquad (j = 1, 2, 3). \end{aligned}$$

If the problem has no feasible program, construct a vector Y which satisfies a system of homogeneous linear inequalities of the form (4.33).

4.10. Find a feasible program to the following problem or construct a solution to a system of homogeneous linear inequalities of the form (4.33):

$$\text{maximize } z = x_1 + x_2 + 3x_3$$

subject to

$$\begin{aligned} 7x_1 + 3x_2 + x_3 &\leq 20, \\ 4x_1 + 3x_2 + 2x_3 &= 24, \\ 7x_1 + x_2 - x_3 &\leq 14, \\ x_j &\geq 0, \qquad (j = 1, 2, 3). \end{aligned}$$

4.11. Use artificial variables to determine if the following problem has a feasible program. If it does not have a feasible program, construct a solution to a system of homogeneous linear inequalities of the form (4.33). If it has a feasible program, determine such a program.

$$\text{minimize } z = -3x_2 + x_5 - 3x_6$$

subject to

$$\begin{aligned} 4x_2 + x_3 \qquad - 2x_5 \qquad &= 11, \\ x_1 - x_2 \qquad + 3x_5 + 2x_6 &= 7, \\ 3x_2 + \qquad x_4 - 3x_5 + 7x_6 &= 9, \\ x_j &\geq 0, \qquad (j = 1, 2, 3, 4), \\ x_5, x_6 &\text{ free.} \end{aligned}$$

4.12. Solve Problem 3.22 by using artificial variables.

4.13. Solve Problem 3.25 by using artificial variables.

4.14. Suppose that in attempting to solve the problem,

$$\text{minimize } z = d + C^T X_1$$

subject to

$$A_1 X_1 = B, \qquad X_1 \geq 0,$$

where A_1 is p by n and X_1 is n by 1, we first solved the Phase (1) problem,

$$\text{minimize } w = E_p^T X_2$$

subject to

$$A_1X_1 + I_pX_2 = B, \qquad X_1 \geq 0, \qquad X_2 \geq 0,$$

and that one learns that the Phase (1) problem has an optimal program,

$$X' = \begin{bmatrix} X_1' \\ X_2' \end{bmatrix} = \begin{bmatrix} x_1' \\ \cdot \\ \cdot \\ x_{p-1}' \\ 0 \\ \cdot \\ \cdot \\ \cdot \\ 0 \end{bmatrix}, \qquad \begin{matrix} x_j' > 0 & \text{for } j = 1, \cdots, p-1, \\ x_j' = 0 & \text{otherwise,} \end{matrix}$$

and optimal basic sequence $S^* = (1, 2, \cdots, p - 1, n + 1)$. What can we conclude about the original problem? Is the Phase (1) problem degenerate? Is the Phase (2) problem degenerate? Discuss.

4.15. Use the Farkas' Lemma of Section 4.4 to prove the following theorem: Given matrices A and B of size p by n and n by 1, respectively, either $AX \geq B$, $X \geq 0$ has a solution or there is a solution to the system

$$\begin{aligned} A^TY &\leq 0 \\ Y &\geq 0 \\ Y^TB &> 0 \end{aligned}$$

where the word, or, is used in its exclusive sense.

Duality and Marginal Analysis

5

5.1. DUALITY AND INTERPRETATIONS: THE SYMMETRIC CASE.

We consider a pair of linear programming problems,

$$\text{maximize } w = P^T V \tag{5.1}$$

subject to

$$DV \leq R, \qquad V \geq 0, \tag{5.2}$$

and

$$\text{minimize } f = R^T U \tag{5.3}$$

subject to

$$D^T U \geq P, \qquad U \geq 0, \tag{5.4}$$

where D is a p by m matrix. We first observe some of the ways in which these problems are related. The matrix of constraints in the second problem is the transpose of that for the first, the vector of coefficients in (5.1) is the transpose of the column vector of constants in (5.4), and the column vector of constants in (5.2) is the transpose of the vector of coefficients in (5.3). Since D is p by m we know that V is m by 1, P is m by 1, and U is p by 1. Roughly speaking, the two problems have the same data inputs (the elements of D, P, and R); the objectives in (5.1) and (5.3) differ as do the decision or program vectors V and U.

For a specific case (in which $p = 3$, $m = 5$), consider the problems

maximize

$$w = -2v_1 - v_2 + 3v_3 - 5v_4 - 8v_5 \tag{5.5}$$

subject to

$$\begin{aligned} 3v_1 - 5v_2 - 9v_3 + v_4 - 2v_5 &\leq 6, \\ 2v_1 + 3v_2 + v_3 + 5v_4 + v_5 &\leq 9, \\ v_1 + v_2 - 5v_3 - 7v_4 + 11v_5 &\leq 10, \\ v_1 &\geq 0, \\ v_2 &\geq 0, \\ v_3 &\geq 0, \\ v_4 &\geq 0, \\ v_5 &\geq 0. \end{aligned} \tag{5.6}$$

minimize

$$f = 6u_1 + 9u_2 + 10u_3 \tag{5.7}$$

subject to

$$\begin{aligned} u_1 &\geq 0, \\ u_2 &\geq 0, \\ u_3 &\geq 0, \\ 3u_1 + 2u_2 + u_3 &\geq -2, \\ -5u_1 + 3u_2 + u_3 &\geq -1, \\ -9u_1 + u_2 - 5u_3 &\geq 3, \\ u_1 + 5u_2 - 7u_3 &\geq -5, \\ -2u_1 + u_2 + 11u_3 &\geq -8. \end{aligned} \tag{5.8}$$

These problems are examples of dual problems in linear programming. For purposes of exposition, the maximizing problem is called the primal and the other is called the dual problem. It will be shown that the dual of the dual is the primal problem, and, moreover, that the problems are even more closely related than the above discussion suggests. We will also discuss duality for a still more general class of linear programs in Section 5.2.

We assume that (5.1) and (5.2) has an optimal feasible solution and introduce slack variables $y_1, \cdots, y_p$ into the problem. This leads to the problem

$$\text{minimize } z = d + C^T X \tag{5.9}$$

subject to

$$AX = B, \qquad X \geq 0, \tag{5.10}$$

which is in standard form with p equations and with $n = p + m$ variables and in which

$$A = [D \ I], \quad X = \begin{bmatrix} V \\ Y \end{bmatrix}, \quad C = \begin{bmatrix} -P \\ 0 \end{bmatrix}, \quad B = R, \quad d = 0. \tag{5.11}$$

Moreover, the properties C1 and C2 of Section 3.2 hold relative to the basic sequence $S = (m + 1, \cdots, m + p)$. Denote the initial tableau as

$$M = \begin{bmatrix} -d & C^T \\ B & A \end{bmatrix} = \begin{bmatrix} 0 & -P^T & 0 \\ R & D & I_p \end{bmatrix} \tag{5.12}$$

and let the final (optimal) tableau be

$$M^* = \begin{bmatrix} -d^* & C^{*T} \\ B^* & A^* \end{bmatrix} = \begin{bmatrix} -d^* & C_G^{*T} & C_S^{*T} \\ B^* & A_G^* & A_S^* \end{bmatrix}. \tag{5.13}$$

Optimality assures that $C_G^{*T} \geq 0$ and $C_S^{*T} \geq 0$, and an optimal solution X^{S^*}, z^{S^*} is given by $X_{S*}^{S*} = B^*, X_{G*}^{S*} = 0, z^{S^*} = d^*$. We want to construct a feasible program for the dual problem and since transposes appear in the statement of the primal and dual problems we might consider looking for such a program in the rows of (5.13). Suppose we try $U_0 = C_S^*$; from optimality $C_S^{*T} \geq 0$ and, moreover, C_S^* has the proper number of components. Also, we recall from Section 4.1 that $M^* = M_S^* M$; specifically,

$$M_S^* M = \begin{bmatrix} 1 & C_S^{*T} \\ 0 & A_S^* \end{bmatrix} \begin{bmatrix} 0 & -P^T & 0 \\ R & D & I_p \end{bmatrix},$$

$$M_S^* M = \begin{bmatrix} C_S^{*T}R & -P^T + C_S^{*T}D & C_S^{*T} \\ A_S^* R & A_S^* D & A_S^* I_p \end{bmatrix}.$$

Now equating appropriate submatrices of M^* and $M_S^* M$ we obtain

$$-d^* = U_0^T R, \tag{5.14}$$

and

$$C_G^{*T} = -P^T + U_0^T D. \tag{5.15}$$

Since $C_S^* \geq 0$ and $C_G^* \geq 0$, we have in view of (5.15)

$$U_0 \geq 0, \qquad D^T U_0 \geq P. \tag{5.16}$$

Therefore U_0 is a feasible program for the dual problem.

Let $X_0 = \begin{bmatrix} V_0 \\ Y_0 \end{bmatrix}$ be an optimal program for the primal problem in the form (5.9) and (5.10). Then

$$d^* = C^T X_0 = [-P^T \quad 0] \begin{bmatrix} V_0 \\ Y_0 \end{bmatrix} = -P^T V_0; \tag{5.17}$$

hence, taking transposes of (5.14),

$$P^T V_0 = R^T U_0 = -d^*. \tag{5.18}$$

For any feasible programs V for (5.1) and (5.2) and U for (5.3) and (5.4) we have

$$R^T U = U^T R \geq U^T D V \geq P^T V. \tag{5.19}$$

From (5.18), (5.19), and the feasibility of V_0 and U_0 it follows that both V_0 and U_0 are optimal. For, we have from (5.19)

$$R^T U \geq P^T V;$$

in particular,

$$P^T V_0 = R^T U_0 \geq P^T V,$$

so that

$$P^T V_0 \geq P^T V$$

for every feasible V. Hence we can write

$$w^0 = P^T V_0 = \text{maximum } w.$$

A similar argument shows that

$$R^T U_0 \leq R^T U$$

for every feasible U, and we have

$$f^0 = R^T U_0 = \text{minimum } f,$$

in other words, for optimal vectors V_0 and U_0, respectively, we have

$$\text{maximum } w = \text{minimum } f.$$

Furthermore, since U_0 appears as part of the initial row of M^*, we see that the simplex algorithm solves both primal and dual problems simultaneously and that an optimal tableau for one of the problems displays optimal programs for both primal and dual problems when such programs exist.

These concepts are easily illustrated by using the problem (5.5) and (5.6) above. After insertion of slack variables this problem can be written, analogous to (5.9) and (5.10), as follows:

$$\text{minimize } z' = 2v_1 + v_2 - 3v_3 + 5v_4 + 8v_5 + 0y_1 + 0y_2 + 0y_3$$

subject to

$$\begin{aligned}
3v_1 - 5v_2 - 9v_3 + v_4 - 2v_5 + y_1 \qquad\qquad &= 6, \\
2v_1 + 3v_2 + v_3 + 5v_4 + v_5 \qquad + y_2 \qquad &= 9, \\
v_1 + v_2 - 5v_3 - 7v_4 + 11v_5 \qquad\qquad + y_3 &= 10, \\
v_j &\geq 0, \qquad (j = 1, \cdots, 5) \\
y_i &\geq 0, \qquad (i = 1, 2, 3).
\end{aligned}$$

Simplex iterations for this problem are displayed in Table 5.1. An optimal program for this problem is

$$X_0 = \begin{bmatrix} V_0 \\ Y_0 \end{bmatrix} = \begin{bmatrix} 0 \\ 0 \\ 9 \\ 0 \\ 0 \\ 87 \\ 0 \\ 55 \end{bmatrix},$$

and the V_0 portion of X_0 (the first five components) is an optimal program for the primal problem (5.5), (5.6); an optimal program for the dual problem (5.7) and (5.8) is

$$U_0 = C_S^* = \begin{bmatrix} 0 \\ 3 \\ 0 \end{bmatrix}.$$

Also, we see that w^0 = maximum $w = 27$ and

$$f^0 = \text{minimum } f = 6(0) + 9(3) + 10(0) = 27.$$

The problem (5.1), (5.2) and its dual (5.3), (5.4) constitute the symmetric case of duality (the problems themselves are also called symmetric duals). The relationship between these problems is a special instance of duality in linear programming; as indicated earlier, a discussion of duality for more general linear programming problems is given in the next section.

Table 5.1. SOLUTION TO PRIMAL PROBLEM.

0	2	1	−3	5	8	0	0	0
6	3	−5	−9	1	−2	1	0	0
9	2	3	1	5	1	0	1	0
10	1	1	−5	−7	11	0	0	1
−27	8	10	0	20	11	0	3	0
87	21	22	0	46	7	1	9	0
9	2	3	1	5	1	0	1	0
55	11	16	0	18	16	0	5	1

$S = (6, 7, 8)$; C_S^{*T} (row −27, last three columns); A_S^* (last three rows, last three columns); A_G^* (last three rows, columns 2–6); C_G^{*T} (row −27, columns 2–6)

We turn now to a consideration of interpretations of symmetric duals and raise the following question: given a problem in either of the forms (5.1), (5.2) or (5.3), (5.4) and an interpretation of it (say the diet problem), we know what the mathematical form of the dual problem is; what practical interpretation can be given to the latter? It turns out that one of the major reasons for the importance of linear programming in applications is that this question not only has an answer, but an answer that is often as important from a practical point of view as knowing an optimal primal program (this is true also of duality in more general linear programming problems). Moreover, we sometimes want to study the interpretations of both problems simultaneously and to investigate the joint effects of manipulation of one or the other or both problems. As we will see later, pursuing the interpretations of interactions between the problems will lead us to consider certain of the given constants as parameters and consequently provide greatly enhanced flexibilities in applications.

Table 5.2. DIETARY DATA.

Nutrient \ *Food*	*Requirement*	*Potatoes*	*Steak*
Carbohydrate	b_1	a_{11}	a_{12}
Vitamins	b_2	a_{21}	a_{22}
Protein	b_3	a_{31}	a_{32}

Let us begin with a brief recapitulation of the diet problem introduced in Section 1.1 but in a slightly more general symbolic form as in Table 5.2. There are two *activities*, eating one unit of potatoes and eating one unit of steak. Also,

$$a_{ij} = \text{number of units of nutrient } i \text{ in one unit of food } j,$$
$$b_i = \text{number of units of nutrient } i \text{ required in the diet,}$$

and we have

$$c_j = \text{cost (price) per unit of food } j$$

and

$$x_j = \text{number of units of food } j \text{ eaten or consumed.}$$

The problem (a special case of (5.3), (5.4)) is

$$\text{minimize } z = c_1x_1 + c_2x_2 \tag{5.20}$$

subject to

$$\begin{aligned} a_{11}x_1 + a_{12}x_2 &\geq b_1, \\ a_{21}x_1 + a_{22}x_2 &\geq b_2, \\ a_{31}x_1 + a_{32}x_2 &\geq b_3, \\ x_j &\geq 0, \qquad (j = 1, 2). \end{aligned} \tag{5.21}$$

The dual of this problem (a special case of (5.1), (5.2)) is

$$\text{maximize } w = b_1y_1 + b_2y_2 + b_3y_3 \tag{5.22}$$

subject to

$$\begin{aligned} a_{11}y_1 + a_{21}y_2 + a_{31}y_3 &\leq c_1, \\ a_{12}y_1 + a_{22}y_2 + a_{32}y_3 &\leq c_2, \\ y_j &\geq 0, \qquad (j = 1, 2, 3). \end{aligned} \tag{5.23}$$

Again it is emphasized that both primal and dual problems have the same given data base, the a_{ij}, b_i, and c_j, so a practical interpretation of the dual, given an interpretation of the primal, is essentially a matter of developing an interpretation of the dual variables y_j.

Consider the first inequality of (5.23). The right-hand term c_1 denotes the cost or price per unit of commodity 1 (food 1 or potatoes); the numbers a_{11}, a_{21}, a_{31} denote the respective numbers of units of nutrients 1, 2, and 3 in food 1. Therefore y_j must be a price assigned for a unit of nutrient $j (j = 1, 2, 3)$. It may be helpful to think of a manufacturer of pills P_1, P_2, P_3 where one pill of type P_j supplies one unit of nutrient j (for example, P_2 is a vitamin pill). Then y_j is the price assigned by the manufacturer per pill P_j and we may introduce the concept of an artificial unit of potatoes, namely a package made up of a_{11} pills of type P_1, a_{21} pills of type P_2, and a_{31} pills of type P_3. This package is equivalent nutritionally to a unit of real potatoes; we require that the cost of this package shall not exceed the cost c_1 of a unit of potatoes. Thus we have:

$$\underbrace{a_{11}y_1 + a_{21}y_2 + a_{31}y_3}_{\substack{\text{price or value of a}\\ \text{package of pills}\\ \text{equivalent nutri-}\\ \text{tionally to a unit of}\\ \text{potatoes}}} \leq \underbrace{c_1}_{\substack{\text{price or}\\ \text{cost per}\\ \text{unit of}\\ \text{potatoes}}}.$$

Note carefully the interpretational significance of the transposition of the a_{ij} in (5.23) as contrasted to (5.21).

A similar examination of the second inequality in (5.23) leads to the statement that the price per unit of a package of pills equivalent to steak must be less than or equal to the cost per unit of steak. Thus a *dual activity* i can be regarded as assigning prices per unit of a pill P_i (whereas a *primal activity* consists of eating a unit of a food). The condition (5.22) requires that the prices y_j assigned should be those that maximize the total returns or revenues from selling pills to meet the nutritional requirements, and (5.23) assures that the values of y_j be such that the seller of pills can avoid unfavorable competition with a seller of the foods (that is, a grocer). The y_j values determined by the dual are thus *competitive prices.*

Now consider a slight generalization of the wood products problem in Section 1.3 as given in Table 5.3. There are three activities in this problem,

Table 5.3. TECHNOLOGY TABLE.

	Units of wood	*Units of labor*	*Price or unit value of output*
Per picture frame	a_{11}	a_{21}	r_1
Per cabinet	a_{12}	a_{22}	r_2
Per figurine	a_{13}	a_{23}	r_3
Available materials	b_1	b_2	$\cdots$

producing one unit of picture frames (one picture frame), producing one cabinet, and producing one figurine; there are two inputs, wood and labor; and there are three possible outputs. Let

a_{ij} = number of units of input i required in the production of one unit of output j,
b_i = number of units of input i available,
r_j = price per unit of output j,
x_j = number of units of output j produced in a given time period.

The problem in linear programming form is to

$$\text{maximize } w = r_1x_1 + r_2x_2 + r_3x_3 \tag{5.24}$$

subject to

$$\begin{aligned} a_{11}x_1 + a_{12}x_2 + a_{13}x_3 &\leq b_1, \\ a_{21}x_1 + a_{22}x_2 + a_{23}x_3 &\leq b_2, \\ x_j &\geq 0, \qquad (j = 1, 2, 3), \end{aligned} \tag{5.25}$$

where w represents the total return from the sale of outputs. This is a *resource allocation* problem, since we are concerned with the question of allocating the scarce resources, which the firm has in amounts limited to b_1 and b_2, among competing production activities so as to maximize total returns.

The dual of this problem is to

$$\text{minimize } z = b_1u_1 + b_2u_2 \tag{5.26}$$

subject to

$$\begin{aligned} a_{11}u_1 + a_{21}u_2 &\geq r_1, \\ a_{12}u_1 + a_{22}u_2 &\geq r_2, \\ a_{13}u_1 + a_{23}u_2 &\geq r_3, \\ u_i &\geq 0, \qquad (i = 1, 2). \end{aligned} \tag{5.27}$$

Examine the second inequality from (5.27),

$$a_{12}u_1 + a_{22}u_2 \geq r_2. \tag{5.28}$$

From the primal statement of the problem we know that r_2 is a price per unit of a commodity (output 2 or cabinets) and we know that a_{12} and a_{22} are the numbers of units of inputs 1 and 2 required to produce one unit of output 2. Therefore the u_1 and u_2 must be expressed in terms of monetary value (price) per unit if (5.28) is to make sense. The first dual activity at unit level consists of assigning price 1 per unit of wood; therefore at level u_1 it consists of assigning price u_1 per unit of wood. Similarly, the second dual activity at level u_2 consists of assigning price u_2 per unit of labor. We see therefore that the left-hand side of (5.28) places a total price $a_{12}u_1 + a_{22}u_2$ on the a_{12} units of wood and a_{22} units of labor required in production of a cabinet. The inequality (5.28) states that, to the firm, values or prices per unit must be assigned (imputed) to the inputs so that if the inputs are used in producing a cabinet the total value of the inputs required cannot be less than the revenue realized from the sale of a cabinet.

One might ask what the consequences of violating the condition (5.28) might be. If we assign prices u_1 and u_2 for which $a_{12}u_1 + a_{22}u_2 < r_2$ then a competing manufacturer of cabinets could attract our inputs of wood and labor from our production facility into his by offering respective prices v_1, v_2 for which $v_1 > u_1$, $v_2 > u_2$ and also $a_{12}v_1 + a_{22}v_2 < r_2$.

Similar interpretations arise from the other inequalities in (5.27), assuring that prices or values per unit u_1 and u_2 will be assigned to inputs so that the total value of inputs used in producing each unit of output will not be less than the revenue realized by the sale of each unit of output. Moreover, (5.26) means that the prices or values will, in addition to satisfying (5.27), be chosen so as to minimize the total value of all inputs required in the firm's production.

The dual of a resource allocation primal, that of assigning unit prices to inputs, is usually not an artificial problem for several reasons. The costs of the inputs may not explicitly enter into accounting profit calculations in a firm because the costs may have been completely amortized or the inputs may have been produced by other divisions of the firm for use in this division. In the latter case there may be no market price per unit for the inputs whatever. If the inputs have a value other than zero, questions of efficient use must arise. For example, suppose the inputs are produced for sale by other firms and by another division of the same firm. The objective (5.26) would enable a manager to determine whether to use resources generated internally in the firm or to buy them from other producers. Alternatively, we could turn the problem around and ask the following question: inputs into the production division are now being used which come from the firm's division X; is the operation of X economical, given that a reliable firm in the market has lately begun producing and selling the commodities we use for inputs?[1]

A somewhat stronger statement can be made concerning the dual. It was established that not only do the primal and dual provide us with optimal vectors X and U, if they exist, satisfying (5.24), (5.25), and (5.26), (5.27) respectively, but also we have

$$\text{total revenue} = \max w = \Sigma r_j x_j = \Sigma u_i b_i = \text{total imputed value.}$$

Therefore the optimal values u_i provide an allocation or *imputation* of all of the total revenue to the inputs; that is, none of the total revenue is left over or is unassigned to inputs. The product $b_i u_i$, in other words, is the total accounting or imputed value to the firm of the available resource or input i. The dual can also be stated in words as: find the smallest imputed

[1]The problem of pricing resources that flow from one division of a firm to another is sometimes called the *transfer pricing problem*, and is a matter of great importance in large-scale, decentralized business operations. The dual of a resource allocation model provides a way of assigning transfer prices; this possibility occurs later in decomposition problems (Chapter 10), where a large-scale linear programming problem (of a certain class) will be broken up or decomposed into a collection of smaller problems, each of which has a *pricing dual*. The decomposed primals can represent decentralized optimal resource allocations, and the corresponding duals represent input valuations in the decentralized divisions. Moreover, the totality of decentralized primal optima represents the firm's optimal resource decision and the totality of dual prices the corresponding optimal assignment of input values throughout the system.

value per unit of input subject to the requirement that all the revenues from the sale of outputs be fully allocated to the inputs available.

An interesting feature of optimality in linear programming is *complementary slackness.* Suppose we return to the symmetric duals (5.1), (5.2) and (5.3), (5.4) and recall that for feasible vectors we have from (5.19)

$$P^T V \leq U^T D V \leq U^T R.$$

Now let V^* and U^* be optimal programs, respectively; then

$$P^T V^* = U^{*T} D V^* = U^{*T} R.$$

Consider

$$U^{*T} R - U^{*T} D V^* = 0$$

or

$$U^{*T}(R - DV^*) = 0; \tag{5.29}$$

from (5.2) and (5.4) we see that the vectors U^* and $R - DV^*$ are nonnegative. Therefore, if the ith component of U^* is positive, then the ith component of the vector $R - DV^*$ must be zero. The latter condition can be written

$$r_i - (d_{i1} v_1^* + \cdots + d_{im} v_m^*) = 0$$

or

$$d_{i1} v_1^* + \cdots + d_{im} v_m^* = r_i,$$

which means that the ith constraint in the primal is satisfied as an equality with respect to V^* if the ith dual variable is positive in an optimal solution U^*. On the other hand, if the ith component of the vector $R - DV^*$ is positive in (5.29), that is, if

$$d_{i1} v_1^* + \cdots + d_{im} v_m^* < r_i,$$

the ith constraint in the primal is satisfied as a strict inequality, then the ith dual variable must be zero in an optimal solution.

Furthermore, from (5.29) we also get

$$U^{*T} D V^* - P^T V^* = 0$$

or

$$(D^T U^* - P)^T V^* = 0 \tag{5.30}$$

and statements analogous to those above can be made about the components of the vector $(D^T U^* - P)^T$ and V^*.

Conversely, if V^*, U^* are feasible programs for which the complementary slackness conditions (5.29) and (5.30) hold, then we have

$$P^T V^* = U^{*T} D V^* = U^{*T} R$$

and hence using (5.19) we conclude that U^* and V^* are optimal. Any two vectors V and U in V_m and V_p, respectively, are said to be *an orthogonal pair* [relative to the problems (5.1), (5.2), (5.3), (5.4)] if they satisfy the complementary slackness conditions (5.29) and (5.30).

We have established the following important result, called the (weak) complementary slackness theorem: two vectors V^* and U^* are optimal primal and dual programs respectively if and only if (1) V^* is primal feasible [satisfies (5.2)], (2) U^* is dual feasible [satisfies (5.4)], and (3) (V^*, U^*) is an orthogonal pair [satisfies (5.29) and (5.30)].

For an example, consider the problem (5.5), (5.6) and its dual (5.7), (5.8). For the pair V_0, U_0 of optimal programs obtained from Table 5.1 we have

$$\begin{bmatrix} V_0 \\ R - DV_0 \end{bmatrix} = \begin{bmatrix} 0 \\ 0 \\ 9 \\ 0 \\ 0 \\ 87 \\ 0 \\ 55 \end{bmatrix}, \qquad \begin{bmatrix} D^T U_0 - P \\ U_0 \end{bmatrix} = \begin{bmatrix} 8 \\ 10 \\ 0 \\ 20 \\ 11 \\ 0 \\ 3 \\ 0 \end{bmatrix}. \tag{5.31}$$

We observe that the complementary slackness conditions hold. Moreover, the first vector in (5.31) is the vector X_0 determined in the usual way from the final tableau M^* of Table 5.1. Note also that in addition to having $U_0 = C_S^*$ we have $D^T U_0 - P = C_G^*$, in other words, the second vector in (5.31) is the transpose of the vector of cost elements in M^*.

The sum of the two vectors in (5.31) has each component positive. This is an illustration of the strong form of the complementary slackness theorem which will be stated and proved in Chapter 8.

Complementary slackness also has natural economic interpretations. Suppose we take the wood products problem above as the primal and suppose that u_1^* is positive in the dual. This means that the imputed value of wood is positive. Then for an optimal solution to the primal, the first constraint is satisfied as an equality so that we would have

$$a_{11}x_1^* + a_{12}x_2^* + a_{13}x_3^* = b_1$$

in (5.25); there would be no unused quantities of wood in the optimal primal program in this case. On the other hand, if the second constraint in the primal were satisfied as a strict inequality with respect to a primal optimal program, then the optimal imputed value of labor, u_2^*, must be zero; we could have supplies of labor unutilized only if the imputed value to the firm of labor were zero.

An analogous complementary slackness property holds for the more general case of duality considered in the next section (see Problem 5.16).

5.2. DUALITY: THE GENERAL CASE.

In order to define duality for more general linear programming problems than those introduced in the previous section, we must first consider some characteristics of a *duality relation* in a more general algebraic setting.

Let Σ be any set of objects. A mapping T of Σ into itself is said to be *involutory* if for every element $a \in \Sigma$, $T[T(a)] = a$. That is, if we map the set Σ into itself by T and then apply the mapping T a second time, the product mapping $T[T(a)]$ must be the identity mapping—the mapping that leaves everything the way it was. The term involutory has its origins in geometry and such concepts as a reflection or a reflection and a rotation of 180 degrees are among the simplest examples. Let Σ be the plane, let T be a reflection of the plane in the 45° line through the origin. Then for any point $a = \begin{bmatrix} a_1 \\ a_2 \end{bmatrix}$,

$$T(a) = T\begin{bmatrix} a_1 \\ a_2 \end{bmatrix} = \begin{bmatrix} a_2 \\ a_1 \end{bmatrix}.$$

If $a = \begin{bmatrix} 5 \\ 2 \end{bmatrix}$, $T(a) = \begin{bmatrix} 2 \\ 5 \end{bmatrix}$ and a is reflected about the 45° line as shown in Figure 5.1. The composite mapping $T[T(a)]$ for any $a \in \Sigma$ is obtained by performing T followed by T. This is clearly the identity transformation, and if $a = \begin{bmatrix} 5 \\ 2 \end{bmatrix}$, $T[T(a)] = \begin{bmatrix} 5 \\ 2 \end{bmatrix}$. Another example is a rotation of the plane

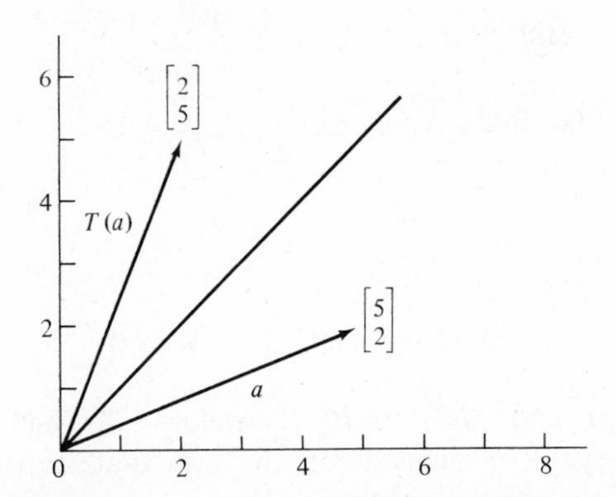

Figure 5.1 An involutory mapping in the plane.

through 180 degrees; the composite or product of two such rotations is the identity mapping. However, the rotation through 90 degrees is not an involutory mapping since its product with itself is not the identity mapping (a discussion of products of functions appears in Appendix A.8).

Let $\sim$ denote an equivalence relation on the set Σ; we say that the mapping T is *equivalence preserving* if $a \sim a'$ implies $T(a) \sim T(a')$ for every $a \in \Sigma$. Also, we say that the mapping T is *meaningful* relative to some relation R on a set Σ if for all $a \in \Sigma$, $aRT(a)$.

For example, let Σ be the set of all linear programming problems each of which has either form (5.1) and (5.2) or the form

$$\text{minimize } S^T U \tag{5.32}$$

subject to

$$KU \geq Q, \qquad U \geq 0, \tag{5.33}$$

and let T be the mapping which sends the primal problem (5.1) and (5.2) into the dual problem (5.3) and (5.4). In more detail, let Σ_1 denote the set of all problems of the form (5.1) and (5.2), let Σ_2 denote the set of all problems of the form (5.32) and (5.33), and let $\Sigma = \Sigma_1 \cup \Sigma_2$. Then T has the effect of interchanging the two subsets Σ_1 and Σ_2 and when applied twice to any problem in Σ gives back that same problem. For example, consider a problem in Σ_1,

$$\text{maximize } P^T V$$

subject to

$$DV \leq R, \qquad V \geq 0.$$

T sends this into the problem in $T(\Sigma_1)$,

$$\text{minimize } R^T U$$

subject to

$$D^T U \geq P, \qquad U \geq 0,$$

a problem of the kind in Σ_2. Applying T to a problem in the set $T(\Sigma_1)$ sends this problem into the problem

$$\text{maximize } P^T V$$

subject to

$$(D^T)^T V \leq (R^T)^T, \qquad V \geq 0,$$

where $(D^T)^T = D$ and $(R^T)^T = R$. Therefore $T[T(\Sigma_1)] = \Sigma_1$, and the involutory property of T depends on the fact that transposing a matrix twice gives back the original matrix.

The mapping T is not only involutory, it is also meaningful with respect to the relation

$$\text{optimal value } a = \text{optimal value } T(a). \tag{5.34}$$

More generally, a mapping T is meaningful if there are useful theorems connecting a and $T(a)$; the duality theorems of linear programming are of this nature.

A mapping T which has all three of the properties discussed above (an involutory, equivalence-preserving, and meaningful mapping) can be regarded as the very "best" type of duality and such a duality is obtained for linear programming under a rather general equivalence relation.

Now consider the set Σ of all linear programming problems in the general form (3.79) through (3.83). We now refer to any such problem as the *primal problem* and in Figure 5.2 we show how to define a new problem in Σ called the *dual problem* related to the given primal. We can characterize the primal problem by means of the $(p_1 + p_2 + 1)$ by $(n_1 + n_2 + 1)$ matrix

$$G = \begin{bmatrix} -d & C_1^T & C_2^T \\ B_1 & A_{11} & A_{12} \\ B_2 & A_{21} & A_{22} \end{bmatrix}$$

together with the sequence of integers (p_1, p_2, n_1, n_2) which indicates the nature of each constraint and variable (we allow for the possibility that some of these integers may be zero). Then the dual problem is characterized by the matrix $-G^T$ together with the sequence (n_1, n_2, p_1, p_2). If the objective in both primal and dual is taken to be one of minimization, there is a change in sign involved and so a negative transposition characterizes the relationship between primal and dual rather than transposition alone as is the case when the primal is taken to be a maximizing problem and the dual a minimizing problem [see, for example, the transposition involved between the primal (5.1), (5.2) and the dual (5.3), (5.4)]. We choose to regard primal and dual as displayed in Figure 5.2 because of its convenience in developing generalizations of duality.

The correspondence between this primal and dual is clearly involutory since $-(-G^T)^T = G$. Each inequality in the primal system corresponds to a nonnegative variable in the dual, and each equation in the primal system corresponds to a free variable in the dual. Corresponding to each nonnegative variable in the primal system is an inequality whose coefficients are the negatives of the coefficients of this variable and whose right-hand side is the negative of the coefficient of the variable in the primal objective function. Each free variable in the primal corresponds to a similarly defined equation in the dual system. The dual objective function has as coefficients the negatives of the right-hand sides of the primal restraints. (The minus sign attached to d in the dual is necessary for uniformity in

Primal	Dual
minimize $C_1^T X_1 + C_2^T X_2 + d = z$	minimize $-B_1^T U_1 - B_2^T U_2 - d = z'$
p_1: $A_{11}X_1 + A_{12}X_2 \geq B_1$	$U_1 \geq 0$
p_2: $A_{21}X_1 + A_{22}X_2 = B_2$	U_2 free
n_1: $X_1 \geq 0$	$-A_{11}^T U_1 - A_{21}^T U_2 \geq -C_1$
n_2: X_2 free	$-A_{12}^T U_1 - A_{22}^T U_2 = -C_2$

Figure 5.2 Dual Systems, general case.

reading relations from the matrix; the initial row represents the equation $z - d = C_1^T X_1 + C_2^T X_2$.)

This duality includes the previous one as a special case if we observe that the problem (5.1) and (5.2) is equivalent to

$$-DV \geq -R, \qquad V \geq 0, \qquad \text{minimize } -P^T V. \tag{5.35}$$

Two special cases of the general linear programming problem are of frequent occurence. The case $p_2 = n_2 = 0$ [compare (5.1) and (5.2)] is called the *symmetric form* and is the one that arises naturally in many applications. The case $p_1 = n_2 = 0$ is called the *standard form* (compare Section 3.2). We recall that reduction to standard form is a first step in preparing the problem for application of the simplex algorithm. The dual of a symmetric problem is again symmetric; however, the dual of a problem in standard form is not in standard form (as the reader should verify).

We saw in Section 3.2 [see (3.32) and (3.33)] how to change from a problem in symmetric form to one in standard form via introduction of slack variables. Thus if we can show that a problem in general form is equivalent to one in symmetric form we will have established that each of the three forms actually includes all problems. To do this we first replace the free vector X_2 by the difference $X_2' - X_2''$ of two nonnegative vectors, and then replace each equation by two oppositely directed inequalities. This yields the symmetric system

$$AX \geq B, \qquad X \geq 0, \qquad \text{minimize } z = d + C^T X \tag{5.36}$$

where

$$A = \begin{bmatrix} A_{11} & A_{12} & -A_{12} \\ A_{21} & A_{22} & -A_{22} \\ -A_{21} & -A_{22} & A_{22} \end{bmatrix},$$

$$B = \begin{bmatrix} B_1 \\ B_2 \\ -B_2 \end{bmatrix}, \qquad C = \begin{bmatrix} C_1 \\ C_2 \\ -C_2 \end{bmatrix}, \qquad X = \begin{bmatrix} X_1 \\ X_2' \\ X_2'' \end{bmatrix}. \tag{5.37}$$

It is easy to verify that the dual of (5.36) is equivalent to the original dual; that is, our duality is equivalence preserving.

More precisely, we might ask in what sense the general primal form is equivalent to (5.36). Clearly any feasible program X for (5.36) defines a feasible program X_1, $X_2 = X_2' - X_2''$ for the original problem and with the same z value. Conversely, if X_1, X_2 is a feasible program for the original problem and if $-t$ is smaller than any component of X_2 then $X_2' = X_2 + t\,E_{p_2}$, $X_2'' = t\,E_{p_2}$ (where for any h, E_h is the vector with h components all equal to one) gives a feasible program X for (5.36) with the same z value. Hence either the two problems have the same optimal value or neither has an optimal program. Alternatively, given X_2 one can find unique X_2', X_2'' with $X_2'^T X_2'' = 0$, $X_2' \geq 0$, $X_2'' \geq 0$, and $X_2 = X_2' - X_2''$. It can be shown, moreover, that the basic programs of the two systems can be put into one to one correspondence; this means that the arguments which justify the simplex method can be extended to show how to obtain all basic solutions of the original problem from the basic solutions of the equivalent problem in standard form.

We give an illustrative example in which $p_1 = 1$, $p_2 = 2$, $n_1 = 3$, $n_2 = 2$.

Primal

$$\begin{aligned}
&\min\ 2x_1 + 3x_2 + 2x_3 - 3x_4 + 7x_5 + 2 = z\\
&\text{subject to}\\
&\quad 2x_1 + 3x_2 - 7x_3 - 4x_4 + x_5 \geq 3,\\
&\quad 3x_1 - x_2 - 3x_3 + 5x_4 + 2x_5 = -7,\\
&\quad 4x_1 + 2x_2 - 6x_3 - 8x_4 - 4x_5 = 2,\\
&\qquad x_1 \geq 0,\\
&\qquad x_2 \geq 0,\\
&\qquad x_3 \geq 0,\\
&\qquad x_4 \text{ free},\\
&\qquad x_5 \text{ free},
\end{aligned}$$

Dual

$$\begin{aligned}
&\min\ -3u_1 + 7u_2 - 2u_3 - 2 = z'\\
&\text{subject to}\\
&\qquad u_1 \geq 0,\\
&\qquad u_2 \text{ free},\\
&\qquad u_3 \text{ free},\\
&\quad -2u_1 - 3u_2 - 4u_3 \geq -2,\\
&\quad -3u_1 + u_2 - 2u_3 \geq -3,\\
&\quad 7u_1 + 3u_2 + 6u_3 \geq -2,\\
&\quad 4u_1 - 5u_2 + 8u_3 = 3,\\
&\quad -u_1 - 2u_2 + 4u_3 = -7.
\end{aligned}$$

The corresponding problem in symmetric form [compare (5.36) and (5.37)] has

$$A = \begin{bmatrix} 2 & 3 & -7 & -4 & 1 & 4 & -1 \\ 3 & -1 & -3 & 5 & 2 & -5 & -2 \\ 4 & 2 & -6 & -8 & -4 & 8 & 4 \\ -3 & 1 & 3 & -5 & -2 & 5 & 2 \\ -4 & -2 & 6 & 8 & 4 & -8 & -4 \end{bmatrix},$$

$$B = \begin{bmatrix} 3 \\ -7 \\ 2 \\ 7 \\ -2 \end{bmatrix}, \qquad C = \begin{bmatrix} 2 \\ 3 \\ 2 \\ -3 \\ 7 \\ 3 \\ -7 \end{bmatrix}.$$

We have shown that our duality is involutory and equivalence preserving so far as transitions from symmetric to standard form are concerned. We conclude this section by stating and proving the general duality theorem for linear programming and showing in the process that duality is meaningful.

We recall that for a general linear programming problem there are three mutually exclusive possibilities: (a) there is a basic optimal feasible program, (b) the problem is feasible but the objective function has no finite lower bound for feasible programs, (c) there is no feasible program. The fundamental duality theorem states:

1. if either primal or dual has an optimal program so does the other;
2. if both primal and dual are feasible then both have optimal programs;
3. if either primal or dual is feasible but with no lower bound for the objective function then the other is infeasible;
4. both primal and dual may be infeasible;
5. if (1) holds then

$$z_{\text{opt}} + z'_{\text{opt}} = 0; \tag{5.38}$$

6. if X_1, X_2 and U_1, U_2 are any feasible programs to the primal and dual, respectively, then

$$z + z' \geq 0. \tag{5.39}$$

We first establish (5.39). We have

$$\begin{aligned} z + z' &= (C_1^T X_1 + C_2^T X_2 + d) - (B_1^T U_1 + B_2^T U_2 + d) \\ &\geq C_1^T X_1 + C_2^T X_2 - U_1^T (A_{11} X_1 + A_{12} X_2) - U_2^T (A_{21} X_1 + A_{22} X_2) \\ &= (C_1^T - U_1^T A_{11} - U_2^T A_{21}) X_1 + (C_2^T - U_1^T A_{12} - U_2^T A_{22}) X_2 \\ &\geq 0, \end{aligned}$$

since the first term is a sum of products of nonnegative factors and the second is 0. Note that we used all of the feasibility conditions for both primal and dual in establishing this chain of inequalities.

Next we assume that the primal problem has an optimal program. More precisely, let

$$M = \begin{bmatrix} -d & C_1^T & C_2^T & 0 \\ B_1 & A_{11} & A_{12} & -I_{p_1} \\ B_2 & A_{21} & A_{22} & 0 \end{bmatrix}$$

be the matrix of the (full) initial tableau and suppose that

$$M^* = \begin{bmatrix} -d^* & C_1^{*T} & C_2^{*T} & C_3^{*T} \\ B_1^* & A_{11}^* & A_{12}^* & A_{13}^* \\ B_2^* & A_{21}^* & A_{22}^* & A_{23}^* \end{bmatrix}$$

is an optimal (final) tableau, where the optimality conditions give us $C_1^{*T} \geq 0$, $C_2^{*T} = 0$, $C_3^{*T} \geq 0$. The equation $C_2^{*T} = 0$ is a consequence of the conclusion concerning Case (b) of Section 3.8. For, consider any free variable x_k. If it is basic in M^*, then $c_k^* = 0$ because of C2 and if it is not basic then $c_k^* = 0$ because of the hypothesis of optimality.

Next let $Q^{-1} = M_{S*}$ be the matrix for which $M = Q^{-1}M^*$, and partition Q into subdivisions the same as for M, say

$$Q = \begin{bmatrix} 1 & Q_{12} & Q_{13} \\ 0 & Q_{22} & Q_{23} \\ 0 & Q_{32} & Q_{33} \end{bmatrix},$$

and let $U_1 = -Q_{12}^T$, $U_2 = -Q_{13}^T$. Then, from $QM = M^*$ and the optimality of M^* we get

$$\begin{aligned} -d - U_1^T B_1 - U_2^T B_2 &= -d^*, \\ C_1^T - U_1^T A_{11} - U_2^T A_{21} &= C_1^{*T} \geq 0, \\ C_2^T - U_1^T A_{12} - U_2^T A_{22} &= C_2^{*T} = 0, \\ U_1^T &= C_3^{*T} \geq 0. \end{aligned} \tag{5.40}$$

From this it follows that U_1, U_2 is a feasible vector for the dual with value of the objective function $z' = -d^* = -z_{\text{opt}}$ and hence

$$z_{\text{opt}} + z' = 0.$$

This together with (5.39) shows that U_1, U_2 is optimal for the dual so (5.38) holds. Since duality is involutory this establishes (1) and (5).

Next we observe that if both problems are feasible then (5.39) shows that neither can fall in Case (b) and this establishes (2). Finally, (3) and (4) together are merely the contrapositive statement of (1) and (2). We give examples to illustrate possibilities (3) and (4). The self-dual problem,

$$\text{minimize } z = -x_1 \text{ subject to the constraints } 0x_1 \geq 1, x_1 \geq 0 \quad (5.41)$$

illustrates (4), and the problem,

$$\text{minimize } z = -x_1 \text{ subject to the constraints } x_1 \geq 1, x_1 \geq 0 \quad (5.42)$$

illustrates (3).

5.3. ACTIVITY ANALYSIS AND DUALITY INTERPRETATIONS.

A more formal development of duality interpretations is sometimes made in activity analysis. One begins with the fundamental concepts of a commodity and an activity. The latter consists of the combination of certain commodities in fixed ratios as inputs to produce certain other commodities as outputs in fixed ratios to inputs. Since an activity is a quantitative concept it has a level and hence a *unit level* or size. An activity can be represented by a vector of input-output coefficients,

$$A_k = \begin{bmatrix} a_{1k} \\ \cdot \\ \cdot \\ \cdot \\ a_{jk} \\ \cdot \\ \cdot \\ \cdot \\ a_{pk} \end{bmatrix}. \quad (5.43)$$

One sometimes begins by defining the a_{jk} in (5.43) as the amount or rate of flow of the jth commodity involved in a *unit level* of the kth activity. Then

$$A_k x_k, \qquad x_k \geq 0$$

represents activity k operated at level x_k. If we assume that all of the activities under consideration are linear (compare Section 2.2) then $A_k x_k$ as well as $A_j + A_k$ may be regarded as representing activities. The activities represented by $A_1, A_2, \cdots, A_n$ are called *fundamental activities* and linear combinations such as $A_k x_k$, $A_j + A_k$, and, more generally,

$$Y = A_1 x_1 + \cdots + A_n x_n = AX,$$

where $A = [A_1 \, A_2 \cdots A_n]$, represent *composite activities*. The vector X whose components x_k represent the levels of the fundamental activities is called the *program vector* or program associated with the composite activity which is represented by Y. Since, in general, we cannot determine an unknown program X uniquely from Y and A a full description of a composite activity requires knowledge both of X and Y. Of course, given the program X, the corresponding activity vector $Y = AX$ is readily calculated. The matrix A is called the technology matrix of the firm.

A sign convention is employed so that the sign of a_{jk} indicates whether the jth commodity is required or consumed in the activity (is an input) or is produced by the activity (is an output). Schematically,

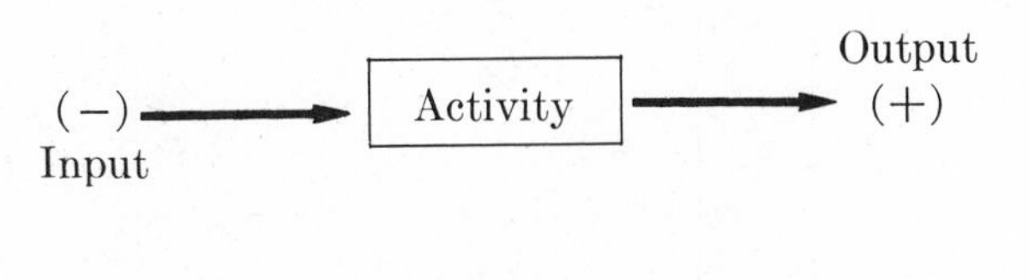

or

$$a_{jk} < 0$$

implies that commodity j is an input for the kth activity,

$$a_{jk} > 0$$

implies that commodity j is an output from activity k, and

$$a_{jk} = 0$$

indicates that commodity j is not involved in the kth activity. Finally, b_j denotes the quantity of the jth commodity available as an input to the activity if $b_j < 0$ and if $b_j > 0$ it denotes the output quantity of the commodity.

We reexamine the transportation problem of Section 2.6 with these definitions and with the sign convention on the components of the activity vectors. Note that the transportation problem is not in symmetric form. It is, however, a special case of the primal given in Figure 5.2, since $p_1 = 0$ and $n_2 = 0$ for this problem. An activity (i, j) consists of shipping one unit of the commodity from origin O_i to destination D_j. For convenience we rewrite the matrix of the equations (2.23) and (2.24) with the appropriate reinterpretations. The results are shown in Table 5.4. Unit level operation of activity (1, 1) can be represented by the vector consisting of the first column of the table. Unit level of this activity consists of shipping one unit from origin O_1 to destination D_1, so that one unit of the commodity is required as an *input* from O_1 into the activity. This results in a -1 in the (1, 1) position. The entry of $+1$ in the $(p + 1, 1)$ position indicates that operating the activity at unit level provides one unit of the commodity to destination D_1 as an *output* (a commodity located at two different places is regarded as two different commodities from the standpoint of this problem). The zeros in the column have an obvious interpretation.

Table 5.4. ACTIVITY ANALYSIS OF TRANSPORTATION PROBLEM.

Activities / *Commodities*	A_{11} $1 \to 1$	A_{12} $1 \to 2$	$\cdots$	A_{1n} $1 \to n$	A_{21} $2 \to 1$	A_{22} $2 \to 2$	$\cdots$	A_{2n} $2 \to n$	$\cdots$	A_{p1} $p \to 1$	A_{p2} $p \to 2$	$\cdots$	A_{pn} $p \to n$
Quantity at O_1	−1	−1	$\cdots$	−1	0	0	$\cdots$	0	$\cdots$	0	0	$\cdots$	0
Quantity at O_2	0	0	$\cdots$	0	−1	−1	$\cdots$	−1	$\cdots$	0	0	$\cdots$	0
.	.	.	.	.	.	.	.	.	.	.	.	.	.
.	.	.	.	.	.	.	.	.	.	.	.	.	.
.	.	.	.	.	.	.	.	.	.	.	.	.	.
Quantity at O_p	0	0	$\cdots$	0	0	0	$\cdots$	0	$\cdots$	−1	−1	$\cdots$	−1
Quantity at D_1	+1	0	$\cdots$	0	+1	0	$\cdots$	0	$\cdots$	+1	0	$\cdots$	0
Quantity at D_2	0	+1	$\cdots$	0	0	+1	$\cdots$	0	$\cdots$	0	+1	$\cdots$	0
.	.	.	.	.	.	.	.	.	.	.	.	.	.
.	.	.	.	.	.	.	.	.	.	.	.	.	.
.	.	.	.	.	.	.	.	.	.	.	.	.	.
Quantity at D_n	0	0	$\cdots$	+1	0	0	$\cdots$	+1	$\cdots$	0	0	$\cdots$	+1

The amount of stock of the commodity at origin O_i is denoted $-b_i$, since it is an input to the system and the amount c_j of the commodity delivered to destination D_j is an output of the system. With this more formal definition of an activity the transportation problem can then be written

$$\text{minimize } z = \sum_{i,j} a_{ij}x_{ij} \tag{5.44}$$

subject to

$$\begin{array}{l}
-x_{11} - x_{12} \cdots - x_{1n} = -b_1, \\
\qquad -x_{21} - x_{22} \cdots - x_{2n} = -b_2, \\
\quad \cdots \\
\qquad\qquad -x_{p1} - x_{p2} \cdots - x_{pn} = -b_p, \\
x_{11} \qquad + x_{21} \qquad + x_{p1} = c_1, \\
\quad x_{12} \qquad + x_{22} \qquad + x_{p2} = c_2, \\
\quad \cdots \\
\qquad x_{1n} \qquad + x_{2n} \qquad + x_{pn} = c_n, \\
\qquad x_{ij} \geq 0 \quad \text{for all } i, j.
\end{array} \tag{5.45}$$

What is the dual of this problem and how is it to be interpreted? Proceeding algebraically, we get as the dual:

$$\text{maximize } -\sum_{i=1}^{p} b_i u_i + \sum_{j=1}^{n} c_j v_j \tag{5.46}$$

subject to

$$\begin{array}{c}
-u_i + v_j \leq a_{ij}, \qquad (i = 1, \cdots, p;\ j = 1, \cdots, n), \\
u_i,\ v_j \text{ free variables.}
\end{array} \tag{5.47}$$

The a_{ij} in (5.47) gives the cost of shipping one unit of the commodity from origin O_i to destination D_j so that the u_i and v_j must be measured in terms of prices per unit and a *dual activity* must be that of assigning prices to a combination of commodities in a way which is operationally equivalent to that of shipping the commodity from O_i to D_j. Suppose then that the manager who is considering problem (5.44) and (5.45) is contacted by a second manager who is in the trucking business and suppose that the latter states that he will buy all the former's stock of the commodity for price u_i per unit at origin O_i and will contract to deliver the commodity to destination D_j in the amount c_j. For this service he will charge a price of v_j per unit delivered at D_j. This is a good deal for the first manager, since

$$v_j \leq a_{ij} + u_i;$$

the trucker now chooses u_i and v_j satisfying (5.47) and which maximize his total returns, that is, which will satisfy (5.46). If both problems have optimal solutions, the trucker maximizes the total return from his activities and the manager is saved the trouble of determining an optimal shipping schedule himself.

Interpretations of duality when the primal has a mixture of constraints (when $p_1 \neq 0$ and $p_2 \neq 0$ in Figure 5.2) are discussed in several places in the literature. An example problem and detailed examination of the use of the dual in decision making in such a problem appears in Reference (20). We do not give further details here.

5.4. DUALITY AND POSTOPTIMALITY ANALYSIS.

After the formulation and solution of a linear programming problem, restraining conditions can change or management may impose additional requirements. The number of units of an input resource may be increased or decreased; for example, in the wood products problem of Chapter 1 a short shipment of wood might be received or an increase in absenteeism might reduce the available supply of labor inputs. In a large firm, the altering of one division's activity may increase the supplies available to other divisions. Quite apart from considerations of these kinds, management may not be willing to accept the full implications of an optimal solution. Again using the wood products problem, an optimal solution to which indicated that no picture frames should be made, management might stipulate that a given number of picture frames must be produced because of product-line and market-penetration considerations. Furthermore, the quality of the data in most applications is rarely what we would like. The model of linear programming requires, strictly speaking, that the costs c_j, the input-output coefficients a_{ij}, and the resource availablities b_i—the givens—be known exactly. This is never fully satisfied. Can we determine which of these givens the model is sensitive to and which givens are associated with low sensitivity? If we knew the answer to this question we could improve our information base for the former and avoid the cost of collecting additional information about the latter. Can we determine a range or interval containing each given such that for any value of the given in the range, an optimal solution (to the dual, say) does not change? For any such change in a primal given, can we quickly recalculate optimal solutions to both primal and dual problems? If such questions can be quickly answered, then the linear programming model can be used by the model builder as a means to refine the model itself; one changes or improves the modeling relative to the sensitive elements, giving less attention to the role of the elements to which the model is relatively insensitive. Answers to such questions could enable a model builder to engage in model refinement of the kind discussed in general terms in Section 2.1.

The procedures we now introduce are part of what is sometimes called postoptimality analysis, marginal analysis, or sensitivity analysis. For the reasons discussed above, postoptimality analysis is fully as important as "optimality analysis" in many applications of linear programming.

It should be clear that one way to engage in postoptimality analysis is to formulate and resolve a modified problem. Substantial economy in time and analysis is often possible if instead we can utilize fully the information available in an optimal solution to the original problem. Moreover, many computer programs now have postoptimality subroutines which enable a user to make many of the changes we will discuss, so the mathematics of postoptimality analysis has practical significance as well.

Our discussion of postoptimality will be grouped into five classes:

1. Changes in an optimal solution as a result of changes made in the level of operation of nonbasic activity k; these changes are further subdivided into cases in which activity k is a slack activity and in which it is a natural or structural activity;
2. Changes in the level of operation of a basic activity;
3. Changes in cost coefficients;
4. Changes in activity coefficients of nonbasic activities;
5. Changes in activity coefficients of basic activities.

We also consider changes in requirements and changes caused by the additional stipulation that a specified amount of a certain output commodity be produced (the latter can be expressed by the addition of a new constraint in the model). Both of these changes will be examined in terms of changes in basic and nonbasic variables.

Suppose we write the linear programming problem introduced in Section 3.1 as

$$\text{minimize } z = d^0 + C^{0T}X \tag{5.48}$$

subject to

$$\begin{aligned} A^0X = A_1^0x_1 + A_2^0x_2 + \cdots + A_n^0x_n &= B^0 \\ x_j &\geq 0 \qquad (j = 1, \cdots, n), \end{aligned} \tag{5.49}$$

where A_j^0 is the jth column of the p by n matrix A^0. Let the problem be nondegenerate, let X^S be an optimal basic program corresponding to the basic sequence $S = (s_1, \cdots, s_p)$. Finally, let M^0 be the initial simplex tableau and M be the final (optimal) tableau, respectively,

$$M^0 = \begin{bmatrix} -d^0 & C^{0T} \\ B^0 & A^0 \end{bmatrix}$$

$$M = \begin{bmatrix} -d & C^T \\ B & A \end{bmatrix}.$$

We use a superscript 0 for the initial tableau and no superscripts for the final tableau in order to simplify the notation required in developing the postoptimality analysis.

The first column of M gives the minimal value of the objective function and the values of the positive coefficients in an optimal program vector, and an optimal solution can be written

$$\begin{aligned} x^S_{s_i} &= b_i, \qquad (i = 1,\cdots, p), \\ x^S_j &= 0, \qquad j \notin S, \\ z^S &= d. \end{aligned} \tag{5.50}$$

We can then reduce (5.49) to the form

$$A^0_{s_1}x^S_{s_1} + A^0_{s_2}x^S_{s_2} + \cdots + A^0_{s_p}x^S_{s_p} = B^0. \tag{5.51}$$

Since $(A^0_{s_1},\cdots, A^0_{s_p})$ is a basis for the vector space V_p, every column of A^0 can be expressed as a linear combination of these vectors or, in the language of activity analysis, every activity in the firm's technology is a combination of these basic activities.

Suppose we now wish to begin a postoptimality analysis by introducing activity k operated at level x'_k, where k is a nonbasic index, into our program and to depart, if necessary, from an optimal solution (5.50). Since k is a nonbasic index, $x_k = 0$ in an optimal solution; this means that no output from activity k is produced. Introducing activity k into a solution can be interpreted as a stipulation by management that some output of activity k be produced because of product line considerations and that departure from optimality is regarded as appropriate, if necessary, in order to do this.

The algebraic effects of this introduction are as follows. We know that

$$A^0_k = \sum_{i=1}^{p} a_{ik}A^0_{s_i},$$

or that

$$A^0_k - \sum_{i=1}^{p} a_{ik}A^0_{s_i} = 0. \tag{5.52}$$

Multiplying (5.52) on both sides by x'_k, adding the result to (5.51), rearranging terms and noting (5.50) results in

$$A^0_{s_1}(b_1 - x'_k a_{1k}) + A^0_{s_2}(b_2 - x'_k a_{2k}) + \cdots + A^0_{s_p}(b_p - x'_k a_{pk}) + x'_k A^0_k = B^0.$$

We obtain a new feasible solution

$$\begin{aligned} x'_{s_i} &= b_i - x'_k a_{ik}, \qquad (i = 1,\cdots, p), \\ x'_j &= 0, \qquad j \notin S, j \neq k, \\ x'_k &\geq 0, \qquad \text{to be selected}, \\ z' &= d + c_k x'_k, \end{aligned} \tag{5.53}$$

provided x'_k is not too large numerically (we will soon identify suitable bounds for x'_k). If the basic sequence $S = (s_1, \cdots, s_p) = (1, \cdots, p)$, it is easy to visualize the old and new solutions as follows,

$$X^S = \begin{bmatrix} b_1 \\ \cdot \\ \cdot \\ \cdot \\ b_p \\ 0 \\ \cdot \\ \cdot \\ \cdot \\ 0 \end{bmatrix}, \qquad X' = \begin{bmatrix} b_1 - x'_k a_{1k} \\ \cdot \\ \cdot \\ \cdot \\ b_p - x'_k a_{pk} \\ 0 \\ \cdot \\ \cdot \\ \cdot \\ x'_k \\ \cdot \\ \cdot \\ \cdot \\ 0 \end{bmatrix}.$$

For any sequence S we have a similar representation except for permutation of components.

It is clear that as activity k is increased by x'_k, the level of activity i must decrease by $x'_k a_{ik}$, $i = 1, \cdots, p$ (note that if $x'_k a_{ik} < 0$, the level of operation of activity i is actually increased). For an arbitrary sequence S, (5.53) gives analagous relationships. The new value z' of z in (5.53) comes about as follows.

$$z' = d + \sum_{i=1}^{p} c_{s_i}(b_i - x'_k a_{ik}) + c_k x'_k;$$

however, $c_{s_i} = 0$ for $i = 1, \cdots, p$, so this becomes

$$z' = d + c_k x'_k,$$

or in view of the last equation of (5.50),

$$z' - z^S = c_k x'_k.$$

Thus c_k is the *marginal cost* corresponding to the introduction of the non-basic activity k at unit level.

In terms of the concepts presented in Section 3.5, c_k, the canonical cost for activity k relative to an optimal basic sequence, is given by $c_k = c_k^0 - z_k^0$ where $z_k^0 = \sum_{i=1}^{p} c_{s_i}^0 a_{ik} = \sum_{i=1}^{p} c_i^0 a_{ik}$ is calculated from the original costs c_i^0

and the final coefficients, here denoted a_{ik}. That is, in terms of the original cost vector C^0 we have

$$z' = d^0 + C^{0T}X' = d^0 + \sum_{i=1}^{p} c_{s_i}^0(b_i - x_k' a_{ik}) + c_k^0 x_k'$$

$$z^S = d^0 + C^{0T}X^S = d^0 + \sum_{i=1}^{p} c_{s_i}^0 b_i$$

$$z' - z^S = C^{0T}X' - C^{0T}X^S = (c_k^0 - \sum_{i=1}^{p} c_{s_i}^0 a_{ik})x_k',$$

and

$$c_k x_k' = (c_k^0 - \sum_{i=1}^{p} c_{s_i}^0 a_{ik})x_k',$$

that is,

$$c_k = c_k^0 - \sum_{i=1}^{p} c_{s_i}^0 a_{ik} = c_k^0 - z_k^0.$$

For each unit of activity k introduced, the ith basic variable is changed from b_i to $x_{s_i}' = b_i - a_{ik}x_k'$, as in (5.53). Therefore the change in cost attributed to the introduction of activity k through its effects on the basic variables x_{s_i} is $-c_{s_i}^0 a_{ik}$. Alternatively, we can regard this as a revenue $c_{s_i}^0 a_{ik}$ and then $z_k^0 = \sum_{i=1}^{p} c_{s_i}^0 a_{ik}$ can be regarded as the *total imputed revenue* attributable to the indirect effects on the basic variables of an introduction of a unit amount of activity k (this is z_k^0). The total effect of this introduction will be $c_k = c_k^0 - z_k^0$, which is the difference in direct unit cost and indirect unit revenue of activity k. When this difference is negative, then the direct cost is less than the imputed revenue; this means that one has a motivation for introducing activity k. This is an economic interpretation of the negative c_k choice criterion in the simplex method developed in Chapter 3. The coefficients a_{ik} are referred to in the economics literature as the marginal rate of substitution of activity k for activity i.

In order to examine the effects of this change further we must consider two cases: Case 1, k is a slack index and Case 2, k is a natural or structural index. Case 2 is simpler so we treat it first.

At this stage of our analysis we find it advantageous to generalize the initial feasibility conditions by no longer requiring the nonnegativity conditions $x_k' \geq 0$ in (5.53). However, natural activities such as building picture frames or eating steak are irreversible so we must still restrict x_k' to nonnegative values for them. An upper bound for x_k' is determined by observing that if any a_{ik} is positive, then x_k' cannot exceed b_i/a_{ik} unless x_{s_i}' becomes negative. Thus we have

$$0 \leq x_k' \leq \min_i \left\{ \frac{b_i}{a_{ik}} \middle| a_{ik} > 0; i = 1, \cdots, p \right\}. \qquad (5.54)$$

If there is no $a_{ik} > 0$, then there is no finite upper bound for x'_k; if this should occur in a practical problem it indicates that some constraining influence may have been overlooked in setting up the model.

For slack activities both positive and negative values of x'_k have meaning. As we noted in Section 1.3, a positive value for a slack variable indicates that not all of some resource has been utilized or that some requirement has been exceeded. A negative value for a slack variable represents an infeasible solution to the original problem but it can be used to account for the effects of a relaxation of constraints in it. For example, to consider what would happen if additional units of some resource should become available we consider negative values for the corresponding slack variable. Thus, if x_k measures labor units then $x_k = -2$ would literally mean a program which exceeded original labor availability by two units and thus (5.53) with $x'_k = -2$ can be used to describe how the optimal program should be altered to take advantage of two added units of labor. If x_k measures excess protein then $x'_k = -2$ would indicate how to adjust an optimal diet best in order to account for a decision to reduce the minimum daily protein requirement by 2 units.

For changes under Case 1, when k is a slack index, the permissible domain for x'_k is given by

$$\max \left\{\frac{b_i}{a_{ik}} | a_{ik} < 0; \quad i = 1, \cdots, p\right\} \leq x'_k \leq \min \left\{\frac{b_i}{a_{ik}} | a_{ik} > 0; i = 1, \cdots, p\right\}. \tag{5.55}$$

Here as in (5.54) we can have either limit infinite. For any x'_k in this interval, (5.53) gives the best adjusted program. In both (5.54) and (5.55) if x'_k reaches its extreme value we have the same result as would be given by a pivotal transformation which brings k into the basic sequence. Any further changes in the kth variable must follow the rule which we develop below for changes in basic variables.

Before discussing changes in basic variables, however, we consider the effect of replacing an initial resource or requirement b_h whose corresponding slack variable is basic by $b_h + \Delta b_h$. If a basic variable x_{s_h} measures the slack in resource h then, clearly, no increase in the initial supply of this resource ($\Delta b_h > 0$) can affect an optimal program or its cost, nor will a decrease ($\Delta b_h < 0$) provided that

$$-\Delta b_h \leq x^S_{s_h}. \tag{5.56}$$

Similar reasoning applies to requirements. For example, if the hth constraint is the minimum protein requirement, then an increase by Δb_h (> 0) in this requirement will have no effect if $\Delta b_h \leq x^S_{s_h}$.

The final tableau matrix M does not indicate the amount of the hth resource that is used in the program. If M^0 denotes the initial tableau matrix, then b^0 is the amount of resource h that was available initially and

$b_h^0 - x_{s_h}^S$ is the amount used in the optimal program. Similarly, for the vitamin illustration, b_h^0 was the original minimum daily requirement and $b_h^0 + x_{s_h}^S$ is the amount of protein actually supplied by the optimal diet.[2]

We now proceed to consider a change in a basic variable. Specifically, let s_h be any basic index and consider the effect of changing $x_{s_h}^S$ by some positive or negative amount τ. According to (5.53) this can be done by choosing some nonbasic index k and letting $\tau = -a_{hk}x_k'$. In this analysis we wish to preserve the original constraints and hence require $x_k' \geq 0$.

For $\tau > 0$ we must choose k such that $a_{hk} < 0$; then the unit cost for increasing x_{s_h} is $-c_k/a_{hk}$ and hence the best k to introduce is given by

$$\frac{-c_k}{a_{hk}} = \min\left\{\frac{-c_j}{a_{hj}} \middle| a_{hj} < 0;\ \ j \in G\right\}. \tag{5.57}$$

Similarly, for τ negative the unit cost and best k are given by

$$\frac{c_k}{a_{hk}} = \min\left\{\frac{c_j}{a_{hj}} \middle| a_{hj} > 0;\ \ j \in G\right\}. \tag{5.58}$$

Whether τ is positive or negative when x_k' takes its upper bound as given by (5.54), say $x_k' = b_s/a_{sk}$, we have for the total increase in cost

$$|c_k\, b_s/a_{sk}|, \tag{5.59}$$

and for further change in x_{s_h} we repeat the entire process beginning with the tableau M^* obtained from M after pivoting at (s, k). We must take into account the fact that M^* is not in optimal form but this does not affect (5.53). The fact that one or more of the c_j^* may be negative gives us no difficulty, provided that we add the requirement $c_k > 0$ in selecting the new k in (5.57) and (5.58).

To analyze effects of changes in costs we observe that the costs in an optimal tableau are related to those in the initial tableau by the equations

$$c_j = c_j^0 - \sum_{i=1}^{p} c_{s_i}^0 a_{ij} = c_j^0 - z_j^0 \qquad (j = 1, \cdots, n),$$

that is, we have $C = C^0 - Z^0$. If C^0 is replaced by the vector $C^0 + \Delta C^0$, where

$$\Delta C^0 = \begin{bmatrix} \Delta c_1^0 \\ \cdot \\ \cdot \\ \cdot \\ \Delta c_n^0 \end{bmatrix}, \tag{5.60}$$

[2]Some authors reserve the term slack variable to denote unused resources and use the term surplus variable to denote the excess over minimal requirements.

then C is replaced by $C + \Delta C$. By the linearity of the problem we have

$$\begin{aligned} C + \Delta C &= (C^0 + \Delta C^0) - (Z^0 + \Delta Z^0) \\ &= C^0 - Z^0 + \Delta C^0 - \Delta Z^0 \\ &= C + \Delta C^0 - \Delta Z^0 \end{aligned}$$

so that

$$\Delta C = \Delta C^0 - \Delta Z^0,$$

that is, for any activity j

$$\Delta c_j = \Delta c_j^0 - \sum_{i=1}^{p} a_{ij}(\Delta c_{s_i}^0) = \Delta c_j^0 - \Delta z_j^0. \tag{5.61}$$

Suppose we investigate the change in cost of a basic activity s_j that would be necessary in order to make the final cost c_k of a nonbasic activity become zero. Such a change in the cost of s_j would then make activity k a candidate to enter a basis. In (5.61) let $\Delta c_k = -c_k$ and $\Delta c_{s_i}^0 = 0$ for $i \neq j$, $i \in S$; then

$$-c_k = -a_{jk}\Delta c_{s_j}^0$$

and

$$\Delta c_{s_j}^0 = \frac{c_k}{a_{jk}}. \tag{5.62}$$

A change in cost of activity s_j equal to c_k/a_{jk} would thus result in a reduction of the final or canonical cost of activity k to zero. Hence if $\Delta c_{s_j}^0$ is in the interval

$$\max\left\{\frac{c_k}{a_{jk}} \middle| a_{jk} < 0,\ k \notin S\right\} \leq \Delta c_{s_j}^0 \leq \min\left\{\frac{c_k}{a_{jk}} \middle| a_{jk} > 0,\ k \notin S\right\} \tag{5.63}$$

an optimal basic sequence S and corresponding optimal program will remain unchanged and z^S will be replaced by $z^S + \Delta c_{s_j}^0 b_j$.

When $\Delta c_{s_j}^0$ is equal to either of the limits in (5.63), two optimal basic programs are available. The second can be obtained by pivoting in column s_j. Any change exceeding the bounds in (5.63) will cause some nonbasic c_k to become negative in the tableau, so that it is no longer optimal. One must then pivot in an appropriate column and proceed to determine a new optimal program.

Changes in c_k^0 for k nonbasic can also be examined by using (5.61). We recall that if c_k^0 is replaced by $c_k^0 + \Delta c_k^0$, then c_k is replaced by $c_k + \Delta c_k^0$. The value $\Delta c_k^0 = -c_k$ is the break-even value and represents the amount by which c_k^0 must be changed to make the kth activity competitive with the basic activities. Thus in the wood products problem of Section 1.3 a decrease of \$5.00 in the cost of a picture frame, in other words, an increase of \$5.00 in the sales price, will be just enough to bring the activity of manufacturing picture frames up to a break-even level with other activities that are being operated at positive levels.

Up to this point we have discussed effects of changing activity levels and of changes in the original requirements (the constant terms b_i^0) as well as changes in the unit costs of activities. We now turn to postoptimality analysis of changes in the activity coefficients a_{ij}^0. Changes in an activity coefficient for a nonbasic activity k have no effect on an optimal program unless c_k becomes negative when we form the matrix product QM^0 to get M [see (4.6)]. The a_{ik} in (5.53) will have to be adjusted, of course, and if c_k becomes negative, further pivots, beginning in column k, will be required to obtain a new optimal program. Changes in a_{ik}^0 for a basic activity k require adjustments in Q and hence possibly in each entry of M; small changes in a_{ik}^0 may not, however, change an optimal basic sequence.

We consider in detail the case of changes in nonbasic activity coefficients. Again we recall that M and $S = (s_1, \cdots, s_p)$ denote optimal tableau and corresponding basic sequence and M^0 and $S^0 = (s_1^0, \cdots, s_p^0)$ denote initial tableau and corresponding basic sequence. In this notation we have $Q = (M_S^0)^{-1}$ and corresponding to (4.15) we have

$$M = (M_S^0)^{-1} M^0, \tag{5.64}$$

where

$$M_S^0 = \begin{bmatrix} 1 & C_S^{0T} \\ 0 & A_S^0 \end{bmatrix}, \qquad (M_S^0)^{-1} = \begin{bmatrix} 1 & -C_S^{0T}(A_S^0)^{-1} \\ 0 & (A_S^0)^{-1} \end{bmatrix}.$$

The element c_k of M in (5.64) is then given by

$$c_k = c_k^0 - C_S^{0T}(A_S^0)^{-1}A_k^0, \qquad k \notin S. \tag{5.65}$$

For k nonbasic in an optimal tableau we see that as A_k^0 changes the matrices A_S^0 and $(A_S^0)^{-1}$ do not change. Specifically, if A_k^0 is replaced by $A_k^0 + D$, then c_k is replaced by $c_k + \Delta c_k$, where

$$\Delta c_k = -C_S^{0T}(A_S^0)^{-1}D; \tag{5.66}$$

a given program remains optimal provided that

$$c_k + \Delta c_k \geq 0$$

and a simplex iteration in which activity k will become basic is called for if

$$c_k + \Delta c_k < 0.$$

If M^0 is in canonical form relative to S^0, then (5.64) can be written

$$M = M_{S0}M^0, \tag{5.67}$$

where, of course,

$$M_{S0} = \begin{bmatrix} 1 & C_{S0}^T \\ 0 & A_{S0} \end{bmatrix}.$$

This means that $-C_S^{0T}(A_S^0)^{-1} = C_{S0}^T$ and $(A_S^0)^{-1} = A_{S0}$; (5.65) is then simplified to

$$c_k = c_k^0 + C_{S0}^T A_k^0 \tag{5.68}$$

and (5.66) becomes

$$\Delta c_k = C_{S0}^T D. \tag{5.69}$$

As we observed earlier, changes in activity coefficients for an activity whose index is in an optimal basic sequence induce changes in $(M_S^0)^{-1}$. It is clear from (5.64) that each entry of the final tableau matrix M may be affected. We do not discuss such changes here. A discussion of changes in one a_{ik}^0 for a basic index k is given in Reference (13, pp. 271–275); a more detailed discussion of changes in activity coefficients appears in Reference (48).

5.5. POSTOPTIMALITY ANALYSIS: ILLUSTRATIVE EXAMPLES.

The algebraic developments of the previous section will now be illustrated by means of simple examples. The XYZ Metal Products Company produces four products as outputs, all made of metal: serving trays, ash trays, bookshelves, and lunch boxes. It uses two commodities as inputs, sheet metal and labor. Input-output relationships, resource availabilities, and sales prices of outputs are shown in Table 5.5.

Table 5.5. DATA FOR XYZ METAL PRODUCTS COMPANY.

Outputs / *Inputs*	*Serving trays*	*Ash trays*	*Book-shelves*	*Lunch boxes*	*Resource (input) availabilities*
Sheet metal	5	1	15	5	100
Labor	3	2	7	5	125
Sales price per unit (dollars)	2	1	10	4	

We make appropriate linearity assumptions and develop the following linear programming model for maximization of sales revenue:

$$\text{maximize } w = 2x_1 + x_2 + 10x_3 + 4x_4$$

subject to

$$\begin{aligned} 5x_1 + x_2 + 15x_3 + 5x_4 &\leq 100, \\ 3x_1 + 2x_2 + 7x_3 + 5x_4 &\leq 125, \\ x_j &\geq 0, \qquad (j = 1, \cdots, 4), \end{aligned}$$

where x_1 = number of serving trays produced, x_2 = number of ash trays produced, x_3 = number of bookshelves produced, and x_4 = number of lunch boxes produced. Table 5.6 shows the initial and final (optimal) tableaus. As usual, we have minimized $z = -w$; x_5 and x_6 are slack variables for metal and labor, respectively.

Table 5.6. INITIAL AND OPTIMAL TABLEAUS.

		x_1	x_2	x_3	x_4	x_5	x_6	
	0	−2	−1	−10	−4	0	0	
M^0	100	5	1	15	5	1	0	$S^0 = (5, 6)$
	125	3	2	7	5	0	1	
	85	8/5	0	2/5	0	3/5	1/5	
M	15	7/5	0	23/5	1	2/5	−1/5	$S = (4, 2)$
	25	−2	1	−8	0	−1	1	

The basic sequence S for the optimal tableau is $S = (4, 2)$, an optimal program is $(X^S)^T = [0\ 25\ 0\ 15\ 0\ 0]$ and minimum $z^S = -85$ so that optimal sales revenue is 85. Activities 1, 2, 3, and 4 are natural or structural activities and relative to the optimal program X^S activities 4 and 2 are basic (the company produces lunch boxes and ash trays) and activities 1 and 3 are nonbasic.

We first consider changes made in an optimal program which take the form of introducing the kth (nonbasic) activity, operated at level x'_k, into our program. From (5.53) we have for a new feasible solution

$$\begin{aligned} x'_{s_i} &= b_i - x'_k a_{ik}, \\ x'_j &= 0, \qquad j \notin S, \qquad j \neq k, \\ x'_k &\quad \text{to be determined}, \\ z' &= d + c_k x'_k. \end{aligned}$$

If the kth natural activity is nonbasic, we have bounds on x'_k from (5.54),

$$0 \leq x'_k \leq \min \left\{ \frac{b_i}{a_{ik}} | a_{ik} > 0; i = 1, \cdots, p \right\}.$$

Suppose we wanted to know how much the level of activity 1 (production of serving trays) could change while still enabling us to determine an alternate solution from (5.53). In this example we obtain

$$0 \leq x'_1 \leq 10\frac{5}{7};$$

so for any increase in level of activity 1 up to and including $x'_1 = 10\ 5/7$ a new feasible solution is given by (5.53); see Section 1.3 for a discussion of fractions in feasible production programs. For example, suppose management states that 10 serving trays must be produced, that is, $x_1' = 10$. Then from (5.53) we get

$$x_4' = 15 - \left(\frac{7}{5}\right)(10) = 1,$$

$$x_2' = 25 - (-2)(10) = 45,$$

$$z' = -85 + \left(\frac{8}{5}\right)(10) = -69.$$

An optimal feasible program under the new conditions is $X'^T = [10 \quad 45 \quad 0 \quad 1 \quad 0 \quad 0]$ and corresponding maximum sales revenue is 69. To put it another way, the stipulation by management that 10 serving trays must be produced causes maximum revenue to be reduced by $(8/5)10 = 85 - 69 = 16$.

It should also be noted that the same result can be obtained by introducing a new constraint of $x_1 = 10$ in the original problem and resolving the resulting (larger) problem. This requires several iterations, even for this simple situation, whereas we obtain the same results quickly, as well as knowing bounds on changes for x_k', by the postoptimality analysis. The economy of the latter is clearly considerable for large problems.

One might ask, what would occur if the level of operation of activity 1 were increased to $x_1' = 10\ 5/7$, its upper bound? Then from (5.53) we get

$$x_4' = 15 - \left(\frac{7}{5}\right)\left(\frac{75}{7}\right) = 0,$$

$$x_2' = 25 - (-2)\left(\frac{75}{7}\right) = 46\frac{3}{7},$$

$$x_1' = 10\frac{5}{7},$$

$$z' = -67\frac{6}{7} \qquad \text{or} \qquad w = 67\frac{6}{7}.$$

When x_1' assumes its upper limit there is a basis change; activity 1 comes into the basis, activity 4 goes out of the basis, there is a new basic sequence $S' = (1, 2)$ and a new basic solution (optimal if $x_1 = 10\ 5/7$ is required)

$$X^{S'} = \begin{bmatrix} 10\frac{5}{7} \\ 46\frac{3}{7} \\ 0 \\ 0 \\ 0 \\ 0 \end{bmatrix}, \qquad z^{S'} = -67\frac{6}{7}.$$

Suppose now that we return to the original optimal tableau in Table 5.6 and consider the effects of changes in activity level x'_k when k is a nonbasic, slack activity. We recall that a positive value for a slack activity level means that the amount of a given resource to be used is less than the supply available and that a negative value means that the requirement for some resource exceeds the supply; alternatively, a negative value for a slack variable can be used to account for an increase in the supply of the corresponding resource. Suppose that a new inventory control system has been installed at XYZ Company and as a result management has learned that the firm has a different number of units of sheet metal available for use in production than was previously thought (i.e., $b_1 \neq 100$). Analyzing this is equivalent to introducing activity 5 (giving x_5 a nonzero value) in accordance with (5.53). The bounds for any nonbasic, slack activity are given by (5.55); for $k = 5$ these bounds are

$$-25 \leq x'_5 \leq \frac{75}{2}.$$

A value $x'_5 = -22$, say, would correspond to the result that the new inventory control system disclosed that there are 22 more units of sheet metal available; an $x'_5 = 15$ would mean that the supply was 15 units less than originally specified. For example, suppose $x'_5 = -10$; then

$$\begin{aligned} x'_5 &= -10, \\ x'_4 &= 15 - \left(\frac{2}{5}\right)(-10) = 19, \\ x'_2 &= 25 - (-1)(-10) = 15, \\ z' &= -85 + \left(\frac{3}{5}\right)(-10) = -91 \qquad \text{or} \qquad w' = 91, \end{aligned}$$

and a new feasible program is $X'^T = [0\ 15\ 0\ 19\ -10\ 0]$. Again, the same result can be obtained by changing b_1 in the original problem from 100 to 110 and resolving the resulting problem.

What would happen if the upper limit of 75/2 on x'_5 were reached (in other words, if this much less sheet metal were available)? We have

$$\begin{aligned} x'_5 &= \frac{75}{2}, \\ x'_4 &= 15 - \left(\frac{2}{5}\right)\left(\frac{75}{2}\right) = 0, \\ x'_2 &= 25 - (-1)\left(\frac{75}{2}\right) = \frac{125}{2}, \\ z' &= -85 + \left(\frac{3}{5}\right)\left(\frac{75}{2}\right) = -62\frac{1}{2}, \end{aligned}$$

and a new basic, optimal program $X'^T = [0\ 125/2\ 0\ 0\ 75/2\ 0]$. In terms of our original problem, this change in resource availabilities means that only ash trays will be produced, that we will have unused sheet metal after executing our optimal program, and optimal profit will be reduced relative to that in the original statement of the problem.

Suppose a change in the activity level for a basic activity is to be evaluated. Let $\tau = -a_{hk}x'_k$ in (5.53) and restrict $x'_k \geq 0$. The change in cost, Δz, is

$$\Delta z = c_k x'_k = c_k \frac{\tau}{-a_{hk}} = -\left(\frac{c_k}{a_{hk}}\right)\tau.$$

The best activity to introduce is given by (5.57) for $\tau > 0$, and by (5.58) for $\tau < 0$; moreover, when x'_k assumes its upper bound from (5.54), the total cost increase is given by (5.59). For example, suppose management specifies that the production of ash trays must be increased by $\tau = 5$ units. Our original optimal solution called for the production of 25 ash trays (see Table 5.5), so we want to increase x_2 from 25 to $x'_2 = 30$. The best activity k to introduce is then determined so that

$$\frac{-c_k}{a_{hk}} = \min\left\{\frac{-8/5}{-2}, \frac{-2/5}{-8}, \frac{-3/5}{-1}\right\}$$
$$= \frac{1}{20},$$

which indicates that $k = 3$. From (5.53) we have

$$x'_{s_i} = b_i - a_{ik}x'_k. \tag{5.70}$$

To interpret this, we recall that the original basic sequence is $S = (4, 2)$, that is, $s_1 = 4$ and $s_2 = 2$. Hence, when $k = 3$, and $s_2 = 2$, (5.70) becomes

$$x'_2 = b_2 - a_{23}x'_3.$$

Since we have $x'_2 = 30$ as stipulated and since a_{23} in the final tableau of Table 5.5 is -8, this can be written

$$30 = 25 - (-8)x'_3$$

or

$$x'_3 = \frac{5}{8}.$$

Similarly, for $k = 2$ and $s_1 = 4$, (5.70) becomes

$$x'_{s_1} = b_1 - x'_k a_{1k}$$

or

$$x'_4 = 15 - \left(\frac{23}{5}\right)x'_3$$

$$x'_4 = 15 - \left(\frac{23}{5}\right)\left(\frac{5}{8}\right) = \frac{97}{8}.$$

The resulting new feasible program is

$$X'^T = [0\ 30\ 5/8\ 97/8\ 0\ 0] \tag{5.71}$$

and $z' = -84\ 3/4$ or $w = 84\ 3/4$. Again we note that the same result would have been obtained had we set $x_2 = 30$ in the initial problem and solved the resulting problem.

The new program (5.71) is feasible but it is not a basic program for the original problem. Suppose management wanted to know how high the level of activity 2 (producing ash trays) could become while producing in addition to this only bookshelves. This corresponds to finding how large x_2 can become while introducing activity 3 into the basis. To put it in still another way, how can we induce a change in the basic sequence $S = (4, 2)$ to $S' = (3, 2)$? A basis change which introduces activity 3 occurs when the value on the right in (5.54) is assumed by x_3', that is, the upper limit of

$$0 \leq x_3' \leq \frac{15}{23/5}.$$

Hence, a basis change occurs when $x_3' = 75/23$. Since x_3 was zero in the initial optimal program, the basis change means that activity 3 is increased by $x_3' = 75/23$. Activity 2 is then adjusted to the new level,

$$x_2' = 25 + 8\left(\frac{75}{23}\right) = \frac{1175}{23}, \tag{5.72}$$

and a new basic program is

$$X'^T = \begin{bmatrix} 0 & \frac{1175}{23} & \frac{75}{23} & 0 & 0 & 0 \end{bmatrix}.$$

The increase in the level of operation of activity 2 is $1175/23 - 25$, of course, since the level of this activity was 25 in the initial optimal program. This difference is $600/23$ and is equal to the value of τ calculated from

$$\tau = -a_{hk}x_k' = -(-8)\left(\frac{75}{23}\right) = \frac{600}{23}.$$

Also, to verify that these changes reduce x_4 to zero (take activity 4 out of the basis), we have from a suitable interpretation of (5.53)

$$x_4' = 15 - \frac{23}{5}x_3',$$

$$x_4' = 0.$$

The new value of z can be calculated to be

$$z = -85 + \left(\frac{2}{5}\right)\left(\frac{75}{23}\right) = -\frac{1925}{23}.$$

Still another way of examining the change in basis from $S = (4, 2)$ to $S' = (3, 2)$ is to return to the tableau matrix M in Table 5.6 and pivot in column 3. Table 5.7 shows the results of this operation performed on M and the new (optimal) tableau matrix M'. It is readily seen from the latter that the new basic (and optimal) solution is the same as that determined in the discussion immediately above.

Table 5.7. ILLUSTRATING BASIS CHANGE FROM S TO S′.

	85	8/5	0	2/5	0	3/5	1/5	
M	15	7/5	0	23/5	1	2/5	−1/5	$S = (4, 2)$
	25	−2	1	−8	0	−1	1	
	1925/23	170/115	0	0	−2/23	65/115	25/115	
M'	75/23	7/23	0	1	5/23	2/23	−1/23	$S' = (3, 2)$
	1175/23	10/23	1	0	40/23	−7/23	15/23	

Let us consider postoptimality analysis of changes in the costs of activities. We begin with changes in cost of a nonbasic activity k. We use the notation of the previous section, where c_j^0 denotes the cost of activity j in the initial tableau and c_j denotes the corresponding cost in the final or optimal tableau. If k denotes a nonbasic activity then $c_k \geq 0$. Also if c_k^0 is replaced by $c_k^0 + \Delta c_k^0$ then the corresponding final cost will be $c_k + \Delta c_k^0$. For a change in original cost of $\Delta c_k^0 = -c_k$, the new (final) cost will be zero. In this case, activity k is a candidate to enter the basis—there is an alternate optimal basic program in which the level of operation of activity k is not zero—and the optimal value of z is unchanged. For $\Delta c_k^0 > -c_k$ our optimal solution is unchanged and for $\Delta c_k^0 < -c_k$, the new value of c_k is negative and one must pivot in column k to obtain a new optimal tableau and new optimal solution.

These comments are easily illustrated by reference to Table 5.6. Suppose c_1^0 is replaced by $c_1^0 + \Delta c_1^0$, where $\Delta c_1^0 = -1$. Since $c_1^0 = -2$ and $c_1 = 8/5$, we have $\Delta c_1^0 = -1 > -8/5$, so our optimal program is unchanged. Replacing $c_1^0 = -2$ with $c_1'^0 = -1$ would produce an optimal tableau identical to that in Table 5.6 except that $c_1' = 8/5 - 1 = 3/5$ instead of 8/5.

Suppose $\Delta c_1^0 = -8/5$ and that $c_1^0 = -2$ is replaced by $-2 - 8/5 = -18/5$. Then $c_1 = 0$ as shown in Table 5.8. Multiple optimal solutions are then indicated and to determine an alternate optimal solution we can pivot in the column corresponding to activity 1 in the first tableau in Table 5.8. There are two basic sequences, $S = (4, 2)$ and $S' = (1, 2)$; corre-

Table 5.8. ILLUSTRATING CHANGE IN C_1^0 OF $\Delta C_1^0 = -8/5$.

85	0	0	2/5	0	3/5	1/5
15	7/5	0	23/5	1	2/5	−1/5
25	−2	1	−8	0	−1	1
85	0	0	2/5	0	3/5	1/5
75/7	1	0	23/7	5/7	2/7	−1/7
325/7	0	1	−10/7	10/7	−3/7	5/7

sponding optimal basic solutions are $(X^S)^T = [0\ 25\ 0\ 15\ 0\ 0]$ and $(X^{S'})^T = [75/7\ 325/7\ 0\ 0\ 0\ 0]$, respectively. The reader can verify that minimum $z = -85$ for each solution. For $\Delta c_1^0 < -8/5$ the new value of c_1 is negative; a pivot operation in column 1 is indicated which will result in a new optimal solution.

For changes in the cost associated with a basic activity, an optimal program will be maintained until the cost change causes the cost of some nonbasic activity to become zero in an optimal tableau. At this point that nonbasic activity can be introduced into a basis. For an example, suppose one wished to determine the bounds (5.63) on changes Δc_2^0 in the cost coefficient of the basic activity 2. In the example of Table 5.6 this cost coefficient is the number -1. We have

$$\max\left\{\frac{8/5}{-2}, \frac{2/5}{-8}, \frac{3/5}{-1}\right\} \leq \Delta c_2^0 \leq \min\left\{\frac{1/5}{1}\right\} \quad \text{or} \quad -\frac{1}{20} \leq \Delta c_2^0 \leq \frac{1}{5}.$$

Thus

$$\min\left\{\frac{c_k}{a_{jk}} \middle| a_{jk} > 0, \quad h \notin S\right\} = \frac{c_6}{a_{26}};$$

this means that a change of 1/5 in the cost $c_2^0 = -1$ will make $c_6 = 0$ in an optimal tableau; activity 6 then becomes a candidate to enter a basis. If this change is made, one optimal program is $X'^T = [0\ 25\ 0\ 15\ 0\ 0]$ and the corresponding new value of the cost function is $z' = z + \Delta c_k^0 b_k = -85 + (1/5)(25) = -80$.

To find a second optimal basic solution, start with the original tableau modified by the insertion of new cost coefficient $c_2^0 = -1 + 1/5 = -4/5$. The optimal basic sequence is $S = (4, 2)$. Since we want to introduce activity 6 into a basis, a new basic sequence will be $S' = (4, 6)$. The necessary modifications are easy to make in the illustration in Table 5.9, since we can pivot in the column corresponding to activity 4 in the modified original tableau. This operation produces a new optimal tableau in Table 5.9 wth basic sequence $S'' = (4, 6)$ as desired. A second optimal program is $X''^T = [0\ 0\ 0\ 20\ 0\ 25]$ and we have $z'' = -80$. Note that although we pivoted in the final tableau, we nonetheless used information from the original tableau of the original problem, specifically, that an original optimal basic sequence was $S = (4, 2)$. It should also be noted that $c_2 = 0$

Table 5.9. ILLUSTRATING EFFECT OF CHANGE IN COST OF BASIC ACTIVITY.

0	−2	−4/5	−10	−4	0	0
100	5	1	15	5	1	0
125	3	2	7	5	0	1
80	2	0	2	0	4/5	0
20	1	1/5	3	1	1/5	0
25	−2	1	−8	0	−1	1

in the final tableau in Table 5.9, so an alternate optimal solution exists. We can calculate it by pivoting in position (2, 2) of this tableau; the resulting optimal program is the vector $X'^T = [0\ 25\ 0\ 15\ 0\ 0]$ that we determined earlier.

We now turn to a postoptimality analysis of nonbasic activity coefficients. From Table 5.6 we see that M^0 is in canonical form relative to S^0. Activity 3, producing bookshelves, is not basic relative to an optimal sequence S. Suppose we regard a_{13}^0 as a parameter and seek to determine a range of values of a_{13}^0 such that it would be profitable to produce bookshelves at the original unit sales price of \$10.00. Activity 3 would become basic after a simplex iteration if $c_3 + \Delta c_3 < 0$; using (5.68) and (5.69) this is indicated when

$$c_3 + \Delta c_3 = -10 + \begin{bmatrix} \frac{3}{5} & \frac{1}{5} \end{bmatrix} \left[\begin{bmatrix} 15 \\ 7 \end{bmatrix} + \begin{bmatrix} d_1 \\ 0 \end{bmatrix} \right] < 0,$$

that is,

$$d_1 < \frac{2}{3}.$$

This means that if the parameter value $a_{13}^0 < 15 - 2/3 = 43/3$ it is then profitable to produce bookshelves.

For another example, suppose both activity coefficients in activity 3 change as specified by the vector

$$D = \begin{bmatrix} d_1 \\ d_2 \end{bmatrix} = \begin{bmatrix} -7 \\ -2 \end{bmatrix}.$$

Is it now profitable to produce bookshelves? We have

$$c_3 + \Delta c_3 = 10 + \begin{bmatrix} \frac{3}{5} & \frac{1}{5} \end{bmatrix} \left[\begin{bmatrix} 15 \\ 7 \end{bmatrix} + \begin{bmatrix} -7 \\ -2 \end{bmatrix} \right] = -\frac{21}{5} < 0,$$

so activity 3 can then be made basic by means of a simplex iteration and bookshelves will be produced.

PROBLEMS

5.1. Write out the dual of Problems 3.4 and 3.6.

5.2. Write out the dual of Problems 3.20 and 4.12.

5.3. Write out and interpret the dual variables and dual activities of Problem 2.3.

5.4. Write out and interpret the dual of Problem 2.12. What are the dual activities and dual variables?

5.5. Write out and interpret the dual of Problem 2.13. What are dual activities and dual variables?

5.6. Can you develop an interpretation of the dual of the blending problem stated in Section 2.3?

5.7. Review the transportation and assignment problems of Section 2.6. Give an interpretation of the dual of each of these problems.

5.8. Give the following information:

$$\text{maximize } z = x_1 + 5x_2$$

subject to

$$\begin{aligned} 5x_1 + 6x_2 &\leq 30, \\ 3x_1 + 2x_2 &\leq 12, \\ -x_1 + 2x_2 &\leq 10, \\ x_j &\geq 0, \qquad (j = 1, 2). \end{aligned}$$

Adding slack variables we get the following tableaus.

0	−1	−5	0	0	0
30	5	6	1	0	0
12	3	2	0	1	0
10	−1	2	0	0	1
25	−7/2	0	0	0	5/2
0	8	0	1	0	−3
2	4	0	0	1	−1
5	−1/2	1	0	0	1/2
25	0	0	7/16	0	19/16
0	1	0	1/8	0	−3/8
2	0	0	−1/2	1	1/2
5	0	1	1/16	0	5/16

(a) What is the dual of the initial problem?
(b) Determine an optimal solution to the dual from the information given above.
(c) Is the primal problem degenerate? Is the dual problem degenerate?
(d) Solve the dual problem and verify that your solution to Problem 5.8(b) is an optimal solution. Does the dual have other optimal solutions?
(e) What is the dual of the problem after slack variables have been added? Is this equivalent to the dual of the initial problem?

5.9. In Problem 3.7 the objective function has no lower bound over the feasible set. Construct the dual problem and verify that the dual has no feasible solutiong.

5.10 The following questions refer to the illustrative example discussed in Section 5.5.

(a) Suppose management stipulates that 5 bookshelves must be produced and that we must, if necessary, depart from an optimal program

$[0 \quad 25 \quad 0 \quad 15 \quad 0 \quad 0]^T$. Determine an alternate program and maximum sales revenue.

(b) Suppose that after determining an optimal solution as in Table 5.6 the cost of activity 1 changed from -2 to $-.75$. Would a change result in the optimal program? Suppose, alternatively, that the cost of activity 4 were to change from -4 to -1. What would happen?

(c) Trace the effects on the optimal program given in Table 5.6 if the input-output coefficients were to change so that a_{11}^0 were to increase to 7 and a_{21}^0 were to be simultaneously reduced to 2.

5.11. In the following tableau activities 1 through 5 are natural activities and activities 6, 7 and 8 are slack activities corresponding to raw materials A, B, and C, respectively. Find a new optimal program and cost in each of the following cases.

200	0	0	0	2	1	9	6	5
12	0	1	0	3/5	8/5	11	−10	3
10	0	0	1	2/5	1	8	14	−12
17	1	0	0	−1	−4	−10	2	15

(a) The supply of raw material A is increased by one unit.
(b) The supply of raw material C is decreased by one unit.
(c) x_4 is to be increased by two units.
(d) x_1 is to be increased by one unit.
(e) x_2 is to be decreased by two units.

5.12. The following represent input-output table and initial and optimal tableaus for a firm.

Outputs / *Inputs*	O_1	O_2	O_3	*Resource (input) availabilities*
Resource X	5	5	10	1,000
Resource Y	10	8	5	2,000
Resource Z	10	5	0	500
Sales revenue per unit of output	100	200	50	

0	−100	−200	−50	0	0	0
1,000	5	5	10	1	0	0
2,000	10	8	5	0	1	0
500	10	5	0	0	0	1
22,500	275	0	0	5	0	35
50	−1/2	0	1	1/10	0	−1/10
950	−7/2	0	0	−1/2	1	−11/10
100	2	1	0	0	0	1/5

(a) If sales revenue per unit of output O_1 increases to 175 what effect will this have on an optimal solution?
(b) Determine a range of values for revenue per unit of O_1 within which the original optimal solution will not change.
(c) Suppose the revenue per unit of O_2 is reduced to 115. What is the effect of this on the original optimal program and optimal profit?
(d) What would occur if an additional supply of 350 units of resource X were to become available to the firm?
(e) What would occur if management were to stipulate that 75 units of output O_1 must be produced?
(f) Suppose the activity vector O_1 were to be replaced by the vector

$$O_1' = \begin{bmatrix} 6 \\ 8 \\ 12 \end{bmatrix};$$

what would happen to the firm's optimal program?

5.13. Construct a linear programming problem such that neither primal nor dual has feasible solutions.

5.14. In the following tableaus activities 1 through 4 are natural activities and activities 5, 6, and 7 are slack activities corresponding to input resources A, B, and C respectively.

0	−20	−50	−80	−50	0	0	0
50	2	4	4	2	1	0	0
25	1	1	3	3	0	1	0
40	2	3	5	3	0	0	1
2000/3	40/3	0	10/3	0	0	0	50/3
35/3	1/3	1	1/3	0	1/2	0	−1/3
25/3	−1/3	0	−4/3	0	1	1	1/3
5/3	1/3	0	4/3	1	−1/2	0	2/3

(a) By how much can the cost of activity 1 change without this activity becoming basic?
(b) If the level of operation of activity 3 is to be increased by 4 units, what happens to optimal program and optimal costs?
(c) Suppose a_{14}^0 changes from 2 to 3; what are the effects of this?
(d) If the supply of resource B is increased by 5 units, what will be a new optimal solution?
(e) Suppose there are no changes at all in the initial problem; can activity 5 be operated at a nonzero level? If so, what is this level?
(f) Suppose the level of operation of activity 2 is to be reduced by 3 units; determine a new optimal solution and optimal cost.

5.15. Verify that in the following problem the objective function has no finite lower bound. Construct the dual of the problem and verify by using the simplex algorithm on the dual that the dual problem is infeasible.

$$\text{Minimize } z = -3x_2 + x_5 + 2x_6$$

subject to

$$\begin{aligned}
4x_2 + x_3 \qquad - 2x_5 \qquad &= 12, \\
x_1 - x_2 \qquad + 3x_5 + 2x_6 &= 7, \\
3x_2 \qquad + x_4 - 4x_5 + 8x_6 &= 10, \\
x_j &\geq 0, \qquad (j = 1, \ldots, 4), \\
x_5, x_6 &\text{ free.}
\end{aligned}$$

5.16. Given primal and dual problems as defined in Figure 5.2, prove that if X^* and U^* are feasible, they are optimal if and only if they satisfy the equations

$$U_1^T(A_{11}X_1 + A_{12}X_2 - B_1) = 0,$$
$$X_1^T(C_1 - A_{11}^T U_1 - A_{21}^T U_2) = 0.$$

5.17. This is a continuation of Problem 3.56.

Oleum Company Ltd. (B)[3]

William Graham, Oleum's managing director responsible for Far Eastern Area operations, had, in August 1964, just received from Charles White, manager of planning, a program for guiding the area's activities in 1965. The plan outlined crude purchases and transportation requirements, suggested refinery crude runs and process conditions, and specified a product distribution pattern within the area and included as well the amount and distribution of imported products required (see Exhibit 1). An accompanying memorandum from White briefly described the new method that had been used in arriving at the plans.[4] The approach was similar to present, relatively successful, detail planning, but it permitted a far more rigorous and comprehensive way of making trade offs between alternative plans and local constraints. Mr. White's memorandum further stated that for the conditions presented, which had been derived by his assistants working with local staff and management, the plan obtained was the most economical one for overall Oleum operations in the area. He suggested that he meet with Mr. Graham to discuss in more detail the strengths and weaknesses of the techniques used to arrive at the plan and to advise Mr. Graham on the many additional studies that could be made easily now that the basic solution model had been formulated.[5]

[3]This case, Oleum Company Limited (B), IM 1821, was prepared by S. W. Yost under the direction of J. L. McKinney of the Harvard University Graduate School of Business Administration as a basis for classroom discussion rather than to illustrate either effective or ineffective handling of administrative situations. Copyright (C) 1964 by the President and Fellows of Harvard College. Used by specific permission.

[4]Described in Oleum Company, Ltd. (A).

[5]Exhibit 2 is a matrix of the model formulated, based on information from Oleum Company, Ltd. (A); Exhibit 3 is computer output of matrix giving optimal solution, as well as cost and right-hand side (RHS) ranging.

The Oleum Company is a large international holding company for a group of oil producing, transporting, refining, and marketing companies located in various countries throughout the world. While the responsibility for day-to-day operations resides with local affiliate management, the Oleum Company headquarters staff, located in London, provides planning guidelines, administers capital budgeting, and, in general, attempts to balance affiliate operations so that the greatest overall benefit accrues to the company.

The following conversation informally summarizes the main topics discussed at the meeting held between Graham and White:

Graham: "Chuck, I'm familiar with the use of linear programming as used by our refinery managers for developing refinery operating programs, but in this case I gather that you've set up an LP model for planning and coordinating the Far Eastern Area's operations. Perhaps you can tell me what kind of answers this adaptation of the LP technique of yours gives—both as to quality and type. By quality, I mean the expected accuracy, sensitivity, and reliability that you feel you have been able to achieve."

White: "Well, to tackle the quality question first, just as with the LP refinery models that we are using, the planning LP model is dependent on careful analysis and inclusion of all the salient factors that can affect the solution. The question of the sensitivity of the outcome to the various coefficients is very important, especially since not all of the coefficients are linear over the range considered. However, we feel that we have been able to reach close approximations in sensitive areas with the help of the local people with whom we worked closely in building the model. In fact, we used their own refinery models to derive process condition constraints and coefficients. I think I can safely say that, *for the given conditions*, the plan developed with the model is realistic and is the best plan for Oleum in the area. I emphasize "for given conditions" because the plan is static and is based on the present circumstances. The plan derived by the LP model is a "physical" one that minimizes costs. By physical, I mean in the sense that it specifies the barrels of crude to be run, the barrels of product to be shipped, where, and so on. So, you see that if during the planning period any of the circumstances change, such as economic, market demand, physical capacities, and so forth, then it may be necessary to revise the whole plan. You can't be sure, of course; this comes back to how sensitive the model is to the particular variable changed. In any case, it is easy to update the model and derive new plans."

Graham: "I understand that, Chuck, but what are the "economic" answers that you mentioned in your memorandum to me?"

White: "The economic answers calculated by the model give you such things as the cost of the last barrel of product delivered or the value of additional crude availability. In other words, these answers are the value of changing a restriction or constraint or limitation presently existing in the operation. However, all these answers apply to the base case solution—the present set of circumstances as described by the model. The great potential for management, as I see it, is the use of the model for the investigation of so-called "side studies." This is accomplished by a technique called parametric linear programming, PLP, where the values of the restrictions or cost inputs are changed in a purposeful manner.

Using this method the sensitivity of the solution to the input information can be determined, answering such questions as how much the data can vary and still reach the same basic conclusion. Remember, we used this technique in establishing the sensitivity of the local refinery cost and process inputs for use in our planning model. For instance, we have been able to determine by this method that the cost of Iranian crude could vary between $1.97 and $2.03 per barrel and the same volume would still be run.

Graham: "Okay, I follow you on that, but looking over the economic aspects of your planning report, it appears to indicate that we could save 6.2 cents per barrel if we processed less Brunei crude. That stuff has very desirable yield characteristics from a refining standpoint. Am I reading this right?"

White: "Yes, that's what it says. It would appear that at the present volume we are losing money. Perhaps at a reduced rate we would not, but we have a contract with NPM in Borneo to purchase 40,000 B/D during 1965 at $2.60 per barrel."

Graham: "Then that contract we negotiated with them, that we thought was favorable, is actually costing us an additional 6.2 cents per barrel to use?"

White: "Well, let's say that a saving of 6.2 cents per barrel would be achieved if we could reduce the amount of oil we are obligated to buy from them. Or, conversely, if a discount of 6.2 cents per barrel could be negotiated, the crude would break even with the operations now planned for 1965. I think that it is important to remember the 6.2 cents represents the marginal value of the last barrel processed. The range of daily usage over which this value applies is likely to be quite small."

Graham: "Chuck, could you run one of your side studies and tell me what daily amount we could reduce at the present price before some other factor becomes limiting? With this information we may want to reopen negotiations with the NPM people. Of course, at the same time I'd like to know what the various effects of any reduction will be on operations throughout the area as distinguished from the present plan you have developed."

The following are excerpts of a meeting that took place between Graham and White several days later:

Graham: "I've been doing a lot of thinking about your planning model and the capability it gives for studying of the effects of possible changes in the area. We have a couple of problems that a lot of discussion has gone into, but no firm decisions have been made, partly because it's so difficult to get a feel for the side effects that the proposed action might have. After talking to you the other day and doing a little "homework" in a linear programming text, I'm pretty sure that these problems can be handled by parametric studies utilizing your planning model, although they involve such things as marketing and capital budgeting rather than operational plans."

Following these comments, Mr. Graham went on to describe the two problems to Mr. White, requesting that the planning group study the situations and prepare recommendations for Mr. Graham's action.

Price/Volume Relationship for
Motor Gasoline at the APL Refinery

Aussi Petroleums, Ltd. (APL) management is considering bidding on a supply contract for motor gasoline for the Australian government. The government lets its mogas supply contracts on an annual basis and in separate increments of five million gallons[6] each to be delivered over the course of the year. The award is placed on the basis of the lowest bid for the particular contract being procured. APL management understands that five mogas contracts will be placed up for bid early in 1965 and is trying to decide how many of the contracts they should bid on and what their bid price should be. I noted that the marginal value of mogas produced by APL under your 1965 plan is $4.27. What effect would the processing of greater amounts of mogas at APL in 1965 have on its marginal cost and is any other information necessary to evaluate the bidding strategy in terms of overall Far Eastern operations?

Market Growth in New Zealand
and the Problem of Acquisition and
Increasing Refinery Capacity

The marketing affiliate in New Zealand predicts a steadily expanding market for all Oleum products during the next decade. Recently they have been given the opportunity to buy out two local marketing competitors. This would result in an immediate increase in their 1965 sales of approximately 30 percent in one case and of 20 percent in the other case, or a total of a 50 percent increase if both competitors were acquired.

I am concerned as to whether supplying this increased demand will prove to be economical in view of the refining capacities in the area and the

[6]In the petroleum industry 42 gallons is equivalent to one barrel.

cost of importing products. Consequently, I am doubtful about advising the acquisition of one or both of the competitors and about whether or not the value of additional refinery capacity is great enough to warrant further investment. I hope that you will be able to give me information that will make it easier to assess the situation from an overall Far Eastern Area standpoint.

Exhibit 1

OLEUM COMPANY, LTD. (B)

Planning Data for Far Eastern Area Operations in 1965

1. *Summary*
   A. Crude transportation
      (1) Owned ships — 6.14 47,000 DWT EQ
      (2) Spot charter — 0.0 47,000 DWT EQ
      (3) Crude cost — 175.44 M$/D
   B. Refining operations
      (1) Crude run
         Brunei crude — 40.000 MB/D[7]
         Iranian crude — 35.721 MB/D
         Total — 75.721 MB/D
      (2) Cost
         Variable cost — 18.01 M$/D
         Total cost — 63.51 M$/D
         Cost/B crude — 0.84 $/B
   C. Distribution operations
      (1) Product import volumes (MB/D)

	Volumes	*Max. vol.*	*Mar. value $/B*
Dist.	4.33	10.000	0.0
Fuel	7.00	7.000	0.71
Total	11.33		

      (2) Product import cost (total) — 17.84 M$/D
      (3) Product transportation cost — 13.97 M$/D

2. *Refinery operations*
   A. APL Refinery (capacity 50.000 MB/D.)
      (1) Crude summary

	Volume MB/D	*Mar. value $/B*
Brunei type	10.000	2.80
Iranian type	35.721	2.62
Total	45.721	

(value of additional capacity $/B/D = 0.0)

[7]Note: Brunei crude volume fixed by contract, incentive to process less = −0.062 $/B/D.

(2) Product summary

	Volume *MB/D*	*Mar. value* *$/B*
Mogas	14.590	4.27
Dist.	11.990	2.70
Fuel	15.680	2.56

(3) Manufacturing cost

Incremental	13.207	M$/D
Total cost	40.708	M$/D
Cost/B crude	.890	$/B

B. NOC Refinery (Capacity 28.600 MB/D nominal)

(1) Crude summary

	Volume *MB/D*	*Mar. value* *$/B*
Brunei type	30.000	2.78
Iranian type	0	2.59
Total	30.000	

(value of additional capacity $/B/D = 0.025)

(2) Product summary

	Volume	*Mar. value*
Mogas	7.770	4.27
Dist.	11.130	3.65
Fuel	9.510	2.76

(3) Manufacturing cost

Incremental	4.800	M$/D
Total cost	22.800	M$/D
Cost	0.760	$/B

3. *Product distribution*

A. Distribution (MB/D)

		Market Areas				
Product	*Source*	*APL*	*NOC*	*MAL*	*PHL*	*NWZ*
Mogas	APL	9.000	–	–	2.500	3.090
	NOC	–	3.000	4.500	–	0.270
Dist.	APL	10.000	–	–	1.990	–
	NOC	–	5.250	4.500	1.380	–
	IMP	–	–	–	0.130	4.200
Fuel	APL	11.000	–	3.240	1.440	–
	NOC	–	6.750	2.760	–	–
	IMP	–	–	–	2.560	4.400

B. Cost (Cost of last barrel–$/B, no fixed costs included)

Product	*APL*	*NOC*	*MAL*	*PHL*	*NWZ*
Mogas	4.27	4.27	4.47	4.42	4.37
Dist.	2.70	2.65	2.85	2.85	2.70
Fuel	2.56	2.76	3.16	2.76	2.61

Exhibit 2

OLEUM COMPANY, LTD. (B)

	MB/D except 47 EQ FLTCP	C B A P L	C B N O C	C I A P L	C I N O C	C C S	B L A P L	B H A P L	I L A P L	I H A P L	B L N O C	B H N O C	I L N O C	I H N O C
COST	*MIN*	2.86	2.84	2.62	2.59	3.20	.12	.28	.15	.30	.16	.34	.20	.39
CBAVL	= 40.	1	1											
CIAVL	≤ 100.			1	1									
FLTCP	≤ 6.3	.050	.045	.120	.110	−1								
APLCB	= 0.0	1					−1	−1						
APLCI	= 0.0			1					−1	−1				
APLCP	≤ 50.						1	1	1	1				
NOCCB	= 0.0		1								−1	−1		
NOCCI	= 0.0				1								−1	−1
NOCCP	≤ 30.0										1	1.05	1	1.05
NOCFQ	≤ 0.0										−.5	−.5	1	1
APLMG	= 9.0						.259	.365	.186	.312				
APLDT	= 10.0						.371	.379	.229	.230				
APLFL	= 11.0						.317	.194	.503	.378				
NOCMG	= 3.0										.259	.350	.186	.300
NOCDT	= 5.25										.371	.378	.229	.230
NOCFL	= 6.75										.317	.210	.503	.390
IMPDT	≤ 10.0													
IMPFL	≤ 7.0													
MALMG	= 4.5													
MALDT	= 4.5													
MALFL	= 6.0													
PHLMG	= 2.5													
PHLDT	= 3.5													
PHLFL	= 4.0													
NWZMG	= 3.36													
NWZDT	= 4.20													
NWZFL	= 4.44													

Coding examples:

APLCI
- Refinery name
- Crude operations
- Crude type (I, Iranian; B, Brunei)

CSC
- Spot charter
- Crude operations

NOCBL
- Refinery name
- Crude type
- Process intensity (L = low; H = high)

Exhibit 2 (continued)

OLEUM COMPANY, LTD. (B)

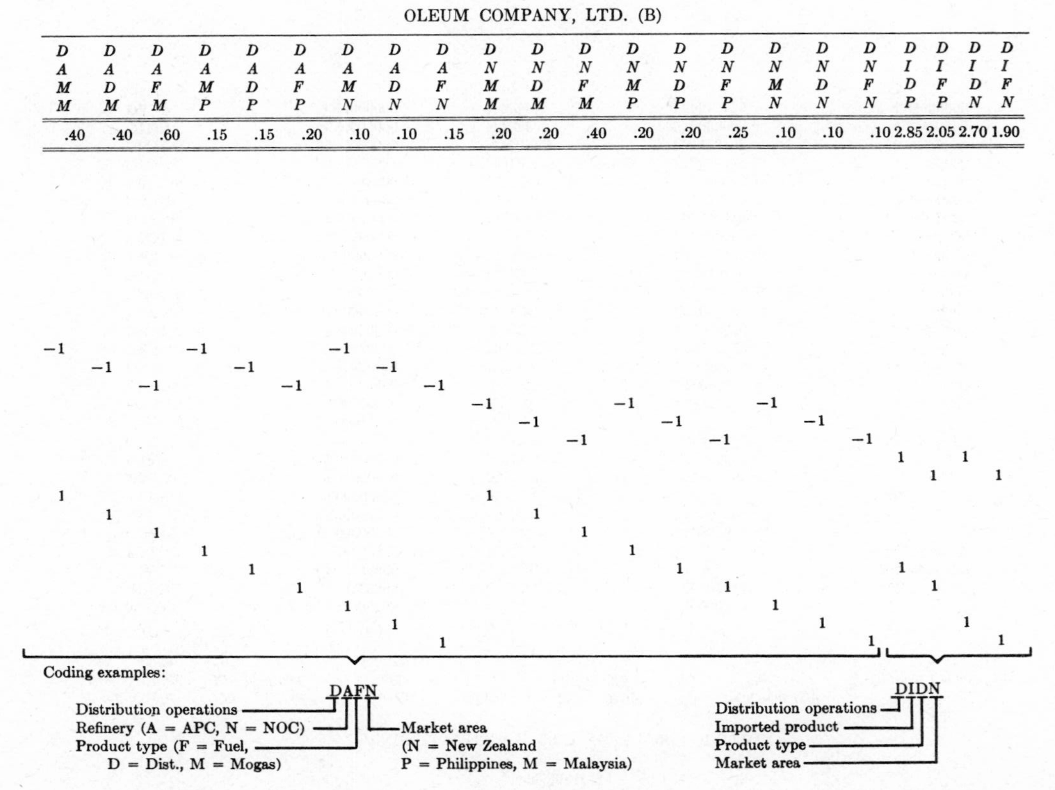

D	*D*	*D*	*D*	*D*	*D*	*D*	*D*	*D*	*D*	*D*	*D*	*D*	*D*	*D*	*D*	*D*	*D*	*D*	*D*	*D*	*D*
A	*A*	*A*	*A*	*A*	*A*	*A*	*A*	*A*	*N*	*N*	*N*	*N*	*N*	*N*	*N*	*N*	*N*	*I*	*I*	*I*	*I*
M	*D*	*F*	*M*	*D*	*F*	*M*	*D*	*F*	*M*	*D*	*F*	*M*	*D*	*F*	*M*	*D*	*F*	*D*	*F*	*D*	*F*
M	*M*	*M*	*P*	*P*	*P*	*N*	*N*	*N*	*M*	*M*	*M*	*P*	*P*	*P*	*N*	*N*	*N*	*P*	*P*	*N*	*N*
.40	.40	.60	.15	.15	.20	.10	.10	.15	.20	.20	.40	.20	.20	.25	.10	.10	.10	2.85	2.05	2.70	1.90
−1			−1			−1															
	−1			−1			−1														
		−1			−1			−1													
									−1			−1			−1						
										−1			−1			−1					
											−1			−1			−1				
																		1		1	
																			1		1
1									1												
	1									1											
		1									1										
			1									1									
				1									1					1			
					1									1					1		
						1									1						
							1									1				1	
								1									1				1

Coding examples:

Exhibit 3

OLEUM COMPANY, LTD. (B) LP/90 SOLUTION

Total Iters	*No. Etas*	*Eta Rec*	*Row Ident.*	*Current Value*	*Chosen Vector*	*Vectr Remvd*	*RHS No.*	*C/V No.*	*Current D/J Theta/Phi*	*Optimal Primal-Dual*
39	*39*	*0*	*Cost*	*−257.21006*			*1*		* * * *	

	J(H)	*Beta (H)*		*Row (I)*	*PI (I)*	*B (I)*
0	00000	257.21006195−		COST	1.00000000	.
	APLBL	1.93207893		CBAVL	.06162614−	40.00000000
+	CIAVL	64.27948600	+	CIAVL		100.00000000
+	FLTCP	.16353832	+	FLTCP	.	6.30000000
	CBAPL	10.00000000		APLCB	2.79837386−	.
	APLIH	35.72051400		APLCI	2.62000000−	.
	APLBH	8.06792107	+	APLCP	.	50.00000000
	DNFM	2.76000000		NOCCB	2.77837386−	.
	DIFP	2.56000000		NOCCI	2.59000000−	.
	CINOC	.	+	NOCCP	.02485000	30.00000000
+	NOCFQ	15.00000000	+	NOCFQ	.	.
	DNDP	1.38000000		APLMG	4.27173780−	9.00000000
	CIAPL	35.72051400		APLDT	2.70000000−	10.00000000
	DADP	1.99026159		APLFL	2.55613176−	11.00000000
	CBNOC	30.00000000		NOCMG	4.27173780−	3.00000000
	DNMN	.27000000		NOCDT	2.65000000−	5.25000000
	NOCBL	30.00000000		NOCFL	2.75613176−	6.75000000
+	IMPDT	5.67026159	+	IMPDT	.	10.00000000
	DIFN	4.44000000	+	IMPFL	.70613176	7.00000000
	DAMN	3.09000000		MALMG	4.47173780−	4.50000000
	DNMM	4.50000000		MALDT	2.85000000−	4.50000000
	DAFM	3.24000000		MALFL	3.15613176−	6.00000000
	DAMP	2.50000000		PHLMG	4.42173780−	2.50000000
+	APLCP	4.27948600		PHLDT	2.85000000−	3.50000000
	DIDP	.12973841		PHLFL	2.75613176−	4.00000000
	DNDM	4.50000000		NWZMG	4.37173780−	3.36000000
	DIDN	4.20000000		NWZDT	2.70000000−	4.20000000
	DAFP	1.44000000		NWZFL	2.60613176−	4.44000000

Exhibit 3 (continued)

OLEUM COMPANY, LTD. (B)

Cost Ranges

Basic Vector	*Beta Value*	*Cost in Problem*	*Lim 1*	*Lim 2*	*Incoming Vector At Lim 1*	*Incoming Vector At Lim 2*
APLBL	1.9320789	.12000000	.09515000	.18467500	+ NOCCP	+ APLIL
+ CIAVL	64.279485		* * * *	.52256415	UNBOUNDED	+ IMPFL
+ FLTCP	.16353832		−3.2000000	4.3547012	CSC	+ IMPFL
CBAPL	9.9999999	2.8600000	2.8351500	* * * *	+ NOCCP	UNBOUNDED
APLIH	35.720514	.30000000	−.22256415	.32792268	+ IMPFL	NOCIL
APLBH	8.0679210	.28000000	.25543985	.35970367	NOCIL	NOCBH
DNFM	2.7600000	.40000000	.25418015	.47839117	NOCIL	+ NOCCP
DIFP	2.5600000	2.0500000	1.9500000	2.7561317	DAFN	+ IMPFL
CINOC		2.5900000	2.5628775	* * * *	NOCIL	UNBOUNDED
+ NOCFQ	15.000000		* * * *	.01808166	UNBOUNDED	NOCIL
DNDP	1.3800000	.20000000	−.05000000	.25000000	DADM	DNDN
CIAPL	35.720514	2.6200000	2.0974359	2.6479227	+ IMPFL	NOCIL
DADP	1.9902616	.15000000	.08301887	.25000000	+ NOCCP	DADN
CBNOC	30.000000	2.8400000	* * * *	2.8648500	UNBOUNDED	+ NOCCP
DNMN	.27000000	.10000000	−.10000000	.15000000	DAMM	DNMP
NOCBL	30.000000	.16000000	* * * *	.18485000	UNBOUNDED	+ NOCCP
+ IMPDT	5.6702615		−1.8801336	* * * *	NOCIL	UNBOUNDED
DIFN	4.44000000	.19000000	* * * *	2.0000000	UNBOUNDED	DAFN
DAMN	3.0900000	.10000000	.05000000	.30000000	DNMP	DAMM
DNMM	4.4999999	.20000000	* * * *	.40000000	UNBOUNDED	DAMM
DAFM	3.2400000	.60000000	.52160884	.74581985	+ NOCCP	NOCIL
DAMP	2.500000	.15000000	* * * *	.20000000	UNBOUNDED	DNMP
+ APLCP	4.2794860		* * * *	.02485000	UNBOUNDED	+ NOCCP
DIDP	.12973841	2.8500000	2.8000000	4.7301336	DNDN	NOCIL
DNDM	4.4999999	.20000000	* * * *	.45000000	UNBOUNDED	DADM
DIDN	4.2000000	2.7000000	* * * *	2.7500000	UNBOUNDED	DNDN
DAFP	1.4400000	.20000000	−.50613176	.30000000	+ IMPFL	DAFN

Exhibit 3 (continued)

OLEUM COMPANY, LTD. (B)

Right Hand Side Ranges

Row Name	Current RHS Val	PI Value	Minimum Value	Maximum Value	Outgoing Vector At Min	Outgoing Vector At Max
CBAVL	40.00000	−.06162614	38.042922	40.724089	APLBL	DIDP
+ CIAVL	100.0000		35.720514	UNBOUNDED	+ CIAVL	
+ FLTCP	6.300000		6.1364616	UNBOUNDED	+ FLTCP	
APLCB		−2.7983738	−.72408942	1.6331597	DIDP	+ FLTCP
APLCI		−2.6200000	−35.720514	1.3628193	CIAPL	+ FLTCP
+ APLCP	50.00000		45.720514	UNBOUNDED	+ APLCP	
NOCCB		−2.7783738	−.72408942	1.7189924	DIDP	+ FLTCP
NOCCI		−2.5900000		1.4867120	CINOC	+ FLTCP
+ NOCCP	30.00000	.02485000	28.957529	31.932079	DNMN	APLBL
+ NOCFQ			−15.000000	UNBOUNDED	+ NOCFQ	
APLMG	9.000000	−4.2717378	7.3257142	9.3250048	APLBH	DIDP
APLDT	10.00000	−2.7000000	9.8702615	11.990262	DIDP	DADP
APLFL	11.00000	−2.5561317	10.514231	11.465052	APLBL	DIDP
NOCMG	3.000000	−4.2717378	1.3257143	3.2700000	APLBH	DNMN
NOCDT	5.250000	−2.6500000	5.1202616	6.6300000	DIDP	DNDP
NOCFL	6.750000	−2.7561317	6.2642308	7.2150520	APLBL	DIDP
+ IMPDT	10.00000		4.3297384	UNBOUNDED	+ IMPDT	
+ IMPFL	7.000000	.70613176	6.5349479	7.4857692	DIDP	APLBL
MALMG	4.500000	−4.4717378	2.8257143	4.7700000	APLBH	DNMN
MALDT	4.500000	−2.8500000	4.3702616	5.8800000	DIDP	DNDP
MALFL	5.999999	−3.1561318	5.5142307	6.4650521	APLBL	DIDP
PHLMG	2.500000	−4.4217378	.82571428	2.8250048	APLBH	DIDP
PHLDT	3.500000	−2.8500000	3.3702616	9.1702616	DIDP	+ IMPDT
PHLFL	4.000000	−2.7561317	3.5142307	4.4650520	APLBL	DIDP
NWZMG	3.360000	−4.3717378	1.6857143	3.6850048	APLBH	DIDP
NWZDT	4.200000	−2.7000000		9.8702615	DIDN	+ IMPDT
NWZFL	4.440000	−2.6061317	3.9542307	4.9050520	APLBL	DIDP

The Assignment Problem and Special Algorithms

6

6.1. INTRODUCTION TO THE ASSIGNMENT PROBLEM.

Suppose that a manager has three jobs numbered 1, 2, and 3 to fill, and three men, identified by A, B, and C, available for assignment to them. He wishes to make assignments so as to maximize total value of performance. Suppose that the productivity rating of each man for each job is known and is as stated in Table 6.1. There are six assignments

Table 6.1. PRODUCTIVITY RATINGS.

		Job		
		1	2	3
Man	A	6	8	12
	B	3	4	10
	C	5	3	8

possible; the value of each assignment is the sum of the ratings in each assignment. For example, the assignment A1, B2, C3 has the value $6 + 4 + 8 = 18$. One procedure for determining the best assignment is to list all assignments, calculate the value of each, and select the one having the largest value. The results are shown in Table 6.2, and the optimal assignment is seen to be C1, A2, B3 with value 23.

The procedure of listing all assignments is easy enough for three jobs, but for 10 jobs it would require a consideration of over three million

(10!) assignments, each involving the addition of 10 ratings. If each assignment were to require one second this would take more than 800 hours. Thus a manager working eight hours per day would take one third of a year to determine the best assignment. If a digital computer is used and the

Table 6.2. VALUE OF ASSIGNMENTS.

1	2	3	
A	B	C	$6+4+8=18$
A	C	B	$6+3+10=19$
B	A	C	$3+8+8=19$
B	C	A	$3+3+12=18$
C	A	B	$5+8+10=23$ (optimal)
C	B	A	$5+4+12=21$

time per assignment is reduced from 1 second to 10 microseconds or 1×10^{-5} seconds then only slightly more than 8/1000 of an hour would be required for a solution. However, if the number of jobs is increased to 15 and the time per assignment is increased to 1.5×10^{-5} seconds the time then needed for a solution exceeds 110 hours and if the computer charge is $200.00 per hour the computer cost for the solution exceeds $22,000.00 and the cost of waiting 110 hours for the solution must be added to this. For 20 jobs under the above assumptions the time exceeds 2.5×10^{8} hours and the computer cost exceeds $\$5.00 \times 10^{10}$; the computer cost of 50 billion dollars would be equal to almost 1/4 of the annual federal budget! Thus it is clear that better solution methods are needed. We develop an algorithm below which has an average of less than 3 seconds of computer time for complete solution of a 16-job assignment (other solution procedures are also available).

We return to our simple three-job example to illustrate some of the ideas which are used in efficient algorithms.

The first modification we consider is the replacement of our original problem by one which calls for minimization rather than for maximization. This can be easily done by changing all signs in the productivity matrix, giving a *cost matrix* denoted by $A^{(1)}$ in which negative costs are interpreted

$$A^{(1)}: \begin{array}{c|ccc|} & 1 & 2 & 3 \\ \hline A & -6 & -8 & -12 \\ B & -3 & -4 & -10 \\ C & -5 & -3 & -8 \\ \hline \end{array}$$

as positive productivities. An equivalent result can be obtained in a way which may have more intuitive appeal. Suppose that 10 represents an ideal productivity for any job performance. Then man A in job 1 will be *deficient* 10 − 6 = 4 units, whereas in job 3 he will be deficient 10 − 12 = −2 units (meaning that his performance exceeds the ideal standard by 2 units). Then the *deficiency matrix* $A^{(2)}$ can be regarded as giving costs of assignments. If we evaluate the six possible assignments for these two

$A^{(2)}$:

	1	2	3
A	4	2	−2
B	7	6	0
C	5	7	2

cost matrices we get costs as given in Figure 6.1. We observe: Property 1, that for each of $A^{(1)}$ and $A^{(2)}$ the minimum cost assignment is C1, A2, B3 which was the optimal assignment for the original problem; Property 2,

1	2	3	$A^{(1)}$	$A^{(2)}$
A	B	C	−18	12
A	C	B	−19	11
B	A	C	−19	11
B	C	A	−18	12
C	A	B	−23	7 (optimal)
C	B	A	−21	9

Figure 6.1

that each assignment cost for $A^{(1)}$ is the negative of the corresponding productivity in the original problem; and Property 3, that for each assignment the cost for $A^{(2)}$ is exactly 30 more than its cost for $A^{(1)}$. Indeed Property 3 alone will guarantee that $A^{(2)}$ will yield the same optimal assignment as $A^{(1)}$, although the cost will differ. We search for more general transformations that will have Property 3 in the general form, original cost minus transformed cost is the same constant for all assignments.

Suppose, for example, we reduce the ideal rating for job 1 by 4 units. This has the effect of replacing $A^{(2)}$ by $A^{(3)}$. Since each assignment uses

$A^{(3)}$:

	1	2	3
A	0	2	−2
B	3	6	0
C	1	7	2

exactly one cost from column 1, the cost of every assignment for $A^{(3)}$ will be

4 less than its cost for $A^{(2)}$. Now subtract 2 from each position in column 2 and add 2 to each position in column 3 and get a new cost matrix $A^{(4)}$ with

$A^{(4)}$:

	1	2	3
A	0	0	0
B	3	4	2
C	1	5	4

no net change in assignment costs from $A^{(3)}$ since the two alterations cancel each other. Next subtract 2 from row 2 and 1 from row 3 to obtain $A^{(5)}$.

$A^{(5)}$:

	1	2	3
A	0	0	0
B	1	2	0
C	0	4	3

This problem will differ by 3 in cost from $A^{(4)}$ for each assignment. Combining several steps we conclude that each assignment for $A^{(5)}$ costs exactly 7 units less than that for the same assignment for $A^{(2)}$. It follows that the best assignment for $A^{(5)}$ will be the same as for $A^{(2)}$ and therefore that the optimal costs for these two matrices will differ by 7.

Now in $A^{(5)}$ the assignment C1, A2, B3 has zero cost and since no cost in $A^{(5)}$ is negative no assignment can cost less than zero; we are able to infer that C1, A2, B3 is the optimal assignment for $A^{(5)}$ without listing all of the six assignment costs. It follows, of course, that C1, A2, B3 is also the optimal assignment for $A^{(2)}$ and that for $A^{(2)}$ the optimal cost is 7 (this can be checked by addition of the proper elements in $A^{(2)}$).

We give another illustrative example. Suppose that there are 10 men and 10 jobs and that the cost of assigning man i to job j is given by the entry a_{ij} in the matrix A. Here we note the following properties:

$$A = \begin{bmatrix} 0 & 2 & 2 & 4 & 3 & 0 & 1 & 6 & 2 & 4 \\ 2 & 0 & 1 & 1 & 2 & 1 & 6 & 2 & 2 & 3 \\ 1 & 4 & 0 & 7 & 3 & 5 & 1 & 4 & 6 & 2 \\ 3 & 5 & 4 & 0 & 3 & 1 & 2 & 3 & 4 & 5 \\ 0 & 3 & 5 & 2 & 0 & 4 & 1 & 3 & 5 & 9 \\ 6 & 1 & 3 & 2 & 1 & 0 & 4 & 3 & 2 & 1 \\ 2 & 5 & 8 & 6 & 4 & 2 & 0 & 2 & 4 & 6 \\ 1 & 3 & 6 & 5 & 4 & 7 & 2 & 0 & 3 & 2 \\ 4 & 7 & 8 & 2 & 1 & 6 & 5 & 8 & 0 & 7 \\ 3 & 3 & 2 & 5 & 1 & 3 & 9 & 4 & 3 & 0 \end{bmatrix}$$

Every cost is nonnegative; (6.1)

There is an assignment with cost zero. (6.2)

The assignment with zero cost is the one which sends man i to job i (the diagonal assignment) and clearly no other assignment can decrease this cost. This conclusion can be reached without the necessity of calculating costs for any of the (over three million) remaining possible assignments.

Given any ten-man, ten-job problem, what one wishes to do is to find an equivalent one with properties (6.1) and (6.2). Algorithms exist for solving assignment problems in which the basic idea is to adjust costs by adding (or subtracting) some number from every position in a row or a column. Some of these algorithms, the so-called *primal algorithms*, achieve (6.2) initially and proceed step by step to achieve (6.1); others, the *dual algorithms*, achieve (6.1) initially and then proceed step by step to achieve (6.2). There are still other algorithms which are neither primal nor dual but have some features of each.

We will give a brief description of a dual algorithm called the Hungarian Method, but our major attention will be devoted to describing a new primal algorithm which appears to be quite efficient.

In general, primal algorithms have an advantage over dual ones since they provide a feasible solution at every step and since any accumulated experience as to what is a good assignment can be used to initiate the calculations with a consequent reduction of total computing time.

In our 10 by 10 matrix A it was relatively easy to identify the optimal assignment. For large problems visual search is not efficient and is not suitable for computer programs. In Section 6.5, we discuss a labeling process which can be used in computer algorithms (or even in hand calculation) as a systematic alternative to visual search.

6.2. THE ASSIGNMENT PROBLEM.

We suppose that there are n positions (jobs) $J_1, \cdots, J_n$ to be filled by n individuals $I_1, \cdots, I_n$; we suppose that a_{ij} is the cost of assigning individual I_i to job J_j, and we call the square matrix

$$A = [a_{ij}], \tag{6.3}$$

the *cost matrix*. An assignment is defined by a matrix

$$X = [x_{ij}], \tag{6.4}$$

where

$$x_{ij} = \begin{cases} 1 \text{ if individual } I_i \text{ is assigned to job } J_j, \\ 0 \text{ otherwise.} \end{cases} \tag{6.5}$$

Since each man must be assigned to exactly one job we require

$$\sum_{j=1}^{n} x_{ij} = 1 \qquad (i = 1, \cdots, n), \tag{6.6}$$

and since each job must be filled we require

$$\sum_{i=1}^{n} x_{ij} = 1 \qquad (j = 1, \cdots, n). \tag{6.7}$$

We also require

$$x_{ij} \geq 0 \qquad (i, j = 1, \cdots, n). \tag{6.8}$$

Consider the linear programming problem defined by

$$\min z = \alpha(X) = \sum_i \sum_j a_{ij} x_{ij} \tag{6.9}$$

subject to the feasibility conditions (6.6), (6.7), and (6.8).

It is a remarkable fact that this linear programming problem will have an optimal feasible program X^0 in which each x_{ij}^0 is either 0 or 1 and hence we need not explicitly require that (6.5) be satisfied. In other words, even without a specific requirement that the x_{ij} be integers it turns out that there are no fractional values for the x_{ij} which will reduce $z = \alpha(X)$ below the value $z^0 = \alpha(X^0)$.

It is this property which makes the assignment problem relatively easy to solve. In contrast, the traveling salesman problem, which does not have this integer property, has so far remained intractable to attempts to find a practical algorithm for solution.

It is helpful to consider the assignment problem in connection with its dual, which is (compare Problems 2.31 and 5.7)

$$\text{maximize } w = \beta(U, V) = \sum_i u_i + \sum_j v_j \tag{6.10}$$

subject to the feasibility conditions

$$u_i \text{ is free} \qquad (i = 1, \cdots, n), \tag{6.11}$$

$$v_j \text{ is free} \qquad (j = 1, \cdots, n), \tag{6.12}$$

and

$$u_i + v_j \leq a_{ij} \qquad (i, j = 1, \cdots, n). \tag{6.13}$$

Suppose that X is a feasible primal vector and that (U, V) is any dual vector (feasible or not), then

$$\begin{aligned} z - w &= \alpha(X) - \beta(U, V) \\ &= \sum_{ij} a_{ij} x_{ij} - \sum_i u_i - \sum_j v_j \\ &= \sum_{ij} a_{ij} x_{ij} - \sum_i u_i \sum_j x_{ij} - \sum_j v_j \sum_i x_{ij} \\ &= \sum_{ij} (a_{ij} - u_i - v_j) x_{ij} \\ &= \sum_{ij} a_{ij}^* x_{ij}, \end{aligned} \tag{6.14}$$

where

$$a_{ij}^* = a_{ij} - u_i - v_j \qquad (i, j = 1, \cdots, n). \tag{6.15}$$

The dual vector (U, V) is feasible if and only if [compare (6.13)]

$$a^*_{ij} \geq 0 \qquad (i, j = 1, \cdots, n). \tag{6.16}$$

Now, if (U, V) is feasible as well as X then from (6.14) and (6.16) we conclude that

$$z - w = \alpha(X) - \beta(U, V) = \textstyle\sum_{ij} a^*_{ij} x_{ij} \geq 0, \tag{6.17}$$

that is, the cost z for any feasible assignment is at least as great as the return w for any feasible dual vector.

6.3. ORTHOGONALITY AND OPTIMALITY.

A dual vector (U, V) is said to be *orthogonal*[1] to a primal vector X if

$$a^*_{ij} x_{ij} = 0 \qquad (i, j = 1, \cdots, n). \tag{6.18}$$

This definition does not require feasibility of either the primal or the dual vector. However, if X^0 is feasible and (U^0, V^0) is orthogonal to X^0 then by (6.14) and (6.18) we conclude that $z^0 = w^0$; if, moreover, (U^0, V^0) is feasible we conclude that both X^0 and (U^0, V^0) are optimal, since, by (6.17), we have $z \geq w^0 = z^0$ for any feasible X and for any feasible (U, V) we have $w^0 = z^0 \geq w$. It follows that our problem is solved if we can find feasible vectors for both the primal and the dual problems which are orthogonal.

It should be noted that introduction of the dual problem provides a convenient mechanism for replacing the initial cost matrix A with a new one A^* which is equivalent to A in the sense that for all feasible vectors X we have

$$\begin{aligned} \alpha(X) &= \textstyle\sum_{ij} a_{ij} x_{ij} = \sum_{ij} (a^*_{ij} + u_i + v_j) x_{ij} \\ &= \textstyle\sum a^*_{ij} x_{ij} + \sum_{ij} u_i x_{ij} + \sum_{ij} v_j x_{ij} \\ &= \alpha^*(X) + \textstyle\sum u_i + \sum v_j \\ &= \alpha^*(X) + \beta(U, V). \end{aligned} \tag{6.19}$$

Therefore, if X^0 minimizes α^* it also minimizes α and vice versa. Moreover, if X^0 is optimal, and if (U^0, V^0) is orthogonal to X^0, (6.19) gives

$$\begin{aligned} \alpha(X^0) &= \alpha^*(X^0) + \beta(U^0, V^0) \\ &= 0 \qquad + \beta(U^0, V^0) \\ &= \textstyle\sum_i u^0_i \ + \sum_j v^0_j. \end{aligned} \tag{6.20}$$

[1]This slight departure from the language of Chapter 5 is made in the interest of conformity with the literature. More precisely, the vectors X and (U,V) constitute an orthogonal pair and it is X and A^* that are orthogonal in the sense of Appendix B.

Thus we may interpret the optimal dual vector (U^0, V^0) as providing a set of numbers, one for each row and one for each column of A, such that when for each i row i is decreased by u_i^0 and for each j column j is next decreased by v_j^0, there results a new cost matrix A^* which satisfies (6.1) and (6.2) and hence provides an optimal assignment. For example, refer to $A^{(2)}$ introduced in Section 6.1,

$A^{(2)}$:		1	2	3
	A	4	2	−2
	B	7	6	0
	C	5	7	2

which has an optimal feasible vector $X^0 = \begin{bmatrix} 0 & 1 & 0 \\ 0 & 0 & 1 \\ 1 & 0 & 0 \end{bmatrix}$ and $\alpha(X^0) = \sum_{ij} a_{ij} x_{ij}^0 = 7$. If $U^0 = \begin{bmatrix} -1 \\ 1 \\ 0 \end{bmatrix}$ and $V^0 = [5 \quad 3 \quad -1]$, then $A^* = \begin{bmatrix} 0 & 0 & 0 \\ 1 & 2 & 0 \\ 0 & 4 & 3 \end{bmatrix}$, which is $A^{(5)}$ of Section 6.1.

6.4. HUNGARIAN METHOD OF SOLUTION.

Consider the following matrix

$$A = \begin{bmatrix} 0 & 3 & 4 & 0 & 2 \\ 1 & 4 & 3 & 0 & 4 \\ 4 & 0 & 0 & 1 & 0 \\ 0 & 2 & 4 & 1 & 2 \\ 0 & 2 & 2 & 3 & 4 \end{bmatrix} \tag{6.21}$$

having eight zero entries; we call the set

$$Z = \{(1, 1), (1, 4), (2, 4), (3, 2), (3, 3), (3, 5), (4, 1), (5, 1)\} \tag{6.22}$$

the *set of zero positions in* A. The three zero positions (1, 1), (2, 4), (3, 5) have the property that no two of these lie in any common row or common column of A; such a subset of Z is called an *independent set of zero positions*, and the number of such positions (three in this case) is called the *order* of the set. An independent set of zero positions is said to be *maximal* if there is no other independent set of zero positions with greater order. For example the three positions listed constitute a maximal independent set of zero positions for A. It should be noted that the set $\{(1, 4), (3, 2), (5, 1)\}$ is another maximal independent set of zero positions for (6.21) and that there are several others.

In general, for any matrix A we denote the set of zero positions by Z, let I denote an independent set of zero positions, and let $m(I)$ denote the order of I. Then we let

$$s = \max \{m(I) | I \text{ an independent set of zero positions}\} \tag{6.23}$$

and call s the *independence order* of Z for the matrix A.

We note that for the matrix A of (6.21) all of the zero positions lie in one of row 3, column 1, or column 4. We say that Z is *covered* by the index set $J = \{3; 1, 4\}$. More generally, for any matrix A and its set Z we call $J = \{i_1, \cdots, i_r; j_{r+1}, \cdots, j_t\}$ a *covering set* for Z if each position (i, j) in Z has either $i \in \{i_1, \cdots, i_r\}$ or $j \in \{j_{r+1}, \cdots, j_t\}$. We denote the order t of J by $m(J)$, in our example $m(J) = 3$, and let

$$g = \min \{m(J) | J \text{ is a covering set for } Z\}. \tag{6.24}$$

We call g the *covering index* of Z. Finally, we say that rows $i_1, \cdots, i_r$ and columns $j_{r+1}, \cdots, j_t$ of A are *covered* (by J).

An important result in combinatorial analysis is known as Koenig's Theorem.

Theorem 6.1. Koenig's Theorem. For any matrix A and its set Z of zero positions

$$\text{Independence order} = \text{Covering index}. \tag{6.25}$$

This theorem can be paraphrased: The maximum number of independent zeros equals the minimum number of covering lines.

Suppose now that an assignment problem with n jobs has cost matrix A with

$$A \geq 0 \tag{6.26}$$

and let g be the covering index of the set Z of zero positions in A. If $g = n$ an optimal assignment (that is, one with zero cost) can be achieved by using the n positions $(1, j_1), \cdots, (n, j_n)$ of a set J with $m(J) = n$ to define an assignment matrix X with

$$\begin{aligned} x_{ij_i} &= 1 \qquad (i = 1, \cdots, n), \\ x_{ij} &= 0 \qquad \text{otherwise}. \end{aligned} \tag{6.27}$$

If $g < n$ we let $J = \{i_1, \cdots, i_r; j_{r+1}, \cdots, j_g\}$ be a minimal covering set and consider

$$u = \min \{a_{ij} | i \notin \{i_1, \cdots, i_r\} \text{ and } j \notin \{j_{r+1}, \cdots, j_g\}\}; \tag{6.28}$$

thus u is the smallest element of A that is neither in a covered row nor a covered column. In our example for $J = \{3; 1, 4\}$ we have $u = 2$. Since all of the zeros are covered and since $A \geq 0$ we know that u is positive. Now let A^* be the matrix obtained from A by adding u to every covered row and subtracting u from every uncovered column.

This transformation can be fitted into the language of primal and dual problems if we define dual vectors U and V by

$$\begin{aligned} u_i &= -u \text{ if row } i \text{ is covered,} \\ u_i &= 0 \text{ if row } i \text{ is not covered,} \\ v_j &= 0 \text{ if column } j \text{ is covered,} \\ v_j &= u \text{ if column } j \text{ is uncovered;} \end{aligned} \tag{6.29}$$

then a^*_{ij} is given by (6.15). We observe that

$$\begin{aligned} a^*_{ij} &= a_{ij} && \text{if exactly one of row } i \text{ and column } j \text{ is covered,} \\ a^*_{ij} &= a_{ij} - u && \text{if neither row } i \text{ nor column } j \text{ is covered,} \\ a^*_{ij} &= a_{ij} + u && \text{if both row } i \text{ and column } j \text{ are covered.} \end{aligned} \tag{6.30}$$

The choice of u thus insures that $A^* \geq 0$. Moreover, for any assignment X we have

$$\begin{aligned} \alpha(X) - \alpha^*(X) &= \textstyle\sum_i u_i + \sum_j v_j \\ &= u(-r + n - (g - r)) = u(n - g), \end{aligned} \tag{6.31}$$

and for the sum of all elements in the cost matrices we have

$$\textstyle\sum_{ij} a_{ij} - \sum_{ij} a^*_{ij} = nu(n - g). \tag{6.32}$$

Now, consider the problem with cost matrix A^* and let g^* be the covering index of A^*. If $g^* < n$ the process can be iterated but because of (6.32) and the fact that $A^* \geq 0$ only a finite number of iterations is possible and the process must terminate. But this can only happen when a cost matrix A^0 is reached having independence order $g^0 = n$ and hence an optimal assignment X^0 with cost $\alpha^0(X^0) = 0$. Clearly X^0 will also be optimal for A and $\alpha(X^0)$ will be the sum of all the right-hand sides of (6.31) and its successors in the iteration.

This process is the essence of the Hungarian Method for solving assignment problems. A complete description would have to include methods for constructing the sets I and J, that is, methods for finding a maximal set of independent zero positions and a minimal set of covering lines. However, for small n the method can be applied based on visual search for I and J.

To apply this method to an arbitrary initial cost matrix A we carry out one preliminary dual transformation to obtain an A^* which is nonnegative and which, moreover, has at least one zero in each row and in each column; such a matrix is said to be *reduced*. The matrix A of (6.21), for example, is reduced.

If we define U, V as follows:

$$\begin{aligned} u_i &= \min\{a_{i1}, \cdots, a_{in}\}, \qquad (i = 1, \cdots, n), \\ v_j &= \min\{a_{1j} - u_1, \cdots, a_{nj} - u_n\}, \qquad (j = 1, \cdots, n), \end{aligned} \tag{6.33}$$

then the matrix A^* given by (6.15) will be reduced.

We illustrate the Hungarian Method by using it to solve the problem defined by (6.21). Here $g = 3$, $J = \{3; 1, 4\}$ is a minimal covering set, and $u = 2$. For bookkeeping purposes we denote successive transformed matrices with superscripts and write the nonzero u_i, v_j adjacent to the matrix. We also indicate the maximal independent set and the covered lines by asterisks as follows:

$$
\begin{array}{cc}
 & \begin{array}{ccccc} * & v_2 = 2 & v_3 = 2 & * & v_5 = 2 \end{array} \\
u_3 = -2^*, \quad A = & \begin{bmatrix} 0^* & 3 & 4 & 0 & 2 \\ 1 & 4 & 3 & 0^* & 4 \\ 4 & 0 & 0 & 1 & 0^* \\ 0 & 2 & 4 & 1 & 2 \\ 0 & 2 & 2 & 3 & 4 \end{bmatrix}
\end{array}
\quad
\begin{aligned} u &= 2 \\ g &= 3 \\ u(n - g) &= 2(5 - 3) \\ &= 4 \end{aligned} \qquad (6.34)
$$

$$
A^{(1)} = \begin{bmatrix} 0^* & 1 & 2 & 0 & 0 \\ 1 & 2 & 1 & 0^* & 2 \\ 6 & 0 & 0 & 3 & 0^* \\ 0 & 0^* & 2 & 1 & 0 \\ 0 & 0 & 0^* & 3 & 2 \end{bmatrix} \quad g = 5. \qquad (6.35)
$$

The optimal assignment has $x_{11} = x_{24} = x_{35} = x_{42} = x_{53} = 1$ with cost for A of 4.

6.5. SOME COMBINATORIAL PRELIMINARIES; THE LABELING PROCESS.

The process which we refer to as (P, Q) labeling is described for general networks by Ford and Fulkerson (Reference 18), is used also by Balinski and Gomory (Reference 3) in their primal algorithm, and is also central in the algorithm which will be developed in the following sections.

A transportation problem with p origins and n destinations can be described analytically by several p by n matrices, p vectors, and n vectors. In our version of the labeling process we do not use the general language of networks but substitute for it terms in matrix language. We associate each origin with a row and each destination with a column of a p by n matrix and associate the route from origin i to destination j with the position (i, j) in the matrix. We use the term *line* to mean either a row or a column of a matrix.

We denote by $N = \{(i, j) | i = 1, \cdots, p; j = 1, \cdots, n\}$ the set of all positions in a p by n matrix. Each such position, of course, represents a route in the corresponding transportation problem. The solution of transportation problems requires that we give attention to various subsets of N, that is, to collections of routes which possess certain properties. These

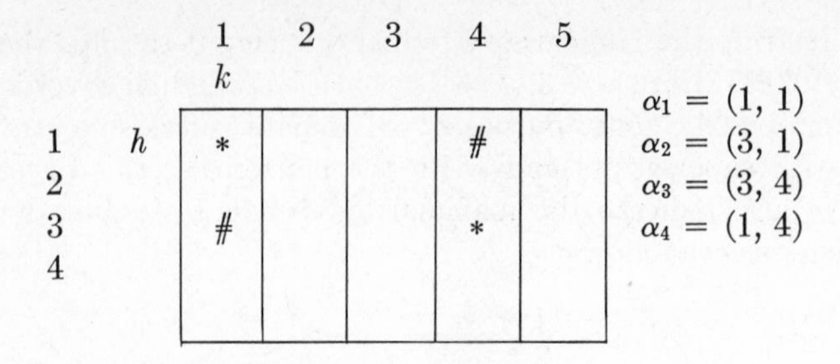

Figure 6.2 A (P, Q) loop with $s = 4$.

subsets ordinarily occur two at a time and in order to handle three distinct situations we now develop a general theory of labeling in terms of two arbitrary subsets P and Q of N.

A sequence C of distinct positions

$$\alpha_1, \cdots, \alpha_s \tag{6.36}$$

is said to be a (P, Q) *chain* if:

L1. each adjacent pair of positions is collinear (lie in the line of the matrix),
L2. the positions belong alternately to P and Q,
L3. no line contains more than two positions.

A (P, Q) chain (6.36) is called a (P, Q) *loop* if in addition:

L4. $s \geq 4$ and α_1 and α_s are collinear.

A chain C is said to *connect* any two of its positions.

In Figure 6.2 and in the illustrations below the symbol * represents a position in P and the symbol # represents a position in Q. Consider the sequence $\alpha_1 = (1, 1)$, $\alpha_2 = (3, 1)$, $\alpha_3 = (3, 4)$ and $\alpha_4 = (1, 4)$ and let $P = \{\alpha_1, \alpha_3\}$ and $Q = \{\alpha_2, \alpha_4\}$. Then we claim that this sequence is both a (P, Q) chain and a (P, Q) loop. To see this we note that the starting position $\alpha_1 = (h, k) = (1, 1) \in P$. Proceeding vertically, the second position $\alpha_2 \in Q$ and is collinear with α_1 since both lie in the first column. Similarly, $\alpha_3 \in Q$ and is collinear with α_2 since both lie in row 3. Next, $\alpha_4 \in Q$ and is collinear with α_3. Therefore properties L1, L2, and L3 are satisfied and hence the sequence is a (P, Q) chain with $s = 4$ elements. Finally, since α_1 and α_4 are collinear and $s \geq 4$, property L4 holds so that the sequence is also a (P, Q) loop. One can verify all this at a glance by observing the configuration of symbols * and # in Figure 6.2. The visual economy made possible by the symbols will be exploited in a number of the figures below.

Now suppose that P^* and Q^* are subsets of N such that $P^* \supset P$ and $Q^* \supset Q$; then the sequence above is also a (P^*, Q^*) chain and a (P^*, Q^*) loop. This condition holds true even if each of the positions α_i is in both P^* and Q^*. For example, the sequence is also a (N, N) loop.

Given any sets P and Q and a position $(h, k) \in P$, the following labeling process can be used to determine if there is a (P, Q) loop with $\alpha_1 = (h, k)$. First, assign the label $\{h\}$ to column k (see Figure 6.3), and then proceed inductively with labeling routines (1) and (2).

(1) If column j has been labeled, if row i has not been labeled, and if $(i, j) \in Q$, then assign label $J(i) = \{j\}$ to row i. [See Figure 6.4 in which we assume that $(3, 1) \in P$ but $(3, 1) \notin Q$ and that the labeling continues from Figure 6.3.]

(2) If row i has been labeled, if column j has not been labeled, and if $(i, j) \in P$, then assign label $I(j) = \{i\}$ to column j. (See Figure 6.5 in which the labeling continues from that in Figure 6.4.)

Eventually we arrive at (a) *breakthrough:* row h receives a label (see

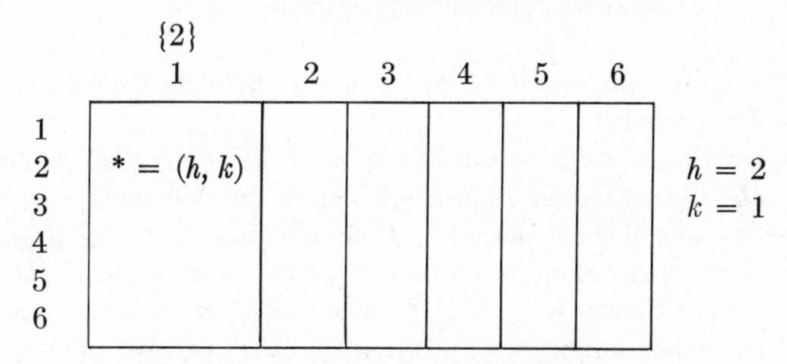

Figure 6.3 Labeling the starting position.

		{2}					
		1	2	3	4	5	6
	1						
	2	* = (h, k)					
	3	*					
{1}	4	#					
	5						
{1}	6	#					

Figure 6.4 Row Labeling.

		{2}		{4}	{6}		{6}
		1	2	3	4	5	6
	1						
	2	* = (h, k)					
	3	*					
{1}	4	#	#	*			
	5						
{1}	6	#		#	*		*

Figure 6.5 Column Labeling.

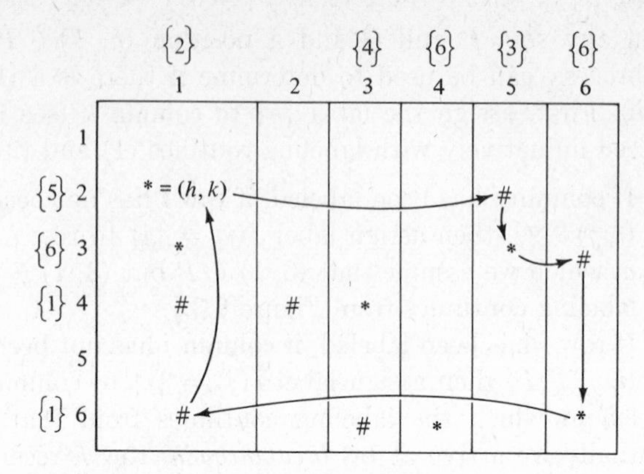

Figure 6.6 Breakthrough example.

Figure 6.6) or (b) *nonbreakthrough:* no further labeling is possible and row h has not been labeled.

Suppose that row i has been labeled and that $(i, j) \in P$. Then, except for one case illustrated in Figure 6.7 below, by following the labels in reverse order we can construct a (P, Q) chain connecting (i, j) to (h, k). For example, given the labeling in Figure 6.6, we can construct a (P, Q) chain C beginning at position $\alpha_1 = (3, 5) \in P$. The row of this starting position has the label $\{6\}$ which implies that position (3, 6) is in Q. We take $\alpha_2 = (3, 6)$ as the second position in C. This column has the label $\{6\}$ so we set $\alpha_3 =$ (6, 6). Proceeding in this way we obtain in turn $\alpha_4 = (6,1)$, and $\alpha_5 = (2, 1) = (h, k)$. Note that the positions in C belong alternately to P and to Q. Of course, if we reverse C we obtain a chain $\alpha_5, \alpha_4, \alpha_3, \alpha_2, \alpha_1$ starting at $(h, k) = (2, 1)$ and ending at $(i, j) = (3, 5)$.

More precisely, if row i has been labeled, if $(i, j) \in P$, and if $i \neq h$, then for some s the sequence

$$(i(1), j(1)), (i(1), j(2)), (i(2), j(2)), \cdots, (i(s), j(s)) \tag{6.37}$$

[where (a): $(i(1), j(1)) = (i, j)$; (b): $j(t) = J(i(t-1)$, $i(t) = I(j(t))$, $t = 2, \cdots, s$; and (c): $(i(s), j(s)) = (h, k)$] will be a (P, Q) chain like that in Figure 6.8, beginning at (i, j) and ending at (h, k). The restriction $i \neq h$ is necessary to avoid a case with three positions in row h as illustrated in Figure 6.7.

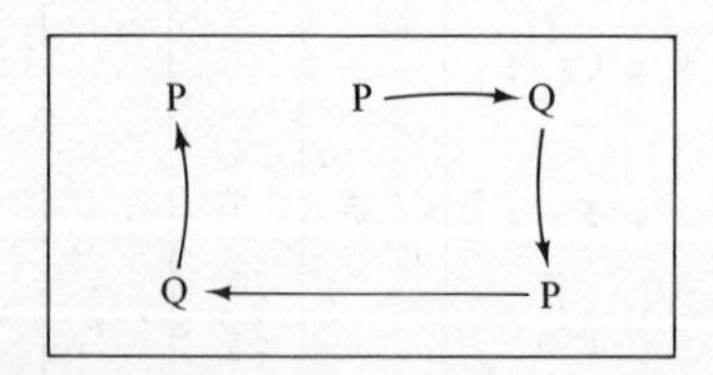

Figure 6.7 The excluded case.

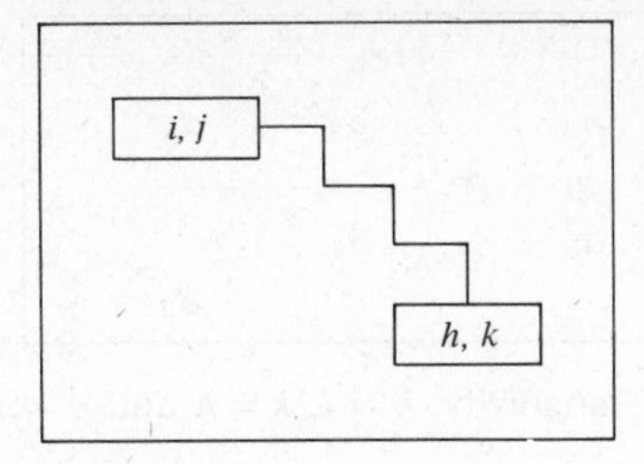

Figure 6.8 First step horizontal, last step vertical.

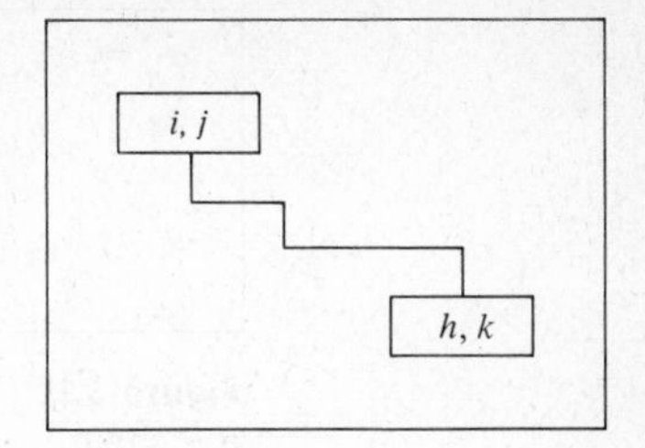

Figure 6.9 First step vertical, last step vertical.

Similarly, if column j has been labeled and if $(i, j) \in Q$, there will be an s for which the (P, Q) chain

$$(i(1), j(1)), (i(2), j(1)), (i(2), j(2)), \cdots, (i(s), j(s)), (i(s+1), j(s)), \quad (6.38)$$

satisfying (a), (b′), and (c′) joins (i, j) to (h, k) as in Figure 6.9. Here (b'): $i(t) = I(j(t-1))$, $t = 2, \cdots, s+1$, $j(t) = J(i(t))$, $t = 2, \cdots, s$, and (c′): $(i(s+1), j(s)) = (h, k)$. In this case if $i = h$, then (6.38) is, moreover, a (P, Q) loop.

The distinction between sequences of type (6.37) and (6.38) may be best expressed by the comment that in (6.37) the first step is horizontal and the last is vertical whereas in (6.38) both the first step and last step are vertical.

In the breakthrough case if row h receives the label $\{j\}$, then by the definition of labeling process $(h, j) \in Q$. We conclude that in the breakthrough case there is a (P, Q) loop beginning with (h, k).

In the nonbreakthrough case we denote by R_L, C_L, R_U, C_U, respectively, the sets of labeled rows, labeled columns, unlabeled rows, unlabeled columns. Then the following statement is valid:

(1) If $(i, j) \in (R_L, C_U)$, then $(i, j) \in P^C = N - P$ (that is, P^C is the complement of P in N), and

(2) If $(i, j) \in (R_U, C_L)$, then either $(i, j) \in Q^C = N - Q$ or $(i, j) = (h, k) \in Q \cap P$. (6.39)

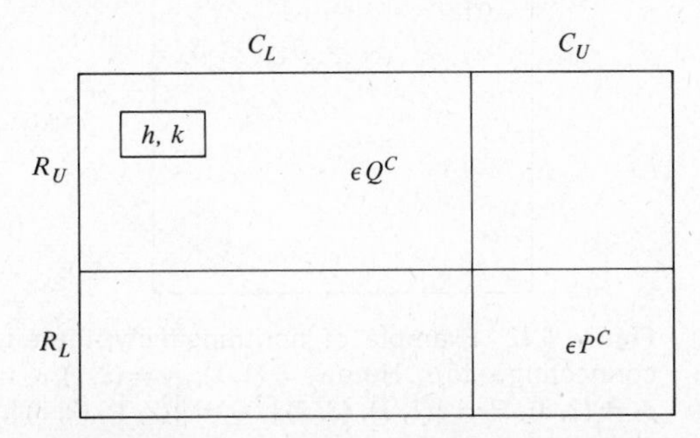

Figure 6.10 Nonbreakthrough configuration.

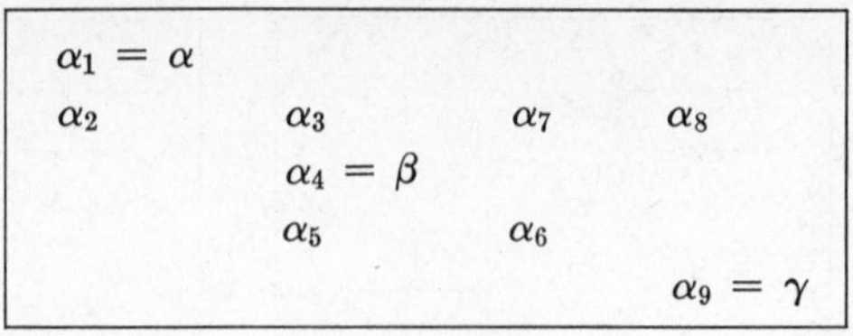

$\alpha_1 = \alpha$			
α_2	α_3	α_7	α_8
	$\alpha_4 = \beta$		
	α_5	α_6	
			$\alpha_9 = \gamma$

Figure 6.11 Transitivity: $h = 2$, $k = 4$ and $\alpha = \alpha_1, \alpha_2, \alpha_8, \alpha_9 = \gamma$ is a P chain.

Figure 6.10 illustrates the nonbreakthrough case when the labeled and unlabeled rows and columns are grouped together. In practice, the starting position (h, k) will frequently be selected as an element of P which is not in Q so that the second alternative in (2) of (6.39) does not arise.

In summary, the labeling process either locates a (P, Q) loop beginning at (h, k) or proves that there is no such loop and provides by means of (6.39) a useful classification of positions.

The case $P = Q$ presents some additional features of interest since in this case chains can be used to set up an equivalence relation in the set P. We abbreviate the notation of (P, P) chain (loop) to P chain (loop). We introduce a binary relation $\sim$ called *equivalence* on P by the definition: $\alpha \sim \beta$ if (a) $\alpha = \beta$, or if (b) there is a P chain connecting α to β. Reflexivity and symmetry of $\sim$ are obvious; to show that $\sim$ is an equivalence relation we need only establish transitivity. If the relations $\alpha \sim \beta$ and $\beta \sim \gamma$ are established by the P chains $\alpha = \alpha_1, \cdots, \alpha_t = \beta$ and $\beta = \alpha_t, \alpha_{t+1}, \cdots, \alpha_s = \gamma$, respectively, then either the sequence $\alpha_1, \cdots, \alpha_s$ is a P chain connecting α to γ or by deletion of some set of consecutive elements $\alpha_{t-h+1}, \cdots, \alpha_{t+k-1}$ with $h \geq 0$, $k \geq 0$ it becomes such a P chain (see Figure 6.11).

To establish this we observe that the sequence $\alpha_1, \cdots, \alpha_s$ will be a P chain unless there are three collinear positions in the sequence. If this happens, the positions involved cannot all belong to either of the original P chains. Hence, there exists an $h > 0$ such that α_{t-h} is the first element

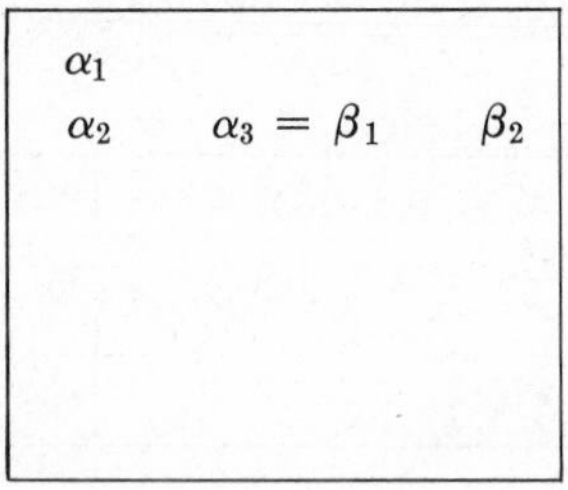

α_1		
α_2	$\alpha_3 = \beta_1$	β_2

Figure 6.12 Example of nontransitivity; there is no (P, Q) chain connecting α_1 to β_2. Here $\alpha_1 = (1, 1)$, $\alpha_2 = (2, 1)$, $\alpha_3 = (2, 2)$, $\beta_1 = (2, 2)$, $\beta_2 = (2, 3)$, $P = \{(1, 1), (2, 2)\}$, $Q = \{(2, 1), (2, 3)\}$.

of the first chain to be involved and then there exists a $k > 0$ such that α_{t+k} is the last element of the second chain collinear with α_{t-h}. Then $\alpha = \alpha_1, \cdots, \alpha_{t-h}, \alpha_{t+h}, \cdots, \alpha_s = \gamma$ is a P chain. See Figure 6.12 for an example which shows that transivity breaks down for (P, Q) chains.

Next, suppose that α and β are two (distinct) elements of P and suppose that

$$\alpha = \alpha_1, \cdots, \alpha_s = \beta \tag{6.40}$$

and

$$\alpha = \beta_1, \cdots, \beta_t = \beta \tag{6.41}$$

are two different P chains connecting α and β. Then the following construction leads to a P loop. Suppose that $\alpha_1 = \beta_1, \cdots, \alpha_h = \beta_h, \alpha_{h+1} \neq \beta_{h+1}$ and let $\alpha_k = \beta_{k'}$ be the equality with smallest $k > h$ and $k' > h$ (see Figure 6.13). Consider the sequence

$$\alpha_{h+1}, \cdots, \alpha_{k-1}, \alpha_k, \beta_{k'-1}, \cdots, \beta_{h+1}, \alpha_h; \tag{6.42}$$

if neither of the pairs $(\alpha_{k-1}, \beta_{k'-1})$, $(\alpha_{h+1}, \beta_{h+1})$ is collinear (see Figure 6.13), then (6.42) is a P loop. If either (see Figures 6.14 and 6.15) or both (see Figure 6.16) of these pairs is collinear, delete, correspondingly, either or both of α_k, α_h from (6.42) and what remains is a P loop.

Recall that in our use of the labeling process for the general case of (P, Q) chains the central question is existence of a (P, Q) loop which starts at some preassigned position (h, k). However, in the case where $P = Q$, we are interested in locating any P loop whatever, and in case no such loops exist, we use the equivalence relation $\sim$ to partition P into its disjoint connected subsets (sometimes called *trees*). Two modifications of the labeling process are required; we refer to the first as *double labeling* and to the second as *reverse labeling*.

To introduce double labeling we modify the labeling routines (1) and (2) as follows:

(1′) If column j has been labeled, and if $(i, j) \in P$, then assign the label $\{j\}$ to row i. If row i, for $i \neq h$, has already been assigned a

1	$\alpha = \alpha_1 = \beta_1$		β_2			
2			β_3			β_4
3		α_4			α_5	
4					α_6	$\beta = \alpha_7 = \beta_5$
5	α_2	α_3				

$h = 1 \quad k = 7 \qquad k' = 5$

Figure 6.13 α_{k-1} is not collinear with $\beta_{k'-1}$ and α_{h+1} is not collinear with β_{h+1}. The P loop is $\alpha_1, \alpha_2, \alpha_3, \alpha_4, \alpha_5, \alpha_6, \alpha_7, \beta_4, \beta_3, \beta_2$.

Row						
1	$\alpha = \alpha_1 = \beta_1$					
2			β_4			β_5
3		α_4			α_5	
4					α_6	$\beta = \alpha_7 = \beta_6$
5	$\alpha_2 = \beta_2$	α_3	β_3			
		$h = 2$	$k = 7$		$k' = 6$	

Figure 6.14 α_{k-1} is not collinear with $\beta_{k'-1}$ and α_{h+1} is collinear with β_{h+1}. The P loop is $\alpha_3, \alpha_4, \alpha_5, \alpha_6, \alpha_7, \beta_5, \beta_4, \beta_3$.

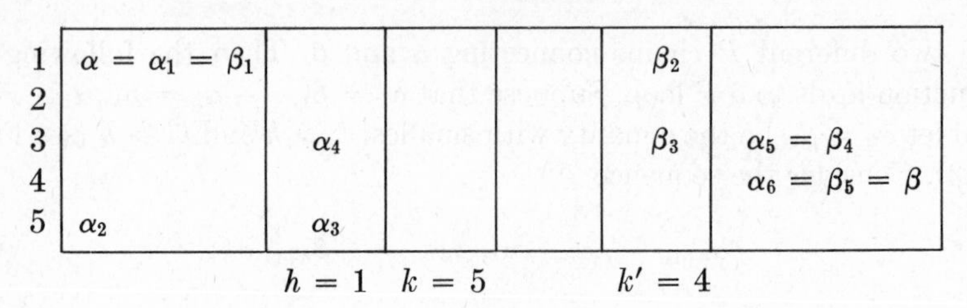

Figure 6.15 α_{k-1} is collinear with $\beta_{k'-1}$ and α_{h+1} is not collinear with β_{h+1}. The P loop is $\alpha_1, \alpha_2, \alpha_3, \alpha_4, \beta_3, \beta_2$.

Row						
1	$\alpha = \alpha_1 = \beta_1$					
2			β_4		β_5	
3		α_4			β_6	$\alpha_5 = \beta_7$
4						$\alpha_6 = \beta_8 = \beta$
5	$\alpha_2 = \beta_2$	α_3	β_3			
		$h = 2$	$k = 5$		$k' = 7$	

Figure 6.16 α_{k-1} is collinear with $\beta_{k'-1}$ and α_{h+1} is collinear with β_{h+1}. The P loop is $\alpha_3, \alpha_4, \beta_6, \beta_5, \beta_4, \beta_3$.

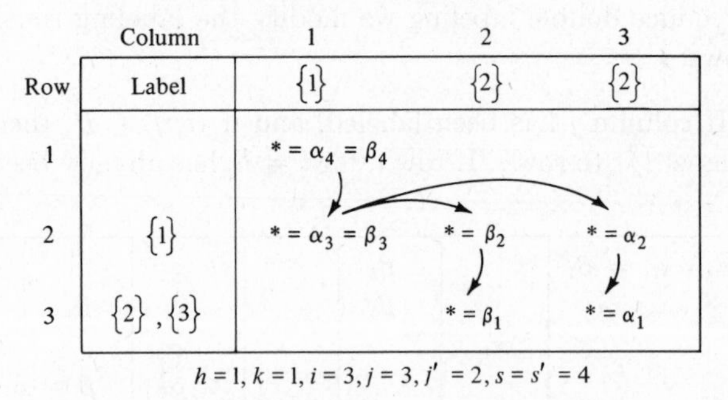

Figure 6.17 A row receives a second label.

label, say $\{j'\}$ (see Figure 6.17), proceed to (3), if $i = h$ proceed to (5), otherwise continue the labeling process with step (2′).

(2′) If row i has been labeled, and if $(i, j) \in P$ then assign the label

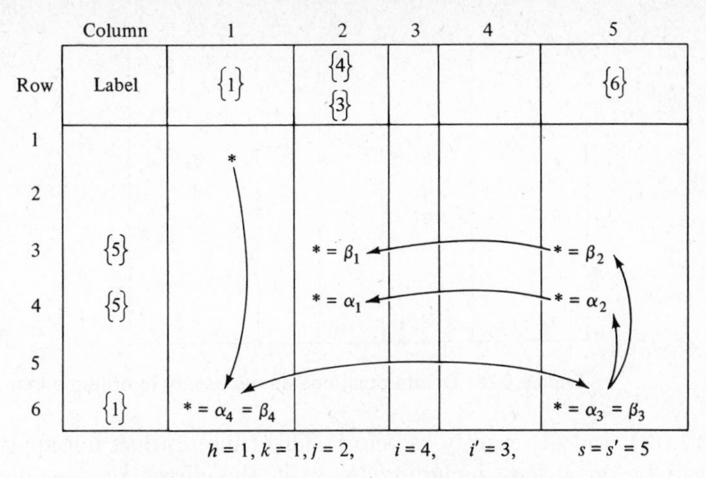

Figure 6.18 A column receives a second label.

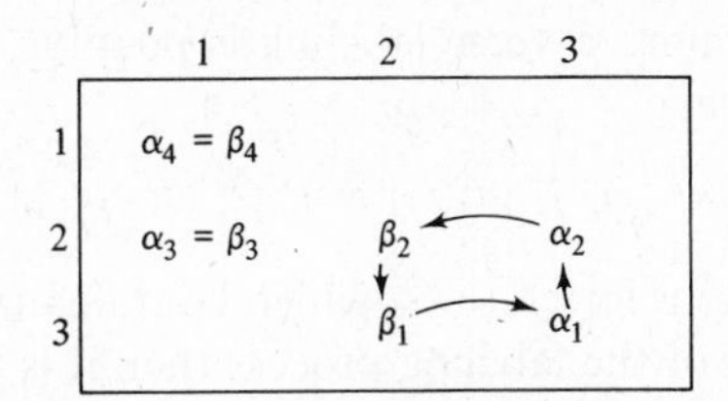

Figure 6.19 Delete positions as necessary from $\alpha_1, \alpha_2, \alpha_3, \alpha_4, = \beta_4, \beta_3, \beta_2, \beta_1$, to obtain the loop $\alpha_1, \alpha_2, \beta_2, \beta_1$.

$\{i\}$ to column j. If column j has already been assigned a label, say $\{i'\}$, proceed to (4) (see Figure 6.18), otherwise continue the labeling process with step (1′).

(3) Let $(i, j) = \alpha_1, \cdots, \alpha_s = (h, k)$ and $(i, j') = \beta_1, \cdots, \beta_{s'} = (h, k)$ be the P chains defined by applying (6.38) beginning, respectively, with (i, j) and (i, j'). Then apply the loop construction process of (6.40), (6.41), (6.42) to the two P chains $\alpha_1, \cdots, \alpha_s$ and $\alpha_1, \beta_1, \cdots, \beta_{s'}$ to determine a P loop (see Figure 6.19).

(4) Let $(i, j) = \alpha_1, \cdots, \alpha_s = (h, k)$ and $(i', j) = \beta_1, \cdots, \beta_{s'} = (h, k)$ be the P chains defined by applying (6.37) beginning, respectively, with (i, j) and (i', j). Then apply the loop construction process of (6.40), (6.41), (6.42) to the two P chains $\alpha_1, \cdots, \alpha_s$ and $\alpha_1, \beta_1, \cdots, \beta_{s'}$ to determine a P loop (see Figure 6.20).

(5) The P chain obtained by applying (6.38) beginning with (h, j) will be a P loop.

The process continues until (a) we construct a P loop, or (b) no more labeling is possible. In case (b) we apply what we call "reverse labeling." We begin again at (h, k) and this time assign label $\{k\}$ to row h. We apply

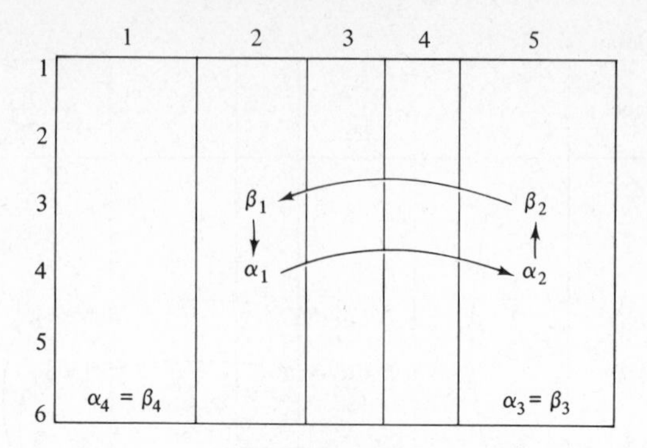

Figure 6.20 Delete positions as necessary to obtain a loop.

(1′), (2′), (3), and (4) exactly as before. This time (5) does not apply since there can be no P loop including (h, k) if the direct process does not lead to one. The reverse labeling continues until (a) we construct a P loop, or (b) until no more reverse labeling is possible.

In Case (b) the set

$$T = \{(i, j) \mid (i, j) \in P \cap (R_L, C_L)\} \tag{6.43}$$

is the equivalence class relative to $\sim$ which contains (h, k) and it is a tree.

By the properties of the labeling process then it is true that no element of P not in T lies in either R_L or C_L, that is, if $(i, j) \in P$ either $(i, j) \in (R_L, C_L)$ or $(i, j) \in (R_U, C_U)$. If $P' = P - T$ is not empty, we repeat the entire labeling process beginning with some position $(h', k') \in P'$. Eventually, we either construct a P loop or decompose P into a union of trees such that no two positions in different trees are collinear.

To verify that T is a tree suppose to the contrary that

$$L = (\alpha_1, \cdots, \alpha_{2s})$$

is a T loop. By our construction we know that (h, k) does not belong to L and that there exists a P chain

$$\beta_1 = (h, k), \beta_2, \cdots, \beta_r = \alpha_j$$

joining (h, k) to L and such that no β_j for $j < r$ is in L. Then by the nature of the labeling process one of the positions adjacent to α_j will be in the same line as β_{r-1} and α_j. By renumbering the elements of L and reversing the sequence if necessary we may assume that $j = 1$ and that β_{r-1}, α_1, α_{2s} are collinear. Then as we continue the labeling process we will construct a P chain

$$\beta_1, \cdots, \beta_{r-1}, \alpha_1, \alpha_2, \cdots, \alpha_q$$

for some $q < 2s$ with the property that either the row or the column of α_{q+1} will be double labeled, once from some β_i with $1 < i < r$ and again

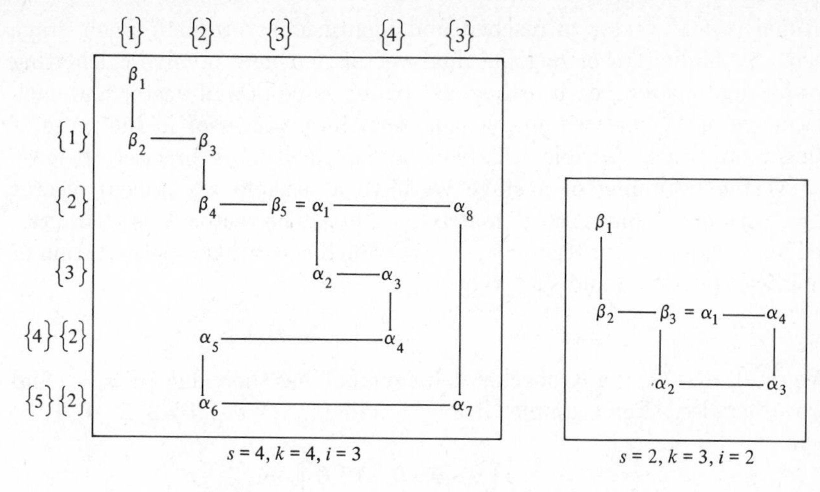

Figure 6.21 Double labeling.

from α_q. We know that $i > 1$ because otherwise (h, k) would belong to a P loop. Two typical situations are illustrated in Figure 6.21. For example, in the illustration on the left in Figure 6.21, the sequence $\beta_1, \beta_2, \beta_3, \beta_4, \beta_5 = \alpha_1, \alpha_2, \alpha_3, \alpha_4$ is a P chain which cannot be extended under the labeling process by adding α_5 because the row containing α_5 has already been labeled from β_3. But the row of α_5 would otherwise have been labeled from α_3, so it is double labeled. The illustration on the right in Figure 6.21 presents a case in which $q = 2s - 1$.

Finally, to show that T contains all positions $(i, j) \sim (h, k)$ suppose that there is a P chain

$$\alpha_1 = (h, k), \alpha_2, \cdots, \alpha_s = (i, j).$$

We first observe that both row h and column k are labeled because we have included both forward and reverse labeling. Moreover, by the nature of the labeling process we know that both the row and column of each position $\alpha_2, \cdots, \alpha_s$ in the chain have been labeled; hence $a_s \in T$. This shows that T is the full equivalence class of (h, k) relative to $\sim$.

6.6. THE GRAVES-THRALL PRIMAL ALGORITHM.

The algorithm which we now describe is adapted to the assignment problem from a more general algorithm which was constructed at the RAND Corporation in 1965 (Reference 22). This algorithm can be initiated with any feasible primal vector and proceeds toward optimality by stages each of which replaces a feasible primal vector X (displayed in matrix form) by a new one X^* with lower cost $[\alpha(X^*) < \alpha(X)]$ until an optimal primal vector is reached and identified as optimal. Each stage begins by finding an orthogonal dual vector and may involve calculating several dual vectors, each orthogonal to the same primal vector and each successive one being an improvement over its predecessor in the sense of being more nearly feasible (this will be described more precisely below).

At the beginning of a stage we have a feasible assignment vector $X = [x_{ij}]$ and a "current cost" matrix $A = [a_{ij}]$. The vector X is characterized by a sequence of integers $(h_1, \cdots, h_n)$ which constitute a permutation of the integers $1, \cdots, n$ and such that

$$x_{h_i j} = \delta_{ij}, \qquad (i, j = 1, \cdots, n). \tag{6.44}$$

(We recall that δ_{ij}, the Kronecker delta symbol, has the value 1 if $i = j$ and zero otherwise.) For example, if $n = 5$ and $(h_1, \cdots, h_5) = (3, 2, 4, 1, 5)$ then

$$X = \begin{bmatrix} 0 & 0 & 0 & 1 & 0 \\ 0 & 1 & 0 & 0 & 0 \\ 1 & 0 & 0 & 0 & 0 \\ 0 & 0 & 1 & 0 & 0 \\ 0 & 0 & 0 & 0 & 1 \end{bmatrix}. \tag{6.45}$$

Next, we define a dual vector (U, V) by letting

$$u_i = 0, \qquad (i = 1, \cdots, n), \qquad v_j = a_{h_j j} \qquad (j = 1, \cdots, n) \tag{6.46}$$

and define an adjusted cost matrix $A^* = [a^*_{ij}]$ by

$$a^*_{ij} = a_{ij} - u_i - v_j, \qquad (i, j = 1, \cdots, n). \tag{6.47}$$

We observe that

$$a^*_{h_j j} = 0, \qquad (j = 1, \cdots, n) \tag{6.48}$$

so that

$$a^*_{ij} x_{ij} = 0, \qquad (i, j = 1, \cdots, n) \tag{6.49}$$

and hence U, V is orthogonal to X. Also $\alpha^*(X) = \Sigma_{ij} a^*_{ij} x_{ij} = 0$ so that

$$\begin{aligned} \alpha(X) &= \alpha^*(X) + \textstyle\sum_i u_i + \sum_j v_j \\ &= 0 + 0 + \textstyle\sum_j v_j. \end{aligned} \tag{6.50}$$

Now, if $A^* \geq 0$, α^* cannot have a negative value for any feasible primal vector so that X is optimal and we are done. Otherwise, there will be some position (h, k) for which $a^*_{hk} < 0$. We define

$$P = \{(i, j) | a^*_{ij} \leq 0\}, \qquad Q = \{(i, j) | x_{ij} = 1\} \tag{6.51}$$

and initiate a (P, Q) labeling at (h, k). Now, either we achieve

(B) Breakthrough,

or

(N) No breakthrough.

In case (B), let L be a (P, Q) loop beginning at (h, k). We let

$$\begin{aligned} L_{\text{odd}} &= \{\text{odd-numbered positions in } L \text{ (counting } (h, k) \text{ as position one)}\}, \\ L_{\text{even}} &= \{\text{even-numbered positions in } L\}. \end{aligned} \tag{6.52}$$

By the nature of the labeling process we note that

$$L_{\text{odd}} \subset P, \; L_{\text{even}} \subset Q. \tag{6.53}$$

We next define a new primal vector X^* by

$$\begin{aligned} x_{ij}^* &= 1 \; (= x_{ij} + 1) \quad \text{if} \quad (i, j) \in L_{\text{odd}}, \\ x_{ij}^* &= 0 \; (= x_{ij} - 1) \quad \text{if} \quad (i, j) \in L_{\text{even}}, \\ x_{ij}^* &= x_{ij} \quad \text{if} \quad (i, j) \notin L. \end{aligned} \tag{6.54}$$

Since each row or column of X either contains no position in L or contains one position in L_{odd} and one in L_{even}, we see that in any line where X is changed there is one increase and one decrease so that each line of X^* sums to 1; hence X^* is primal feasible. Now, since by (6.48) $a_{ij}^* = 0$ for all $(i, j) \in Q$, the only possible nonzero terms in the sum for $\alpha^*(X^*)$ are those for positions (i, j) in L_{odd}. Each nonzero summand will be negative by the definition of P and at least one summand, namely the term $a_{hk}^* x_{hk}^*$, is negative. Hence

$$\alpha^*(X^*) = \textstyle\sum_{(i,j) \in L_{\text{odd}}} a_{ij}^* < 0 \tag{6.55}$$

so that

$$\begin{aligned} \alpha(X^*) &= \alpha^*(X^*) + \textstyle\sum_i u_i + \sum_j v_j \\ &= \alpha^*(X^*) + \alpha(X) < \alpha(X); \end{aligned} \tag{6.56}$$

that is, X^* is better than X. We are now ready to initiate a new state with X^*.

In case (N), we let

$$\epsilon = \min \{-a_{hk}^*, \min \{a_{ij}^* \mid (i, j) \in (R_L, C_U)\}\}. \tag{6.57}$$

By (6.39), if $(i, j) \in (R_L, C_U)$ then $(i, j) \in P^C$ whence by (6.51), $a_{ij}^* > 0$ and therefore,

$$\epsilon > 0. \tag{6.58}$$

Now we define new dual vectors $(\Delta U^*, \Delta V^*)$, and (U^*, V^*) as follows:

$$\begin{aligned} \Delta u_i^* &= 0 && \text{if} \quad i \in R_U, \\ \Delta u_i^* &= \epsilon && \text{if} \quad i \in R_L, \\ \Delta v_j^* &= 0 && \text{if} \quad j \in C_U, \\ \Delta v_j^* &= -\epsilon && \text{if} \quad j \in C_L; \end{aligned} \tag{6.59}$$

and let

$$\begin{aligned} a_{ij}^{**} &= a_{ij}^* - \Delta u_i^* - \Delta v_j^* \\ &= a_{ij} - u_i^* - v_j^*, \end{aligned} \tag{6.60}$$

where

$$u_i^* = u_i + \Delta u_i^*, \qquad v_j^* = v_j + \Delta v_j^* \qquad (i, j = 1, \cdots, n). \tag{6.61}$$

Then

$$\begin{aligned} a_{ij}^{**} &= a_{ij}^* && \text{if} \quad (i, j) \in (R_L, C_L) \cup (R_U, C_U) \\ &= a_{ij}^* + \epsilon && \text{if} \quad (i, j) \in (R_U, C_L) \\ &= a_{ij}^* - \epsilon && \text{if} \quad (i, j) \in (R_L, C_U). \end{aligned} \tag{6.62}$$

We note the following: (1) that the only decreases in costs are at positions in (R_L, C_U) and by the definition of ϵ no such position becomes negative, and (2) that no infeasible position (one for which $a_{ij}^* < 0$) becomes less feasible for (U^*, V^*) than for (U, V) and that since $a_{hk}^{**} > a_{hk}^*$ at least one position, namely (h, k), is numerically closer to being feasible for (U^*, V^*) than for (U, V). We describe this situation by saying that (U^*, V^*) is *more feasible* than (U, V).

We now have three possibilities:

N1. $\qquad A^{**} \geq 0,$

N2. $\qquad a_{hk}^{**} = 0 \quad \text{but} \quad A^{**} \ngeq 0,$

N3. $\qquad a_{hk}^{**} < 0.$

In Case N1, X^* is now identified as optimal and we are done.

In Case N2, we substitute A^{**} for A^*, (U^*, V^*) for (U, V), return to (6.51) and initiate a new labeling at any position where the ** cost is negative.

In Case N3, we could proceed as in Case N2 and again initiate the labeling at (h, k). However, we can achieve an equivalent result by continuing our existing labeling with a modified P defined by A^{**} instead of A^*. By the definition of ϵ and the fact that in Case N3, $\epsilon < -a_{hk}^*$, there is at least one position (i, j) in the previous set (R_L, C_U) which was initially in P^C but which is now in P (with $a_{ij}^{**} = 0$). Hence, the continued labeling will add at least one column to the labeled set. We continue labeling until we again reach either (B) or (N) and proceed as described above for these cases.

We contend that the algorithm must eventually terminate with a nonnegative adjusted cost matrix and hence with an optimal solution. To establish this it is sufficient to show that T1, T2, and T3 hold.

T1. There can be only a finite number of stages (that is, only a finite number of primal improvements),

T2. There can be only a finite number of occurrences of N2 in any stage, and

T3. There can be only a finite number of consecutive occurrences of N3 in any stage.

To establish T1 we observe that there are only a finite number of assignment vectors X and no assignment can appear at two stages since the cost decreases at every stage. To establish T2 we observe (1) that there are only finitely many positions (i, j) with $a^*_{ij} < 0$, (2) that no new such negative positions are introduced within any stage, and (3) that each occurrence of N2 reduces (by at least one) the total number of negative positions. Hence N2 can occur only a finite number of times within any stage. To establish T3 we observe that each occurrence of N3 increases the number of labeled columns by at least one and hence must either reach N2 or label the column j for which $x_{hj} = 1$ and so achieve a breakthrough.

One important feature of the algorithm is that once the first adjusted cost matrix A^* is defined, no transformation from one cost matrix to another either introduces any new negative position or decreases any cost which is negative. Moreover, for each transformation at least one negative position has increased cost. We may summarize these three properties of the algorithm by saying that each labeling pass leads to a new adjusted cost matrix which is uniformly closer to optimal than its predecessor.

We can initiate the algorithm with any feasible assignment; for example with the *diagonal assignment* $x_{ij} = \delta_{ij}$ $(i, j = 1, \cdots, n)$. However, some care in selection of the initial assignment may reduce computing time considerably. The Vogel approximation discussed below for the general transportation problem provides an excellent starting assignment for hand calculation but requires too many search operations to be effective in a computer algorithm.

The starting assignment used in our computer algorithm defines the initial sequence of integers $(h_1, \cdots, h_n)$ by

$$\begin{aligned} a_{h_1 1} &= \min \{a_{i1}\}, \\ a_{h_2 2} &= \min \{a_{i2} \mid i \neq h_1\}, \\ a_{h_3 3} &= \min \{a_{i3} \mid i \neq h_1, h_2\}, \\ &\cdots \\ a_{h_j j} &= \min \{a_{ij} \mid i \neq h_1, h_2, \cdots, h_{j-1}\}, \\ &\cdots \\ a_{h_n n} &= \min \{a_{in} \mid i \neq h_1, h_2, \cdots, h_{n-1}\}. \end{aligned} \tag{6.63}$$

In words, we let $a_{h_1 1}$ be the smallest cost in the first column; then cover the row and column of $a_{h_1 1}$ and let $a_{h_2 2}$ be the smallest cost in the first column of the remaining matrix, and proceed in this manner until finally $a_{h_n n}$ is the only remaining position. For example, suppose

$$A = \begin{bmatrix} 2 & 3 & 5 & 3 & 1 \\ 3 & 2 & 1 & 4 & 2 \\ 1 & 3 & 5 & 1 & 4 \\ 4 & 4 & 1 & 2 & 2 \\ 6 & 4 & 3 & 5 & 4 \end{bmatrix}, \tag{6.64}$$

our initial assignment X has $(h_1, \cdots, h_5) = (3, 2, 4, 1, 5)$ and is given by (6.45) with corresponding cost $z = 11$. Continuing with the algorithm we write the components of U and V as given by (6.46) adjacent to the appropriate rows and columns of A. This facilitates the computation of A^*; thus we have

$$\begin{array}{c|ccccc|} & 1 & 2 & 1 & 3 & 4 \\ \hline 0 & 2 & 3 & 5 & 3^* & 1 \\ 0 & 3 & 2^* & 1 & 4 & 2 \\ 0 & 1^* & 3 & 5 & 1 & 4 \\ 0 & 4 & 4 & 1^* & 2 & 2 \\ 0 & 6 & 4 & 3 & 5 & 4^* \\ \hline \end{array} \qquad \text{Cost} = 11 \tag{6.65}$$

where the asterisks indicate the assignments given by X. The transformed matrix A^* is then

$$\begin{array}{c|ccccc|} & 1 & 2 & 1 & 3 & 4 \\ \hline 0 & 1 & 1 & 4 & 0^* & -3 \\ 0 & 2 & 0^* & 0 & 1 & -2 \\ 0 & 0^* & 1 & 4 & -2 & 0 \\ 0 & 3 & 2 & 0^* & -1 & -2 \\ 0 & 5 & 2 & 2 & 2 & 0^* \\ \hline \end{array} \tag{6.66}$$

and $\Sigma v_j = 11 = \alpha(X)$.

Here we have five negative positions, and

$$P = \{(1, 4), (1, 5), (2, 2), (2, 3), (2, 5), (3, 1), (3, 4), (3, 5), (4, 3), (4, 4), (4, 5), (5, 5)\},$$
$$Q = \{(3, 1), (2, 2), (4, 3), (1, 4), (5, 5)\}.$$

We choose $(h, k) = (1, 5)$ as starting position for our labeling because it has the largest negative adjusted cost. The labels are placed to the left of the u_i and above the v_j. The labeling stops after row 5 has been labeled.

$$\begin{array}{cc|ccccc|} & & & & & & -2 \\ & & & & & & \{1\} \\ & & 1 & 2 & 1 & 3 & 4 \\ \hline & 0 & 1 & 1 & 4 & 0^* & \boxed{-3} \\ & 0 & 2 & 0^* & 0 & 1 & -2 \\ & 0 & 0^* & 1 & 4 & -2 & 0 \\ & 0 & 3 & 2 & 0^* & -1 & -2 \\ 2\ \{5\} & 0 & 5 & 2 & 2 & 2 & 0^* \\ \hline \end{array} \qquad \begin{array}{l} (h, k) = (1, 5) \\ \text{Cost} = 11. \end{array} \tag{6.67}$$

We have

$$\epsilon = \min \{-(-3), \min \{5, 2, 2, 2\}\} = 2; \tag{6.68}$$

we are in case (N3) since $a_{hk}^{**} = a_{15}^{**} = -3 + \epsilon = -3 + 2 = -1 < 0$. Now $\Delta v_5^* = -2$, $\Delta u_5^* = 2$ (these values are placed adjacent to the labels in (6.67)) and the new matrix is shown in (6.69).

						−1
			{5}	{5}	{5}	{1}
		1	2	1	3	2
{4}	0	1	1	4	0*	[−1]
{2}	0	2	0*	0	1	0
	0	0*	1	4	−2	2
{3}	0	3	2	0*	−1	0
{5}	2	3	0	0	0	0*

$(h, k) = (1, 5)$
New cost $= 11 - 1 = 10$. (6.69)

Note that we now have only three negative positions and that the three new zeros in row five enable us to continue the labeling and obtain a breakthrough with the loop $L = \{(1, 5), (1, 4), (5, 4), (5, 5)\}$. We change to the new assignment having $x_{31}^* = x_{22}^* = x_{43}^* = x_{54}^* = x_{15}^* = 1$ and adjust the costs to obtain orthogonality by selecting $\Delta v_5^* = -1$.

It should be observed that we keep a cumulative record of the dual vector changes so that at any stage we could add the u_i, v_j to the current costs and recover the initial matrix A of (6.64) and also calculate the new cost of assignment.

This time we start our labeling at $(h, k) = (3, 4)$;

				−1	−1	−1	
				{5}	{5}	{3}	
			1	2	1	3	1
		0	1	1	4	0	0*
1	{2}	0	2	0*	0	1	1
		0	0*	1	4	[−2]	3
1	{3}	0	3	2	0*	−1	1
1	{4}	2	3	0	0	0*	1

$(h, k) = (3, 4)$
Cost $= 10$. (6.70)

We are again in Case (N3) with $\epsilon = 1$.

				−1	−1	−1	−1
				{5}	{5}	{3}	{5}
			1	1	0	2	1
1	{5}	0	1	2	5	1	0*
1	{2}	1	1	0*	0	1	0
		0	0*	2	5	[−1]	3
1	{3}	1	2	2	0*	−1	0
1	{4}	3	2	0	0	0*	0

$(h, k) = (3, 4)$
Cost $= 10$. (6.71)

This time we are in Case N2 with $\epsilon = 1$.

$$
\begin{array}{cc|ccccc|l}
 & & & & & -1 & & \\
 & & & \{5\} & \{5\} & \{4\} & \{5\} & \\
 & & 1 & 0 & -1 & 1 & 0 & \\
\hline
 & 1 & 0 & 2 & 5 & 1 & 0^* & \\
 & 2 & 0 & 0^* & 0 & 1 & 0 & (h, k) = (4, 4) \\
 & 0 & 0^* & 3 & 6 & 0 & 4 & \text{New cost} = 10 - 1 = 9. \\
\{3\} & 2 & 1 & 2 & 0^* & \boxed{-1} & 0 & \\
\{4\} & 4 & 1 & 0 & 0 & 0^* & 0 & \\
\hline
\end{array}
\qquad (6.72)
$$

This time we reach breakthrough and the resulting solution is optimal and has cost $\Sigma u_i + \Sigma v_j = 9 + 0 = 9$.

$$
\begin{array}{c|ccccc|l}
 & 1 & 0 & -1 & 0 & 0 & \\
\hline
1 & 0 & 2 & 5 & 2 & 0^* & \\
2 & 0 & 0^* & 0 & 2 & 0 & \\
0 & 0^* & 3 & 6 & 1 & 4 & \text{Optimal cost} = 9. \\
2 & 1 & 2 & 0 & 0^* & 0 & \\
4 & 1 & 0 & 0^* & 1 & 0 & \\
\hline
\end{array}
\qquad (6.73)
$$

As a check we add the final u_i, v_j to their rows and columns and obtain the initial cost matrix with optimal assignment indicated.

$$
\begin{array}{c|ccccc|}
 & 1 & 0 & -1 & 0 & 0 \\
\hline
1 & 2 & 3 & 5 & 3 & 1^* \\
2 & 3 & 2^* & 1 & 4 & 2 \\
0 & 1^* & 3 & 5 & 1 & 4 \\
2 & 4 & 4 & 1 & 2^* & 2 \\
4 & 6 & 4 & 3^* & 5 & 4 \\
\hline
\end{array}
\qquad (6.74)
$$

6.7. SOME COMPUTATIONAL EXPERIENCE WITH THE GRAVES-THRALL ALGORITHM.

Appendix C gives a general and a detailed flow chart for the Graves-Thrall Assignment Algorithm. To test the efficiency of the algorithm a set of ten problems whose size and costs were randomly generated were solved by (A) the Graves-Thrall capacitated transportation algorithm (which is described in the next chapter), (B) the Graves-Thrall assignment algorithm, and (C) the Balinşki-Gomory assignment algorithm.

Table 6.3. SUMMARY OF SOLUTIONS FOR RANDOMLY GENERATED ASSIGNMENT PROBLEMS.

		Graves-Thrall algorithm for capacitated transportation problem			*Adaptation of Graves-Thrall algorithm for assignment problem*			*Same assignment routine but with labelling set of Balinski-Gomory*		
Problem number	n *(size)*	*Loops*	*Dual improvements*	*Solution time (sec.)*	*Loops*	*Dual improvements*	*Solution time (sec.)*	*Loops*	*Dual improvements*	*Solution time (sec.)*
1	16	3	21	5.13	3	17	2.63	4	26	2.90
2	5	3	4	4.83	2	5	0.35	2	5	0.35
3	14	4	14	3.97	5	10	2.02	5	28	2.65
4	9	4	14	1.55	3	13	0.88	1	11	0.80
5	14	3	10	2.43	2	15	2.07	2	13	1.92
6	4	1	3	0.33	1	3	0.27	1	3	0.27
7	10	3	7	1.58	3	6	1.12	3	7	1.15
8	14	3	12	2.80	2	11	1.88	4	17	2.17
9	16	4	14	4.27	4	13	2.43	5	23	2.82
10	14	3	15	3.57	4	15	2.13	5	20	2.33
Totals	116	31	114	30.46	29	108	15.78	32	153	17.36

The results are summarized in Table 6.3. (Since these experiments were performed there have been a number of programming improvements which reduce the time required in using each of the three algorithms. However, experience has shown that the relative times have not changed significantly.)

The purpose of the comparison between (A) and (B) was to determine what might be gained in computer time by eliminating some routines in the more general purpose algorithm (A) that were not needed for the simple case of the assignment problem. For the ten test problems the total time needed for (A) was almost twice that of (B).

Algorithm C is a relatively new primal assignment algorithm and was one of the first to compete on even terms in the theoretical bound for the maximum number of operations needed for a solution with the bound given some years ago by Munkries for a dual algorithm. The Graves-Thrall algorithm was inspired by the Balinski-Gomory one and differs from it in essentially only one feature, namely in the definition of P. In the Graves-Thrall algorithm

$$P = \{(i, j) \mid a^*_{ij} \leq 0\} \tag{6.75}$$

whereas in the Balinski-Gomory algorithm one selects a single position (h, k) with $a^*_{hk} < 0$ and then defines P by

$$P = \{(i, j) \mid a^*_{ij} = 0\} \cup \{(h, k)\}. \tag{6.76}$$

This apparently minor change has three important consequences: (1) in the Graves-Thrall algorithm each dual transformation (a) guarantees improvement at (h, k), (b) introduces no new dual infeasibilities, and (c) increases no current dual infeasibility, whereas for the Balinski-Gomory algorithm (a) and (b) hold but not (c); (2) in the Graves-Thrall algorithm there can be contributions to improvement in cost from any number of positions in a (P, Q) loop, whereas in the Balinski-Gomory algorithm only position (h, k) can contribute; and (3) the greater size of P makes breakthrough more likely.

The following example

$$\begin{bmatrix} 0^* & 1 & 2 & 3 & 4 & 5 & -1 \\ -1 & 0^* & 1 & 2 & 3 & 4 & 5 \\ 5 & -1 & 0^* & 1 & 2 & 3 & 4 \\ 4 & 5 & -1 & 0^* & 1 & 2 & 3 \\ 3 & 4 & 5 & -1 & 0^* & 1 & 2 \\ 2 & 3 & 4 & 5 & -1 & 0^* & 1 \\ 1 & 2 & 3 & 4 & 5 & -1 & 0^* \end{bmatrix} \tag{6.77}$$

illustrates consequences (1), (2), and (3). If we start with $(h, k) = (2, 1)$ the Graves-Thrall algorithm obtains breakthrough with the loop

$$L = ((2, 1), (1, 1), (1, 6), (6, 6), (6, 5), (5, 5), (5, 4), (4, 4), (4, 3), (3, 3), (3, 2,) (2, 2)) \tag{6.78}$$

with new assignments at

$$L_{\text{odd}} = ((2, 1), (1, 6), (6, 5), (5, 4), (4, 3), (3, 2)) \tag{6.79}$$

and cost improvements of 6. On the other hand the Balinski-Gomory algorithm requires 6 labeling passes and transformations before breakthrough is obtained and at each pass until the last some negative cost becomes more negative (that is, some dual infeasibility is increased).

The ten test problems show ten percent more computer time for the Balinski-Gomory algorithm than for the Graves-Thrall algorithm, and thirty-five percent more labeling passes and consequent transformations. Since the total computer time included a number of operations common to all three algorithms the number of transformations may be a better comparative measure than computer time. The remarkable fact is that even the poorest of the three averaged only three seconds per problem, and the two better ones averaged under two seconds and did 16-man assignments in less than 3 seconds. For algorithm (B) the equation $t = (n^2 + 10)/100$ gives a fairly good fit for computer time versus problem size. Whether a computer library should stock only a more general purpose algorithm such as (A) or whether it should have both (A) and (B) depends on the range of problems expected. Unless quite a few rather large assignment problems are expected the computer time is so minor that even a fifty percent reduction in computer time is of questionable value.

PROBLEMS

6.1. Suppose the following is a cost matrix for an assignment problem. Write out the dual problem and the complementary slackness equations.

$$\begin{bmatrix} 2 & 8 & 7 & 2 \\ 2 & 7 & 1 & 5 \\ 9 & 1 & 3 & 6 \\ 6 & 2 & 8 & 6 \end{bmatrix}$$

6.2. Let

$$A^* = \begin{bmatrix} 2 & 0 & 0 \\ 0 & 1 & 4 \\ 1 & 3 & 0 \end{bmatrix}, \quad U = \begin{bmatrix} 2 \\ 1 \\ -1 \end{bmatrix}, \quad V = \begin{bmatrix} 0 \\ 1 \\ 5 \end{bmatrix}.$$

(a) Determine an optimal assignment vector.
(b) What is the cost of this assignment?
(c) Calculate the initial cost matrix A.

6.3. Suppose an assignment problem has cost matrix A, where

$$A = \begin{bmatrix} 0 & 1 & 3 & 2 & 2 \\ 1 & 2 & 0 & 5 & 1 \\ 4 & 0 & 1 & 3 & 0 \\ 2 & 2 & 1 & 0 & 4 \\ 0 & 1 & 2 & 1 & 3 \end{bmatrix},$$

and let Z be the set of zero positions in A.

(a) List the elements in Z.
(b) Find a maximal independent set of zero positions.
(c) What is the independence order of Z?
(d) Find a minimal covering set for Z.
(e) What is the covering index of Z?

6.4. Using the Hungarian method, solve the assignment problem whose cost matrix is given in Problem 6.3.

6.5. Let * represent a P position and # a Q position below. Use the labeling process to find a (P, Q) loop beginning at position (1, 1).

Row \ Column	1	2	3	4	5
1	*				#
2	#	*		*	#
3					
4		#	*		*
5			#		*

6.6. Let * represent a P position below.

Row \ Column	1	2	3	4	5
1			*		*
2		*	*		
3	*	*			*
4	*	*			
5					*

(a) Find a P loop beginning at position (1, 3).
(b) Using the double-labeling procedure, begin labeling at position (5, 5); can you find a P loop?

6.7. Given the cost matrix

$$A = \begin{bmatrix} 4 & 2 & 3 & 1 & 5 \\ 1 & 2 & 1 & 1 & 9 \\ 6 & 1 & 8 & 6 & 2 \\ 2 & 4 & 2 & 2 & 1 \\ 1 & 3 & 4 & 1 & 3 \end{bmatrix}.$$

(a) Find a starting assignment X using the rule (6.63).
(b) Find a dual vector (U, V) orthogonal to X.
(c) Compute the adjusted cost matrix A^*.

6.8. The matrix below is the cost matrix for an assignment problem. An assignment is indicated by asterisks. What is the set P? What is the set Q?

$$\begin{bmatrix} -2 & 4 & 2 & 0^* & 1 \\ 2 & 0^* & 0 & 6 & -4 \\ -2 & 3 & -6 & 3 & 0^* \\ 4 & 1 & 0^* & 0 & 3 \\ 0^* & -3 & 1 & 6 & 4 \end{bmatrix}$$

6.9. The cost matrix for an assignment problem is given by

$$A = \begin{bmatrix} 5 & 0^* & 5 & -1 & -3 \\ 6 & 6 & 0^* & 0 & 1 \\ 5 & 4 & 0 & 3 & 0^* \\ 0^* & 0 & 1 & 2 & 2 \\ -4 & 2 & 3 & 0^* & 1 \end{bmatrix}$$

and an assignment vector X is indicated by asterisks.
(a) List the elements of the sets P and Q.
(b) Find a (P, Q) loop beginning at position (5, 1).

6.10. The cost matrix A and vectors U and V are shown below (* indicates an assignment).

	−3	2	4	2	6
0	9	0	5	0*	−3
0	10	6	0*	1	1
0	9	4	0	4	0*
0	4	0*	1	3	2
0	0*	2	3	1	1

(a) List the elements of the sets P and Q.
(b) Begin labeling at position (1, 5) and either find a better assignment X^* or find a dual vector (U^*, V^*) more feasible than (U, V).

For each of the following three problems, the adjusted cost matrix A^*, the vectors U and V, and an assignment vector X (indicated by asterisks) are given. Begin labeling at the position enclosed in a box and carry out two complete steps of the Graves-Thrall algorithm.

6.11.

U \ V	4	0	2	3
0	5	0*	2	4
0	3	$\boxed{-2}$	0*	2
0	1	6	−1	0*
0	0*	4	7	0

6.12.

U \ V	2	5	3	4
0	3	3	0*	0
0	0*	2	−1	2
0	−1	0*	$\boxed{-3}$	1
0	7	−1	6	0*

6.13.

U \ V	2	4	6	7
0	0*	−2	−5	$\boxed{-7}$
0	3	5	−2	0*
0	5	2	0*	−2
0	3	0*	−3	−5

6.14. Solve the following assignment problem, given the initial assignment and orthogonal vectors. Show value of minimum cost.

$U \; \boxed{A^{*X}}$ with V =

	2	3	4	3
0	0^1	0	−1	2
0	−1	3	0^1	0
0	3	0^1	2	−2
0	6	4	1	0^1

6.15. Solve the following assignment problem, where the numbers in the matrix represent productivity ratings.

Man \ Job	1	2	3	4
A	3	1	6	4
B	4	4	5	1
C	2	6	1	6
D	7	1	4	8

6.16. Solve the following assignment problem, where the numbers in the box represent the cost of assigning a man to a job.

		Job 1	2	3	4	5
Man	A	8	2	8	3	2
	B	3	9	2	1	1
	C	7	9	7	7	3
	D	5	3	4	5	1
	E	1	1	4	4	9

6.17. Solve the following assignment problem, where the numbers in the box represent the cost of assigning a man to a job.

		Job 1	2	3	4	5
Man	A	5	3	7	3	4
	B	5	6	12	7	8
	C	2	8	3	4	5
	D	9	6	10	5	6
	E	3	2	1	4	5

6.18. Solve Problem 6.17 if the numbers in the box represent the value of a man in a job.

The Capacitated Transportation Problem

7

7.1. THE CAPACITATED TRANSPORTATION PROBLEM AND ITS DUAL.

Transportation problems and their generalization, network flow problems, have occupied an important place in mathematical programming, both from the point of view of important applications and an interesting theoretical framework. In these notes we consider the standard transportation problem with the modification that there may be finite shipping capacities over some or all of the routes.

The algorithm described below was developed (Reference 22) for the still more general class of problems in which the cost of shipping over a given route is given by a convex, piecewise linear function rather than being limited to a single linear function. We believe that this algorithm will be useful even for noncapacitated transportation problems. This belief is based on our computational experience as well as on certain theoretical features which we examine below.

The algorithm employs many features of Fulkerson's out-of-kilter algorithm (Reference 18, p. 162 ff.) and also borrows heavily from the Balinski-Gomory primal algorithm for transportation problems (Reference 3). The motivation in developing this new algorithm is to see how far one can exploit the trade off between generality of formulation and computational efficiency. The labeling process described for general networks by Ford and Fulkerson (Reference 18, pp. 17–19) and also used by Balinski and Gomory in the algorithm just cited was presented in Section 6.5 in the form in which it is used in the algorithm we will develop.

The simple form of the subalgorithm given below for constructing a feasible solution or proving infeasibility for a capacitated transportation problem is derived from a feasibility algorithm for circulations in more general networks given by Ford and Fulkerson (Reference 18, pp. 52–53), and is also closely related to Beale's algorithm (Reference 5).

The significance of upper-bound constraints in network flows requires little discussion. In many applications of the transportation model these constraints have been neglected with the intent of simplifying the solution and with the hope that so few of them would be violated in an otherwise optimal feasible solution that patch-up techniques could be used to restore feasibility and still have a good solution even if not an optimal one. In the present algorithm there is no need for this avoidance of true constraints since they are handled with very little more work than is needed for noncapacitated problems.

We begin with a transportation problem with p origins and n destinations; let b_i be the amount to be shipped from origin i, let c_j be the amount needed at destination j, and define a shipping program $X = [x_{ij}]$ by specifying the amount x_{ij} to be shipped from origin i to destination $j (i = 1, \cdots, p; j = 1, \cdots, n)$. We assume that supply and demand are in balance, so that

$$\textstyle\sum_i b_i = \sum_j c_j, \tag{7.1}$$

and we assume that a_{ij} is the known unit shipping cost from origin i to destination j. We wish to minimize the total shipping cost

$$z = \alpha(X) = \textstyle\sum_{ij} a_{ij} x_{ij} \tag{7.2}$$

subject to the feasibility requirements

$$\textstyle\sum_j x_{ij} = b_i, \qquad (i = 1, \cdots, p), \tag{7.3}$$

$$\textstyle\sum_i x_{ij} = c_j, \qquad (j = 1, \cdots, n), \tag{7.4}$$

$$x_{ij} \geq 0, \qquad (i = 1, \cdots, p; j = 1, \cdots, n), \tag{7.5}$$

$$x_{ij} \leq e_{ij}, \qquad (i = 1, \cdots, p; j = 1, \cdots, n), \tag{7.6}$$

where e_{ij} represents the capacity or upper bound for shipments from origin i to destination j. We allow the possibility of infinite capacity ($e_{ij} = \infty$) and otherwise assume that the b_i and c_j are positive integers, that the e_{ij} are nonnegative integers, and that the a_{ij} are arbitrary real numbers (positive, negative, or zero).

The problem of existence of a feasible solution is treated in Section 7.2 where an algorithm is given which either leads to a feasible program or establishes infeasibility.

The dual problem is to maximize

$$w = \beta(U, V, W) = \textstyle\sum_i b_i u_i + \sum_j c_j v_j - \sum_{ij} e_{ij} w_{ij} \tag{7.2'}$$

subject to the feasibility requirements

$$u_i \text{ free}, \qquad (i = 1, \cdots, p), \tag{7.3'}$$
$$v_j \text{ free}, \qquad (j = 1, \cdots, n), \tag{7.4'}$$
$$u_i + v_j - w_{ij} \leq a_{ij}, \qquad (i = 1, \cdots, p; j = 1, \cdots, n), \tag{7.5'}$$
$$w_{ij} \geq 0, \qquad (i = 1, \cdots, p; j = 1, \cdots, n). \tag{7.6'}$$

Let X and (U, V, W) be any primal and dual vectors (feasible or not) and consider the difference of the objective functions; by some simple algebraic manipulations we can show that

$$\begin{aligned} z - w &= \alpha(X) - \beta(U, V, W) \\ &= \textstyle\sum_{ij}(a_{ij} - u_i - v_j + w_{ij})x_{ij} + \\ &\textstyle\sum_i((\sum_j x_{ij}) - b_i)u_i + \sum_j((\sum_i x_{ij}) - c_j)v_j + \sum_{ij}(e_{ij} - x_{ij})w_{ij}. \end{aligned} \tag{7.7}$$

The reader may wish to observe that (7.7) is analogous to the manipulations in (6.14). Also, as in (6.15) let

$$a_{ij}^* = a_{ij} - u_i - v_j, \qquad (i = 1, \cdots, p; j = 1, \cdots, n). \tag{7.8}$$

The primal and dual vectors are said to be *orthogonal* if each individual summand in (7.7) is zero, that is, if

$$(a_{ij}^* + w_{ij})x_{ij} = 0, \qquad (i = 1, \cdots, p; j = 1, \cdots, n), \tag{7.9}$$
$$((\textstyle\sum_j x_{ij}) - b_i)u_i = 0, \qquad (i = 1, \cdots, p), \tag{7.10}$$
$$((\textstyle\sum_i x_{ij}) - c_j)v_j = 0, \qquad (j = 1, \cdots, n), \tag{7.11}$$

and

$$(e_{ij} - x_{ij})w_{ij} = 0, \qquad (i = 1, \cdots, p; j = 1, \cdots, n). \tag{7.12}$$

Clearly, $\alpha(X) = \beta(U, V, W)$ if the vectors are orthogonal. If X is feasible (7.10) and (7.11) are satisfied for all dual vectors; thus for feasible X, (7.9) and (7.12) are sufficient conditions for orthogonality.

Next, we observe that if both X and (U, V, W) are feasible each term in (7.9) and (7.12) is nonnegative so that

$$z - w = \alpha(X) - \beta(U, V, W) \geq 0 \tag{7.13}$$

for all feasible X and (U, V, W). Suppose that X_0 and (U_0, V_0, W_0) are both feasible and let $z_0 = \alpha(X_0)$, $w_0 = \beta(U_0, V_0, W_0)$. Then it follows from (7.13) that $z_0 = w_0$ if and only if X_0 and (U_0, V_0, W_0) are orthogonal. Moreover, it is true that $z_0 = w_0$ implies that both X_0 and (U_0, V_0, W_0) are optimal. For let X and (U, V, W) be any feasible vectors. Then by (7.13), $z \geq w_0$ and $z_0 \geq w$; now since $w_0 = z_0$ this implies that

$$z \geq z_0 = w_0 \geq w \tag{7.14}$$

which shows that $z_0 = w_0$ is the minimum value of α and the maximum value of β, so that both X_0 and (U_0, V_0, W_0) are optimal.

These basic relationships between the primal and dual problems can be exploited in a variety of ways in the construction of computer algorithms. The general outline of the Graves-Thrall algorithm which we describe in detail in the three following sections involves three subsidiary algorithms or subalgorithms as follows.

Feasibility. This subalgorithm either establishes that the primal problem is infeasible or constructs a feasible primal vector. Once feasibility is established all primal vectors considered will be feasible.

Orthogonality. This subalgorithm begins with a feasible primal vector and either constructs a dual vector orthogonal to it or constructs a better primal vector (one with smaller cost).

Optimality. This subalgorithm begins with a feasible primal vector and a dual orthogonal vector and at each stage either establishes optimality of the primal vector or constructs a better one. Within a given stage there may be a substage in which the primal vector remains unchanged but a new orthogonal dual vector is constructed which is less infeasible than the initial one.

It will be proved that if the primal problem is feasible then after a finite number of steps the algorithm terminates with optimal primal and dual vectors. Finally, this is described as a *primal algorithm* since it maintains primal feasibility throughout (once feasibility is established). The orthogonality guarantees $z = w$ and the optimality algorithm uses dual infeasibilities as guides for step by step changes each yielding either a less costly feasible primal or a new orthogonal dual with decreased infeasibility. A by-product of the orthogonality subalgorithm is a proof that if the primal problem is feasible and if each b_i and c_j is an integer then it has an optimal feasible solution with integer components; the algorithm itself deals exclusively with integral primal vectors.

7.2. FEASIBILITY SUBALGORITHM FOR CAPACITATED TRANSPORTATION PROBLEM.

We consider the feasibility of the capacitated transportation problem defined by (7.1) through (7.6). The nature of the cost function is irrelevant since we are now only interested in feasibility. In the ordinary transportation problem the capacity constraints (7.6) are not present, and there are well known methods for obtaining a feasible solution. In the capacitated problem we say that a primal vector X is *semifeasible* if it satisfies the ordinary feasibility conditions (7.3), (7.4), and (7.5). We first give the details of several methods for obtaining a semifeasible primal vector X (this problem is, of course, the same as that of obtaining a feasible primal vector for the ordinary transportation problem). A subalgorithm is then developed which, starting with a semifeasible vector, either (1) demonstrates infeasibility or (2) constructs a feasible vector.

One can construct an initial semifeasible primal vector in the following way (called the northwest-corner rule). We begin in the upper left-hand corner (northwest corner) of the X matrix by taking

$$x_{11} = \min\{b_1, c_1\}.$$

If $b_1 < c_1$, we next take $x_{21} = \min\{b_2, c_1 - x_{11}\}$, and proceed in this way down the first column until we have

$$x_{11} = b_1,\ x_{21} = b_2, \cdots, x_{h1} = c_1 - b_1 - b_2 \cdots - b_{h-1} < b_h.$$

Then proceed to the right along the hth row assigning as much as possible to each succeeding position until b_h is exhausted. Next, proceed down the column thus reached and so on until, finally, $x_{pn} = \min\{b_p, c_n\}$ is determined. All other x_{ij} are set equal to zero.

If $c_1 < b_1$, the process begins with the first row until b_1 is exhausted and then alternates between rows and columns until x_{pn} is reached. The general schematic picture of X is,

$$\begin{bmatrix} * & * & & & & & \\ & * & * & * & & & \\ & & & * & & & \\ & & & * & * & * & * \\ & & & & & & * \\ & & & & & & * \\ & & & & & & * & * & * & * \end{bmatrix}, \tag{7.15}$$

that is, there are zeros everywhere except for a descending stair running from the upper left- to the lower right-hand corner. By construction, the row sums and the column sums satisfy (7.3) and (7.4); hence X is a semifeasible vector. Note that this construction yields an X with at most $p + n - 1$ nonzero entries.

We introduce an example which we will use to illustrate various points in our discussion. Let

$$A = \begin{bmatrix} 10 & 20 & 5 & 9 & 10 \\ 2 & 10 & 8 & 30 & 6 \\ 1 & 20 & 7 & 10 & 4 \end{bmatrix}, \quad X = \begin{bmatrix} x_{11} & x_{12} & x_{13} & x_{14} & x_{15} \\ x_{21} & x_{22} & x_{23} & x_{24} & x_{25} \\ x_{31} & x_{32} & x_{33} & x_{34} & x_{35} \end{bmatrix},$$

$$B = \begin{bmatrix} 24 \\ 5 \\ 14 \end{bmatrix}, \quad C = \begin{bmatrix} 3 \\ 6 \\ 4 \\ 9 \\ 21 \end{bmatrix}; \tag{7.16}$$

then the northwest-corner solution

$$X = \begin{bmatrix} 3 & 6 & 4 & 9 & 2 \\ & & & & 5 \\ & & & & 14 \end{bmatrix} \begin{matrix} 24 \\ 5 \\ 14 \end{matrix}$$
$$\begin{matrix} 3 & 6 & 4 & 9 & 21 \end{matrix}$$

has just one bend (here and later, unfilled entries of X are assumed to be zero).

If

$$B = \begin{bmatrix} 6 \\ 8 \\ 2 \\ 4 \end{bmatrix} \quad \text{and} \quad C = \begin{bmatrix} 5 \\ 4 \\ 6 \\ 3 \\ 2 \end{bmatrix}$$

then the northwest-corner solution is

$$X = \begin{bmatrix} 5 & 1 & & & \\ & 3 & 5 & & \\ & & 1 & 1 & \\ & & & 2 & 2 \end{bmatrix} \begin{matrix} 6 \\ 8 \\ 2 \\ 4 \end{matrix}$$
$$\begin{matrix} 5 & 4 & 6 & 3 & 2 \end{matrix}$$

and has six bends (there is a bend downward at the position occupied by the integer 1 in the first row, then a bend to the right at the position occupied by the integer 5 in the second row, and so on).

If for some h and k, $b_1 + \cdots + b_h = c_1 + \cdots + c_k$, we continue the process at position $(h + 1, k + 1)$ just as we did initially at position $(1, 1)$; that is, we set $x_{h+1,k+1} = \min \{b_{h+1}, c_{k+1}\}$ and proceed in whichever direction there is something left over to assign. Such ties indicate degeneracy but require no special treatment in the algorithm. For example, if

$$B = \begin{bmatrix} 6 \\ 6 \\ 3 \\ 5 \end{bmatrix} \quad \text{and} \quad C = \begin{bmatrix} 5 \\ 4 \\ 6 \\ 3 \\ 2 \end{bmatrix}$$

then the northwest-corner vector is

$$X = \begin{bmatrix} 5 & 1 & & & \\ & 3 & 3 & & \\ & & 3 & & \\ & & & 3 & 2 \end{bmatrix} \begin{matrix} 6 \\ 6 \\ 3 \\ 5 \end{matrix};$$
$$\begin{matrix} 5 & 4 & 6 & 3 & 2 \end{matrix}$$

note that X has $7 < p + n - 1 = 8$ nonzero entries.

Vogel's approximation method (Reference 35) is another method for providing an initial semifeasible vector. In essence, this method consists of assigning initial values x_{ij} with some reference to the costs, whereas the costs are not taken into consideration in the northwest-corner solution. One would like to make all assignments to the smallest element in each row and column, but this may not be feasible. The Vogel approximation measures the difference between the two lowest cost entries in a row or column and thus indicates where departure from the smallest cost will yield the largest loss.

We illustrate with our example of (7.16). Writing out the cost matrix and bordering it above and to the left with the b_i and c_j (see Figure 7.1). At the end of each row and at the bottom of each column we write the absolute value of the difference of the two smallest costs. The largest of these differences is the 10 in the second column. We then assign as much as possible to the smallest entry in this column. In this case, $x_{22} = 5$. We enter this assignment as a superscript in position (2, 2) and subtract 5 from a_2 and b_2. Next, we calculate the differences leaving out the second row from consideration since it is already saturated by the first assignment. The largest difference now is 9 in the first column; we assign $x_{31} = 3$ and again adjust the bordering entries. This process continues until finally only one line (in this case, row 1) is left and fill in the remaining requirements in this row. The x_{ij} are indicated as superscripts in the cost matrix. If both a row and a column are saturated simultaneously, both are dropped from consideration for later assignment. The vector X given by Vogel's approximation method is semifeasible. In general, use of the Vogel approximation method may save computation in contrast with use of the northwest-corner method. This can be more than ample compensation for its greater complexity.

Still another procedure that will always generate a semifeasible vector X is that of making assignments one at a time, each of which saturates either a row or a column requirement (the Vogel method is but one of several useful methods which utilize this procedure). For example, one could select the (i, j) with the smallest cost among all positions not in a

			3	6	4	9	21		
			0	1			10		
24			10	20^{1}	5^{4}	9^{9}	10^{10}	4	
5	0		2	10^{5}	8	30	6	4	2
14	11	0	1^{3}	20	7	10	4^{11}	3	
			1	10	2	1	2		
			9	0			6		

Figure 7.1 Illustration of Vogel's approximation.

saturated row or column. However, this method shares with the Vogel method the disadvantage of requiring many scans of the cost matrix. The method used in the following algorithm makes the cheapest possible assignments in column 1 until c_1 is used up, then proceeds to column 2 and so on across the columns in order. More recently (Lee, Reference 30) methods for generating semifeasible vectors have been developed which give some weight to route capacities as well as to costs.

For example, for the problem of (7.16) the above rule yields the semifeasible vector X indicated in (7.17) by the nonzero superscripts on the costs.

$$
\begin{array}{c|ccccc|}
 & 3 & 1 & 4 & 9 & 10 \\
\hline
24 & 10 & 20^{1} & 5^{4} & 9^{9} & 10^{10} \\
50 & 2 & 10^{5} & 8 & 30 & 6 \\
14 & 1^{3} & 20 & 7 & 10 & 4^{11} \\
\hline
\end{array}
\tag{7.17}
$$

It is merely a coincidence that this is the same X as was given by the Vogel approximation.

We are now ready to introduce the subalgorithm (of the Graves-Thrall algorithm) which starts with a semifeasible X and terminates either with a feasible vector or with a demonstration of infeasibility. We are going to be concerned with the relationships, for each position (i, j), that can exist between the lower bound 0, the upper bound e_{ij}, and the assignment x_{ij}. The following five mutually exclusive situations exhaust all the possibilities:

$$
\begin{aligned}
0 &\leq e_{ij} < x_{ij}, \\
0 &< x_{ij} = e_{ij}, \\
0 &< x_{ij} < e_{ij}, \\
0 &= x_{ij} < e_{ij}, \\
0 &= x_{ij} = e_{ij}.
\end{aligned}
$$

We introduce a notation to indicate the sets of positions corresponding to each of these possibilities:

$$
\begin{aligned}
S^{+}(X) &= \{(i, j) \mid 0 \leq e_{ij} < x_{ij}\}, \text{ above capacity;} \\
S^{c}(X) &= \{(i, j) \mid 0 < x_{ij} = e_{ij}\}, \text{ at capacity;} \\
S(X) &= \{(i, j) \mid 0 < x_{ij} < e_{ij}\}, \text{ active (can increase or decrease);} \\
S^{0}(X) &= \{(i, j) \mid 0 = x_{ij} < e_{ij}\}, \text{ zero level activity;} \\
S^{00}(X) &= \{(i, j) \mid 0 = x_{ij} = e_{ij}\}, \text{ excluded activity.}
\end{aligned}
\tag{7.18}
$$

Note that these five sets partition N. When only one X is under discussion below we may abbreviate the notation for one of the sets above by deleting the symbols (X).

S^{+} consists of the set of positions where x_{ij} is over capacity and therefore must be reduced to attain feasibility. We introduce the labeling sets

$P = S \cup S^c \cup S^+$ and $Q = S \cup S^0$. Note that P contains all the positions where $x_{ij} > 0$ (and therefore can be decreased), Q contains all of the positions where $x_{ij} < e_{ij}$ (and therefore can be increased). Clearly, if S^+ is empty the vector X is feasible. If S^+ is not empty, let $(h, k) \in S^+$ and initiate a (P, Q) labeling process at (h, k). If a breakthrough occurs, we have a loop L beginning at (h, k) for which $L_{\text{odd}} \subset P$ and $L_{\text{even}} \subset Q$. Let

$$\delta = \min\{\delta^-, \delta^+\} \tag{7.19}$$

where

$$\delta^- = \min\{x_{ij} \mid (i, j) \in L_{\text{odd}}\}$$

and

$$\delta^+ = \min\{e_{ij} - x_{ij} \mid (i, j) \in L_{\text{even}}\}.$$

Note that δ^+ may be ∞ and that by definition of P and Q, $\delta > 0$. Now the vector X' defined by

$$\begin{aligned} x'_{ij} &= x_{ij} - \delta, && (i, j) \in L_{\text{odd}}, \\ x'_{ij} &= x_{ij} + \delta, && (i, j) \in L_{\text{even}}, \\ x'_{ij} &= x_{ij}, && (i, j) \notin L, \end{aligned} \tag{7.20}$$

is semifeasible and is uniformly more feasible than X in the sense that

(1) $S^+(X') \subset S^+(X)$,
(2) $x'_{ij} \leq x_{ij}$ for all $(i, j) \in S^+(X)$,
(3) $x'_{hk} < x_{hk}$.

In practice a smaller δ than the one given by (7.19) might be preferable. Let

$$\delta^{--} = \min\{x_{ij} - e_{ij} \mid (i, j) \in L_{\text{odd}} \text{ and } x_{ij} - e_{ij} > 0\}$$

and let

$$\delta' = \min\{\delta^-, \delta^+, \delta^{--}\}. \tag{7.21}$$

Then each (integer) δ'' with $\delta' \leq \delta'' \leq \delta$ will yield a new semifeasible vector X'' satisfying (7.21).

The computer algorithm for which the flow chart is given in Appendix C below uses

$$\delta'' = \min\{\delta, x_{hk} - e_{hk}\}. \tag{7.19'}$$

This selection is convenient to program and is adapted to the method used for computing an initial semifeasible X in the sense that it tends to preserve as much use of favorable cost routes as is feasible.

If no breakthrough occurs, let

$$B_2 = \sum b_i, \quad i \in R_L, \qquad C_1 = \sum c_j, \quad j \in C_L, \tag{7.22}$$

and for any p by n matrix $G = [g_{ij}]$ let

$$\begin{aligned} G_{11} &= \sum g_{ij}, & (i, j) &\in (R_U, C_L), \\ G_{12} &= \sum g_{ij}, & (i, j) &\in (R_U, C_U), \\ G_{21} &= \sum g_{ij}, & (i, j) &\in (R_L, C_L), \\ G_{22} &= \sum g_{ij}, & (i, j) &\in (R_L, C_U). \end{aligned} \tag{7.23}$$

For example, this notation means that for G_{11} we sum over g_{ij} for all $i \in R_U$ and $j \in C_L$. For any semifeasible primal vector Y we have

$$Y_{11} + Y_{21} = C_1, \qquad Y_{21} + Y_{22} = B_2, \tag{7.24}$$

and hence

$$Y_{11} = C_1 - B_2 + Y_{22} \geq C_1 - B_2. \tag{7.25}$$

Since [compare (6.39)] for $(i, j) \in (R_U, C_L)$ we cannot have $(i, j) \in Q$, we conclude that $x_{ij} \geq e_{ij}$, and since $x_{hk} > e_{hk}$, we conclude that

$$X_{11} > E_{11}. \tag{7.26}$$

Next, since for $(i, j) \in (R_L, C_U)$ we cannot have $(i, j) \in P$, we conclude that $x_{ij} = 0$, and hence

$$X_{22} = 0. \tag{7.27}$$

Finally, from (7.25) for X instead of Y, and using (7.26) and (7.27) we have

$$Y_{11} \geq C_1 - B_2 = X_{11} > E_{11}, \tag{7.28}$$

so that Y is not feasible. We conclude that the labeling process fails to reach breakthrough only if the problem is infeasible. Now, since each breakthrough leads to an integer reduction in the total amount of infeasibility,

$$\sum x_{ij} - e_{ij}, \qquad (i, j) \in S^+(X) \tag{7.29}$$

we conclude that after a finite number of labeling passes we either demonstrate infeasibility or produce a feasible vector.

The argument concerning infeasibility is a variant of a circulation theorem due to Alan Hoffman (Reference 24; see also Reference 18, p. 50). We may interpret the inequality $E_{11} < C_1 - B_2$ as stating that the amount required at labeled destinations minus the total amount available at labeled origins exceeds the total capacity over routes from unlabeled origins to labeled destinations; and, hence, that there can be no feasible vector.

7.3. ORTHOGONALITY SUBALGORITHM.

The purpose of the orthogonality subalgorithm is to provide a process whose input is a feasible primal vector and whose output is a feasible primal vector at least as good as the input one and a dual vector (not necessarily feasible) orthogonal to it.

The concept of "basic", which plays a central role in the simplex algorithm, appears here under the name "loopless" as a sufficient condition for the existence of a dual vector. Before continuing with the algorithm itself we insert some remarks on solvability of equations of the type involved in the construction of the dual vector.

Let M be any subset of the set N of all positions in a p by n matrix and let $H = [h_{ij}]$ be any real p by n matrix. It is well known (Reference 13, p. 303) that the system of equations

$$u_i + v_j = h_{ij} \qquad \text{for all } (i, j) \in M \tag{7.30}$$

has a solution (U, V) if there are no M loops. Conversely, if there exists an M loop there will be choices of h_{ij} for which the system (7.30) has no solution.

We say that a set M is *loopless* if there are no M loops. In Section 6.5 we proved that if M is loopless, then for some $r \geq 1$, M is the union of disjoint trees $T_1, \cdots, T_r$ and no two positions from distinct trees lie in the same row or column of N. Hence the solution of (7.30) for all positions in M is equivalent to solving the r systems

$$u_i + v_j = h_{ij} \qquad \text{for all } (i, j) \in T_\lambda,\ \lambda = 1, \cdots, r. \tag{7.31}$$

The equivalence follows from the fact that no variable u_i or v_j can appear (with nonzero coefficient) in more than one of the r systems. Therefore, to establish theoretical solvability of (7.30) for all loopless M it is sufficient to consider the case where M is a tree. Practically, we solve for general loopless M by considering its trees one at a time.

Let (h, k) be any fixed reference position in M and let (i, j) be any other position in M. Then since M is a tree there is a unique M chain,

$$\alpha_1 = (i, j),\ \alpha_2, \cdots, \alpha_s = (h, k) \tag{7.32}$$

connecting (i, j) with (h, k). We define a partial order called a *precedence* on M by the statement (i', j') precedes (i, j), written $(i', j')\ p\ (i, j)$, if (i', j') appears in (7.32). Now set $u_h = 0$ and arrange the equations $u_i + v_j = g_{ij}$ for $(i, j) \in M$ in such an order that equation (i, j) follows equation (i', j') if $(i', j')\ p\ (i, j)$. This system of equations is triangular; when equation (i, j) is reached, it will be the first appearance of $v_j(u_i)$ and $u_i(v_j)$ will already have been determined by an earlier equation if row i (column j) is labeled prior to column j (row i).

Thus when M is a tree the Equations (7.30) will have a unique solution with $u_h = 0$. More generally, if $M = T_1 \cup \cdots \cup T_r$ is a partitioning of M into r trees with tree T_i initiating from position $(h(i), k(i))$, then there will be a unique solution of (7.30) having $u_{h(1)} = \cdots = u_{h(r)} = 0$ and having u_i or v_j zero for all rows i or columns j not occupied by any element of M.

This argument provides a constructive proof of the existence of a solution to the system (7.30). It also shows that if M contains m positions then the system (7.30) has rank m.

For example, suppose M consists of the positions indicated by the symbol $*$ in the matrix shown in Table 7.1; there are $r = 2$ trees and $m = 12$ positions in M. From (7.31) after a suitable permutation of rows and columns, the resulting system is shown in Table 7.2. If we set u_1 and u_5 equal to zero, the resulting system of equations is triangular.

Table 7.1. MATRIX DEFINING M WITH $r = 2$.

Table 7.2. TRIANGULAR SYSTEM WHEN $u_1 = u_5 = 0$.

v_1	v_2	v_3	u_2	u_3	u_4	v_4	v_5	u_6	v_7	u_7	v_6	u_1	u_5	$= 1$
1												1		h_{11}
	1											1		h_{12}
		1										1		h_{13}
	1		1											h_{22}
	1			1										h_{32}
		1			1									h_{43}
					1	1								h_{44}
							1						1	h_{55}
							1	1						h_{65}
								1	1					h_{67}
							1			1				h_{75}
										1	1			h_{76}

Returning now to our algorithm, let X be a primal feasible vector with a corresponding set $S(X)$ of positions which we call *active*. We say that X is *loopless* if there are no $S(X)$ loops. If X is loopless we now show how to find a dual vector (U, V, W) orthogonal to X but not necessarily feasible. Since X is feasible we need consider only (7.9) and (7.12). We first apply the theorem represented by (7.30) to obtain (U, V) for which

$$u_i + v_j = a_{ij} \qquad \text{for all } (i, j) \in S(X). \tag{7.33}$$

Using the $a^*_{ij} = a_{ij} - u_i - v_j$ defined by (U, V) we then determine W by the conditions

$$\begin{aligned} w_{ij} &= 0 \qquad \text{for all } (i, j) \in S^0(X) \cup S(X) \\ w_{ij} &= -a^*_{ij} \qquad \text{for all } (i, j) \in S^C(X) \\ w_{ij} &= \max \{0, -a^*_{ij}\} \text{ for all } (i, j) \in S^{00}(X). \end{aligned} \tag{7.34}$$

Table 7.3 indicates how our selection of the dual vector satisfies the orthogonality conditions for each of the subsets into which N is partitioned relative to X.

Suppose next that X is not loopless. We now show how to construct a new loopless vector X' with $S(X') \langle S(X)$ and $\alpha(X') \leq \alpha(X)$. The same

Table 7.3. ORTHOGONALITY ANALYSIS.

	$S^0(X)$	$S(X)$	$S^C(X)$	$S^{00}(X)$
(7.9) $(a^*_{ij} + w_{ij})x_{ij} = 0$	$x_{ij} = 0$	$a^*_{ij} = 0$	$w_{ij} = -a^*_{ij}$	$x_{ij} = 0$
(7.12) $(e_{ij} - x_{ij})w_{ij} = 0$	$w_{ij} = 0$	$w_{ij} = 0$	$e_{ij} = x_{ij}$	$e_{ij} = 0$

routine that establishes the looplessness of X' can be used as an aid in the actual solution of the corresponding dual orthogonality equations.

Since X is not loopless there exists an $S(X)$ loop, call it L. By a change in the starting position for L, if necessary, we may assume that

$$\theta = (\sum a_{ij},\ (i,\ j) \in L_{\text{odd}}) - (\sum a_{ij},\ (i,\ j) \in L_{\text{even}}) \geq 0. \tag{7.35}$$

Next, define δ by (7.19) and X' by (7.20); then

$$\alpha(X) - \alpha(X') = \theta\lambda \geq 0. \tag{7.36}$$

By definition of δ, there will be either an $(i, j) \in L_{\text{odd}}$ for which $\delta = x_{ij}$ or an $(i, j) \in L_{\text{even}}$ for which $\delta = e_{ij} - x_{ij}$. This (i, j) will not belong to $S(X')$. By construction of X' there are no positions in $S(X')$ that are not in $S(X)$. We conclude that $S(X')$ is a proper subset of $S(X)$. By iterations of this process we must eventually reach a loopless vector X'' for which $\alpha(X'') \leq \alpha(X)$ and $S(X'') \langle S(X)$.

Note that the labeling process can be modified so that it simultaneously tests for loops and constructs a dual vector orthogonal to X. The nature of the labeling process guarantees that if T is a tree in $S(X)$ and if (U, V) satisfies the orthogonality equations on T, the components of (U, V) thereby determined need not be changed for any later vector X' obtained from X by the loop-reduction process. Moreover, later labeling processes will never enter the rows and columns occupied by elements of T. This property adds greatly to the efficiency of the algorithm since scanning an entire large matrix can be quite time consuming as compared with scanning smaller and smaller submatrices.

In practice if there are rows and columns not occupied by elements of $S(X)$, we may possibly accelerate the computation by replacing the passive choice of zero for the corresponding u_i and v_j by a more active one which guarantees that in each row and column at least one of the numbers a^*_{ij} will be zero and that none of them will be negative.

In programming this part of the algorithm, it may be desirable to consider the following point. The reason we wished to get X loopless was so that we could solve the dual equations (7.33). Now, if a loop L is found for which (7.35) gives $\theta = 0$, then it is not really necessary to change X since those equations of (7.33) corresponding to this loop will be solvable. Hence, we need only apply (7.22) when $\theta > 0$. With this modification we can strengthen the inequality in (7.36) to $>$, and strengthen the conclusion to read: Either (i) we can solve the system (7.33) or (ii) we can iterate the orthogonality subroutine until we find a vector X'' for which (a') $\alpha(X'') < \alpha(X)$, (b) $S(X'') \langle S(X)$, and (c) equations (7.33) formed for X'' have a solution.

We note that solution of (7.33) is the critical step in constructing the dual vector; because if (U, V) satisfies (7.33) W is explicitly determined by (7.34).

Next, we observe that if X is loopless then it is integral (consists of integer coordinates only). In the first place any x_{ij} which is at capacity must be integral by our hypothesis on the e_{ij}. Next, let $b_i'(c_j')$ be $b_i(c_j)$ reduced by the sum of all x_{ij} which are at capacity in row i (column j). Then X must satisfy the equations

$$\begin{aligned} \sum_{j \epsilon S_i} x_{ij} &= b_i' \qquad (i = 1, \cdots, p), \\ \sum_{i \epsilon S^j} x_{ij} &= c_j' \qquad (j = 1, \cdots, n), \end{aligned} \tag{7.37}$$

where

$$S_i = \{j \mid (i, j) \in S(X)\}$$

and

$$S^j = \{i \mid (i, j) \in S(X)\}.$$

If $b_i' = 0$ then $x_{ij} = 0$, $(j = 1, \cdots, n)$ and if $c_j' = 0$ then $x_{ij} = 0$, $(i = 1, \cdots, p)$. Moreover, no tree in $S(X)$ will have elements in such a row or column. If we delete from (7.37) all equations having zero right-hand side, the coefficient matrix for the residual system is the transpose of that of system (7.33); hence it has rank equal to the number s of variables [the number of positions in $S(X)$] and will become triangular when r redundant equations [where r is the number of trees in $S(X)$] are removed. But, since all of the coefficients are $+1$, -1, or 0 and since the constant terms are integers, it follows that the unique solution X is integral.

If, for example, M is again given by Table 7.1, then equations (7.37) with suitable ordering of equations and variables are given in Table 7.4. We observe that

$$c_1' + c_2' + c_3' - b_2' - b_3' - b_4' + c_4' = b_1'$$

and

$$c_5' - b_6' + c_7' - b_7' + c_6' = b_5';$$

hence the last two equations are redundant and when they are deleted the remaining system is triangular. Clearly the x_{ij} are integral if the b'_i and c'_j are.

Table 7.4. TRIANGULAR SYSTEM AFTER DELETION OF TWO REDUNDANT EQUATIONS.

x_{11}	x_{12}	x_{13}	x_{22}	x_{32}	x_{43}	x_{44}	x_{55}	x_{65}	x_{67}	x_{75}	x_{76}	$= 1$
1												c'_1
	1		1	1								c'_2
		1			1							c'_3
			1									b'_2
				1								b'_3
					1	1						b'_4
						1						c'_4
							1	1		1		c'_5
								1	1			b'_6
									1			c'_7
										1	1	b'_7
											1	c'_6
1	1	1										b'_1
							1					b'_5

What we have proved above, stated in other terms, says that adding a new feasibility condition,

$$x_{ij} \text{ is an integer for all } (i, j), \tag{7.38}$$

to the defining conditions (7.1) through (7.6), although it may exclude some previously optimal feasible vectors that are not integral, will still leave as feasible any loopless vector X and hence will not change the optimal cost in any feasible problem. This serendipitous property of the transportation problem differentiates it sharply in difficulty from genuine integer programs such as the traveling salesman problem.

It is interesting to note that the coefficient matrices in Tables 7.2 and 7.4 are transposes of each other. The triangularity of the first of these matrices enables us to establish the key result that if X is basic (loopless), then there exists an orthogonal dual solution. The second matrix plays a key role in establishing that the components of all basic solutions are integral.

7.4. OPTIMALITY SUBALGORITHM.

We enter the optimality subalgorithm with a feasible vector X and a dual vector (U, V, W) orthogonal to X and with W satisfying (7.34). The dual vector will be feasible and hence both vectors optimal if

$$u_i + v_j - w_{ij} \leq a_{ij} \tag{7.5'}$$

and

$$w_{ij} \geq 0 \tag{7.6'}$$

are satisfied for all $(i, j) \in N$. We recall that we introduced an auxiliary cost matrix $A^* = [a^*_{ij}]$ by setting

$$a^*_{ij} = a_{ij} - u_i - v_j \tag{7.8}$$

for all $(i, j) \in N$. In view of (7.8) we can rewrite (7.5′) in the form

$$a^*_{ij} + w_{ij} \geq 0 \tag{7.39}$$

for all $(i, j) \in N$.

We observe that all feasible X

$$S^{00}(X) = S^{00} = \{(i, j) \mid e_{ij} = 0\}$$

and that the selection $w_{ij} = \max \{0, \ -a^*_{ij}\}$ in (7.34), which may have appeared arbitrary when introduced, satisfies both (7.6′) and (7.39) for all $(i, j) \in S^{00}$. Next, for $(i, j) \in S(X)$ we have $w_{ij} = a^*_{ij} = 0$ and hence both (7.6′) and (7.39) are satisfied as equalities. Finally, we observe that (7.6′) holds as an equality for $(i, j) \in S^0(X)$, so that (7.6′) can fail to hold only if

$$(i, j) \in S^C(X) \qquad \text{and} \qquad a^*_{ij} > 0 \tag{7.40}$$

and (7.39) can fail to hold only if

$$(i, j) \in S^0(X) \qquad \text{and} \qquad a^*_{ij} < 0. \tag{7.41}$$

This suggests a classification of the positions in N according to the sign of a^*_{ij}; indeed N can be partitioned into the four sets

$$\begin{aligned} R^+(A^*) &= \{(i, j) \mid a^*_{ij} > 0, e_{ij} > 0\}, \\ R^0(A^*) &= \{(i, j) \mid a^*_{ij} = 0, e_{ij} > 0\}, \\ R^-(A^*) &= \{(i, j) \mid a^*_{ij} < 0, e_{ij} > 0\}, \\ S^{00} &= \{(i, j) \mid e_{ij} = 0\}. \end{aligned} \tag{7.42}$$

Now because of (7.33) we have $S(X) \subset R^0(A^*)$. We can restate our conclusions about dual feasibility in the following form: $w_{ij} \geq 0$ if and only if

$$Q_0 = S^C(X) \cap R^+(A^*) \tag{7.43}$$

is empty and (7.39) holds if and only if

$$P_0 = S^0(X) \cap R^-(A^*) \tag{7.44}$$

is empty. We call P_0 and Q_0 the *nonoptimality sets* (or *dual infeasibility sets*). The optimality subalgorithm uses the labeling process, beginning at any dual infeasible position, to construct either a new primal X' with $\alpha(X') < \alpha(X)$ in case of breakthrough or a new dual vector (U', V', W'), still orthogonal to X, but uniformly more feasible than (U, V, W).

To make the algorithm effective we must choose labeling sets P and Q so that if a (P, Q) loop is generated then alternate additions and subtractions of some amount δ to the x_{ij} around the loop will decrease the cost at one or more positions and increase it at no position. Suppose that (i, j) is a position where x_{ij} is to be increased; then we must have $x_{ij} < e_{ij}$ and hence $(i, j) \in S^0(X) \cup S(X)$. If the cost is not to be increased we must also have $(i, j) \in R^-(A^*) \cup R^0(A^*)$. This suggests that we choose

$$P = (S^0(X) \cup S(X)) \cap (R^-(A^*) \cup R^0(A^*)). \tag{7.45}$$

Positions in P can be characterized as those in which *one can and is not unwilling to increase current assignments.* Similarly,

$$Q = (S^c(X) \cup S(X)) \cap (R^+(A^*) \cup R^0(A^*)) \tag{7.46}$$

consists of positions where *one can and is not unwilling to decrease current assignments.*

A simple calculation shows that

$$P = P_0 \cup S(X) \cup (S^0(X) \cap R^0(A^*))$$

and

$$Q = Q_0 \cup S(X) \cup (S^c(X) \cap R^0(A^*)),$$

so the nonoptimality sets P_0 and Q_0 satisfy the inclusion relations

$$P_0 \subset P, \qquad Q_0 \subset Q. \tag{7.47}$$

Clearly, if a (P, Q) loop L contains any element of $P_0 \cup Q_0$, there is a cost reduction when the positions in L are alternately increased and decreased by $\delta > 0$.

The most significant feature of the algorithm is that, given the particular choices of P and Q, if there is no breakthrough there is a uniformly better dual vector (U', V', W') available with the properties (1) (U', V', W') is orthogonal to X, and (2) either $P_0' \cup Q_0' < P_0 \cup Q_0$ or a (P, Q) labeling pass starting at the same position will label at least one new row or column. Property (2) guarantees finiteness of the dual-improvement part of the algorithm since the number of labeled lines cannot increase indefinitely without reaching breakthrough and if $P_0' \cup Q_0'$ is empty, then dual feasibility is established.

For an example of some of these concepts, consider the array ${}^EA^{*X}$,

u_i \ v_j	2	5	3	4
0	${}^{7}0^{3}$	${}^{4}5^{4}$	${}^{8}-3$	${}^{3}4$
3	${}^{5}0^{2}$	${}^{2}0^{2}$	${}^{4}0^{2}$	${}^{0}5$
5	${}^{3}3$	0^{3}	${}^{5}-5$	${}^{2}-3^{2}$
−1	${}^{3}4^{3}$	${}^{0}-4$	${}^{10}0^{7}$	0^{6}

,

which was derived from the capacities and cost matrices E and A and the supply and demand vectors B and C where

$$E = \begin{bmatrix} 7 & 4 & 8 & 3 \\ 5 & 2 & 4 & 0 \\ 3 & & 5 & 2 \\ 3 & 0 & 10 & \end{bmatrix} \quad \text{(blank spaces indicate capacity } \infty\text{)},$$

$$\begin{array}{cc} & & b_i \\ A = & \begin{bmatrix} 2 & 10 & 0 & 8 \\ 5 & 8 & 6 & 12 \\ 10 & 10 & 3 & 6 \\ 5 & 0 & 2 & 3 \end{bmatrix} & \begin{matrix} 7 \\ 6 \\ 5 \\ 16 \end{matrix} \\ c_j & \begin{matrix} 8 & 9 & 9 & 8 \end{matrix} & \end{array}$$

We assume that an initial solution X has been given, where

$$X = \begin{bmatrix} 3 & 4 & 0 & 0 \\ 2 & 2 & 2 & 0 \\ 0 & 3 & 0 & 2 \\ 3 & 0 & 7 & 6 \end{bmatrix},$$

and $S(X)$ is loopless. Now from ${}^{E}A^{*X}$ we have the following,

$$\begin{aligned} S^{00} &= \{(2, 4), (4, 2)\} \\ S^{0}(X) &= \{(1, 3), (1, 4), (3, 1), (3, 3)\} \\ S(X) &= \{(1, 1), (2, 1), (2, 3), (3, 2), (4, 3), (4, 4)\} \\ S^{C}(X) &= \{(1, 2), (2, 2), (3, 4), (4, 1)\}. \end{aligned}$$

Also,

$$\begin{aligned} R^{-}(A^*) &= \{(1, 3), (3, 3), (3, 4)\} \\ R^{0}(A^*) &= \{(1, 1), (2, 1), (2, 2), (2, 3), (3, 2), (4, 3), (4, 4)\} \\ R^{+}(A^*) &= \{(1, 2), (1, 4), (3, 1), (4, 1)\}, \end{aligned}$$

and

$$\begin{aligned} P_0 &= S^{0}(X) \cap R^{-}(A^*) = \{(1, 3), (3, 3)\} \\ Q_0 &= S^{C}(X) \cap R^{+}(A^*) = \{(1, 2), (4, 1)\}; \end{aligned}$$

finally,

$$\begin{aligned} P &= P_0 \cup S(X), \text{ since } S^{0}(X) \cap R^{0}(A^*) = \phi \\ Q &= Q_0 \cup S(X) \cup \{(2, 2)\}, \end{aligned}$$

since $S^{C}(X) \cap R^{0}(A^*) = \{(2, 2)\}$.

The optimality subalgorithm is subdivided into two principal cases:

Case (1) P_0 is nonempty;
Case (2) P_0 is empty but Q_0 is nonempty.

This subdivision has the feature that Case 2 is vacuous if all capacities are infinite, that is, for the classical transportation problem.

We examine Case 1 in detail. A (P, Q) labeling is initiated at a position $(h, k) \in P_0$. Then we either get breakthrough (B) or nonbreakthrough (N). For (B) we define a new primal vector X' by

$$
\begin{aligned}
x'_{ij} &= x_{ij} + \delta, & (i, j) &\in L_{\text{odd}}, \\
x'_{ij} &= x_{ij} - \delta, & (i, j) &\in L_{\text{even}}, \\
x'_{ij} &= x_{ij} & (i, j) &\notin L,
\end{aligned}
\tag{7.48}
$$

where

$$\delta = \min \{\delta^-, \delta^+\} \tag{7.49}$$

and

$$
\begin{aligned}
\delta^- &= \min \{x_{ij} \mid (i, j) \in L_{\text{even}}\} \\
\delta^+ &= \min \{e_{ij} - x_{ij} \mid (i, j) \in L_{\text{odd}}\}.
\end{aligned}
$$

By our choice of P and Q we have $\delta > 0$. The change in cost will be

$$\alpha(X') - \alpha(X) = \delta(\textstyle\sum_{L_{\text{odd}}} a^*_{ij} - \sum_{L_{\text{even}}} a^*_{ij}). \tag{7.50}$$

Since $L_{\text{odd}} \subset P \subset R^-(A^*) \cup R^0(A^*)$ and $L_{\text{even}} \subset Q \subset R^+(A^*) \cup R^0(A^*)$, each term in the first summand in (7.50) is nonpositive and each term in the second summand is nonnegative. Moreover, the first term a^*_{hk} in the first summand is negative so we conclude that

$$\alpha(X') < \alpha(X); \tag{7.51}$$

that is, X' is better than X. We then return to the orthogonality subalgorithm and enter it with X replaced by X'.

In the nonbreakthrough case (N) we are in the situation pictured by Figure 6.10 above. From (6.39) and (7.45) we have (recall that $P^C = N - P$ and $Q^C = N - Q$)

$$(R_L, C_U) \subset N - P = S^C(X) \cup R^+(A^*) \cup S^{00}, \tag{7.52}$$

and similarly, using (7.46) and that $(h, k) \notin Q$ we have

$$(R_U, C_L) \subset N - Q = S^0(X) \cup R^-(A^*) \cup S^{00}. \tag{7.53}$$

Using the fact that $S(X) \subset R^0(A^*)$ these relations can be restated in the equivalent forms:

$$
\begin{aligned}
&\text{if } (i, j) \in (R_L, C_U), \text{ then either (1) } a^*_{ij} > 0 \text{ and } e_{ij} > x_{ij} = 0, \\
&\qquad \text{or (2) } x_{ij} = e_{ij} > 0, \text{ or (3) } e_{ij} = 0;
\end{aligned}
\tag{7.54}
$$

and

$$
\begin{aligned}
&\text{if } (i, j) \in (R_U, C_L), \text{ then either (1) } a^*_{ij} < 0 \text{ and } x_{ij} = e_{ij} > 0, \\
&\qquad \text{or (2) } e_{ij} > x_{ij} = 0, \text{ or (3) } e_{ij} = 0.
\end{aligned}
\tag{7.55}
$$

We define

$$\epsilon = \min \{\epsilon^+, \epsilon^-, -a^*_{hk}\}, \tag{7.56}$$

where

$$\epsilon^+ = \min \{a^*_{ij} \mid (i, j) \in S^0(X) \cap (R_L, C_U)\} \tag{7.57}$$

and

$$\epsilon^- = \min \{ -a^*_{ij} \mid (i, j) \in S^C(X) \cap (R_U, C_L) \}. \tag{7.58}$$

It follows from (7.54) and (7.55) that ϵ^+ and ϵ^- are each positive; by hypothesis $a^*_{ij} < 0$, so $\epsilon > 0$.

We also define a new dual vector (U', V', W') and a new adjusted-cost matrix $A'^* = [a'^*_{ij}]$ by

$$\begin{aligned} u'_i &= u_i + \epsilon && \text{for } i \in R_L, \\ u'_i &= u_i && \text{for } i \in R_U, \\ v'_j &= v_j - \epsilon && \text{for } i \in C_L, \\ v'_j &= v_j && \text{for } i \in C_U, \end{aligned} \tag{7.59}$$

and

$$a'^*_{ij} = a_{ij} - u'_i - v'_j \text{ for all } (i, j) \in N, \tag{7.60}$$

and W' is defined as in (7.34) except that we use a'^*_{ij} in place of a^*_{ij}. It then follows from (7.8) and (7.59) that

$$\begin{aligned} a'^*_{ij} &= a^*_{ij} \text{ for all } (i, j) \in (R_L, C_L) \cup (R_U, C_U), \\ a'^*_{ij} &= a^*_{ij} - \epsilon \text{ for all } (i, j) \in (R_L, C_U), \\ a'^*_{ij} &= a^*_{ij} + \epsilon \text{ for all } (i, j) \in (R_U, C_L). \end{aligned} \tag{7.61}$$

It can be seen from (7.52), (7.53) and (7.61) that

$$\begin{aligned} &\text{(a) } a'^*_{ij} \geq a^*_{ij} \text{ for all } (i, j) \in P, \\ &\text{(b) } a'^*_{ij} \leq a^*_{ij} \text{ for all } (i, j) \in Q. \end{aligned} \tag{7.62}$$

Also, since $P_0 \subset P$ and $Q_0 \subset Q$, we see that for no infeasible position (i, j) is the amount of infeasibility increased. Moreover, the choices of ϵ^+ and ϵ^-, respectively, guarantee that

$$\begin{aligned} &\text{(a) } P'_0 \subset P_0, \\ &\text{(b) } Q'_0 \subset Q_0, \end{aligned} \tag{7.63}$$

that is, that the infeasibility sets are not enlarged. In addition, since $\epsilon > 0$ we have

$$0 \geq a'^*_{hk} > a^*_{hk}, \tag{7.64}$$

that is, (7.62 (a)) holds with strict inequality for at least the position $(h, k) \in P_0$. We summarize these properties by the statement that (U', V', W') is uniformly more feasible than (U, V, W).

Finally, we claim that if (*i*) $\epsilon = -a^*_{hk}$, then $(h, k) \notin P'_0$ and hence $P'_0 \cup Q'_0 < P_0 \cup Q_0$; if (*ii*) $\epsilon = \epsilon^+ < -a^*_{hk}$, then we can start a (P', Q') labeling process at (h, k) and obtain $R'_L \supset R_L$, $C'_L > C_L$; and if (*iii*) $\epsilon = \epsilon^- < -a^*_{hk}$, we obtain, similarly, $R'_L > R_L$, $C'_L \supset C_L$. To establish (*ii*) and (*iii*) we observe that $(h,k) \in P'_0$ so that (h,k) is eligible as a starting position for a (P', Q') labeling pass. Next, we observe that because of (7.61), $P \cap (R_L, C_L) = P' \cap (R_L, C_L)$ and $Q \cap (R_L, C_L) = Q' \cap (R_L, C_L)$;

now by the nature of the labeling process all labeled columns except column k derive their labels from positions in P that are in rows already labeled and all labeled rows are labeled from positions Q in columns already labeled. Hence for either case (*ii*) or (*iii*) all previously labeled lines will be labeled again (and with the same labels). Now if $\epsilon = \epsilon^+ = a^*_{ij}$ for $(i, j) \in S^0(X) \cap (R_L, C_U)$, then $a'^*_{ij} = 0$ so that $(i, j) \in P'$ and hence column j is labeled in the (P', Q') labeling so that $C'_L > C_L$. Similarly, in case (*iii*) at least one new row is labeled.

The algorithm now proceeds by reentering the optimality subalgorithm with the same X and with (U', V', W') instead of (U, V, W).

For an example, consider the array ${}^EA^{*X}$ given above. Since the set P_0 is nonempty, we could begin our (P, Q) labeling at position (1, 3) or (3, 3). We choose the latter since the cost is more negative for this position. The first and several subsequent steps are shown below.

			−3		−3	−3
			{2}		{3}	{4}
			2	5	3	4
3	{1}	0	$^{7}0^{3}$	$^{4}5^{4}$	$^{8}-3$	$^{3}4$
3	{3}	3	$^{5}0^{2}$	$^{2}0^{2}$	$^{4}0^{2}$	$^{0}5$
		5	$^{3}3$	0^{3}	$^{5}-5$	$^{2}-3^{2}$
3	{3}	−1	$^{3}4^{3}$	$^{0}-4$	$^{10}0^{7}$	0^{6}

Nonbreakthrough: $\epsilon = \min\{-a^*_{hk} = 5, \epsilon^+ = \infty, \epsilon^- = 3\} = 3$.

				2		
			{2}		{3}	{4}
			−1	5	0	1
	{1}	3	$^{7}0^{3}$	$^{4}2^{4}$	$^{8}-3$	$^{3}4$
	{3}	6	$^{5}0^{2}$	$^{2}-3^{2}$	$^{4}0^{2}$	$^{0}5$
−2	{4}	5	$^{3}6$	0^{3}	$^{5}-2^{\delta}$	$^{2}0^{2-\delta}$
	{3}	2	$^{3}4^{3}$	$^{0}-7$	$^{10}0^{7-\delta}$	$0^{6+\delta}$

Breakthrough: $\delta = 2$, X' is loopless.

			3			
			{2}	{3}	{1}	{4}
			−1	7	0	1
−3	{1}	3	$^{7}0^{3-\delta}$	$^{4}0^{4}$	$^{8}-3^{\delta}$	$^{3}4$
−3	{3}	6	$^{5}0^{2+\delta}$	$^{2}-5^{2}$	$^{4}0^{2-\delta}$	$^{0}5$
	{3}	3	$^{3}8$	0^{3}	$^{5}0^{2}$	$^{2}2$
	{3}	2	$^{3}4^{3}$	$^{0}-9$	$^{10}0^{5}$	0^{8}

Breakthrough: $\delta = 2$, X' loopless, $(h, k) = (1, 3)$.

We are now in Case 2 and the example will be completed after we introduce a method for dealing with this case.

Case 2 is essentially dual to Case 1. We interchange the roles of P and Q [defined by (7.45) and (7.46)] and initiate a (Q, P) labeling at some position $(h, k) \in Q_0$. If we obtain breakthrough and a (Q, P) loop L, we can renumber the positions in L so that (h, k) is last instead of first; then $(h, k) \in L_{\text{even}}$. We again define a new primal vector by (7.48) and (7.49) and the argument used in Case 1 to obtain (7.51) requires only the modification that the guaranteed improvement in $\alpha(X')$ follows from the positive term a^*_{hk} in the second summand in the right-hand side of (7.50). As in the breakthrough situation of Case 1, we then return to the orthogonality subalgorithm and enter it with X' instead of X.

In the nonbreakthrough case everything holds with the roles of P and Q interchanged. We use primes to designate corresponding equations and relations and we omit supporting discussions.

$$(R_L, C_U) \subset N - Q = S^0(X) \cup R^-(A^*) \cup S^{00}, \tag{7.52'}$$

$$(R_U, C_L) \subset N - P = S^C(X) \cup R^+(A^*) \cup S^{00}. \tag{7.53'}$$

$$\text{If } (i, j) \in (R_L, C_U), \text{ then either (1), } a^*_{ij} < 0 \text{ and } x_{ij} = e_{ij} > 0, \text{ or (2) } e_{ij} > x_{ij} = 0, \text{ or (3) } e_{ij} = 0. \tag{7.54'}$$

$$\text{If } (i, j) \in (R_U, C_L), \text{ then either (1) } a^*_{ij} > 0 \text{ and } e_{ij} > x_{ij} = 0, \text{ or (2) } x_{ij} = e_{ij} > 0, \text{ or (3) } e_{ij} = 0. \tag{7.55'}$$

$$\epsilon = \min \{\epsilon^+, \epsilon^-, a^*_{hk}\}, \tag{7.56'}$$

$$\epsilon^+ = \min \{ -a^*_{ij} \mid (i, j) \in S^C(X) \cap (R_L, C_U)\}, \tag{7.57'}$$

$$\epsilon^- = \min \{a^*_{ij} \mid (i, j) \in S^0(X) \cap (R_U, C_L)\}, \tag{7.58'}$$

$$\begin{aligned} u'_i &= u_i - \epsilon && \text{for } i \in R_L, \\ u'_i &= u_i && \text{for } i \in R_U, \\ v'_j &= v_j + \epsilon && \text{for } j \in C_L, \\ v'_j &= v_j && \text{for } j \in C_U, \end{aligned} \tag{7.59'}$$

$$a'^*_{ij} = a_{ij} - u'_i - v'_j \quad \text{for all } (i, j) \in N, \quad \text{(same as (7.60))} \tag{7.60'}$$

$$\begin{aligned} a'^*_{ij} &= a^*_{ij} && \text{for all } (i, j) \in (R_L, C_L) \cup (R_U, C_U), \\ a'^*_{ij} &= a^*_{ij} + \epsilon && \text{for all } (i, j) \in (R_L, C_U), \\ a'^*_{ij} &= a^*_{ij} - \epsilon && \text{for all } (i, j) \in (R_U, C_L), \end{aligned} \tag{7.61'}$$

$$\begin{aligned} &\text{(a) } a'^*_{ij} \geq a^*_{ij} && \text{for all } (i, j) \in P, \\ &\text{(b) } a'^*_{ij} \leq a^*_{ij} && \text{for all } (i, j) \in Q, \end{aligned} \quad \text{(same as (7.62))} \tag{7.62'}$$

$$\begin{aligned} &\text{(a) } P'_0 \subset P_0, \\ &\text{(b) } Q'_0 \subset Q_0, \end{aligned} \quad \text{(same as (7.63))} \tag{7.63'}$$

$$a^*_{hk} > a'^*_{hk} \geq 0. \tag{7.64'}$$

If $\epsilon = a^*_{hk}$ then $(h, k) \notin Q_0$ and hence $P'_0 \cup Q'_0 < P_0 \cup Q_0$; if $\epsilon < a^*_{hk}$, then $R'_L \cup C'_L > R_L \cup C_L$. As in Case 1 for nonbreakthrough we return to the optimality subalgorithm with the same X and with (U', V', W') instead of (U, V, W).

This can be illustrated by continuing the example introduced above. We observed that we were in Case 2; the remaining steps are as follows.

				3		
				{1}	{3}	
			2	7	0	1
	{3}	0	${}^{7}0^{1}$	${}^{4}3^{4-\delta}$	$0^{2+\delta}$	${}^{3}7$
		3	${}^{5}0^{4}$	${}^{2}-2^{2}$	${}^{4}3$	${}^{0}8$
−3	{2}	3	${}^{3}5$	$0^{3+\delta}$	${}^{5}0^{2-\delta}$	${}^{2}2$
	{3}	2	${}^{3}1^{3}$	${}^{0}-9$	${}^{10}0^{5}$	0^{8}

Breakthrough: $\delta = 2$, $(h, k) = (1, 2)$, X' is loopless.

		{4}		{1}	{1}
		2	10	0	1
{1}	0	${}^{7}0^{1+\delta}$	${}^{4}0^{2}$	${}^{8}0^{4-\delta}$	${}^{3}7$
{1}	3	${}^{5}0^{4}$	${}^{2}-5^{2}$	${}^{4}3$	${}^{0}8$
{2}	0	${}^{3}8$	0^{5}	${}^{5}3$	${}^{2}5$
{3}	2	${}^{3}1^{3-\delta}$	${}^{0}-12$	${}^{10}0^{5+\delta}$	0^{8}

Breakthrough: $\delta = 3$, $(h, k) = (4, 1)$.

	2	10	0	1
0	${}^{7}0^{4}$	${}^{4}0^{2}$	${}^{8}0^{1}$	${}^{3}7$
3	${}^{5}0^{4}$	${}^{2}-5^{2}$	${}^{4}3$	${}^{0}8$
0	${}^{3}8$	0^{5}	${}^{5}3$	${}^{2}5$
2	${}^{3}1$	${}^{0}-12$	${}^{10}0^{8}$	0^{8}

Optimal solution is attained.

To establish that the algorithm terminates in a finite number of steps we observe that each nonbreakthrough in Case 1 or in Case 2 either shrinks the set of infeasible positions $P_0 \cup Q_0$ or expands the set (R_L, C_L) of labeled lines so that ultimately we must either reach optimality or breakthrough (the algorithm as programmed in Appendix C is based upon a somewhat sharper variant of this argument). There can be only a finite number of breakthrough steps since (1), each breakthrough leads (via the orthogonality subalgorithm if necessary) to an improved loopless feasible primal vector, and (2) the set of all such vectors is finite. No such vector can

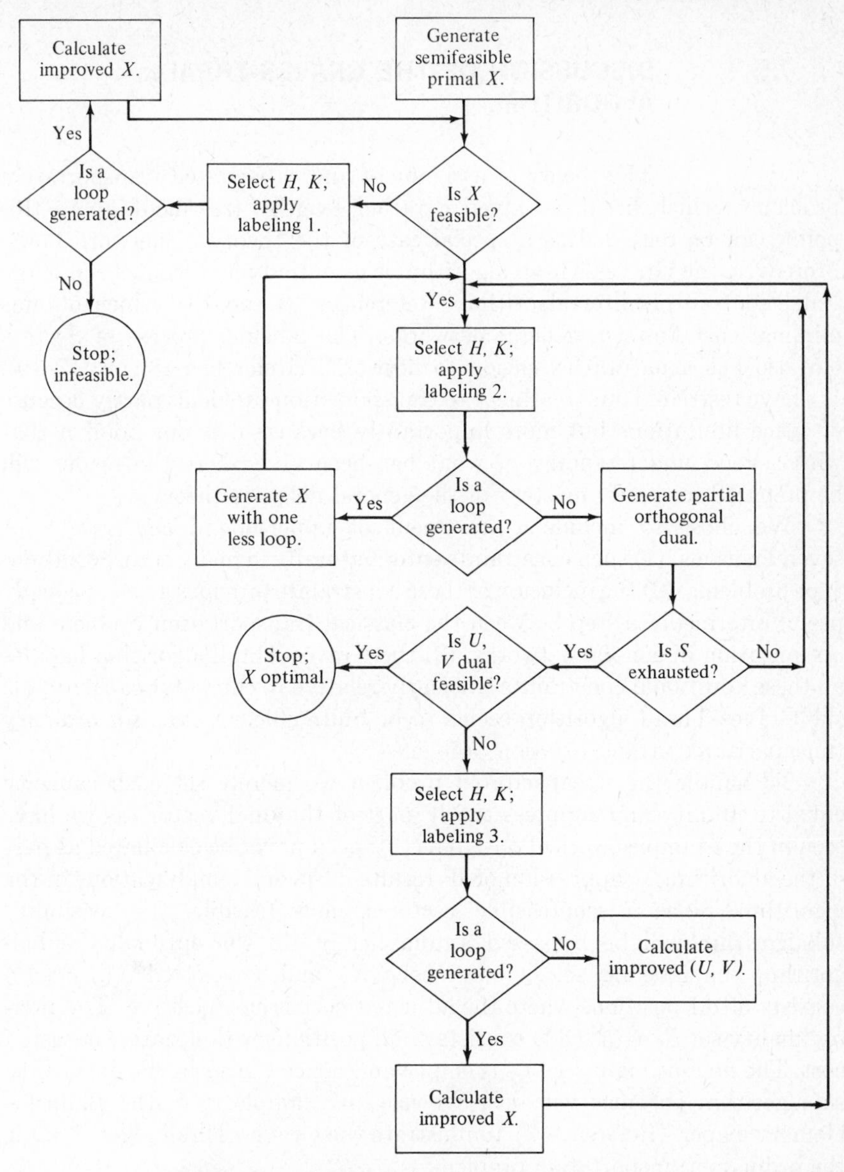

Figure 7.2 Flowchart for Graves-Thrall capacitated transportation algorithm (labelings 1, 2, and 3 are those appearing in Sections 7.2, 7.3, and 7.4, respectively).

appear twice because of (1); hence by (2) the process must terminate, but this can only happen when optimal primal and dual vectors have been found. Figure 7.2 gives a flow chart illustrating the general plan of the Graves-Thrall algorithm and Appendix C provides a more detailed flow chart for the entire algorithm.

7.5. DISCUSSION OF THE GRAVES-THRALL ALGORITHM.

The theory of assignment and capacitated transportation problems, which has been given a rather lengthy treatment up to this point, can be regarded as a special case of the theory of network flows. Moreover, the Graves-Thrall algorithm is essentially a specialization of the Fulkerson out-of-kilter algorithm (References 18 and 19), which obtains minimal cost flows in general networks. The labeling process of Section 6.5 also has a natural extension to networks (Reference 18, pp. 17–19). We have restricted our treatment to transportation problems partly because of space limitations but more importantly because it is our opinion that with a good understanding of what has been given here the reader will be prepared for ready mastery of the general network theory.

We chose to include a treatment of upper-bound constraints for several reasons: (1) such constraints enter naturally in many transportation-type problems, (2) the inclusion of these constraints provides a pedagogically useful intermediate step between the classical transportation problem and optimization in general networks, (3) the Graves-Thrall algorithm handles all these additional constraints efficiently relative to computer capacity, (4) the Graves-Thrall algorithm seems to be quite efficient even for ordinary (uncapacitated) transportation problems.

To handle the uncapacitated problem we merely set each capacity equal to infinity and suppress the W part of the dual vector (as we have seen in the examples worked out above, W need never be calculated as part of the algorithm). Suppression of W results in several simplifications in the algorithm. Since a semifeasible vector is now feasible, the feasibility subalgorithm will be bypassed automatically. In the optimality subalgorithm the labeling set Q becomes $S(X)$ and $P = R^-(A^*) \cup R^0(A^*)$ consists of all positions where the adjusted cost is not positive. The nonoptimality set $P_0 = R^-(A^*)$ consists of all positions with negative adjusted cost. The nonoptimality set Q_0 is empty and hence Case 2 of the optimality subalgorithm becomes vacuous. We use an example from the Balinski-Gomory paper (Reference 3) to illustrate the Graves-Thrall algorithm for the ordinary transportation problem.

$$A = \begin{array}{|ccccc|}\hline 3 & 6 & 3 & 1 & 1 \\ 2 & 4 & 3 & 2 & 7 \\ 1 & 1 & 2 & 1 & 2 \\ \hline \end{array} \qquad \begin{array}{c} \text{Supplies } b_i \\ 4 \\ 5 \\ 6 \end{array}$$

$$\text{Demands } c_j \quad 2 \quad 2 \quad 3 \quad 4 \quad 4$$

We begin with the same loopless feasible X and orthogonal (U, V) used by Balinski-Gomory.

	$u \backslash v$	{1} 3	{3} 6	{1} 3	{1} 2	{1} 3
{2}	0	0^2	$0^{2-\delta}$	0	-1	-2^δ
	0	-1	-2	0^3	0^2	4
{5}	-1	-1	$\boxed{-4^\delta}$	0	0^2	$0^{4-\delta}$

$$
\begin{aligned}
Q &= \{(1, 1), (1, 2), (2, 3), (2, 4), (3, 4), (3, 5)\} = S(X), \\
P &= Q \cup \{(1, 3), (1, 4), (1, 5), (2, 1), (2, 2), (3, 1), (3, 2), (3, 3)\}, \\
P_0 &= \{(1, 4), (1, 5), (2, 1), (2, 2), (3, 1), (3, 2)\}, \\
(h, k) &= (3, 2).
\end{aligned}
$$

Breakthrough: $\delta = 2$, $\theta = 6$, and the new solution X is loopless.

$\Delta u \backslash \Delta v$	0	-6	-2	-2	-2
0	0^2	0	0	-1	-2^2
2	-1	-2	0^3	0^2	4
2	-1	-4^2	0	0^2	0^2

Calculation of new orthogonal (U, V) is then shown.

	$u \backslash v$	{3} 3	0	1	0	{1} 1
{1}	0	$0^{2-\delta}$	6	2	1	$0^{2+\delta}$
	2	-3	2	0^3	0^2	4
{5}	1	$\boxed{-3^\delta}$	0^2	0	0^2	$0^{2-\delta}$

$(h, k) = (3, 1)$.

Breakthrough: $\delta = 2$, $\theta = 3$, the new X is again loopless.

$\Delta v \backslash \Delta u$	-3	0	0	0	0
0	0	6	2	1	0^4
0	-3	2	0^3	0^2	4
0	-3^2	0^2	0	0^2	0

Calculation of the new (U, V) shows that the solution is optimal.

$v \backslash u$	0	0	1	0	1
0	3	6	2	1	0^4
2	0	2	0^3	0^2	4
1	0^2	0^2	0	0^2	0

For the ordinary transportation problem the main difference between the Balinski-Gomory algorithm and the Graves-Thrall algorithm is in the greater size of the labeling set P in the latter. The difference in the size of P has several consequences:

1. The probability of breakthrough is somewhat higher for Graves-Thrall than for Balinski-Gomory.
2. Since improvement may occur at more than one position in a loop, the improvement obtained with breakthrough tends to be greater for Graves-Thrall than for Balinski-Gomory.
3. Computation of a new orthogonal vector in the Graves-Thrall algorithm may require some labeling passes not involved in the Balinski-Gomory algorithm, but this is compensated for by the primal improvement accompanying each such pass.
4. When a new orthogonal dual vector is obtained, the Balinski-Gomory algorithm guarantees (a) definite improvement toward feasibility at one dual infeasible position, and (b) that the set of infeasible positions does not grow, whereas the Graves-Thrall algorithm guarantees (a), (b), and also the uniformity condition (c) that the size of the infeasibility is not increased at any infeasible position.

We give an additional example using an ordinary transportation problem. In this example loops and labels are calculated by inspection. Suppose that

$$X = \begin{bmatrix} 2 & 4 & 0 & 0 \\ 0 & 6 & 0 & 3 \\ 0 & 0 & 1 & 2 \end{bmatrix}, \qquad A = \begin{bmatrix} 5 & 6 & 4 & 3 \\ 2 & 1 & 5 & 2 \\ 4 & 1 & 3 & 4 \end{bmatrix}$$

give an initial feasible primal vector X and cost matrix A. These matrices are combined by writing the nonzero x_{ij} as superscripts on the corresponding a_{ij}. Thus we write

$$A^X = \begin{bmatrix} 5^2 & 6^4 & 4 & 3 \\ 2 & 1^6 & 5 & 2^3 \\ 4 & 1 & 3^1 & 4^2 \end{bmatrix}.$$

We then also place the u_i and v_j as they are determined, each at the beginning of its row or column. Thus, for example, we have

$$\begin{array}{cc} u_i & v_j \quad 5 \quad\ 6 \quad\ 6 \quad\ 7 \\ \begin{array}{r} 0 \\ -5 \\ -3 \end{array} & \begin{bmatrix} 5^2 & 6^4 & 4 & 3 \\ 2 & 1^6 & 5 & 2^3 \\ 4 & 1 & 3^1 & 4^2 \end{bmatrix}. \end{array}$$

Next, we calculate A^* and retain the x_{ij} as superscripts; we get

$$\begin{array}{c} \\ 0 \\ -5 \\ -3 \end{array} \begin{array}{c} \begin{array}{cccc} 5 & 6 & 6 & 7 \end{array} \\ \begin{bmatrix} 0^2 & 0^4 & -2 & -4 \\ 2 & 0^6 & 4 & 0^3 \\ 2 & -2 & 0^1 & 0^2 \end{bmatrix} \end{array};$$

the orthogonality of X and (U, V) is now obvious, since each nonzero x_{ij} appears as superscript to a zero a^*_{ij}. The (U, V) is retained, since in later cycles of the algorithm it is more convenient to work with A^* than with A and to determine the new dual vector (U', V') and corresponding A'^* as modifications of the current (U, V) and A^* rather than starting from scratch. Here $P_0 = \{(1, 3), (1, 4), (3, 2)\}$ and since $a^*_{14} = -4$ is the largest negative entry we take $(h, k) = (1, 4)$ and find the P loop $((1, 4), (3, 4), (3, 2), (1, 2))$ with $\theta = -6$, $\delta = 2$. The matrix $A^{*X'}$ is then

$$A^{*X'} = \begin{array}{c} \\ 0 \\ -5 \\ -3 \end{array} \begin{array}{c} \begin{array}{cccc} 5 & 6 & 6 & 7 \end{array} \\ \begin{bmatrix} 0^2 & 0^2 & -2 & -4^2 \\ 2 & 0^6 & 4 & 0^3 \\ 2 & -2^2 & 0^1 & 0 \end{bmatrix} \end{array}.$$

This X' has a loop $((1, 2), (1, 4), (2, 4), (2, 2))$ with $\theta = 4$, $\delta = -2$ giving a new primal vector which is loopless,

$$\begin{array}{c} \\ 0 \\ -5 \\ -3 \end{array} \begin{array}{c} \begin{array}{cccc} 5 & 6 & 6 & 7 \end{array} \\ \begin{bmatrix} 0^2 & 0 & -2 & -4^4 \\ 2 & 0^8 & 4 & 0^1 \\ 2 & -2^2 & 0^1 & 0 \end{bmatrix} \end{array}.$$

Now, take $\Delta v_4 = -4$, $\Delta u_2 = 4$, $\Delta v_2 = -4$, $\Delta u_3 = 2$, $\Delta v_3 = -2$ and we get orthogonality and have

$$\begin{array}{c} \\ 0 \\ -1 \\ -1 \end{array} \begin{array}{c} \begin{array}{cccc} 5 & 2 & 4 & 3 \end{array} \\ \begin{bmatrix} 0^2 & 4 & 0 & 0^4 \\ -2 & 0^8 & 2 & 0^1 \\ 0 & 0^2 & 0^1 & 2 \end{bmatrix} \end{array}.$$

Two more applications of the optimality subalgorithm lead to the following optimal solution,

$$\begin{array}{c} \\ 0 \\ -3 \\ -3 \end{array} \begin{array}{c} \begin{array}{cccc} 5 & 4 & 4 & 3 \end{array} \\ \begin{bmatrix} 0 & 2 & 0^1 & 0^5 \\ 0^2 & 0^7 & 4 & 2 \\ 2 & 0^3 & 2 & 4 \end{bmatrix} \end{array}.$$

7.6. UNBALANCED TRANSPORTATION PROBLEMS.

Suppose that the balance condition (7.1) is not satisfied. The problem is then infeasible. However, if instead of (7.1) we have

$$\textstyle\sum_i b_i - \sum_j c_j = d > 0, \tag{7.65}$$

then we can create a new, fictitious destination $n + 1$, set $c_{n+1} = d$, $a_{i,n+1} = 0$, $i = 1, \cdots, p$. The resulting problem having p origins and $n + 1$ destinations is then balanced. In any optimal program for this new problem the final column of the solution matrix indicates how the available surplus should be allocated among the origins.

Similarly, if $d < 0$ we create a new, fictitious origin $p + 1$, set $b_{p+1} = -d$, $a_{p+1,j} = 0$ for $j = 1, \cdots, n$ and again we have a balanced problem. The final row of the solution matrix indicates how the shortage is spread among the destinations. Nonzero values for the $a_{p+1,j}$ can be used if we wish to place different weights on shortages at the several destinations.

More generally, suppose that (7.65), (7.5), (7.6) hold and that (7.3) and (7.4) are replaced by

$$\textstyle\sum_j x_{ij} \leq b_i, \qquad (i = 1, \cdots, p) \tag{7.66}$$

and

$$\textstyle\sum_i x_{ij} \geq c_j, \qquad (j = 1, \cdots, n) \tag{7.67}$$

and (7.2) is retained as the objective function. Then if we introduce nonnegative slack variables $x_{i,n+1}$, $y_{p+1,j}$ we replace (7.66) and (7.67) by

$$\textstyle\sum_j x_{ij} + x_{i,n+1} = b_i, \qquad (i = 1, \cdots, p) \tag{7.66$'$}$$

$$\textstyle\sum_i x_{ij} - y_{p+1,j} = c_j \qquad (j = 1, \cdots, n). \tag{7.67$'$}$$

This is not in the form of a transportation problem. However using (7.65), (7.66′), and (7.67′) we get

$$d = \textstyle\sum_i b_i - \sum_j c_j = \displaystyle\sum_{i=1}^{p} x_{i,n+1} + \sum_{j=1}^{n} y_{p+1,j} \tag{7.68}$$

and hence each slack variable is bounded by d. Next, let

$$x_{p+1,j} = d - y_{p+1,j}, \qquad x_{p+1,n+1} = d - \sum_{i=1}^{p} x_{i,n+1}. \tag{7.69}$$

Then from (7.68), (7.69) and the nonnegativity of the slack variables we have

$$0 \leq x_{p+1,j} \leq d, \qquad (j = 1, \cdots, n + 1). \tag{7.70}$$

Now let $c_{n+1} = d$, $b_{p+1} = nd$, and $a_{p+1,j} = a_{i,n+1} = 0$ for $j = 1, \cdots, n + 1$, $i = 1, \cdots, p + 1$. We now have a balanced transportation problem with $p + 1$ origins and $n + 1$ destinations and the capacity constraints of (7.6) together with the new ones

$$x_{p+1,j} \leq d, \qquad (j = 1, \cdots, n). \tag{7.71}$$

Any optimal program X' of this new problem will on deletion of the last row and column yield an optimal program of the original unbalanced problem and with the same value for the objective function.

The presence of the capacity constraints (7.70) in the new problem illustrates once again the practical importance of having a solution algorithm designed to handle capacity constraints.

As an example suppose that

$$A = \begin{bmatrix} 5 & -4 & 3 \\ 3 & 2 & 2 \end{bmatrix}, \qquad B = \begin{bmatrix} 10 \\ 20 \end{bmatrix}, \qquad C = \begin{bmatrix} 13 \\ 7 \\ 4 \end{bmatrix};$$

then $d = 6$ and the equivalent balanced problem has

$$^{E}A = \begin{bmatrix} 5 & -4 & 3 & 0 \\ 3 & 2 & 2 & 0 \\ {}^{6}0 & {}^{6}0 & {}^{6}0 & 0 \end{bmatrix}, \qquad B = \begin{bmatrix} 10 \\ 20 \\ 18 \end{bmatrix}, \qquad C = \begin{bmatrix} 19 \\ 13 \\ 10 \\ 6 \end{bmatrix}.$$

An optimal solution is given by

$$\begin{array}{ccc} & V & \\ U & [{}^{E}A^{*X}] = & \end{array} \begin{array}{c} \\ -4 \\ 0 \\ 0 \end{array} \begin{array}{c} \begin{array}{cccc} 3 & 0 & 2 & 0 \end{array} \\ \begin{bmatrix} 6 & 0^{10} & 7 & 4 \\ 0^{13} & 2 & 0^{4} & 0^{3} \\ {}^{6}-3^{6} & {}^{6}0^{3} & {}^{6}-2^{6} & 0^{3} \end{bmatrix} \end{array}.$$

Observe that the total amount actually shipped of 27 lies between $\sum c_j = 24$ and $\sum b_i = 30$.

7.7. INTERPRETATION OF THE TERMINAL COST MATRIX; MULTIPLE OPTIMAL SOLUTIONS.

Let A^* and X represent the final cost matrix and optimal assignment, respectively, for a capacitated transportation problem. A position (h, k) with $a^*_{hk} = 0$ and either $0 = x_{hk} < e_{hk}$ or $0 < x_{hk} = e_{hk}$ indicates a possible additional optimal basic feasible solution. Let

$$\begin{aligned} P &= [S(X) \cup S^0(X)] \cap R^0(A^*), \\ Q &= [S(X) \cup S^C(X)] \cap R^0(A^*); \end{aligned} \tag{7.72}$$

now if $0 = x_{hk} < e_{hk}$ and there is a (P, Q) loop beginning at (h, k), then we proceed to construct, as in the optimality subalgorithm, a new optimal feasible solution with $x_{hk} > 0$. If no such loop exists, then there is no optimal feasible solution with $x_{hk} > 0$. Similarly, if $0 < x_{hk} = e_{hk}$, we look for a (Q, P) loop beginning at (h, k). Next, consider a position $(h, k) \in R^+(A) \cap S^0(X)$ at which we wish to make a positive assignment x'_{hk}. Suppose there is a (P, Q) loop beginning at (h, k), where

$$\begin{aligned} P &= \{(h, k)\} \cup [(S^0(X) \cup S(X)) \cap R^0(X^*)], \\ Q &= [S^C(X) \cup S(X)] \cap R^0(A^*). \end{aligned} \tag{7.73}$$

Then for $0 < x'_{hk} \leq \delta$, where δ is defined by (7.49), and by adding and subtracting x'_{hk} around the loop, we can construct a new X' with the prescribed x'_{hk} and such that $\alpha(X') - \alpha(X) = x'_{hk}a^*_{hk}$. Furthermore, no other solution X'' with $x''_{hk} = x'_{hk}$ can have a lower cost.

If no (P, Q) loop exists, the set $S(X)$ is not a tree of order $p + n - 1$; the primal optimal solution is then degenerate and the dual has multiple optimal solutions. For one of these solutions the set $R^0(A^{*\prime})$ may have a (P', Q') loop beginning at (h, k). We can adapt the nonbreakthrough construction from Section 7.4 to construct such a loop if one exists. On the other hand, there may be no such loop and this too will be detected by the nonbreakthrough construction and is indicated by $\epsilon^- = \epsilon^+ = \infty$. We do not provide a full algorithmic discussion here; a few examples are introduced which illustrate typical cases.

Example 1. Position $(h, k) = (1, 1)$, breakthrough;

$$
\begin{array}{ccc}
 & V & \begin{array}{cc} 0 & \quad 3 \end{array} \\
U & [{}^{E}A^{*X}] = \begin{array}{c} 2 \\ 5 \end{array} & \begin{bmatrix} 3^{\delta} & {}^{8}0^{6-\delta} \\ 0^{5-\delta} & 0^{2+\delta} \end{bmatrix}.
\end{array}
$$

Cost of assigning δ at $(1, 1)$ is 3δ for $0 < \delta \leq 5$.

Example 2. Position $(h, k) = (1, 1)$, nonbreakthrough;

$$
\begin{array}{cccc}
 & & & \begin{array}{cc} \{1\} & \quad \{2\} \end{array} \\
 & V & & \begin{array}{cc} 0 & \quad 3 \end{array} \\
U & [{}^{E}A^{*X}] = & \begin{array}{c} 2 \\ \{1\} \;\; 5 \end{array} & \begin{bmatrix} 3 & {}^{8}-4^{8} \\ 0^{5} & 0^{2} \end{bmatrix};
\end{array}
$$

$\epsilon = 4$. $(h, k) = (1, 1)$ leads to breakthrough,

$$
\begin{array}{cc}
 & \begin{array}{cc} 0 & \quad 3 \end{array} \\
\begin{array}{c} -2 \\ 5 \end{array} & \begin{bmatrix} 7^{\delta} & {}^{8}0^{8-\delta} \\ 0^{5-\delta} & 0^{2+\delta} \end{bmatrix}.
\end{array}
$$

Cost of assigning δ at $(1, 1)$ is 7δ for $0 < \delta \leq 5$. Note that in a degenerate problem the initial a^*_{hk} may not measure the unit cost of assignment.

Example 3. Position $(h, k) = (1, 1)$,

$$
\begin{array}{cc}
 & \begin{array}{cc} \{1\} & \quad \end{array} \\
[{}^{E}A^{X}] = \begin{array}{c} \\ \{1\} \end{array} & \begin{bmatrix} 3 & 0^{2} \\ 0^{3} & {}^{7}0^{7} \end{bmatrix}.
\end{array}
$$

Here $\epsilon^+ = \epsilon^- = \infty$ (each is the minimum of a function over an empty set) and clearly no assignment can be made at $(1, 1)$ without exceeding the upper-bound constraint at $(2, 2)$.

Example 4. Position $(h, k) = (1, 1)$,

$$[{}^{E}A^{X}] = \begin{array}{c} \\ \{1\} \end{array} \overset{\{1\}}{\begin{bmatrix} 3 & 0^2 & 0^4 \\ 0^3 & {}^{7}0^{7} & 5 \end{bmatrix}}.$$

Then $\epsilon^+ = 5$, $\epsilon^- = \infty$ leads to

$$\begin{array}{c} \\ 5 \end{array} \overset{-5}{\begin{bmatrix} 8^{\delta} & 0^2 & 0^{4-\delta} \\ 0^{3-\delta} & {}^{7}-5^{7} & 0^{\delta} \end{bmatrix}},$$

and we have breakthrough with cost 8δ for $0 < \delta \leq 3$.

PROBLEMS

7.1. Let

$$A = \begin{bmatrix} 3 & 2 & 8 & 7 & 1 \\ 1 & 2 & 9 & 2 & 5 \\ 10 & 1 & 2 & 1 & 8 \\ 6 & 7 & 3 & 1 & 4 \end{bmatrix}, \quad B = \begin{bmatrix} 10 \\ 5 \\ 8 \\ 3 \end{bmatrix}, \quad C = \begin{bmatrix} 7 \\ 7 \\ 6 \\ 3 \\ 3 \end{bmatrix};$$

find a semifeasible vector X using

(a) the northwest-corner rule,
(b) Vogel's approximation method,
(c) the method suggested in the text (making cheapest assignments in column 1 until c_1 is used up, and so on).

7.2. Let

$$X = \begin{bmatrix} 0 & 0 & 0 & 4 & 1 \\ 4 & 0 & 0 & 0 & 2 \\ 0 & 6 & 2 & 0 & 0 \\ 0 & 1 & 4 & 0 & 0 \end{bmatrix}$$

and

$$E = \begin{bmatrix} \infty & 0 & \infty & 6 & 0 \\ 4 & \infty & \infty & \infty & \infty \\ \infty & 4 & 2 & \infty & \infty \\ \infty & \infty & \infty & \infty & 7 \end{bmatrix};$$

list the elements of each of the sets $S^+(X)$, $S^C(X)$, $S(X)$, $S^0(X)$, $S^{00}(X)$.

7.3. Let X and E be given as in Problem 7.2 and let the labeling sets P and Q be defined as in Section 7.2 of the text.

(a) List the elements of P and Q.
(b) Begin a (P, Q) labeling process at position (3, 2) and find a (P, Q) loop. What are the values of δ, δ', δ''?

7.4. Given:

$$A = \begin{bmatrix} 2 & 3 & 5 \\ 1 & 2 & 7 \end{bmatrix}, \qquad B = \begin{bmatrix} 5 \\ 8 \end{bmatrix}, \qquad C = \begin{bmatrix} 4 \\ 6 \\ 3 \end{bmatrix},$$

$$E = [e_{ij}] = \begin{bmatrix} \infty & 4 & 2 \\ \infty & 3 & \infty \end{bmatrix},$$

find a feasible solution (if one exists) by using the feasibility subalgorithm for the capacitated transportation problem.

7.5. Let A be as given in Problem 7.1 above and let

$$X = \begin{bmatrix} 4 & 0 & 6 & 2 & 0 \\ 0 & 7 & \boxed{2} & 0 & 1 \\ 0 & 0 & 3 & 2 & 0 \\ 5 & 3 & 0 & 0 & 7 \end{bmatrix}.$$

Also suppose that $e_{13} = 6$, $e_{15} = 0$, $e_{55} = 7$ and that all other capacities are infinite.

(a) What are the elements of the set $S(X)$?
(b) Initiate an S labeling at position (2, 3) and locate an S loop if one exists.

7.6. Data for a transportation problem are given as follows:

$$[{}^{E}A^{X}] = \begin{bmatrix} 2^{3} & 1^{1} & 1 & {}^{3}6^{3} \\ {}^{2}1^{2} & 3^{4} & 4 & 2^{4} \\ 4 & 3 & 2^{6} & 8 \end{bmatrix}.$$

Find a dual vector (U, V, W) orthogonal to X.

7.7. Consider the following transportation problem

$$[{}^{E}A^{X}] = \begin{bmatrix} 3^{4} & {}^{2}2^{2} & 8^{1} & 3^{2} \\ 6^{6} & 1^{1} & 2 & 2 \\ 1 & 4^{3} & {}^{0}6 & 1^{7} \end{bmatrix}.$$

(a) Find an S loop beginning at position (1, 1).
(b) Find a vector X' such that $\alpha(X') < \alpha(X)$.

7.8. Given the following capacitated transportation problem, find a feasible program with orthogonal u_i and v_j. Capacities are upper left superscripts.

	Costs				*Requirements*
	${}^{4}2$	5	6	${}^{2}1$	10
	5	${}^{5}2$	3	5	5
	${}^{5}1$	3	1	3	6
	3	${}^{2}1$	4	2	7
Resources	8	6	5	9	

7.9. Let

$$[{}^{E}A^{*X}] = \begin{bmatrix} 2 & {}^{2}1^{2} & -3 & {}^{0}2 & 0^{6} \\ {}^{7}0^{5} & 0^{4} & 2 & -4 & 4 \\ 1 & 5 & 0^{3} & 0^{4} & 0^{2} \end{bmatrix}.$$

(a) List the elements of each of the following sets: $S^{00}(X)$, $S^{0}(X)$, $S(X)$, $S^{C}(X)$, $R^{+}(A^{*})$, $R^{0}(A^{*})$, $R^{-}(A^{*})$.
(b) Find the nonoptimality sets P_0 and Q_0.
(c) Find the sets P and Q as defined by (7.45) and (7.46).

7.10. In the following two (uncapacitated) transportation problems the optimality subalgorithm is to be applied. In each case list the elements of the sets P and Q, then begin labeling at the position indicated by the asterisk and find a (P, Q) loop if possible. Otherwise, find a vector (U', V') more feasible than (U, V).

(a)

$$\begin{array}{c|c|} & V \\ \hline U & A^{*X} \\ \hline \end{array} = \begin{array}{r|rrr|} & 2 & -1 & 1 \\ \hline 0 & 0^{3} & 4 & 0^{4} \\ 2 & -1^{*} & 0^{2} & 0^{5} \\ -1 & 0^{6} & 6 & 5 \\ \hline \end{array}$$

(b)

$$\begin{array}{c|c|} & V \\ \hline U & A^{*X} \\ \hline \end{array} = \begin{array}{r|rrr|} & 2 & 8 & 0 \\ \hline 0 & 6 & 0^{7} & 4 \\ 1 & 0^{3} & 2 & 0^{8} \\ 3 & -2 & 0^{5} & -4^{*} \\ \hline \end{array}$$

7.11. A (P, Q) labeling has resulted in the following array [initial position: $(h, k) = (5, 3)$]:

$$\begin{array}{c|rrrrr|} & & & \{5\} & \{2\} & \\ \hline \{4\} & 2 & 5 & 3 & {}^{5}2^{5} & 2 \\ \{3\} & {}^{2}3 & 2 & 0^{5} & 0^{1} & 5 \\ & {}^{5}{-3}^{5} & 1 & {}^{1}{-1}^{1} & 0 & 0^{1} \\ & 0 & 0^{4} & 0 & 2 & 0^{1} \\ & 6 & 1 & -3 & 3 & 0^{4} \\ \hline \end{array};$$

find ϵ.

7.12. A (Q, P) labeling has resulted in the following array [initial position: $(h, k) = (4, 4)$]:

$$\begin{array}{c|rrrrr|} & \{1\} & & \{5\} & \{4\} & \{1\} \\ \hline \{4\} & 0^{3} & 5 & 5 & {}^{5}0^{4} & {}^{4}0^{3} \\ \{3\} & {}^{2}1 & 2 & 0^{3} & 3 & 3 \\ \{4\} & {}^{5}{-3}^{5} & 1 & -2 & 0^{1} & 0 \\ & 1 & 0^{4} & 2 & {}^{2}2^{2} & 3 \\ \{4\} & 1 & 1 & 0^{2} & 0^{2} & 2 \\ \hline \end{array};$$

find ϵ.

7.13. For each of the following three problems carry out one complete step of the Graves-Thrall algorithm indicating P, Q sets and phases of the algorithm used (orthogonality, optimality). Start labeling at position denoted by *.

(a)

	3	4	2
2	${}^{3}-3^{3}$	-2^{*}	0^{2}
1	4	${}^{1}0^{1}$	2
0	0^{1}	0^{3}	-1

(b)

	3	-1	3
2	4	0^{3}	2
4	0^{2}	-1	0^{1}
1	${}^{3}-2^{3}$	2	-1^{*}

(c)

	0	1	2
3	${}^{2}1$	0^{2}	1
1	2	${}^{3}2^{3*}$	0^{1}
2	0^{4}	3	-3

7.14. Solve the following capacitated transportation problem. Display the solution in tableau form with assignments, capacities, orthogonal u_i, v_j, and adjusted costs. If capacity restrictions are removed, how much would minimum cost decrease?

	2	1	3
1	${}^{3}-1$	0^{4}	-2
-1	0^{1}	1	0^{1}
-2	${}^{3}0^{3}$	4	0^{1}

7.15. (a) Solve the following transportation problem:

					Origin supply
	5	8	3	6	30
	4	5	7	4	50
	6	2	4	5	40
Destination requirement	30	20	40	30	

(b) Solve the following capacitated problem:

5	8	3	6	30
${}^{20}4$	5	7	4	50
6	2	4	5	40
30	20	40	30	

7.16. Solve the following transportation problem:

					Surplus at origin
5	6	6	12	9	5
5	4	3	14	11	10
9	7	3	7	12	8
7	8	9	7	3	7
6	3	7	9	5	

Deficiency at destination (bottom row)

7.17. Solve the following transportation problem:

						Surplus at origin
5	3	7	3	8	5	3
5	6	12	5	7	11	4
2	8	3	4	8	2	2
9	6	10	5	10	9	8
3	3	6	2	1	2	

Deficiency at destination (bottom row)

7.18. The M Steel Company has three open-hearth furnaces and five rolling mills. Transportation costs (dollars per ton) for shipping steel from furnaces to rolling mills are shown in the following table.

		Mills					
		M_1	M_2	M_3	M_4	M_5	*Capacities (tons)*
Furnaces	F_1	4	2	3	2	6	8
	F_2	5	4	5	2	1	12
	F_3	6	5	4	7	3	14
Requirements (tons)		4	4	6	8	8	

(a) What is an optimal shipping schedule?
(b) Suppose that shipping restrictions make it possible to ship a maximum of 2 tons from F_2 to M_5. What is an optimal schedule now?

7.19. A company manufactures freezers in three factories. It is necessary to supply four warehouses from these factories. Unit freight costs are given in the following table. The bordering numbers indicate factory capacities and warehouse requirements for a given period. Only 200 freezers can be shipped from factory 3 to warehouse 1 and only 100 can be shipped from factory 1 to warehouse 4. Find an optimal shipping plan. How should the surplus be allocated?

		Warehouses				
		W_1	W_2	W_3	W_4	*Capacities*
Factories	F_1	15	46	54	5	500
	F_2	47	17	25	55	250
	F_3	4	25	5	24	350
Requirements		350	300	200	200	

7.20. Seven patients R, S, T, U, V, W, X require blood transfusions. Suppose there are four blood types A, B, AB, and O, where AB is a type for a universal recipient and O is a type for a universal donor. The blood supply and cost situation are as follows:

Blood type	*Supply (pint)*	*Cost per pint (dollars)*
A	7	1.00
B	4	4.00
AB	6	2.00
O	5	5.00

The data for the patients are as follows:

Patient	*Type*	*Requirement*
R	A	2
S	AB	3
T	B	1
U	O	2
V	A	3
W	B	2
X	AB	1

The problem is to minimize replacement cost of blood subject to the requirement that each patient receives the required amount and type of blood.

(a) Show that this problem can be formulated as a transportation problem.

(b) Show that the optimal transfusion schedule is

	R	S	T	U	V	W	X	D
A	2	1			3		1	
B			1			2		1
AB		2						4
O				2				3

Hint: Since total demand is 14 pints and total supply is 22, introduce a dummy patient D with a demand of 8 pints.

7.21. For the following transportation problem an optimal program X is shown.

$$\begin{array}{cccc} & & & 0 \quad 2 \quad 1 \\ & V & & \\ U & [{}^{E}A^{*X}] & = & \begin{array}{c} 3 \\ 1 \\ 2 \end{array} \begin{bmatrix} 0^{2} & 3 & 0 \\ 1 & {}^{4}0^{4} & 2 \\ 0 & 0^{2} & 0^{2} \end{bmatrix} \end{array}$$

Find another optimal program X'.

7.22. Consider the array:

$$\begin{array}{cc} & V \\ U & [^E A^{*X}] \end{array} = \begin{array}{c} \\ 1 \\ -1 \\ 2 \end{array} \begin{array}{c} \begin{array}{cccc} 3 & 4 & 1 & 0 \end{array} \\ \begin{bmatrix} 0 & 2 & 0^1 & 0^4 \\ 1 & 2 & 0^2 & 2 \\ 0^2 & 0^3 & 3 & 0^1 \end{bmatrix} \end{array}$$

What is the cost of assigning one unit at (2, 2)?

7.23. A transportation problem is specified by the array:

$$\begin{array}{cc} & V \\ U & [^E A^{*X}] \end{array} = \begin{array}{c} \\ 3 \\ 2 \\ 1 \end{array} \begin{array}{c} \begin{array}{ccc} 3 & 2 & -1 \end{array} \\ \begin{bmatrix} 0^2 & 3 & 0 \\ 1 & 0^4 & 2 \\ 10^1 & 0^2 & 0^2 \end{bmatrix} \end{array}.$$

What is the cost of sending one unit from origin 1 to destination 3?

7.24. Solve the following capacitated transportation problem:

$$A = \begin{bmatrix} 2 & 5 & 6 & 1 & 1 \\ 5 & 2 & 3 & 5 & 6 \\ 1 & 3 & 1 & 3 & 3 \\ 3 & 1 & 4 & 2 & 2 \\ 4 & 2 & 2 & 1 & 4 \end{bmatrix}, \quad B = \begin{bmatrix} 10 \\ 5 \\ 6 \\ 7 \\ 4 \end{bmatrix}, \quad C = \begin{bmatrix} 8 \\ 6 \\ 5 \\ 9 \\ 4 \end{bmatrix},$$

$$E = [e_{ij}] = \begin{bmatrix} \infty & \infty & \infty & 5 & 4 \\ 2 & \infty & \infty & \infty & \infty \\ 5 & \infty & \infty & \infty & \infty \\ \infty & \infty & \infty & 2 & \infty \\ \infty & \infty & \infty & \infty & \infty \end{bmatrix}.$$

7.25. Given the following cost matrix of a transportation problem

11	6	15	3
7	8	4	13
22	17	16	31

and a feasible program

0	0	0	11
1	0	4	0
5	10	0	4

,

determine an optimal program and minimum cost.

7.26. Solve the production planning model given in Section 2.7 (formulated as a transportation problem).

7.27. Given the array

$$\begin{array}{|ccc|} {}^{\infty}1 & {}^{2}{-1}^{2} & {}^{1}2^{1} \\ {}^{7}0^{5} & {}^{0}3 & {}^{6}0^{3} \end{array}$$

do each of the following.

(a) Determine the sets P and Q.

(b) Verify that no loop is encountered if one begins the algorithm at position (1, 3).

(c) Calculate ϵ and make the dual transformation.

(d) Complete the solutions.

7.28. A corporation has decided to produce three new products. Five branch plants now have excess production capacity. The unit manufacturing cost of the first product would be $25.00, $28.00, $24.00, $30.00, $27.00 in plants 1, 2, 3, 4, and 5 respectively. The unit manufacturing cost of the second product would be $30.00, $33.00, $28.00, $32.00 and $31.00 in the plants respectively. The unit manufacturing cost of the third product would be $41.00, $43.00, and $39.00 in plants 1, 2, and 3 respectively, whereas the plants 4 and 5 do not have the capability for producing this product. Sales forecasts indicate that 320, 215, and 450 units of products 1, 2, and 3 should be produced, respectively. Plants 1, 2, 3, 4, and 5 have the capacity to produce 250, 400, 200, 500, and 300 units daily, regardless of the product or combination of products involved. Assume that any plant having the capability and capacity can produce any combination of the products in any quantity. Management wishes to know how to allocate the new products to the plants in order to minimize the total manufacturing cost.

(a) Formulate this as a transportation problem by constructing the appropriate cost and requirements table.

(b) Indicate initial assignment and final assignment.

7.29. A department store wishes to purchase the following quantities of order forms:

Order-form type	*A*	*B*	*C*
Quantity (in thousands)	500	300	200

Three suppliers indicate that they can supply not more than the total quantities listed below (each supplier can supply all three order-form types).

Supplier	*X*	*Y*	*Z*
Quantity (in thousands)	300	300	800

The unit cost (per thousand) of each type of order form from each supplier are given in the table below. How should the orders be placed and what is the cost of an optimal program? (*Hint:* add a dummy order-form type to represent the unused capacity of the suppliers.)

	A	*B*	*C*
X	11	14	18
Y	10	13	17
Z	12	14	20

7.30. (a) An airline is planning to introduce a new engine-maintenance plan whereby aircraft are periodically removed from service and replaced by reconditioned (or new) engines, and the aircraft returned to service. This plan will require purchase of some new engines to use as replacements at a cost of $10,000.00 each. Each engine removed for maintenance can be reconditioned by one of the two methods: the normal method which requires two months of time and costs $3000 per engine, and the fast method, which takes 1 month and costs $5000 per engine. Any number of engines can be handled simultaneously by either method. The air line is planning for a period of 5 months into the future and the number of aircraft to be serviced monthly is 80, 100, 140, 88, and 116, respectively. Formulate this as a linear programming problem where the objective is to determine for each month how many aircraft should be purchased, reconditioned by normal method, and reconditioned by fast method, so as to minimize the total cost over the given 5 month period. Define the variables carefully.

(b) This problem can be formulated as a transportation problem. Do this, identifying the factories, warehouses, and unit cost matrix.

(c) Solve the transportation problem.

7.31. (Merrill-Tobin) Suppose you are given the following unbalanced transportation problem:

Minimize

$$z = \sum_{ij} a_{ij} x_{ij}$$

subject to

$$\sum_j x_{ij} \leq b_i \qquad (i = 1,\cdots,p)$$

$$\sum_i x_{ij} \geq c_j \qquad (j = 1,\cdots,n)$$

$$0 \leq x_{ij} < \infty \qquad (i = 1,\cdots,p;\ j = 1,\cdots,n)$$

with

$$\sum_i b_i - \sum_j c_j = d > 0.$$

Solve this problem by formulating it as a balanced transportation problem with no upper bounds by adding only a fictitious destination $n + 1$, if:

(a) $$a_{ij} \geq 0 \qquad (i = 1,\cdots,p;\ j = 1,\cdots,n).$$

Hint: You should show that there exists an optimal program X^*, such that

$$\sum_i x_{ij} = c_j.$$

(b) Solve when at least one

$$a_{ij} < 0 \qquad (i = 1,\cdots,p;\ j = 1,\cdots,n).$$

Hint: The variable $x_{i,n+1}$ will have a different cost, $a_{i,n+1}$, if $a_{ij} < 0$ for some $j \leq n$ than it will if $a_{ij} \geq 0$ for all $j = 1,\cdots,n$.

Schema, Duality, and Complementary Slackness

8

8.1. SCHEMA AND DUALITY.

We now introduce and utilize a new and more compact notation[1] for the simplex algorithm which, among other things, will aid us to examine duality more deeply. The *schema*

$$
S: \quad \begin{array}{c} \\ y \\ y^* \\ \\ \end{array} \begin{array}{c} \begin{array}{cc} x & x^* \end{array} \\ \boxed{\begin{array}{cc} \alpha & \beta \\ \gamma & \delta \end{array}} \\ \begin{array}{cc} = v & = v^* \end{array} \end{array} \begin{array}{c} \\ = -u \\ = -u^* \\ \\ \end{array} \tag{8.1}
$$

conveniently represents the following "dual" systems of linear equations:

$$
\text{Row system} \left\{ \begin{array}{ccccc} \vdots & & \vdots & & \vdots \\ \cdots + \alpha x + \cdots & + & \beta x^* + \cdots & = & -u \\ \vdots & & \vdots & & \vdots \\ \cdots + \gamma x + \cdots & + & \delta x^* + \cdots & = & -u^* \\ \vdots & & \vdots & & \vdots \end{array} \right. \tag{8.2}
$$

[1]This treatment is based on ideas in Tucker (References 39, 42, 43), Balinski and Gomory (Reference 2), and Balinski (Reference 1).

$$
\text{Column system} \quad \left\{ \begin{array}{ccc}
\cdot & \cdot & \cdot \\
\cdot & \cdot & \cdot \\
\cdot & \cdot & \cdot \\
\cdots + y\alpha + & \cdots + y^*\gamma + & \cdots = v \\
\cdot & \cdot & \cdot \\
\cdot & \cdot & \cdot \\
\cdot & \cdot & \cdot \\
\cdots + y\beta + & \cdots + y^*\delta + & \cdots = v^* \\
\cdot & \cdot & \cdot \\
\cdot & \cdot & \cdot \\
\cdot & \cdot & \cdot
\end{array} \right. \tag{8.3}
$$

Each equation of (8.2) is obtained from (8.1) by forming the inner product of a coefficient row with the label row at the top of the schema and then equating this inner product to the label at the right of that coefficient row. Each equation of (8.3) is obtained from (8.1) by forming the inner product of a coefficient column of (8.1) with the label column at the left margin and then equating the inner product to the label below that coefficient column.

In the row (column) system, u and u^* (v and v^*) are considered to be basic (dependent) labels and are expressed as linear combinations of the nonbasic (independent) x and x^* (y and y^*) labels. The term "label" is used to include both variables and constants.

First, we interchange the roles of x and u in (8.2) by solving the $-u$ equation for x and substituting the resulting value of x in the $-u^*$ equation. This gives the new system:

$$
\begin{array}{ccc}
\cdot & \cdot & \cdot \\
\cdot & \cdot & \cdot \\
\cdot & \cdot & \cdot \\
\cdots + \alpha^{-1}u + \cdots + & \alpha^{-1}\beta x^* & + \cdots = -x \\
\cdot & \cdot & \cdot \\
\cdot & \cdot & \cdot \\
\cdot & \cdot & \cdot \\
\cdots - \gamma\alpha^{-1}u + \cdots + & (\delta - \gamma\alpha^{-1}\beta)x^* & + \cdots = -u^* \\
\cdot & \cdot & \cdot \\
\cdot & \cdot & \cdot \\
\cdot & \cdot & \cdot
\end{array} \tag{8.4}
$$

Next, we interchange the roles of y and v in (8.3) giving

$$\begin{array}{ccccccc} & \vdots & & \vdots & & \vdots \\ \cdots + & v\alpha^{-1} & + \cdots - & y^*\gamma\alpha^{-1} & + \cdots = & y \\ & \vdots & & \vdots & & \vdots \\ \cdots + & v\alpha^{-1}\beta & + \cdots + & y^*(\delta - \gamma\alpha^{-1}\beta) & + \cdots = & v^* \\ & \vdots & & \vdots & & \vdots \end{array} \tag{8.5}$$

The following tableau represents the two new systems in the same manner as (8.1) represents the initial systems (8.2) and (8.3). Note

$$\begin{array}{c|cc|c} & u & x^* & \\ \hline v & \alpha^{-1} & \alpha^{-1}\beta & = -x \\ y^* & -\gamma\alpha^{-1} & \delta - \gamma\alpha^{-1}\beta & = -u^*. \\ \hline & = y & = v^* & \end{array} \tag{8.6}$$

the new positions of the labels as well as the new coefficients. We say that (8.6) is obtained from (8.1) by a *pivot* on the (y, x) position. We call the coefficient α in the pivot position the *pivot element*. The pivot process can be described as follows:

(1) replace the pivot element α by its reciprocal α^{-1}.
(2) replace each other element β in the pivot row by $\alpha^{-1}\beta$, the product of the reciprocal of the pivot element and the given element.
(3) replace each other element γ in the pivot column by $-\gamma\alpha^{-1}$, the negative product of the given element and the reciprocal of the pivot element.
(4) replace each element δ not in either the pivot row or the pivot column by $\delta - \gamma\alpha^{-1}\beta$, the element minus the triple product of (a) the element γ in its row and the pivot column, (b) the reciprocal α^{-1} of the pivot element, and (c) the element β in its column and the pivot row.
(5) interchange the labels for the pivot row and pivot column in both primal and dual systems and leave all other labels unchanged.

We illustrate by an example. The schema

$$\begin{array}{c|ccc|c} & x_1 & x_2 & x_3 & \\ \hline y_1 & \textcircled{3} & -1 & -2 & = -u_1 \\ y_2 & 2 & -2 & 1 & = -u_2 \\ y_3 & 4 & 1 & 0 & = -u_3 \\ \hline & = v_1 & = v_2 & = v_3 & \end{array}$$

represents the two systems

$$\text{Primal} \qquad\qquad\qquad\qquad \text{Dual}$$

$$\begin{aligned} 3x_1 - x_2 - 2x_3 &= -u_1, \\ 2x_1 - 2x_2 + x_3 &= -u_2, \\ 4x_1 + x_2 &= -u_3, \end{aligned} \qquad (8.7) \qquad\qquad \begin{aligned} 3y_1 + 2y_2 + 4y_3 &= v_1, \\ -y_1 - 2y_2 + y_3 &= v_2, \\ -2y_1 + y_2 &= v_3. \end{aligned} \qquad (8.8)$$

A pivot transformation on these two systems with pivot element 3 in position (1, 1) solves (8.7) for $-x_1$, $-u_2$, $-u_3$ in terms of u_1, x_2, x_3 and (8.8) for y_1, v_2, v_3 in terms of v_1, y_2, y_3. The arithmetic in full detail is:

$$x_1 - \frac{1}{2}x_2 - \frac{2}{3}x_3 = -\frac{1}{3}u_1, \qquad y_1 + \frac{2}{3}y_2 + \frac{4}{3}y_3 = \frac{1}{3}v_1,$$

$$2x_1 - 2x_2 + x_3 = -u_2, \qquad -y_1 - 2y_2 + y_3 = v_2,$$

$$4x_1 + x_2 = -u_3, \qquad -2y_1 + y_2 = v_3,$$

$$x_1 - \frac{1}{3}x_2 - \frac{2}{3}x_3 = -\frac{1}{2}u_1, \qquad y_1 + \frac{2}{3}y_2 + \frac{4}{3}y_3 = \frac{1}{3}v_1,$$

$$0 - \frac{4}{3}x_2 + \frac{7}{3}x_3 = -u_2 + \frac{2}{3}u_1, \qquad 0 - \frac{4}{3}y_2 + \frac{7}{3}y_3 = v_2 + \frac{1}{3}v_1,$$

$$0 + \frac{7}{3}x_2 + \frac{8}{3}x_3 = -u_3 + \frac{4}{3}u_1, \qquad 0 + \frac{7}{3}y_2 + \frac{8}{3}y_3 = v_3 + \frac{2}{3}v_1,$$

$$\frac{1}{3}u_1 - \frac{1}{3}x_2 - \frac{2}{3}x_3 = -x_1, \qquad \frac{1}{3}v_1 - \frac{2}{3}y_2 - \frac{4}{3}y_3 = y_1,$$

$$-\frac{2}{3}u_1 - \frac{4}{3}x_2 + \frac{7}{3}x_3 = -u_2, \qquad -\frac{1}{3}v_1 - \frac{4}{3}y_2 + \frac{7}{3}y_3 = v_2,$$

$$-\frac{4}{3}u_1 + \frac{7}{3}x_2 + \frac{8}{3}x_3 = -u_3, \qquad -\frac{2}{3}v_1 + \frac{7}{3}y_2 + \frac{8}{3}y_3 = v_3.$$

On the other hand, by applying the pivot rule directly to the schema, we obtain:

$$\begin{array}{c|ccc|c} & u_1 & x_2 & x_3 & \\ \hline v_1 & 1/3 & -1/3 & -2/3 & = -x_1 \\ y_2 & -2/3 & -4/3 & 7/3 & = -u_2 \\ y_3 & -4/3 & 7/3 & 8/3 & = -u_3 \\ \hline & = y_1 & = v_2 & = v_3 & \end{array}$$

It is clear that this new schema simultaneously represents both the new row system and the new column system.

If we regard (8.2) as giving a primal system and (8.3) as giving its dual, we conclude that a pivot operation (including the appropriate permutation of labels) leads to a transformation tableau whose primal is equivalent to the original primal in the sense that every solution

$(\cdots, x, \cdots, x^*, \cdots, u, \cdots, u^*, \cdots)$ of either is also a solution of the other. The old and new dual systems are equivalent in the same sense. When we add the "trimmings" needed to make (8.2) and (8.3) into a full linear programming problem together with its dual we will have established, through the use of schema, the invariance of duality under pivotal transformations and hence under the simplex algorithm when it is also presented in schema form.

We first consider a primal problem in symmetric form and its dual. We have three reasons for this initial limitation to symmetric form. First, as remarked earlier, a substantial proportion of the known applications of linear programming is formulated most naturally in symmetric form so that restriction to this case does not represent a serious loss of generality. Second, solution for this special case presents considerably less notational and conceptual difficulty than for the general case. Third, the algorithm for this special case will be an important subroutine in the algorithm for the general case.

What we now propose to do is to interpret the procedures and results of the earlier sections in schema form. This goal helps explain what might otherwise seem to be an unnatural notation for the dual linear programming problems we next introduce.

The notation below is chosen also to minimize the use of minus signs in the algebraic description of the schema; the numerical entries themselves may, of course, have any sign. Consider the dual problems:

Primal

$$\begin{aligned} AX + B &\leq 0, \\ X &\geq 0, \\ \min u &= d + C^T X, \end{aligned} \tag{8.9}$$

Dual

$$\begin{aligned} Y &\geq 0, \\ A^T Y + C &\geq 0, \\ \max v &= d + B^T Y, \end{aligned} \tag{8.10}$$

where A is p by n, B is p by 1, C is n by 1. Note that there are no sign restrictions on the A or B or C.

The components of X and Y may be labeled in such a way that, after adding slack variables to (8.9) and (8.10) the dual problems become:

$$d + c_1 x_1 + \cdots + c_n x_n = u \qquad \text{(to be minimized)}$$

$$\left.\begin{aligned} b_1 + a_{11}x_1 + \cdots + a_{1n}x_n &= -x_{n+1} \\ b_2 + a_{21}x_1 + \cdots + a_{2n}x_n &= -x_{n+2} \\ \cdot \qquad\qquad\qquad & \cdot \\ \cdot \qquad\qquad\qquad & \cdot \\ \cdot \qquad\qquad\qquad & \cdot \\ b_p + a_{p1}x_1 + \cdots + a_{pn}x_n &= -x_{n+p} \end{aligned}\right\} \text{constraining equations} \tag{8.9$'$}$$

$$x_i \geq 0, \qquad (i = 1, \cdots, n + p)$$

$$
\left.
\begin{aligned}
&d + b_1 y_{n+1} + \cdots + b_p y_{n+p} = v \quad \text{(to be maximized)}\\
&c_1 + a_{11} y_{n+1} + \cdots + a_{p1} y_{n+p} = y_1\\
&c_2 + a_{12} y_{n+1} + \cdots + a_{p2} y_{n+p} = y_2\\
&\quad \vdots \\
&c_n + a_{1n} y_{n+1} + \cdots + a_{pn} y_{n+p} = y_n
\end{aligned}
\right\} \text{constraining equations} \qquad (8.10')
$$

$$y_i \geq 0, \qquad (i = 1, \cdots, n + p).$$

These two systems can be conveniently represented by the single schema S,

$$
S: \quad
\begin{array}{c|c|cccc|l}
 & 1 & x_1 & x_2 & & x_n & \\
\hline
1 & d & c_1 & c_2 & \cdots & c_n & = u \text{ (min)} \\
\hline
y_{n+1} & b_1 & a_{11} & a_{12} & \cdots & a_{1n} & = -x_{n+1} \leq 0 \\
y_{n+2} & b_2 & a_{21} & a_{22} & \cdots & a_{2n} & = -x_{n+2} \leq 0 \\
 & \vdots & \vdots & & & \vdots & \\
y_{n+p} & b_p & a_{p1} & a_{p2} & \cdots & a_{pn} & = -x_{n+p} \leq 0 \\
\hline
 & = v & = y_1 & = y_2 & & = y_n & \\
 & & \geq & \geq & & \geq & \\
 & \text{(max)} & 0 & 0 & & 0 &
\end{array}
\qquad (8.11)
$$

For notational convenience we sometimes write $d = a_{00}$, $b_i = a_{i0}$, $c_j = a_{0j}$ $(i = 1, \cdots, p; j = 1, \cdots, n)$. Because of the special nature and role of the elements in the zero row and column, we are led to call row zero the *distinguished row*, column zero the *distinguished column*, and the element $d = a_{00}$ the *distinguished element*.

Our first main result is the following theorem: given a pair of dual problems with labeled schema (8.11), there is a (not necessarily unique) finite sequence of pivot transformations which transforms (8.11) into one of the following mutually exclusive forms ($\oplus$, $\ominus$ denote nonnegative and nonpositive entries respectively).

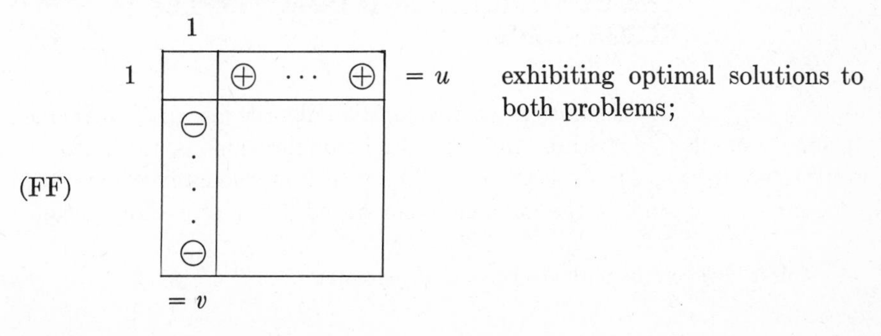

exhibiting optimal solutions to both problems;

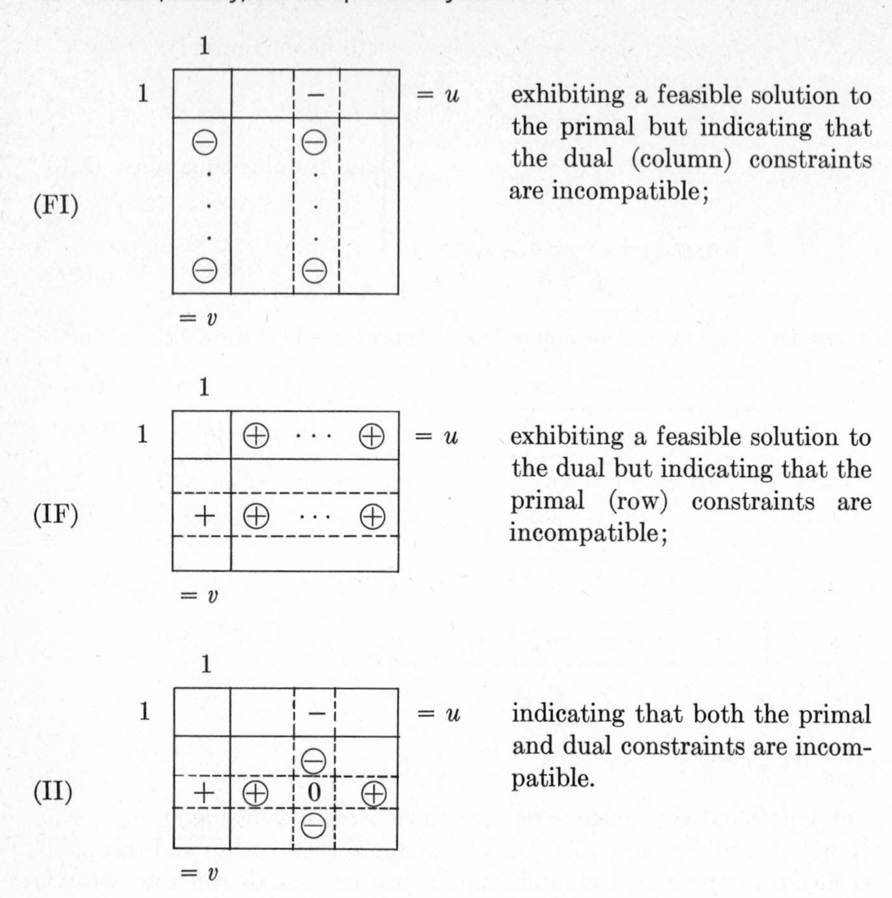

exhibiting a feasible solution to the primal but indicating that the dual (column) constraints are incompatible;

exhibiting a feasible solution to the dual but indicating that the primal (row) constraints are incompatible;

indicating that both the primal and dual constraints are incompatible.

The proof of this theorem, as given in Section 8.2, is constructive and utilizes the results of Chapter 3. Somewhat shorter proofs can be found in Tucker (Reference 42) and Balinski (Reference 1).

8.2. THE CONTRACTED AND DUAL SIMPLEX ALGORITHMS.

In this section we develop two algorithms, the contracted (primal) simplex algorithm and the dual simplex algorithm[2] (also in contracted form). These algorithms, important in themselves, are key elements in our proof of the main theorem stated in the preceding section.

[2]This is based on the work of Lemke (Reference 31).

Any schema of the form (8.11) in which $b_i \le 0$, $(i = 1, \cdots, p)$ is said to be in *row-feasible form*, and any schema of the form (8.11) in which $c_i \ge 0$, $(i = 1, \cdots, n)$ is said to be in *column-feasible form*. Clearly, any schema which is in both row- and column-feasible form is of the form FF (feasible, feasible) and exhibits optimal programs for both problems. For, if the schema is in row-feasible form, the $(p + n)$ vector $[0 \cdots 0 \ -b_1 \cdots -b_p]^T$ is a basic feasible program for the primal problem. If it is in column-feasible form, the vector $[c_1 \cdots c_n \ 0 \cdots 0]^T$ is a basic feasible program for the dual problem.

Suppose that (8.11) or some schema S obtained from it by a sequence of pivotal transformations is in row-feasible form. We associate with any $i(1 \le i \le p)$ the row vector $S(i) = [-b_i \ f_{i,n+1} \cdots f_{i,n+p} \ f_{i1} \cdots f_{in}]$ where f_{ij} is the coefficient of x_j in the ith equation. We observe that we do not ordinarily have $f_{ij} = a_{ij}$ since x_j, if nonbasic, need not be the label for column j and if basic, will have coefficient 1 or 0 according as it is or is not the label for row i. This particular definition of the f_{ij} (1) makes $S(i)$ be the same as the ith row of the corresponding ordinary simplex tableau for the same sequence of pivotal transformations, and (2) makes each vector $S(i)$ for (8.11) lexicographically positive. We define $S(0)$ analogously from row zero of S.

The contracted primal simplex algorithm is initiated with a schema in primal feasible form. The (primal) pivot selection rule is as follows. If S is in row-feasible form but not in either of the terminal forms FF or FI then pivot at position (s, t) where:

> The integer t is any index for which $c_t < 0$. (There is at least one such t lest S be in the form FF.) (8.12)
>
> The integer s is chosen so that $S(s)/a_{st}$ is lexicographically smallest among the set of vectors $S(i)/a_{it}$ with $a_{it} > 0$ and $1 \le i \le p$. (There is at least one such i lest S be in the form FI.) (8.13)

We can replace (8.13) by the simpler rule

$$\frac{b_s}{a_{st}} = \max\left\{\frac{b_i}{a_{it}} \mid a_{it} > 0\right\} \tag{8.13$'$}$$

in case this gives a unique s.

Let S' denote the schema obtained by a pivotal transformation with this choice of pivot element. Then the hypothesis that each vector $S(i)$, $(i = 1, \cdots, p)$ is lexicographically positive leads to the two conclusions: (1) each vector $S'(i)$, $(i = 1, \cdots, p)$ is lexicographically positive and (2) $S'(0) - S(0)$ is lexicographically positive.

Property (2) guarantees that there can be no circling (no recurrence of the set of indices in an earlier basic sequence) in any sequence of pivot operations which satisfy (8.12) and (8.13). Now, since there are only finitely many basic sequences for a given n and p the process must terminate either in the form FF or FI.

Since the rule (8.13) is rather cumbersome, we usually use (8.13′) with the agreement that ties shall be broken randomly. This completes the description and proof of convergence of the contracted primal simplex algorithm.

In certain problems, the diet problem, for example, the initial schema is in column-feasible form. One might ask if there is an algorithm which would preserve such column feasibility and use primal infeasibilities to determine a pivot row. The dual simplex algorithm does just this. We state it in contracted form and handle degeneracy by the random tie-breaking rule.

Suppose that S is in column-feasible form but not in either of the forms FF or IF. Then there exists an integer s for which $b_s > 0$ and an integer j such that $a_{sj} < 0$. We then choose t so that

$$\frac{c_t}{a_{st}} = \max\left\{\frac{c_j}{a_{sj}} \mid a_{sj} < 0\right\} \tag{8.14}$$

where to obtain a definite t any ties are broken by some random device.

An example of one cycle of the dual simplex algorithm is given in Table 8.1 and a small problem is solved in Table 8.2.

Table 8.1. THE DUAL SIMPLEX ALGORITHM, $(s, t) = (3, 2)$.

S:

	1	x_1	x_2	x_3	x_4	
1	3	8	1	0	2	u
y_5	1	0	1	−1	−3	$-x_5$
y_6	0	2	0	−2	5	$-x_6$
y_7	2	−1	[−1]	2	−1	$-x_7$
y_8	−1	8	3	2	0	$-x_8$
	v	y_1	y_2	y_3	y_4	

S':

	1	x_1	x_7	x_3	x_4	
1	5	7	1	2	1	u
y_5	3	−1	1	1	−4	$-x_5$
y_6	0	2	0	−2	5	$-x_6$
y_2	−2	1	−1	−2	1	$-x_2$
y_8	5	5	3	8	−3	$-x_8$
	v	y_1	y_7	y_3	y_4	

We now return to the reduction proof of the main theorem. If S is not initially in row-feasible form, we proceed as in Section 3.6 and either increase the number of feasible relations and thus eventually obtain row feasibility or demonstrate row infeasibility, that is, find an S with a row (not the initial one) of the form $[+ \oplus \cdots \oplus]$.

Table 8.2. EXAMPLE OF DUAL SIMPLEX ALGORITHM.

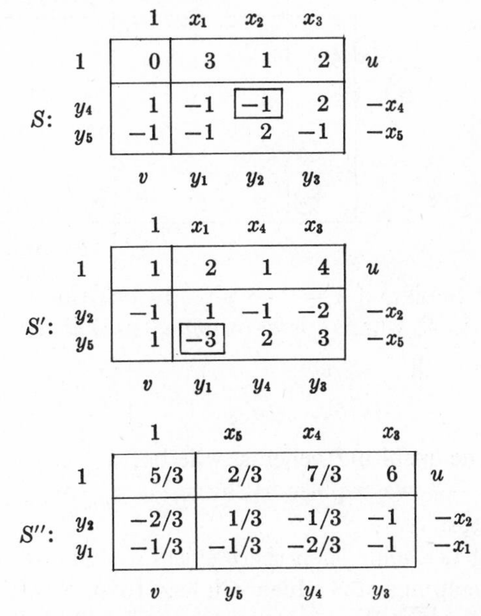

S:

	1	x_1	x_2	x_3	
1	0	3	1	2	u
y_4	1	−1	[−1]	2	$-x_4$
y_5	−1	−1	2	−1	$-x_5$
	v	y_1	y_2	y_3	

S':

	1	x_1	x_4	x_3	
1	1	2	1	4	u
y_2	−1	1	−1	−2	$-x_2$
y_5	1	[−3]	2	3	$-x_5$
	v	y_1	y_4	y_3	

S'':

	1	x_5	x_4	x_3	
1	5/3	2/3	7/3	6	u
y_2	−2/3	1/3	−1/3	−1	$-x_2$
y_1	−1/3	−1/3	−2/3	−1	$-x_1$
	v	y_5	y_4	y_3	

We have already handled the primal feasible case so we now complete the proof that every linear programming schema is equivalent to one of the four terminal forms by showing how to proceed from primal infeasibility to one of the forms IF, II. In this final part of the proof we make use of the dual forms of those results already established. For notational convenience we assume that the last row of S is the one of form $[+ \oplus \cdots \oplus]$. Suppose that there are t zeros ($t \geq 0$) in this row and that they occur at the right-hand end of the row; then S has the following form:

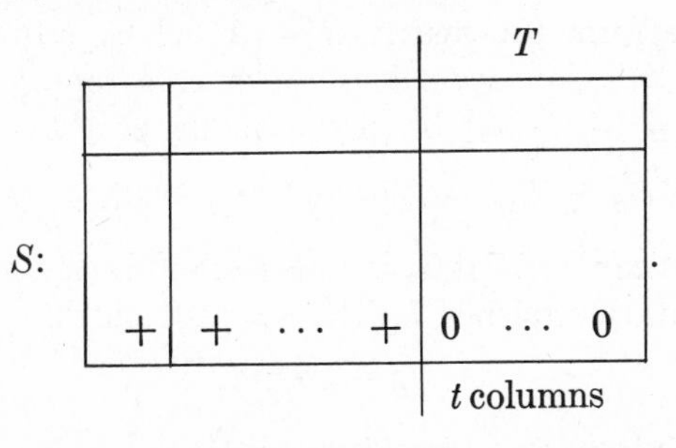

We define T to be the schema consisting of the last t columns of S. (Note that T has no distinguished column). Now, either (1) T is dual feasible or (2) T is dual infeasible. If (1) holds, then there exists a sequence of pivots on positions in the last t columns of S which will lead to S' of the form

S':

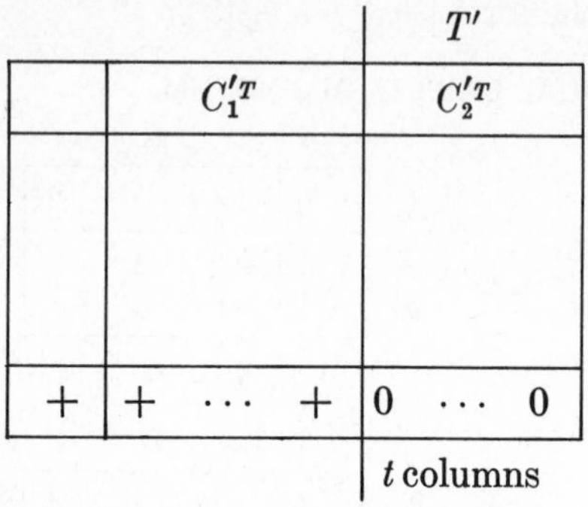

	$C_1'^T$	$C_2'^T$ (T')
$+$	$+ \cdots +$	$0 \cdots 0$

(t columns), where $C_2' \geq 0$.

In particular, if $t = 0$, S already has this form. Next, pivot on position (p, k) in S' where k is determined so that

$$\frac{c_k'}{a_{pk}'} = \min \left\{ \frac{c_j'}{a_{pj}'} \mid a_{pj}' > 0 \right\},$$

and the resulting schema will be in the form IF. Since for $j \neq k$ and $j < n - t$ we replace c_j' by $c_j'' = a_{pj}' \; (c_j'/a_{pj}' - c_k'/a_{pk}')$ which is clearly nonnegative.

If (2) holds then there exists a sequence of pivots on positions in the last t columns of S which will lead to an S' with one of its last t columns in the form $[\ominus]$. But since the last row of T is zero the same will be true of its counterpart T' in S' since all pivot positions are in the last t columns. Hence S' has the form II.

8.3. COMBINATORIAL EQUIVALENCE AND BLOCK PIVOTING[3].

Let A be any p by n matrix and let I be the identity matrix of order p. We form the matrix $G = [A \quad I]$ by adjoining I to A. Let $S = (s_1, \cdots, s_p)$ be a sequence of distinct elements from $P_{n+p} = \{1, \cdots, n, \cdots, n+p\}$, let $T = (t_1, \cdots, t_n)$ be any ordering of $P_{n+p} - S$, and consider

$$G_S = [G_{s_1} \cdots G_{s_p}] \quad \text{and} \quad G_T = [G_{t_1} \cdots G_{t_n}]. \tag{8.15}$$

We say that S is *admissible* if G_S is nonsingular. Each admissible sequence S and complementary sequence T defines a new matrix

$$A^* = G_S^{-1} G_T \tag{8.16}$$

[3]This section is based on Tucker (Reference 41).

which is said to be *combinatorially equivalent* to A. We write in this case $A^* \sim A$. For example, let

$$A = \begin{bmatrix} 6 & 3 & 1 & 2 \\ 1 & 4 & 0 & 1 \\ 2 & 2 & 1 & 1 \end{bmatrix}; \tag{8.17}$$

then

$$G = [A \quad I] = \begin{bmatrix} 6 & 3 & 1 & 2 & 1 & 0 & 0 \\ 1 & 4 & 0 & 1 & 0 & 1 & 0 \\ 2 & 2 & 1 & 1 & 0 & 0 & 1 \end{bmatrix}. \tag{8.18}$$

The sequence $S = (5, 4, 3)$ is admissible; let $T = (7, 1, 6, 2)$. Then

$$G_S = \begin{bmatrix} 1 & 2 & 1 \\ 0 & 1 & 0 \\ 0 & 1 & 1 \end{bmatrix}, \qquad G_T = \begin{bmatrix} 0 & 6 & 0 & 3 \\ 0 & 1 & 1 & 4 \\ 1 & 2 & 0 & 2 \end{bmatrix}$$

and

$$A^* = \begin{bmatrix} 1 & -1 & -1 \\ 0 & 1 & 0 \\ 0 & -1 & 1 \end{bmatrix} \begin{bmatrix} 0 & 6 & 0 & 3 \\ 0 & 1 & 1 & 4 \\ 1 & 2 & 0 & 2 \end{bmatrix}$$

$$= \begin{bmatrix} -1 & 3 & -1 & -3 \\ 0 & 1 & 1 & 4 \\ 1 & 1 & -1 & -2 \end{bmatrix}.$$

We now consider three properties of the relation of combinatorial equivalence.

(1) For every matrix A we have $A \sim A$. For, choose $S = (n + 1, \cdots, n + p)$ and $T = (1, \cdots, n)$.

(2) $A^* \sim A$ implies $A \sim A^*$. Since $A^* \sim A$, $A^* = G_S^{-1} G_T$ for some choice of S and T. Now let $G^* = [A^* \quad I]$; then the equation

$$G_S G^* = G_S[A^* \quad I] = [G_T G_S] = G' \tag{8.19}$$

shows that $G' = G_S G^*$ is obtained by permuting the columns of G. Hence for suitable sequences S^*, T^* we have

$$G_S G^*_{S^*} = I_p \qquad \text{and} \qquad G_S G^*_{T^*} = A, \tag{8.20}$$

and, therefore,

$$G^{*-1}_{S^*} G^*_{T^*} = A \tag{8.21}$$

so that $A \sim A^*$.

(3) $A^* \sim A$ and $A^{**} \sim A^*$ imply $A^{**} \sim A$. Let

$$G = [A \quad I], \qquad G^* = [A^* \quad I], \qquad G^{**} = [A^{**} \quad I], \tag{8.22}$$

and suppose that

$$A^* = G_S^{-1} G_T \qquad \text{and} \qquad A^{**} = G^{*-1}_{S^*} G^*_{T^*}; \tag{8.23}$$

then $G_{S*}^* G^{**} = [G_{T*}^* \; G_{S*}^*]$ is a permutation of the columns of G^* and $G_S G^*$ is a permutation of the columns of G. Hence $G_S G_{S*}^* G^{**}$ is a permutation of the columns of G. It follows that for suitable sequences S', T' we have

$$G_S G_{S*}^* = G_{S'} \qquad \text{and} \qquad G_{T'} = G_S G_{S*}^* A^{**}, \tag{8.24}$$

hence $A^{**} = G_{S'}^{-1} G_{T'} \sim A$.

Statements (1), (2), and (3) prove that combinatorial equivalence is an equivalence relation and therefore partitions the set of all matrices into disjoint classes. Since for given p and n there are only a finite number of choices for S and T, there must be only a finite number of matrices combinatorially equivalent to any matrix A. Hence our equivalence classes are each of finite order.

In the schema version of the simplex algorithm let the initial schema matrix be [compare (8.11)]

$$M(S) = \begin{bmatrix} d & C^T \\ B & A \end{bmatrix}, \tag{8.25}$$

and for a later schema S^*, let

$$M(S^*) = \begin{bmatrix} d^* & C^{*T} \\ B^* & A^* \end{bmatrix}. \tag{8.26}$$

Since we never pivot in row 0 or column 0 and since we do not permute either this row or column in rearranging rows and columns it follows both that $A^* \sim A$ and $M(S^*) \sim M(S)$, and hence we see again that there are only finitely many schema matrices for a given problem.

The duality theory also shows that there is an equivalent dual definition of combinatorial equivalence involving rows and postmultiplication. It turns out that the dual definition resembles the primal one except that a few added minus signs appear from time to time; this phenomenon is related so the fact that the dual problem is equivalent to the primal problem with initial schema matrix $-(M(S))^T$ and hence it is $-I$ rather than I which is adjoined at each step.

Suppose that S^* is the schema immediately following S and that the pivot element is $m_{st} (= a_{st})$. Then $M(S^*)$ is obtained from $M(S)$ by the *pivotal transformation*

$$\begin{aligned} m_{st}^* &= \frac{1}{m_{st}}, \\ m_{sj}^* &= \frac{m_{sj}}{m_{st}}, \qquad j \neq t \\ m_{it}^* &= \frac{-m_{it}}{m_{st}}, \qquad i \neq s \\ m_{ij}^* &= m_{ij} - m_{it}\frac{m_{sj}}{m_{st}}, \qquad i \neq s, \quad j \neq t. \end{aligned} \tag{8.27}$$

Of course, in the simplex algorithm we always have $s > 0$ and $t > 0$. We state without proof that if $A \sim A^*$ then there exists a sequence of pivotal transformations leading from A to a matrix A' from which A^* can be obtained by some permutation of rows and columns. In the schema form of the simplex algorithm this is the way $M(S^*)$ is obtained from $M(S)$ (using pivots each having $s > 0$ and $t > 0$).

If A^* is obtained from A by a pivotal transformation using position (s, t) as pivot position, then we can give two alternate procedures for obtaining A^*. The first involves direct application of (8.16) with $S = (n + 1, \cdots, n + s - 1, t, n + s + 1, \cdots, n + p)$ and $T = (1, \cdots, t - 1, n + s, t + 1, \cdots, n)$. Then G_S^{-1} is the same as the identity matrix except that column s is replaced by

$$\frac{(1 + a_{st})U_s - A_t}{a_{st}} \tag{8.28}$$

where U_s is the sth unit vector and A_t is the tth column of A.

The dual alternative procedure is described as follows. Let Q be the matrix obtained from I_n by replacing its tth row U_t^T by

$$\frac{(a_{st} - 1)U_t^T - (A^T)_s}{a_{st}} \tag{8.29}$$

where $(A^T)_s$ is the sth row of A. Then the product

$$\begin{bmatrix} A \\ -I_n \end{bmatrix} Q$$

will have $-I_n$ as a submatrix and A^* as the submatrix consisting of the remaining p rows.

Suppose that beginning with a schema

$$\begin{array}{cccc} & X & & \\ Y & \boxed{\quad A \quad} & = -U & \\ & = V & & \end{array} \tag{8.30}$$

a sequence of pivots results in a new schema

$$\begin{array}{cccc} & X' & & \\ Y' & \boxed{\quad A' \quad} & = -U' & \\ & = V' & & \end{array} \tag{8.31}$$

in which, of course, A' is combinatorially equivalent to A; suppose further that exactly r of the new (primal) basic variables u_i' were initially nonbasic.

Then also exactly r of the new nonbasic variables were initially basic. We may suppose the notation is chosen so that

$$X = \begin{bmatrix} X_1 \\ X_2 \end{bmatrix}, \quad U = \begin{bmatrix} U_1 \\ U_2 \end{bmatrix}, \quad Y = \begin{bmatrix} Y_1 \\ Y_2 \end{bmatrix}, \quad V = \begin{bmatrix} V_1 \\ V_2 \end{bmatrix},$$

$$X' = \begin{bmatrix} U_1 \\ X_2 \end{bmatrix}, \quad U' = \begin{bmatrix} X_1 \\ U_2 \end{bmatrix}, \quad Y' = \begin{bmatrix} V_1 \\ Y_2 \end{bmatrix}, \quad V' = \begin{bmatrix} Y_1 \\ V_2 \end{bmatrix}$$

and partition A and A' correspondingly so that the schema (8.30) and (8.31) take the respective forms

$$\begin{array}{cccc} & X_1 & X_2 & \\ Y_1 & \alpha & \beta & = -U_1 \\ Y_2 & \gamma & \delta & = -U_2 \\ & = V_1 & = V_2 & \end{array} \tag{8.32}$$

and

$$\begin{array}{cccc} & U_1 & X_2 & \\ V_1 & \alpha' & \beta' & = -X_1 \\ Y_2 & \gamma' & \delta' & = -U_2 \\ & = Y_1 & = V_2 & \end{array} \tag{8.33}$$

We determine $\alpha', \beta', \gamma', \delta'$ using the fact that (8.33) is obtainable from (8.32) by solving the system

$$\begin{aligned} \alpha X_1 + \beta X_2 &= -U_1 \\ \gamma X_1 + \delta X_2 &= -U_2 \end{aligned} \tag{8.34}$$

for X_1, U_2 in terms of U_1, X_2. Clearly, this cannot be done unless α is nonsingular and if this is true we get

$$\begin{aligned} \alpha^{-1}U_1 + \qquad \alpha^{-1}\beta X_2 &= -X_1 \\ -\gamma\alpha^{-1}U_1 + (\delta - \gamma\alpha^{-1}\beta)X_2 &= -U_2. \end{aligned} \tag{8.35}$$

We conclude that

$$A' = \begin{bmatrix} \alpha^{-1} & \alpha^{-1}\beta \\ -\gamma\alpha^{-1} & \delta - \gamma\alpha^{-1}\beta \end{bmatrix}, \tag{8.36}$$

a result which closely resembles (8.6). Indeed (8.6) comes from the special case $r = 1$ of (8.36) [except that (8.6) is expressed in scalar rather than in matrix form].

The process leading from (8.32) to (8.33) is called a *block pivot* and the matrix α is called the *pivot matrix* or the *kernel*. Of course the kernel need not be in the upper left-hand corner but can be any nonsingular submatrix α of A. We can apply the concept of block pivot to combinatorial equivalence and obtain the following result. A matrix A^* is combinatorially equivalent to a given matrix A if and only if A^* can be obtained by permuting rows and columns of a matrix A' which in turn can be obtained from A by a block pivot.

For example, suppose S is the following schema:

$$
\begin{array}{lll|cc|ccc|ll}
 & & & \multicolumn{2}{c|}{X_2} & \multicolumn{3}{c|}{\overbrace{\hphantom{xxxxxxxxxx}}^{X_1}} & & \\
 & & & 1 & x_1 & x_2 & x_3 & x_4 & & \\
\hline
 & & & 0 & -1 & -2 & -4 & -1 & = u & \\
\hline
 & & y_5 & -3 & 0 & 2 & 1 & 0 & -x_5 & \\
S: & Y_1 & y_6 & -4 & 1 & 1 & 3 & 0 & -x_6 & -U_1 \\
 & & y_7 & -3 & 1 & 0 & 1 & 4 & -x_7 & \\
\hline
 & & & = v & = y_1 & = y_2 & = y_3 & = y_4 & & U_2 \text{ empty} \\
Y_2 \text{ empty} & & & \multicolumn{2}{c|}{V_2} & \multicolumn{3}{c|}{\underbrace{\hphantom{xxxxxxxxxx}}_{V_1}} & &
\end{array}
$$

where α, the pivot matrix, is the 3 by 3 matrix in the lower right-hand corner, and where

$$
\beta = \begin{bmatrix} -3 & 0 \\ -4 & 1 \\ -3 & 1 \end{bmatrix}, \qquad \delta = [0 \quad -1], \qquad \gamma = [-2 \quad -4 \quad -1].
$$

Then we have

$$
\alpha^{-1} = \begin{bmatrix} 3/5 & -1/5 & 0 \\ -1/5 & 2/5 & 0 \\ 1/20 & -1/10 & 1/4 \end{bmatrix}, \qquad \alpha^{-1}\beta = \begin{bmatrix} -1 & -1/5 \\ -1 & 2/5 \\ -1/2 & 3/20 \end{bmatrix}
$$

$$
-\gamma\alpha^{-1} = -[-9/20 \quad -11/10 \quad -1/4], \qquad \delta - \gamma\alpha^{-1}\beta = [-13/2 \quad 7/20],
$$

and

$$
\begin{array}{lll|cc|ccc|ll}
 & & & \multicolumn{2}{c|}{X_2} & \multicolumn{3}{c|}{\overbrace{\hphantom{xxxxxxxxxx}}^{U_1}} & & \\
 & & & 1 & x_1 & x_5 & x_6 & x_7 & & \\
\hline
 & & & -13/2 & 7/20 & 9/20 & 11/10 & 1/4 & = u & \\
\hline
 & & y_2 & -1 & -1/5 & 3/5 & -1/5 & 0 & x_2 & \\
S': & Y_2 & y_3 & -1 & 2/5 & -1/5 & 2/5 & 0 & x_3 & X_1 \\
 & & y_4 & -1 & 3/20 & 1/20 & -1/10 & 1/4 & x_4 & \\
\hline
 & & & = v & = y_1 & = y_5 & = y_6 & = y_7 & & \\
 & & & \multicolumn{2}{c|}{V_2} & \multicolumn{3}{c|}{\underbrace{\hphantom{xxxxxxxxxx}}_{Y_1}} & &
\end{array}
$$

8.4. SCHEMA AND THE GENERAL LINEAR PROGRAMMING PROBLEM.

The general linear programming problem can be written in the form:

$$\begin{aligned} &\text{minimize} \quad u = d + C_1^T X_1 + C_2^T X_2 \\ &\text{subject to} \\ &B_1 + A_{11}X_1 + A_{12}X_2 \le 0, \qquad (p_1 \text{ constraints}) \\ &B_2 + A_{21}X_1 + A_{22}X_2 = 0, \qquad (p_2 \text{ constraints}) \\ &X_1 \ge 0, \qquad (n_1 \text{ variables}) \\ &X_2 \text{ free}, \qquad (n_2 \text{ variables}). \end{aligned} \tag{8.37}$$

The dual of this problem then can be written in the form:

$$\begin{aligned} &\text{maximize} \quad v = d + B_1^T U_1 + B_2^T U_2 \\ &\text{subject to} \\ &U_1 \ge 0, \qquad (p_1 \text{ variables}) \\ &U_2 \quad \text{free}, \qquad (p_2 \text{ variables}) \\ &C_1 + A_{11}^T U_1 + A_{21}^T U_2 \ge 0, \qquad (n_1 \text{ constraints}) \\ &C_2 + A_{12}^T U_1 + A_{22}^T U_2 = 0, \qquad (n_2 \text{ constraints}). \end{aligned} \tag{8.38}$$

Both of these systems can be represented by the following schema, where Y_1 and V_1 represent slack variables:

S:		1	$X_1^T \ge 0$	X_2^T free		
	1	d	C_1^T	C_2^T	$= u$ (min)	
	$U_1 \ge 0$	B_1	A_{11}	A_{12}	$= -Y_1 \le 0$	(8.39)
	U_2 free	B_2	A_{21}	A_{22}	$= -Y_2 = 0$	
		$=$	$=$	$=$		
		v(max)	$V_1^T \ge 0$	$V_2^T = 0$		

The new feature we face here is the presence of zero and of free labels. A basic zero primal label cannot appear in a basic feasible program unless the corresponding constant is also zero, whereas a nonbasic zero label presents no difficulty and as we shall see merely indicates that its column plays no role in the primal problem. A basic free primal label means that the corresponding constraint can be satisfied regardless of the values assigned to the remaining nonzero labels. Our goal is to perform preliminary pivot operations which will permit a reduction of the general problem to consideration of one in symmetric form.

To achieve this goal we carry out pivot operations which will (1) increase the number of nonbasic primal constant labels (and therefore increase the number of basic dual free labels) in the schema and (2) increase the number of basic primal free labels (and therefore increase the number of nonbasic dual constant labels) in the schema.

This can be accomplished most efficiently by first pivoting on a position in the block occupied by A_{22}, because constant and free labels will be interchanged by such an operation. If $A_{22} = 0$, a pivot position may be selected in either A_{21} or A_{12}, achieving either goal (1) or (2) of the preceeding paragraph. If $p_2 = n_2 = 0$ then (8.37) is already in the symmetric form of Section 8.1 for which we have given an algorithm.

Proceeding inductively, suppose we have reached a schema of the form:

		1	$X_1'^T \geq 0$	$X_2'^T = 0$	$X_3'^T$ free	
	1	h	G_1^T	G_2^T	G_3^T	$= u(\min)$
	$U_1' \geq 0$	F_1	E_{11}	E_{12}	E_{13}	$= -Y_1' \leq 0$
S^*:	$U_2' = 0$	F_2	E_{21}	E_{22}	E_{23}	$= -Y_2'$ free
	U_3' free	F_3	E_{31}	E_{32}	E_{33}	$= -Y_3' = 0$
		$=$	$=$	$=$	$=$	
		$v(\max)$	$V_1'^T \geq 0$	$V_2'^T$ free	$V_3'^T = 0$	(8.40)

If there is a nonzero entry in E_{33} we may pivot on the corresponding position and thus simultaneously achieve both goals 1 and 2 above. If $E_{33} = 0$ but either E_{31} or $E_{13} \neq 0$ we may pivot on the position of any nonzero entry and achieve either goal 1 or 2. We can continue thus until a schema S^* is attained for which E_{31}, E_{13}, E_{33} all have only zero entries. Of course S^* merely represents another system of primal and dual equations which have exactly the same solutions as (8.37) and (8.38).

Now, if $F_3 \neq 0$ the primal system is inconsistent and if $G_3 \neq 0$ then either the primal system is not feasible or the primal objective function has no finite lower bound.

Dually, $G_3 \neq 0$ implies dual infeasibility and $F_3 \neq 0$ implies either dual infeasibility or the dual objective function has no finite upper bound. The infeasibilities which arise here indicate that the equations in (8.37) or (8.38) are inconsistent and would remain so even if the restrictions $X_1 \geq 0$ or $U_1 \geq 0$ were removed.

We conclude that if either the primal or the dual is to have an optimal solution we can always pivot until a schema S^* of the form is obtained:

	1	$X_1'^T \geq 0$	$X_2'^T = 0$	$X_3'^T$ free	
	h	G_1^T	G_2^T	0	$= u(\min)$
$U_1' \geq 0$	F_1	E_{11}	E_{12}	0	$= -Y_1' \leq 0$
$U_2' = 0$	F_2	E_{21}	E_{22}	E_{23}	$= -Y_2'$ free
U_3' free	0	0	E_{32}	0	$= -Y_3' = 0$
	$=$	$=$	$=$	$=$	
	$v(\max)$	$V_1'^T \geq 0$	$V_2'^T$ free	$V_3'^T = 0$	(8.41)

The columns with second subscript 2 can be dropped from the primal equations since they enter with coefficient $X_2' = 0$. Now, since $F_2 + E_{21}X_1' + E_{22}(0) + E_{23}X_3' = -Y_2'$ merely expresses the *free* vector $-Y_2'$ in terms of X_1' and X_3', and since Y_2' does not occur in the objective function u, this equation can be set aside and used to determine Y_2' in terms of X_3' and X_1' after the remaining subproblem (8.42) has been solved. Thus

T:		1	$X_1'^T(\geq 0)$		
	1	h	G_1^T	$= u$	(8.42)
	U_1' (≥ 0)	F_1	E_{11}	$= -Y_1'(\leq 0)$	
		$=$	$=$		
		v	$V_1'^T(\geq 0)$		

is a subschema of S^* which represents a primal problem (essentially) equivalent to (8.37) and which is symmetric.

The deletions made to arrive at (8.42) from (8.41) can be justified also for the dual problem and so the dual problem represented by T is equivalent to (8.38).

The schema formulation makes possible a more complete statement of the fundamental duality theorem for linear programming than our earlier one. We partition the collection of all primal and dual problems into five classes and give the possible connections between them in Table 8.3 (in this table P indicates a possible pairing and N indicates a pairing that cannot occur).

Class 1. Problem has an optimal feasible program.

Class 2. Problem has a basic feasible program but the cost function is unbounded because there is a nonbasic free variable with nonzero cost and which appears only in equations whose associated basic variables are free.

Class 3. Problem has a feasible solution with no free variable in corresponding canonical cost function, but has no optimal program because cost function is unbounded.
Class 4. Problem is infeasible because equations are inconsistent.
Class 5. Equations are consistent but problem is infeasible.

These five classes can be described analytically in the notation of (8.40). We set E_{13}, E_{31}, E_{33} all equal to zero; then the primal problem is in one of the following classes whenever the corresponding condition is satisfied.

Class 1: if $F_3 = 0$, $F_1 \leq 0$, $G_3 = 0$, $G_1 \geq 0$.
Class 2: if $F_3 = 0$, $F_1 \leq 0$, $G_3 \neq 0$.
Class 3: if $F_3 = 0$, $F_1 \leq 0$, $G_3 = 0$, $\begin{bmatrix} G_1^T \\ E_{11} \end{bmatrix}$ has a column $[\ominus]$.
Class 4: if $F_3 \neq 0$.
Class 5: if $F_3 = 0$; $[F_1 \quad E_{11}]$ has a row $[+\oplus]$.

Of course, the dual classes are described by interchanging the roles of the F_i and the G_i with reversal of inequalities.

Table 8.3. POSSIBLE CLASSES FOR PRIMAL AND DUAL.

primal in class \ Dual in class		Feasible 1	Feasible 2	Feasible 3	Infeasible 4	Infeasible 5
Feasible	1	P	N	N	N	N
	2	N	N	N	P	N
	3	N	N	N	N	P
Infeasible	4	N	P	N	P	P
	5	N	N	P	P	P

Table 8.4 illustrates the nine possible cases. They are expressed in the notation of (8.41) with the additional assumption that each of the submatrices has one row and one column. In all cases we set h, G_2, F_2 and all of the E_{ij} equal to 0. Let (i, j) describe the primal problem in class i, dual in class j.

Table 8.4. ILLUSTRATIONS OF NINE POSSIBLE CASES.

Case	Illustration
(1, 1)	$F_3 = G_3 = 0, G_1 = -F_1 = 1$
(2, 4)	$F_3 = G_1 = 0, G_3 = -F_1 = 1$
(3, 5)	$F_3 = G_3 = 0, F_1 = G_1 = -1$
(4, 2)	$F_3 = G_1 = 1, F_1 = G_3 = 0$
(5, 3)	$F_3 = G_3 = 0, F_1 = G_1 = 1$
(4, 4)	$F_3 = G_3 = 1, F_1 = G_1 = 0$
(4, 5)	$F_1 = G_3 = 0, F_3 = -G_1 = 1$
(5, 4)	$F_3 = G_1 = 0, F_1 = G_3 = 1$
(5, 5)	$F_3 = G_3 = 0, F_1 = -G_1 = 1$

8.5. COMPLEMENTARY SLACKNESS THEOREMS.

In this section we establish some stronger duality theorems which throw additional light on the nature of components of feasible and optimal programs. We find it convenient to introduce several related primal, dual pairs of problems. Let, as before, A, B, and C be matrices of respective degrees p by n, p by 1, and n by 1; let

$$X = \begin{bmatrix} X_1 \\ X_2 \end{bmatrix}, \quad Y = \begin{bmatrix} Y_1 \\ Y_2 \end{bmatrix}, \quad U = \begin{bmatrix} U_1 \\ U_2 \end{bmatrix}, \quad V = \begin{bmatrix} V_1 \\ V_2 \end{bmatrix} \tag{8.43}$$

be vectors of degree $m = n + p$, each partitioned into a first part of degree n and a second part of degree p.

Consider first the dual pair of problems,

Primal 1

$$\begin{aligned} &\min z = d + C^T X_1, \\ &\text{subject to } AX_1 + B \leq 0, \\ &\qquad X_1 \geq 0, \end{aligned} \tag{8.44}$$

Dual 1

$$\begin{aligned} &\max w = d + B^T Y_2, \\ &\text{subject to } Y_2 \geq 0, \\ &\qquad A^T Y_2 + C \geq 0, \end{aligned} \tag{8.45}$$

and the corresponding pair of problems obtained by adjoining slack variables X_2 and Y_1,

Primal 2

$$\begin{aligned} &\min z = d + C^T X_1, \\ &\text{subject to } AX_1 + X_2 + B = 0, \\ &\qquad X \geq 0. \end{aligned} \tag{8.46}$$

Dual 2

$$\begin{aligned} &\max w = d + B^T Y_2, \\ &\text{subject to } Y \geq 0, \\ &\qquad -Y_1 + A^T Y_2 + C = 0. \end{aligned} \tag{8.47}$$

Consider next the homogeneous systems related to Primal 2 and Dual 2,

Primal 3

$$\begin{aligned} AU_1 + U_2 &= 0, \\ U &\geq 0. \end{aligned} \tag{8.48}$$

Dual 3

$$\begin{aligned} V &\geq 0, \\ -V_1 + A^T V_2 &= 0. \end{aligned} \tag{8.49}$$

Theorem 1 (Strong complementary slackness theorem). If Primal 1 and Dual 1 are both feasible then there exist optimal programs X_1^0, Y_2^0 such that

$$\begin{aligned} X_1^0 + (C + A^T Y_2^0) > 0, &\qquad X_1^{0T}(C + A^T Y_2^0) = 0, \\ -(B + AX_1^0) + Y_2^0 > 0, &\qquad (B + AX_1^0)^T Y_2^0 = 0, \end{aligned} \tag{8.50}$$

or, equivalently, optimal programs X^0, Y^0 such that

$$X^0 + Y^0 > 0, \qquad X^{0T} Y^0 = 0. \tag{8.51}$$

Proof [due to Balinski (Reference 1, pp. 59, 60)]: Consider the schema

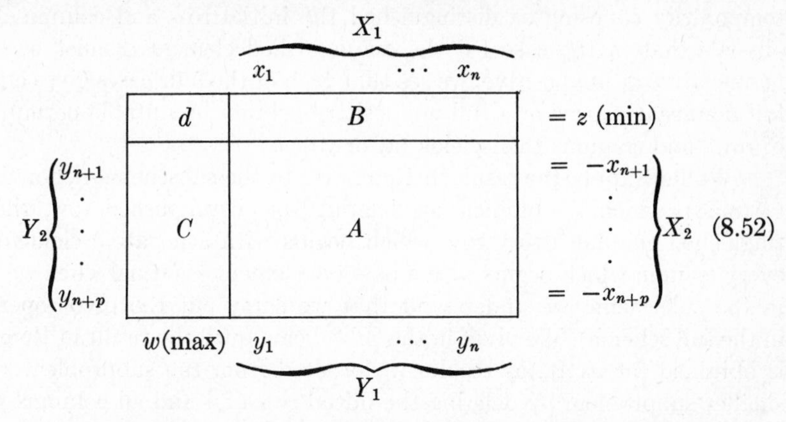

(8.52)

By the theorem in Section 8.1 we must be able to reach the form FF. We assume that rows and columns are permuted so that the positive components in the corresponding basic programs are last as shown in the schema below.

	0 ··· 0	+ ··· +
0 · · · 0	*	
− · · · −		

(8.53)

Remark 1. If a position is in a row beginning with a zero and a column beginning with a zero, then the distinguished row and column are unchanged by a pivot operation on this position. More generally, the procedure we develop below guarantees that a pivot position will be chosen so as to leave unaltered a certain number of initial rows and a certain number of initial columns. This is the case because every element in the intersection of one of these rows or columns and the pivot column or row, respectively, will be zero.

Remark 2. If there is no distinguished row or column, the theorem of Section 8.1 implies that one can pivot and permute rows and columns, if necessary, until one has either: (a) the initial row consisting of zeros followed by positive elements, or (b) the initial column consisting of zeros followed by negative elements. To establish this remark we apply the theorem, temporarily choosing as distinguished the initial row and column. If we observe that in the case FF the distinguished element d must be either nonnegative or nonpositive, we see that each of the four cases gives either a full nonnegative row or a full nonpositive column. A suitable permutation of rows and columns then yields (a) or (b).

We now apply the result in Remark 2 to the subschema denoted by $*$ in (8.53), which is obtained by deleting the distinguished row, the distinguished column, every row which begins with a negative element and every column which begins with a positive element (as usual when we pivot in the subschema we understand that we carry out the pivot operation in the full schema). We pivot in the subschema until the result in Remark 2 is obtained for it. If (a) results we proceed from the subproblem $*$ to a smaller subproblem by deleting the initial row of $*$ and all columns which have a positive element in the first row of $*$. We proceed dually in case (b). We continue to apply Remark 2 until we have either no rows or no columns remaining. According to Remark 1, none of the initial rows or columns that have been deleted will be changed by subsequent pivots or rearrangements. At the conclusion of the process we will have reached a schema of the form:

$$
\begin{array}{ll|c|c|c|l}
 & & \multicolumn{2}{c|}{\overbrace{x_1 \quad \cdots}^{n-r+1}} & \overbrace{\cdots \qquad x_n}^{r} & \\
\hline
 & & d & & 0 \ \cdots \ 0 \quad + \ \cdots \ + & \\
\hline
p-s+1 \left\{ \right. & y_{n+1} & & & \cdots \ 0 \ + \cdots + & = -x_{n+1} \\
 & & 0 & 0 & & \\
 & & & & + & \\
\hline
 & & 0 & 0 \qquad - & & \cdot \\
 & & \cdot & \cdot & & \cdot \\
 & & \cdot & \cdot & & \cdot \\
 & & \cdot & \cdot & & \\
 & & & 0 & & \\
 & & & - & & \\
 & & & \cdot & & \\
 & & & \cdot & & \\
 & & & \cdot & & \\
s \left\{ \right. & & 0 & - & & \\
\cline{3-4}
 & & - & & & \\
 & & \cdot & & & \\
 & & \cdot & & & \\
 & & \cdot & & & \\
 & y_{n+p} & - & & & = -x_{n+p} \\
\hline
 & & \multicolumn{2}{c|}{y_1 \quad \cdots} & \cdots \qquad y_n &
\end{array}
\tag{8.54}
$$

The last s rows of this schema are lexicographically negative, the last r columns are lexicographically positive, and the upper left-hand corner [a $(p - s + 1)$ by $(n - r + 1)$ matrix] is zero everywhere except possibly for the distinguished element d. Moreover, among the lexicographically negative rows the number of zeros preceding the first nonzero entry in any row is at least as great as the number in the row following and similarly for the lexicographically positive columns.

We now choose vectors X^0 and Y^0 so that $X^0 + Y^0 > 0$ and $X^{0T}Y^0 = 0$. There is no loss in generality in the assumption that the initial schema is in optimal form; denote by X' and Y' the corresponding (optimal) basic programs:

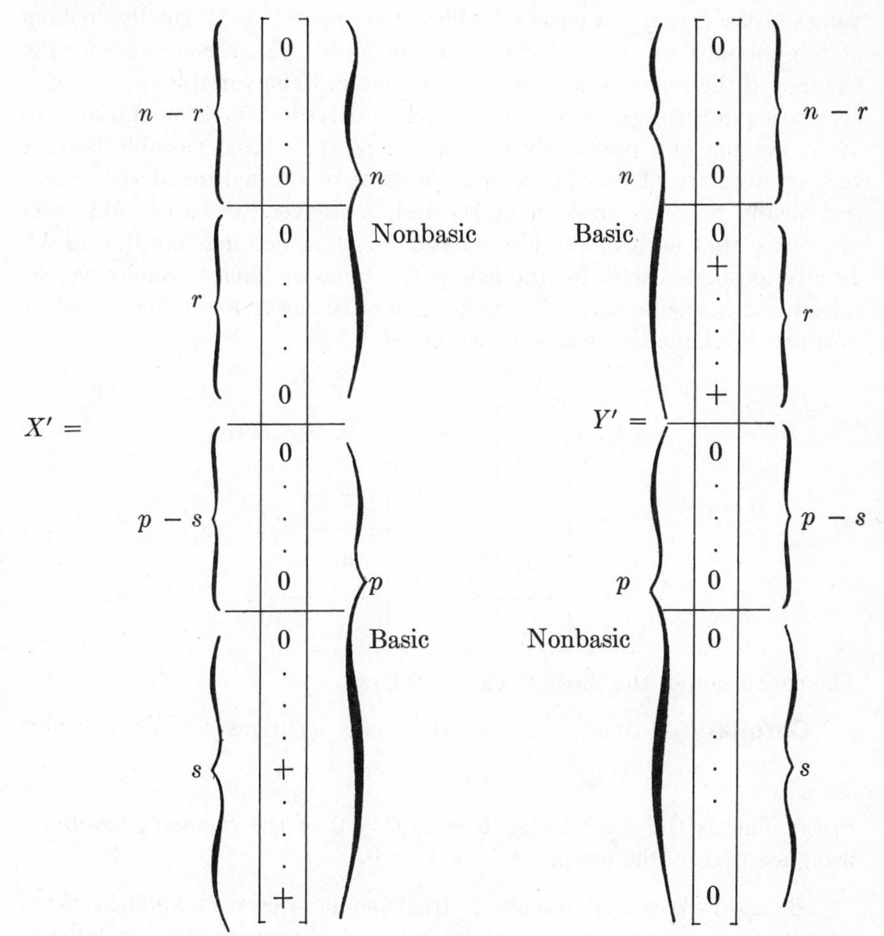

To get X^0 and Y^0 we introduce positive values for the first $n - r$ nonbasic (primal) variables $x_1^0, \cdots, x_{n-r}^0$ and positive values for the first $p - s$ nonbasic (dual) variables $y_{n+1}^0, \cdots, y_{n+p-s}^0$. The values of the remaining nonbasic variables will be kept at zero. At this stage we have the following:

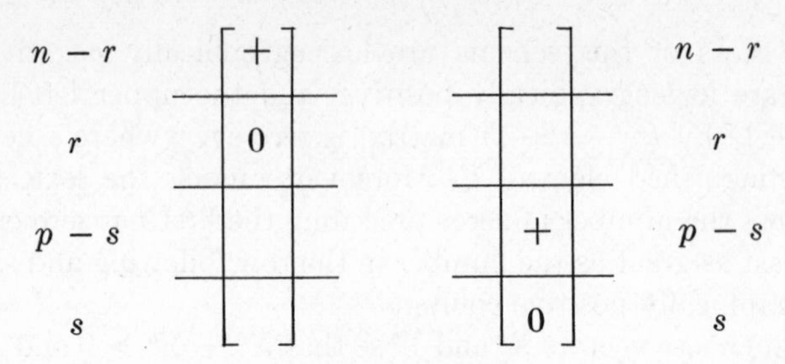

We now want to choose the positive values of the nonbasic variables in X^0 so as to make each of the last s variables have a positive value. We note that assigning positive values to the selected nonbasic variables leaves the values of the first $p - s$ basic variables at zero [see (8.54)]. Dually, looking at the columns we see that the values of $y_1^0, \cdots, y_{n-r}^0$ remain zero for the changes in the selected nonbasic dual variables. The variables $x_1^0, \cdots, x_{n-r}^0$ are chosen in turn beginning with x_1^0 each positive but small enough so as to avoid making any previously established positive basic variable become zero or negative. This can be done because of the nature of the lexicographically negative rows in (8.54) and, moreover, we can in this way guarantee that each of the last s basic variables becomes positive in X^0. Dually, positive values for the first $p - s$ nonbasic dual variables can be introduced in such a way as to make each of the last r basic dual variables positive. We have the results shown below:

$$X^0 = \begin{bmatrix} + \\ \hline 0 \\ \hline 0 \\ \hline + \end{bmatrix} \qquad Y^0 = \begin{bmatrix} 0 \\ \hline + \\ \hline + \\ \hline 0 \end{bmatrix}$$

The conclusion of the theorem clearly follows.

Corollary. Primal 3 and Dual 3 have solutions U^0, V^0 for which

$$U^0 + V^0 > 0, \qquad U^{0T} V^0 = 0.$$

Proof: This is the special case $B = 0$, $C = 0$ of the theorem, feasibility being assured by the programs $U = V = 0$.

Balinski's paper (Reference 1) treats some interesting applications of strong complementary slackness; the following theorem, given by Williams (Reference 47), incorporates in an elegant form some of his own recent results together with some known previously (references to earlier results are given in (Reference 47)).

Theorem 2 (Williams). Suppose that Primal 2 and Dual 2 are both feasible. Then: (1) for each j, exactly one of $\{x_j, y_j\}$ is unbounded for feasible programs; (2) for each j, exactly one of $\{x_j, y_j\}$ is zero for all optimal programs; (3) for each j, x_j (or y_j) is unbounded in the primal (or dual) optimal set if and only if y_j (or x_j) is zero for all dual (or primal) feasible programs; (4) for each j, x_j (or y_j) is bounded in the primal (or dual) optimal set if and only if y_j (or x_j) is positive for some dual (or primal) feasible program.

Proof:

(1) Let X', Y' be any feasible programs for Primal 2 and Dual 2, respectively, and let U^0, V^0 satisfy the corollary above; then for all $t, s \geq 0$

$$X = X' + tU^0 \qquad \text{and} \qquad Y = Y' + sV^0$$

are also feasible. Now, from $u_j^0 + v_j^0 > 0$ it follows that at least one of x_j, y_j is unbounded for feasible programs. To show that not both can be unbounded we note from the double description theorem of Section 3.10 that x_j is unbounded for feasible X if and only if there exists a feasible U for Primal 3 with $u_j > 0$. Now

$$\begin{aligned} U^T V^0 &= U_1^T V_1^0 + U_2^T V_2^0 \\ &= U_1^T(A^T V_2^0) + (-AU_1)^T V_2^0 = 0. \end{aligned}$$

Hence $u_j > 0$ implies $v_j^0 = 0$ and therefore $u_j^0 > 0$. Dually, $v_j > 0$ for feasible V implies that $v_j^0 > 0$ and hence x_j (or y_j) is unbounded if and only if u_j^0 (or v_j^0) is positive. The complementary slackness condition $u_j^0 v_j^0 = 0$ thus implies that not both x_j and y_j can be unbounded for feasible programs.

(2) This is essentially the content of the strong complementary slackness theorem, for the programs X^0, Y^0 given by it show that for no j can we have both primal and dual components zero for all optimal programs. On the other hand, if X is any optimal (primal) program, then $X^T Y^0 = 0$ and therefore $y_j^0 > 0$ implies $x_j = 0$ and dually $x_j^0 > 0$ implies $y_j = 0$ for all optimal Y.

(3) Let E_j be the jth unit vector in V_{p+n} and consider the two dual programs:

Primal 4_j	Dual 4_j
min $C^T U_1$,	max $E_j Y (= y_j)$,
subject to U_1 free,	subject to $Y_1 - A^T Y_2 = C$,
$\begin{bmatrix} I \\ -A \end{bmatrix} U_1 \geq E_j.$	$Y \geq 0.$

Let $U_2 = -AU_1$ and let $U = \begin{bmatrix} U_1 \\ U_2 \end{bmatrix}$ as in Primal 3. Then the feasibility conditions for Primal 4_j reduce to

$$\begin{aligned} U &\geq E_j \\ AU_1 + U_2 &= 0. \end{aligned}$$

Since $E_j \geq 0$ the feasible programs for Primal 4_j are just those feasible programs for Primal 3 which have $u_j \geq 1$. Hence x_j is unbounded for feasible X if and only if Primal 4_j is feasible. Moreover, min $C^T U_1 = 0$ is equivalent to having x_j unbounded for optimal feasible X. Now by the duality theorem of linear programming min $C^T U_1 = 0$ if and only if max $E_j^T Y = 0$. Since the feasibility sets for Dual 2 and Dual 4 are identical and since $E_j^T Y = y_j$ we conclude that x_j is unbounded for optimal X if and only if $y_j = 0$ for all feasible Y.

(4) This is the contrapositive statement of (3). Note that whereas (1) refers only to feasible programs and (2) only to optimal programs for both primal and dual, in (3) and (4) we have statements connecting optimality for one problem with feasibility for its dual.

PROBLEMS

8.1. Write out in inequality form the dual linear programming problems represented by the following schema.

	1	x_1	x_2	x_3	
1	2	-1	3	4	$= u(\min)$
y_4	6	1	0	-2	$= -x_4 \leq 0$
y_5	-2	-3	1	2	$= -x_5 \leq 0$
	$= v$ (max)	$= y_1 \geq 0$	$= y_2 \geq 0$	$= y_3 \geq 0$	

8.2. Consider the primal linear programming problem

$$\text{minimize } u = -3 + x_1 + x_2 - 3x_3 + x_4$$

subject to

$$\begin{aligned} x_1 + x_2 \qquad\qquad - x_4 &\geq -1 \\ -x_1 \qquad + 2x_3 \qquad &\leq \ 2 \\ -x_1 \qquad + 4x_3 + x_4 &\geq \ 1 \\ x_i &\geq \ 0, \qquad (i = 1, \ldots, 4). \end{aligned}$$

(a) Represent this problem and its dual in schema form.
(b) Pivot on the position (2, 3) to obtain a new schema S^* (the pivot element is 2).
(c) What can you conclude from S^* about the existence of feasible solutions to the primal problem? The dual problem?

8.3. The schema

	1	x_1	x_2	x_3	
1	0	-1	-2	6	u
y_4	-4	-2	1	3	$-x_4$
y_5	-12	4	0	7	$-x_5$
y_6	-3	2	0	1	$-x_6$
	v	y_1	y_2	y_3	

is in row-feasible form. Use the contracted primal simplex algorithm to find optimal solutions to the corresponding primal and dual linear programming problems.

8.4. The schema

	1	x_1	x_2	
1	2	2	3	u
y_3	1	1	-3	$-x_3$
y_4	3	-1	-1	$-x_4$
y_5	-1	1	1	$-x_5$
	v	y_1	y_2	

is in column-feasible form. Apply the dual simplex algorithm to obtain a schema in one of the forms FF or IF.

8.5. Solve the following problem by the contracted primal simplex algorithm,

$$\text{minimize } u = -3x_1 - 4x_2 - x_3 - x_4$$

subject to

$$\begin{aligned} x_1 + 2x_2 \qquad\quad + x_4 &\leq 6, \\ 2x_1 + x_2 \qquad\qquad\quad &\leq 2, \\ x_2 + 3x_3 + x_4 &\leq 3, \\ x_j &\geq 0, \qquad (j = 1, 2, 3, 4). \end{aligned}$$

8.6. Solve the following by the dual simplex algorithm,

$$\text{minimize } u = 2x_1 + 2x_2 + 4x_3$$

subject to

$$\begin{aligned} 2x_1 + x_2 + x_3 &\geq 1, \\ x_1 - 2x_2 + x_3 &\geq 2, \\ x_j &\geq 0, \qquad (j = 1, 2, 3). \end{aligned}$$

8.7. In each of the following problems first obtain a schema in row-feasible form using the method of Section 3.6 and then use the contracted primal simplex algorithm to obtain a schema in one of the forms FF or FI.

(a)

	1	x_1	x_2	x_3	
1	0	6	−2	4	u
y_4	3	−3	1	0	x_4
y_5	0	−4	0	3	x_5
y_6	−3	1	0	1	x_6
	v	y_1	y_2	y_3	

(b)

	1	x_1	x_2	x_3	
1	0	2	−2	3	u
y_4	−1	−1	−1	1	x_4
y_5	−2	−3	2	−1	x_5
y_6	3	−1	0	2	x_6
	v	y_1	y_2	y_3	

8.8. The primal is infeasible in each of the following. In each case make appropriate pivots to obtain a schema in one of the forms IF or II.

(a)

	1	x_1	x_2	x_3	x_4	
1	5	−6	2	−4	−3	u
y_5	1	5	−2	5	2	$-x_5$
y_6	0	3	−1	2	1	$-x_6$
y_7	2	1	3	0	0	$-x_7$
	v	y_1	y_2	y_3	y_4	

(b)

	1	x_1	x_2	x_3	
1	2	4	−1	−1	u
y_4	1	−1	2	−1	$-x_4$
y_5	3	2	0	0	$-x_5$
y_6	−1	1	−3	1	$-x_6$
	v	y_1	y_2	y_3	

8.9. Let

$$A = \begin{bmatrix} -1 & 2 & 1 \\ 0 & 1 & -3 \\ 2 & -1 & 0 \end{bmatrix}$$

and suppose that A^* is obtained from A by pivoting on position (1, 2). Find G_S^{-1} and compute A^* using (8.16).

8.10. Suppose A is a nonsingular p by n matrix; show that A is combinatorially equivalent to A^{-1}.

8.11. Consider the schema

$$
\begin{array}{c|ccc|l}
 & x_1 & x_2 & x_3 & \\
\hline
y_1 & 1 & 1 & 2 & = -u_1 \\
y_2 & 1 & 2 & -1 & = -u_2; \\
y_3 & 3 & -2 & 5 & = -u_3 \\
\hline
 & = v_1 & = v_2 & = v_3 &
\end{array}
$$

by an appropriate block pivot obtain a new schema of the form

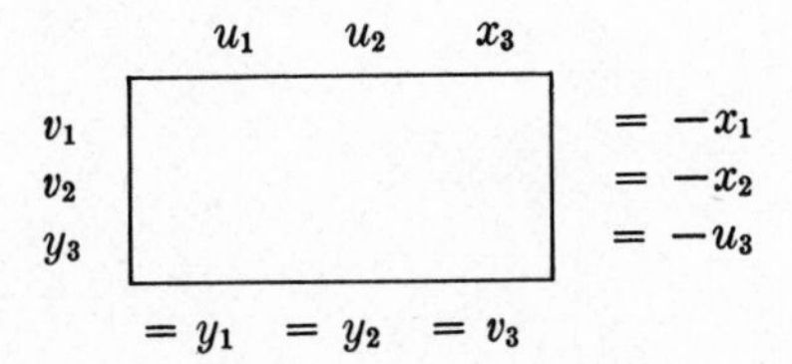

8.12. Suppose A is combinatorially equivalent to A^*; prove that $-A^{*T}$ is combinatorially equivalent to $-A^T$.

8.13. Solve the following linear programming problem,

$$
\begin{array}{r|c|ccc|l}
 & & \geq 0 & \geq 0 & \text{free} & \\
 & 1 & x_1 & x_2 & x_3 & \\
\hline
1 & 0 & 0 & 1 & -1 & = u(\min) \\
\hline
u_1 \geq 0 & 1 & 1 & -4 & -1 & = -y_1 \leq 0 \\
u_2 \geq 0 & 1 & 3 & -1 & 0 & = -y_2 \leq 0 \\
u_3 \text{ free} & -2 & 2 & 3 & 1 & = -y_3 = 0 \\
\hline
 & = & = & = & = & \\
 & v(\max) & v_1 \geq 0 & v_2 \geq 0 & v_3 = 0 &
\end{array}
$$

8.14. Represent the following linear programming problem in schema form and solve,

$$\text{minimize } u = 2x_1 - x_2 + x_3 - x_4$$

subject to

$$
\begin{aligned}
-x_1 - 2x_2 \phantom{{}+x_3} - 3x_4 &\leq 3, \\
x_1 \phantom{{}-2x_2} - x_3 + 2x_4 &\geq 4, \\
- x_2 + x_3 + x_4 &= 1, \\
x_1 + 3x_2 - 2x_3 - 2x_4 &= 1, \\
2x_1 + 3x_2 - x_3 - x_4 &= 5, \\
x_j &\geq 0, \qquad (j = 1, 2, 3). \\
x_4 &\text{ free.}
\end{aligned}
$$

Game Theory[1]

9

9.1. INTRODUCTION.

The theory of games arises in an attempt to provide a mathematical model of conflict situations. Although forerunners of the theory date back many years, the first attempt to build a mathematical theory was made in 1921 by Emile Borel. The theory was first firmly established by John von Neumann in 1928 with his proof of the minimax theorem, but it was not until the appearance of *Theory of Games and Economic Behavior* by John von Neumann and Oskar Morgenstern (Reference 46) in 1944 that the topic received general notice. Since then many facets of game theory have received attention both from mathematicians who have been challenged by its many fascinating problems and from others who have seen in this theory an important model for understanding and resolving problems of economic, military, and political conflict.

Early attention was focused on ramifications of the two-person zero-sum game, finite or infinite; more recently, primary emphasis has shifted to general games where there are still many important unsolved problems and an abundance of opportunities for useful refinements of the basic definitions. In particular, the entire theory of many person games has been given a new turn by the recent results of W. Lucas who constructed a ten-person game

[1]Portions of this chapter have been adapted from a chapter by one of the authors in Machol *et al.*, *System Engineering Handbook*, copyright © 1965 by McGraw-Hill Inc.

(Reference 32) which upset a major conjecture of von Neumann and Morgenstern concerning existence of solutions.

The reader is referred to the References for a selected list of readings on game theory and in turn to the works cited for more complete bibliographies.

9.2. DEFINITION OF A GAME.

A game in extensive form is described by a set of players and a set of rules which specify (1) the choices of action that are open to each player under all possible circumstances and (2) the payoff to each player at the end of any "play." Games are classified in terms of: (1) the number of players, (2) the number of moves, (3) the nature of the payoff, and (4) other characteristics of the rules. A *finite* game is one in which the total number of choices of actions (over all moves) is finite. We shall use the word *game* to mean *finite game*.

Before giving a formal definition of a game we develop some useful auxiliary concepts from graph theory. Some of these concepts were presented in less general form in Chapter 6. Let $A = \{a_1, \cdots, a_p\}$ be a finite set of elements called *nodes* and let R be a symmetric relation in the Cartesian product set $A \times A$ (see Sections A.6 and A.10). We call the elements of R *arcs*. The two sets A and R define a graph H. Intuitively, a graph is a set of nodes with certain pairs of nodes joined by arcs. A *path* in a graph is a set of nodes $b_1, \cdots, b_s$ such that (b_i, b_{i+1}) is an arc in H for $i = 1, \cdots, s - 1$. A graph is said to be *connected* if for each pair of nodes a, b in A there is a path beginning with a and ending with b. A path is said to be a *loop* if (a) $s > 3$, (b) $b_1 = b_s$, and (c) $b_i = b_j$ for $i \neq j$ holds only for the terminal nodes. A *tree* is a connected graph having no loops.

A tree is said to be *rooted* if it has one distinguished node. The analogy with botanical trees is obvious if we identify the distinguished node with the foot of the trunk and disregard the root structure. Several trees are illustrated in Figure 9.1. Tree (*a*) has 27 nodes and 26 arcs; tree (*b*) has 7 nodes

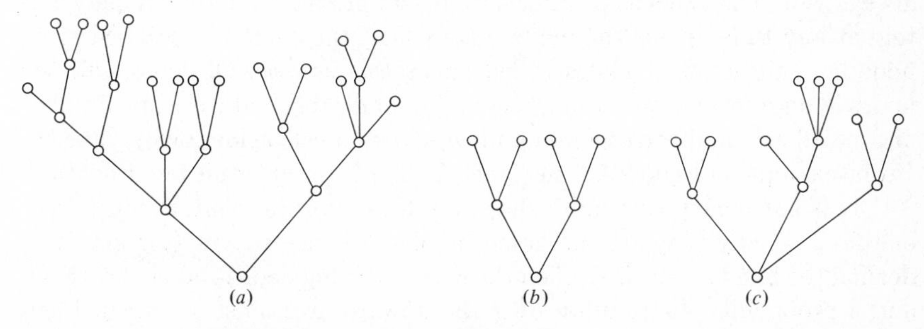

Figure 9.1 Examples of trees.

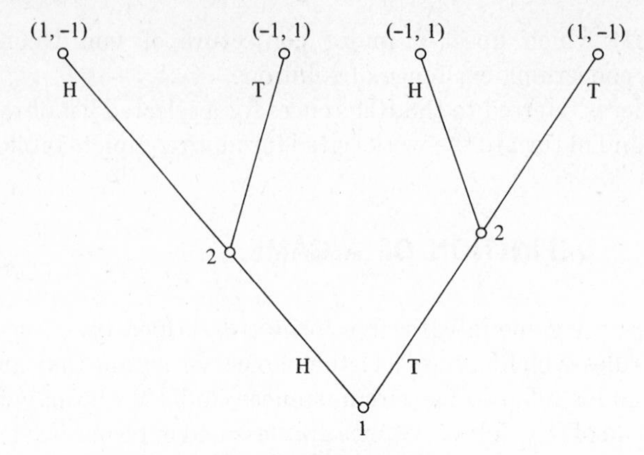

Figure 9.2 Matching pennies tree.

and 6 arcs; tree (*c*) has 13 nodes and 12 arcs. In a tree the number of arcs is always one less than the number of nodes. A node in a tree is said to be *terminal* if it lies on only one arc. Tree (*a*), for example, has 15 terminal nodes.

A game can be described in terms of a tree in which each nonterminal node represents a move by some player and a move consists of a selection of one of the arcs going upward from the node. A play begins at the distinguished node and continues until a terminal node is reached. If player i is to move at a certain stage, the corresponding node is labeled with the symbol i. A path starting at the distinguished node and ending at a terminal node is called a *play* of the game. Note that there is a one-to-one correspondence between plays and terminal nodes. The tree for chess is extremely complex; tree (*b*) belongs to the game of matching pennies; the labels and meanings of the arcs are given in Figure 9.2.

Suppose now that player 1 wishes to match player 2. He then wins in the plays whose arcs are labeled H, H or T, T whereas player 2 wins in the remaining cases H, T and T, H. The components of the vectors at the top of each play indicate the payoffs to the players (ordered by their labels). Note that if player 2 knew just at which node he was making his choice he could always win. The concept of information sets provides for this. A player is told at any time he has the move merely that the game has reached some node in a given set of nodes called an *information set*. Of course, all the nodes in an information set must have the same label and the same number of upward arcs and corresponding upward arcs must be identically labeled. In the example in Figure 9.2, the labels H and T provide this identification.

It is customary to require that no information set contain more than one node of any play. If the game involves chance moves (for example, dealing the hands in bridge), then the corresponding node is labeled with a 0 and a probability distribution over the upward arcs must be given. Each information set belonging to player 0 consists of exactly one node.

Each play has associated with it an *outcome* which is a vector indicating the return (or payoff) to each player accruing from that play. It is customary to label each terminal node with the associated outcome. If for every terminal node the sum of the payoffs is zero, the game is said to be *zero sum*. For example, a parlor game in which all of the payoffs are between the players is zero sum, whereas a labor-management game with the possibility of strikes and lockouts is not zero sum. A zero-sum game thus bears some resemblance to a conservative system in mechanics.

A *pure strategy* for a player is a selection of one alternative (or a set of corresponding upward arcs) from each of his information sets. A *mixed strategy* is a probability distribution over the set of all pure strategies.

In a (finite) game the number of pure strategies for each player is finite although even for familiar parlor games such as chess, poker, and bridge this number is so large as to be beyond the handling capacity of even our largest computers. For example, a pure strategy in chess would consist of selecting a move for each (legal) position of pieces on the chessboard (for complete accuracy one should include as part of a position the sequence of past moves which led to it so as to take account of special rules concerning draws through repetition of positions). Details of each pure strategy could theoretically be recorded in a (large) book and a play of chess would consist of each player selecting a pure strategy and letting a referee determine the outcome by reference to the book.

In this sense every finite game can be reduced to a single decision for each player; if player i, for $i > 0$, has N_i pure strategies, his decision consists of selecting an integer x_i from the set $I(N_i) = \{1, \cdots, N_i\}$. A referee then carries out the chance moves of player 0 and announces the outcome vector.

A game in *normalized* form is thus characterized by:

(1) A set of integers $N_0, N_1, \cdots, N_n$,
(2) A function F on the product set $I(N_0) \times I(N_1) \times \cdots \times I(N_n)$ to Euclidean n space,
(3) A probability distribution p_0 over the set $\{1, \cdots, N_0\}$.

Thus, $F(x_0, x_1, \cdots, x_n)$ is the outcome vector corresponding to the selections of strategy x_i by player $i (i = 0, \cdots, n)$. We denote by $F_i(x_0, x_1, \cdots, x_n)$ the ith component of F. Then, the expected return to player i for given choices $x_1, \cdots, x_n$, is

$$G_i(x_1, \cdots, x_n) = \sum_{x_0=1}^{N_0} p_0(x_0) \, F_i(x_0, x_1, \cdots, x_n) \tag{9.1}$$

where $p_0(x_0)$ is the probability that player 0 chooses alternative x_0.

In applications of game theory the components of the outcome vector represent utilities to the individual players. Since most games involve chance moves and since a player has no control over them, it is a consequence of Formula (9.1) that these utilities should conform to certain assumptions concerning probability mixtures.

Consider two events A and B and let their utilities to an individual be respectively u and v. We construct a new event C called a *probability mixture* of A and B which consists of event A with probability p and event B with probability $1 - p$. Let w be the utility of C; if

$$w = pu + (1 - p)v$$

we say that the utility is linearly adapted to chance events.

More formally, let $A_1, \cdots, A_r$ be any events, let $u(A_i)$ be the utility of A_i and let $\Pi = (p_1, \cdots, p_r)$ be a probability vector (a vector whose components are nonnegative real numbers which have sum one). We introduce a new event A called the Π *mixture* of $A_1, \cdots, A_r$ and designated by $A = [A_1, \cdots, A_r, \Pi]$. We say that u is a *linear utility function* if

$$u(A) = u([A_1, \cdots, A_r, \Pi]) = p_1 u(A_1) + \cdots + p_r u(A_r) \qquad (9.2)$$

for all finite collections $A_1, \cdots, A_r$ and probability vectors Π. Axiomatic characterizations of linear utility functions are examined in (Reference 33). In what follows we shall assume that the numbers in each outcome vector arise from linear utility functions. Thus, formulas such as (9.1) are meaningful, and $G_i(x_1, \cdots, x_n)$ represents the utility to player i of the simultaneous strategy selections $x_1, \cdots, x_n$, and we need no longer consider player 0.

For a game in normalized form a mixed strategy for player i is a probability vector Π_i with N_i components $\Pi_i(1), \cdots, \Pi_i(N_i)$. The unit vectors correspond to pure strategies and the set of all mixed strategies can be regarded as an $(N_i - 1)$-dimensional simplex. The cases for dimensions 2 and 3 are illustrated in Figure 9.3.

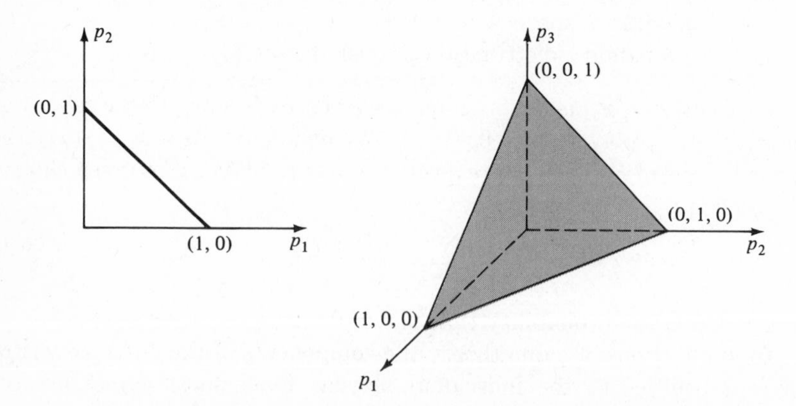

Figure 9.3 Probability simplices for dimensions 2 and 3.

9.3. TWO-PERSON, ZERO-SUM GAMES.

Whereas the analysis of general games is still far from complete, for the case of two-person, zero-sum games the situation is much more satisfactory. There are: (1) an acceptable concept of solution; (2) an existence theorem guaranteeing a solution; (3) a classification of solutions; (4) several reasonably practical methods for determining solutions; and (5) a collection of applications of the theory. Since the computation of solutions is limited by the capacity of computers there is still some interest in studying these games in extensive form in the hope of reducing a larger class to manageable size. However, in the present section we consider only games in normal form.

A two-person game in normal form is characterized by two p by n matrices A and B, where A gives the payoffs for player one, B gives them for player two, and $N_1 = p$, $N_2 = n$. If the game is also zero-sum then $B = -A$ and we thus speak of a *matrix game* defined by a p by n matrix $A = [a_{ij}]$ in which a_{ij} represents what player one is paid by player two when they choose respective strategies i and j. A negative entry indicates a positive payment from player one to player two. We now consider a few examples.

Example 1. *The Debaters.* Two candidates for the legislature have agreed to hold a public debate somewhere in their district and are negotiating over the location for it. Each wishes to have as friendly an audience as possible, and the twelve possible locations vary considerably in the balance between Republicans and Democrats. Each of the twelve possible sites lies on the intersection of an East-West and a North-South highway, as pictured in Figure 9.4.

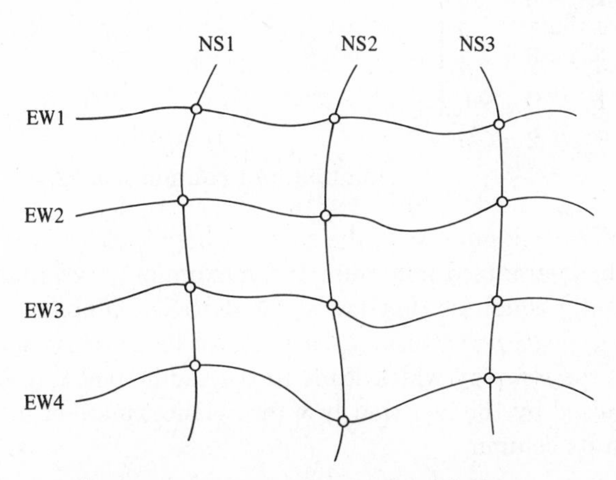

Figure 9.4 Possible sites for the debaters.

The excess (measured in thousands) of Democrats over Republicans at the EWi and NSj intersection is a_{ij} where

$$A = \begin{bmatrix} 7 & -3 & 3 \\ -4 & -1 & 5 \\ 3 & 2 & 4 \\ 1 & 0 & -6 \end{bmatrix}.$$

An elder statesman in the district suggests that the candidates settle the question by turning it into a game in which the Democrat chooses an EW route, the Republican a NS route, and then holding the debate at the intersection of these routes. The candidates agree to this. The Democrat now studies what is the worst that can happen to him if he chooses any route. Thus for $i = 1$, he would like $j = 1$ and $j = 3$ which would give him, respectively, advantages of 7 and 3; however, in the worst case $j = 2$ and he would lose 3. Thus he writes -3 at the end of the first row, and similarly, copies the smallest (minimum) entry of each row at its end,

$$\begin{bmatrix} 7 & -3 & 3 \\ -4 & -1 & 5 \\ 3 & 2 & 4 \\ 1 & 0 & -6 \end{bmatrix} \quad \begin{matrix} -3 \\ -4 \\ 2 \\ -6 \end{matrix} \qquad \begin{matrix} i = 3, \\ \text{maximum of row minima} = 2. \end{matrix}$$

A conservative plan (strategy) would thus be to choose $i = 3$ which guarantees him an advantage of at least two. This is called a *maximin strategy.*

Similarly, the Republican looks for the largest entry in each column (largest rather than smallest because his interests are exactly opposite to those of the Democrat) and chooses the column with the smallest maximum, that is, he follows a *minimax strategy,*

$$\begin{matrix} \begin{bmatrix} 7 & -3 & 3 \\ -4 & -1 & 5 \\ 3 & 2 & 4 \\ 1 & 0 & -6 \end{bmatrix} \\ \begin{matrix} 7 & \quad 2 & \quad 5 \end{matrix} \end{matrix} \qquad \begin{matrix} j = 2, \\ \text{minimum of column maxima} = 2. \end{matrix}$$

Since maximin = minimax = 2, the solution here is clear.

When the guaranteed minimum and maximum payoffs for two opponents are exactly equal (as they are in the debaters' problem) the game is said to have a *saddle point;* clearly, each player has nothing to gain by departing from the strategy which leads to the saddle point. A saddle point can be recognized by the fact that it is the smallest number in its row and the largest in its column.

Example 2. *The Debaters Ten Years Later.* This problem is the same as that in Example 1, except that now the population has changed in city (4, 2) and the matrix A is replaced by

$$B = \begin{bmatrix} 7 & -3 & 3 \\ -4 & -1 & 5 \\ 3 & 2 & 4 \\ 1 & 6 & -6 \end{bmatrix}.$$

Following the previous process, we get

$$\begin{array}{llll} & & & \text{Row minima} \\ & \left[\begin{array}{rrr} 7 & -3 & 3 \\ -4 & -1 & 5 \\ 3 & 2 & 4 \\ 1 & 6 & -6 \end{array}\right] & & \begin{array}{rl} -3 & \\ -4 & \\ 2 & \text{Maximin} \\ -6 & \end{array} \\ \begin{array}{l}\text{Column} \\ \text{maxima}\end{array} & \begin{array}{rrr} 7 & 6 & 5 \\ & & \text{Minimax} \end{array} & & \end{array}$$

We no longer have minimax = maximin, and there is no longer a clear-cut solution. To handle such problems, we employ a *mixed strategy* as illustrated by the following example.

Example 3. *Matching Pennies.* This game is familiar to everyone. We have two players, Red and Blue, and assume that Red is trying to match Blue. Each has two strategies: Heads and Tails. The game can be shown schematically in Figure 9.5.

There is no saddle point here, and if either lets the other know his choice in advance, he is certain to lose. However, suppose that Red determines his choice by flipping an honest coin. Then the expected return to Blue if he chooses Heads is

$$\frac{1}{2}(1) + \frac{1}{2}(-1) = 0,$$

and if he chooses Tails is

$$\frac{1}{2}(-1) + \frac{1}{2}(1) = 0.$$

		Blue	
		H	*T*
Red	H	1	−1
	T	−1	1

Figure 9.5 Matching-pennies game.

		Blue			
		Scissors	*Paper*	*Stone*	*Row minima*
	Scissors	0	1	−1	−1
Red	Paper	−1	0	1	−1
	Stone	1	−1	0	−1
	Column maxima	1	1	1	

Figure 9.6 Game of scissors, paper, stone.

Thus by adopting a mixed strategy (that is, a probability distribution over the two pure strategies) Red can protect himself against loss, at least at the level of mathematical expectation. A similar mixed strategy is available to Blue.

Example 4. *Scissors, Paper, Stone.* Another classical game frequently played by children is scissors, paper, stone. Two players, Red and Blue, simultaneously name one of the three objects. If both name the same object, the game is a draw. Otherwise, the winner is determined by the jingle, "Scissors cut paper, stone breaks scissors, and paper covers stone." The payoff matrix is shown in Figure 9.6. One sees easily that a mixed strategy which chooses each pure strategy with probability 1/3 protects each player against expected loss, and it can be shown that no other mixed strategy has this property.

We now return to the general case. It is clear that when the payoff matrix is obtained, the context of the problem is no longer relevant to the process of solution. As we have seen, a two-person, zero-sum game in normalized form is given by a matrix A. The entry a_{ij} is the payment made by Blue (the column player) to Red (the row player) if Red chooses row i and Blue chooses column j. For example, a 5 by 4 matrix game can be displayed as in Figure 9.7.

		Blue				
		1	2	3	4	*Minimum value in ith row*
	1	a_{11}	a_{12}	a_{13}	a_{14}	a_1
	2	a_{21}	a_{22}	a_{23}	a_{24}	a_2
Red	3	a_{31}	a_{32}	a_{33}	a_{34}	a_3
	4	a_{41}	a_{42}	a_{43}	a_{44}	a_4
	5	a_{51}	a_{52}	a_{53}	a_{54}	a_5
Maximum value in *j*th column		b_1	b_2	b_3	b_4	

Figure 9.7 A 5 by 4 matrix game.

Now let $a_i = \min_j a_{ij}$ = minimum value in the ith row,

$a = \max_i a_i$ = maximum of all the a_i,

$b_j = \max_i a_{ij}$ = maximum value in the jth column,

$b = \min_j b_j$ = minimum of all the b_j;

then
$$a = \max_i \min_j a_{ij},$$
$$b = \min_j \max_i a_{ij}.$$

We have the following theorem.

Theorem 9.1.

$$\min_j \max_i a_{ij} \geq \max_i \min_j a_{ij}.$$

Proof: We must show $a \leq b$. We have $a = \max_i a_i$, say $a = a_p$, and $a = a_p = \min_j a_{pj}$, say $a = a_p = a_{pq}$. Similarly, let $b = b_h = a_{hk}$. Then we have $a = a_{pq} \leq a_{pk} \leq a_{hk} = b$; that is, maximin $\leq$ minimax. This general statement is true for all functions of two variables when both sides of the inequality are defined. If maximin = minimax, we say that the *minimax theorem holds* for pure strategies. The first form of the debaters' problem is an example as are all matrices having saddle points.

Earlier, we saw how to represent chess as a (very large) matrix game. It can be proved that this game has a saddle point in pure strategies, although it is not known whether the result is a win, a loss, or a draw for the first player.

A game in extensive form is said to have *perfect information* if each information set contains exactly one node. It can be proved (Reference 8, pp. 126 ff.) that if a two-person, zero-sum game in extensive form has perfect information then the same game in normal form has a saddle point in pure strategies. Ticktacktoe, checkers, and chess are games with perfect information. Games involving chance moves or in which any player has information about past moves which is not shared by others are not games with perfect information.

Consider a game with matrix $\begin{bmatrix} a & b \\ c & d \end{bmatrix}$ and having no saddle point, where the elements are arranged in such a way that $a \leq b$, $a \leq c$, $a \leq d$. If $b < d$, then the game has a saddle point; therefore, $b \geq d$. If $c < d$, then d is a saddle point; therefore, $c \geq d$. Consider the example in Figure 9.8.

		Blue 1	Blue 2	
Red	1	3	6	3
	2	5	4	4
		5	6	

Figure 9.8 A game having no saddle point.

We have maximin = 4 and minimax = 5, hence there is no saddle point, and we try a mixed strategy. We would like to decide on some ratio for playing the two strategies. The determination of the strategy to be played would then be made in a random fashion with some preassigned probability.

Suppose Red chooses strategy 2 with probability p and strategy 1 with probability $1 - p$, then when Blue chooses strategy 1, Red's expectation is

$$z_1 = 3(1 - p) + 5p. \tag{9.3}$$

When Blue chooses strategy 2, Red's expectation is

$$z_2 = 6(1 - p) + 4p. \tag{9.4}$$

We want to choose p so that the smaller of z_1 and z_2 is as large as possible, that is, we want to find

$$\max_p \min_i \{z_i\}.$$

Graphically, we find that this point is at the intersection of the two lines (9.3) and (9.4) and occurs when (see Figure 9.9)

$$p = \frac{3}{4} \quad \text{and} \quad 1 - p = \frac{1}{4} \quad \text{and} \quad z_1 = z_2 = 4\frac{1}{2}.$$

By using this mixed strategy, the first player guarantees himself an expectation of at least $4\frac{1}{2}$. Similarly, there is a mixed strategy for the second

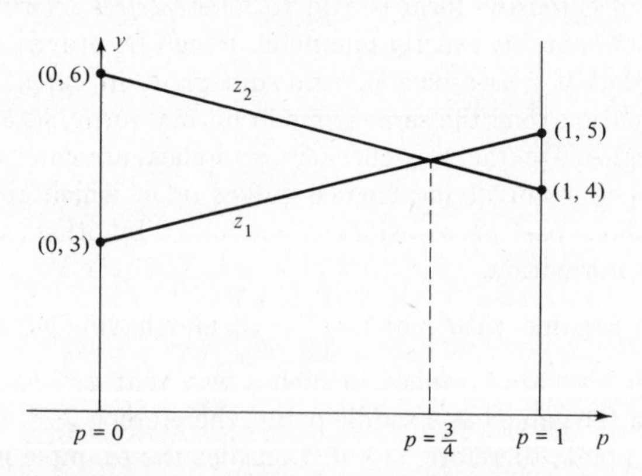

Figure 9.9 Saddle point with mixed strategy.

player which guarantees that his losses will not be more than $4\frac{1}{2}$. Using pure strategies, all Red can guarantee is that he will net a minimum gain of 4, and all Blue can guarantee is that he will suffer a maximum loss of 5.

The number $4\frac{1}{2}$ is the *value of the game;* that is, using a mixed strategy, the first player can guarantee a gain of at least $4\frac{1}{2}$ and the second player can guarantee a loss not exceeding $4\frac{1}{2}$. The ratio of the number of times strategy 1 is played to the number of times strategy 2 is played is 1 : 3. The numbers 1 and 3 are called *oddments.*

In a 2 by 2 game without a saddle point, the oddments for a good mixed strategy for Red are found by subtracting the first column from the second, interchanging the order of the pair of numbers and using their absolute values. Thus $(|d - c|, |b - a|) = (1, 3)$ are the oddments. For Blue, subtract the first row from the second, interchanging the entries and use their absolute values.

9.4. MINIMAX THEOREM FOR MATRIX GAMES.

Consider, again, a general matrix game Γ defined by a p by n matrix $A = [a_{ij}]$. A mixed strategy for the row player (player one) is a probability vector $P = (p_1, \cdots, p_p)$; similarly, a mixed strategy for the column player (player two) is a probability vector $Q = (q_1, \cdots, q_n)$. The goal of the row player is to select a mixed strategy P which maximizes the smallest component of $A^T P$. We introduce a new free variable v_0; given P, finding the smallest component of the vector $A^T P$ is equivalent to maximizing v_0 subject to $A^T P \geq v_0 E_n$, where E_n denotes the column vector of n components, each equal to 1. Then the problem for the row player is to select P and v_0 which maximize v_0 subject to the conditions

$$\begin{aligned} A^T P &\geq v_0 E_n, \\ E_p^T P &= 1, \\ P &\geq 0. \end{aligned} \tag{9.5}$$

It is useful to observe that the verbal definition of v_0 given above is equivalent to

$$\max v_0 = \max_P \min_j P^T A_j$$

where A_j is the jth column of the matrix A. Moreover, since every convex combination of a sequence of integers is at least as large as the minimum of these integers, which is itself a convex combination, it follows that

$$\max v_0 = \max_P \min_Q P^T A Q.$$

Thus the row player can achieve a maximin.

Dually, the column player wishes to select Q and a value of a second free variable v^0 which minimizes v^0 subject to

$$\begin{aligned} AQ &\leq v^0 E_p, \\ E_n^T Q &= 1, \\ Q &\geq 0. \end{aligned} \tag{9.6}$$

By the same reasoning as above we can characterize

$$\min v^0 = \min_Q \max_P P^T A Q$$

and the column player can achieve a minimax.

	Row player (player I)	*Column player (player II)*
n:	$A^T P - v_0 E_n \geq 0,$	$Q \geq 0,$
1:	$-E_p^T P = -1,$	v^0 free,
p:	$P \geq 0,$	$AQ - E_p v^0 \leq 0,$
1:	v_0 free,	$-E_n^T Q = -1,$
	minimize $-v_0$	maximize $-v^0$

Figure 9.10 Duality of games for players I and II.

Both of the problems shown in Figure 9.10 are feasible, since for any probability vectors P and Q we can always choose v_0 sufficiently small (negative if necessary) and v^0 sufficiently large so that P, v_0 and Q, v^0 satisfy each of the problems, respectively. Moreover, the general form of the duality theorem of linear programming can be applied and for optimal vectors P^0 and Q^0 we have

$$\max v_0 = \min v^0 = v. \tag{9.7}$$

We call v the *value* of the game and the vectors P^0 and Q^0 are called *optimal strategies*, respectively, for the row and column players. The result (9.7) is the celebrated minimax theorem of game theory.

We observe that if Γ is a game with matrix A and Γ' has matrix $A' = A + cE_pE_n^T$, then given any strategies P and Q we have $P^TA'Q = P^TAQ + c$. It follows that the values $v(\Gamma')$ of Γ' and $v(\Gamma)$ of Γ are related by

$$v(\Gamma') = v(\Gamma) + c, \tag{9.8}$$

and moreover that the sets of optimal strategies for Γ' are the same as those for Γ. Thus there is no loss of generality in assuming each $a_{ij} > 0$ and, therefore, also that $v(\Gamma) > 0$.

We now exploit the transformation given in the preceding paragraph to obtain a second formulation of a game as a pair of linear programs. Assume $A > 0$; we can then also require that v_0 and v^0 be positive. We substitute $P = v_0 Y$, $Q = v^0 X$ in (9.5) and (9.6), respectively, and can derive the dual systems shown in (9.9) and (9.10):

Row player

$$A^T Y \geq E_n, \quad Y \geq 0, \quad \text{minimize } \frac{1}{v_0} = E_p^T Y. \tag{9.9}$$

Column player

$$X \geq 0, \quad AX \leq E_p, \quad \text{maximize } \frac{1}{v^0} = E_n^T X. \tag{9.10}$$

Both problems are clearly feasible and by the symmetric form of the duality theorem for linear programming we have

$$\max \frac{1}{v^0} = \min \frac{1}{v_0} = \frac{1}{v}. \tag{9.11}$$

This gives a second approach to the minimax theorem.

This formulation of a game as a linear programming problem was given by Dantzig (Reference 11) in 1951; the formulation displayed in Figure 9.10 was exploited later by Tucker (Reference 40) and is developed in greater detail below.

The simplex method can be used to determine optimal strategies for the game. Instead of starting with (9.9), which is not readily put into canonical form, one begins with the column player's program (9.10) where insertion of slack variables immediately yields a problem in canonical form. The initial tableau M and final tableau M^* are displayed in Figure 9.11. An optimal program X^0 is obtained as usual as a basic program from M^*, an optimal program Y^0, given by the duality theorem, is C_2^*, and $-d^* = 1/v$. Optimal strategies for the row and column players are $P^0 = vY^0$ and $Q^0 = vX^0$, respectively.

To illustrate the use of the simplex method in solving games, consider the game with matrix

$$A = \begin{bmatrix} 2 & 1 & 6 \\ 5 & 8 & 3 \end{bmatrix}.$$

For this matrix we consider (9.10), introduce slack variables x_4, x_5, and

$M =$

0	$-E_n^T$	0
E_p	A	I_p

$M^* =$

$-d^*$	C_1^{*T}	C_2^{*T}
B^*	A_1^*	A_2^*

Figure 9.11 Initial and optimal tableaus.

proceed to the simplex solution. From the final tableau (Table 9.1) we find an optimal program has $x_1 = 1/8$, $x_2 = 0$, $x_3 = 1/8$, $d = -1/v = -1/4$, and from the columns of the slack variables we obtain an optimal vector $y_1 = 1/12$, $y_2 = 1/6$ for the dual problem (9.9). Thus the game has value 4; an optimal strategy for the column player is $q_1 = 1/2$, $q_2 = 0$, $q_3 = 1/2$, and an optimal strategy for the row player is $p_1 = 1/3$, $p_2 = 2/3$.

Table 9.1. SIMPLEX SOLUTION TO A GAME.

0	−1	−1	−1	0	0
1	2	1	6	1	0
1	5	8	3	0	1
1/5	0	3/5	−2/5	0	1/5
3/5	0	−11/5	24/5	1	−2/5
1/5	1	8/5	3/5	0	1/5
1/4	0	5/12	0	1/12	1/6
1/8	0	−11/24	1	5/24	−1/12
1/8	1	15/8	0	−1/8	1/4

A converse procedure for passing from a linear programming problem to an equivalent matrix game is also available and provides a proof of the fundamental duality theorem for linear programming based on the minimax theorem (Reference 8, pp. 419–423).

We return now to the Tucker normalization and employing the schema form of the simplex algorithm as described in Chapter 8 we obtain from Figure 9.10 (after adjoining slack variables) the following schema representation of a matrix game:

	1	q_1		q_k		q_n	v^0	
1	0	0	$\cdots$	0	$\cdots$	0	1	$= v^0$ min
p_1	0	a_{11}	$\cdots$	a_{1k}	$\cdots$	a_{1n}	-1	$= -u_1 \leq 0$
	$\vdots$	$\vdots$		$\vdots$		$\vdots$	$\vdots$	
p_j	0	a_{j1}	$\cdots$	a_{jk}	$\cdots$	a_{jn}	-1	$= -u_j \leq 0$ (9.12)
	$\vdots$	$\vdots$		$\vdots$		$\vdots$	$\vdots$	
p_p	0	a_{p1}	$\cdots$	a_{pk}	$\cdots$	a_{pn}	-1	$= -u_p \leq 0$
v_0	1	-1	$\cdots$	-1	$\cdots$	-1	0	$= 0$
	$= v_0$ max	$= v_1 \geq 0$		$= v_k \geq 0$		$= v_n \geq 0$	0	

We initiate the algorithm with two preliminary pivots which will yield an equivalent schema with no basic zero labels and with no nonbasic free variables.

First, we pivot on the jth position of the last column with pivot element -1. The result is:

	1	q_1		q_k		q_n	u_j	
1	0	a_{j1}	$\cdots$	a_{jk}	$\cdots$	a_{jn}	1	$= v^0$ min
p_1	0	$a_{11} - a_{j1}$	$\cdots$	$a_{1k} - a_{jk}$	$\cdots$	$a_{1n} - a_{jn}$	-1	$= -u_1 \leq 0$
$\vdots$	$\vdots$	$\vdots$		$\vdots$		$\vdots$	$\vdots$	$\vdots$
0	0	$-a_{j1}$	$\cdots$	$-a_{jk}$	$\cdots$	$-a_{jn}$	-1	$= -v^0$
$\vdots$	$\vdots$	$\vdots$		$\vdots$		$\vdots$	$\vdots$	$\vdots$
p_p	0	$a_{p1} - a_{j1}$	$\cdots$	$a_{pk} - a_{jk}$	$\cdots$	$a_{pn} - a_{jn}$	-1	$= -u_p \leq 0$
v_0	1	-1	$\cdots$	-1	$\cdots$	-1	0	$= 0$
	$= v_0$ max	$= v_1 \geq 0$		$= v_k \geq 0$		$= v_n \geq 0$	$= p_j \geq 0$	

(9.13)

If we now pivot on the kth position in the last row with pivot element -1, the result is:

	1	q_1		0		q_n	u_j	
1	a_{jk}	$a_{j1} - a_{jk}$	$\cdots$	a_{jk}	$\cdots$	$a_{jn} - a_{jk}$	1	$= v^0$ min
p_1	$a_{1k} - a_{jk}$	a_{11}^*	$\cdots$	$a_{1k} - a_{jk}$	$\cdots$		-1	$= -u_1 \leq 0$
$\vdots$	$\vdots$	$\vdots$		$\vdots$		$\vdots$	$\vdots$	
0	$-a_{jk}$	$a_{jk} - a_{j1}$	$\cdots$	$-a_{jk}$	$\cdots$	$a_{jk} - a_{jn}$	-1	$= -v^0$
$\vdots$	$\vdots$	$\vdots$		$\vdots$		$\vdots$	$\vdots$	
p_p	$a_{pk} - a_{jk}$		$\cdots$	$a_{pk} - a_{jk}$	$\cdots$		-1	$= -u_p \leq 0$
v_k	-1	1	$\cdots$	-1	$\cdots$	1	0	$= -q_k \leq 0$
	$= v_0$ max	$= v_1 \geq 0$		$= v_0$		$= v_n \geq 0$	$= p_j \geq 0$	

(9.14)

Here for $i \neq j$, $h \neq k$, the new element in position (i, h) is $a_{ih}^* = a_{ih} - a_{jh} - a_{ik} + a_{jk}$; in particular, $a_{11}^* = a_{11} - a_{j1} - a_{1k} + a_{jk}$.

Now note that the jth row of this schema represents exactly the same equation as the zero row, and so may be deleted since the corresponding dual variable is 0. The same remark holds for the kth column.

Moreover, if a_{jk} is a saddle point, that is, $a_{jk} \le a_{jq}$ $(q = 1, \cdots, n)$ and $a_{jk} \ge a_{ik}$, $(i = 1, \cdots, p)$, and we choose the pivot row and column accordingly, the schema will be in optimal form and we are done.

If a problem does *not* have a saddle point we may still select a_{jk} to satisfy $a_{jk} \ge a_{ik}$, $(i = 1, \cdots, p)$ and still achieve row feasibility in (9.14).

Example 1. *A game with a saddle point.* Let

$$A = \begin{bmatrix} -6 & 0 & 1 \\ 5 & -1 & -4 \\ 3 & -3 & 7 \\ 4 & \textcircled{2} & 3 \end{bmatrix}; \tag{9.15}$$

here the circled entry is a saddle point, so the pivot row and column are determined and the solution is as follows:

M:

	1	q_1	q_2	q_3	v^0	
1	0	0	0	0	1	v^0
p_1	0	-6	0	1	-1	$-u_1$
p_2	0	5	-1	-4	-1	$-u_2$
p_3	0	3	-3	7	-1	$-u_3$
p_4	0	4	2	3	$\textcircled{-1}$	$-u_4$
v_0	1	-1	-1	-1	0	0
	v_0	v_1	v_2	v_3	0	

M':

	1	q_1	q_2	q_3	u_4	
1	0	4	2	3	1	v^0
p_1	0	-10	-2	-2	-1	$-u_1$
p_2	0	1	-3	-7	-1	$-u_2$
p_3	0	-1	-5	4	-1	$-u_3$
0	0	-4	-2	-3	-1	$-v^0$
v_0	1	-1	$\textcircled{-1}$	-1	0	0
	v_0	v_1	v_2	v_3	p_4	

M'':

	1	q_1	0	q_3	u_4	
1	2	2	2	1	1	v^0
p_1	−2	−8	−2	0	−1	$-u_1$
p_2	−3	5	−3	−4	−1	$-u_2$
p_3	−5	4	−5	9	−1	$-u_3$
0	−2	−2	−2	−1	−1	$-v^0$
v_2	−1	1	−1	−1	0	$-q_2$
	v_0	v_1	v_0	v_3	p_4	

Example 2. *A game with no saddle point.* Let

$$A = \begin{bmatrix} -6 & 0 & 1 \\ 5 & -1 & -4 \\ 3 & -3 & 7 \\ 4 & \textcircled{5} & -3 \end{bmatrix} \quad \left.\begin{matrix} -6 \\ -4 \\ -3 \\ -3 \end{matrix}\right\} \text{Row minima.} \tag{9.16}$$

$$\underbrace{\begin{matrix} 5 & 5 & 7 \end{matrix}}_{\text{Column maxima}}$$

Clearly, there is no saddle point so a_{jk} could be chosen arbitrarily. However, we have chosen the circled entry in order to obtain row feasibility in (9.14). The solution follows:

M:

	1	q_1	q_2	q_3	v^0	
1	0	0	0	0	1	v^0
p_1	0	−6	0	1	−1	$-u_1$
p_2	0	5	−1	−4	−1	$-u_2$
p_3	0	3	−3	7	−1	$-u_3$
p_4	0	4	5	−3	(−1)	$-u_4$
v_0	1	−1	−1	−1	0	0
	v_0	v_1	v_2	v_3	0	

M':

	1	q_1	q_2	q_3	u_4	
1	0	4	5	−3	1	v^0
p_1	0	−10	−5	4	−1	$-u_1$
p_2	0	1	−6	−1	−1	$-u_2$
p_3	0	−1	−8	10	−1	$-u_3$
0	0	−4	−5	3	−1	$-v^0$
v_0	1	−1	(−1)	−1	0	0
	v_0	v_1	v_2	v_3	p_4	

M'':

	1	q_1	0	q_3	u_4	
1	5	7	5	−8	1	v^0
p_1	−5	−14	−5	9	−1	$-u_1$
p_2	−6	2	−6	5	−1	$-u_2$
p_3	−8	−11	−8	18	−1	$-u_3$
0	−5	−7	−5	8	−1	$-v^0$
v_2	−1	1	−1	1	0	$-q_2$
	v_0	v_1	v_0	v_3	p_4	

Thus we consider:

T:

	1	q_1	q_3	u_4	
1	5	7	−8	1	v^0
p_1	−5	−14	9	−1	$-u_1$
p_2	−6	2	5	−1	$-u_2$
p_3	−8	−11	(18)	−1	$-u_3$
v_2	−1	1	1	0	$-q_2$
	v_0	v_1	v_3	p_4	

One pivot gives optimality,

T^*:

	1	q_1	u_3	u_4	
1	13/9	19/9	4/9	5/9	v^0
p_1	−1	−17/2	−1/2	−1/2	$-u_1$
p_2	−34/9	91/18	−5/18	−13/18	$-u_2$
v_3	−4/9	−11/18	1/18	−1/18	$-q_3$
v_2	−5/9	29/18	−1/18	1/18	$-q_2$
	v_0	v_1	p_3	p_4	

Hence $P = [0, 0, 4/9, 5/9]^T$ and $Q = [0, 5/9, 4/9]^T$ and $u_{\min} = v_{\max} = 13/9$.

The Tucker normalization has two desirable features: (1) it does not require that each of the elements of the game matrix be positive, and (2) it uses the p_i, q_j, and v as its variables.

9.5. KERNEL OF A GAME.

We recall that a position (i, j) gives a saddle point of a game if a_{ij} is the smallest element in its row and the largest element in its column. Not all games have saddle points, and we have seen that a solution is easily obtained for those that do.

In this section we wish to generalize the notion of a saddle point and obtain a proof of a basic theorem in game theory due to Shapley and Snow (Reference 36).

We begin with the formulation (9.9) and (9.10) but change the sign of the objective function in each program so that the column player is minimizing and the row player is maximizing. Then the programs can be expressed in a single schema as in Figure 9.12.

(a)

	1	x_1	$\cdots$	x_n	
1	0	−1	$\cdots$	−1	(min) $= -\frac{1}{v^0}$
y_1	−1		A		$= -u_1$
$\cdot$	$\cdot$				
$\cdot$	$\cdot$				
$\cdot$	$\cdot$				
y_p	−1				$= -u_p$
	(max) $= -\frac{1}{v_0}$	$= v_1$	$\cdots$	$= v_n$	

(b)

	1	X^T	
1	0	$-E_n^T$	(min)
Y	$-E_p$	A	$-U$
	(max)	V^T	

Figure 9.12 Schema for dual programs.

Applying the simplex method in schema form to the problem in Figure 9.12(a) we get optimal vectors X^*, Y^* and value v in a finite number of iterations. Moreover, the results of Chapter 8 imply the existence of an r by r submatrix A_{11} of A such that a single matrix pivot operation on A_{11} yields a schema which displays optimal vectors. Such a submatrix is called a *kernel* of A. It may or may not be unique, depending on whether there are or are not multiple optimal vectors to either the primal or the dual program. For convenience in displaying our results we assume that rows and columns have been permuted if necessary so that the kernel is in the upper left-hand corner of A.

Of course, we were originally interested in finding optimal strategies P^* and Q^* so if one uses this new formulation, one must calculate P^* and Q^* from the formulas

$$P^* = v^*Y^*$$
$$Q^* = v^*X^*$$

where $v^* = 1/E_n^TX^*$.

We indicate the matrix pivot transformation in Figure 9.13, where

$$B^* = \begin{bmatrix} B_1^* \\ B_2^* \end{bmatrix} \leq 0,$$

$$C^{*T} = [C_1^{*T}C_2^{*T}] \geq 0,$$

and where $r + s = n$ and $r + t = p$.

From Chapter 8 we find that some of the formulas governing this matrix pivot transformation are:

$$d^* = -E_r^TA_{11}^{-1}E_r, \tag{9.17}$$
$$C_1^{*T} = E_r^TA_{11}^{-1}, \tag{9.18}$$
$$C_2^{*T} = -E_s^T + E_r^TA_{11}^{-1}A_{12}, \tag{9.19}$$
$$B_1^* = -A_{11}^{-1}E_r, \tag{9.20}$$
$$B_2^* = -E_t + A_{21}A_{11}^{-1}E_r. \tag{9.21}$$

We know that

$$d^* = \min(-E_n^TX),$$
$$= -E_n^TX^* = -\frac{1}{v^*},$$

	1	X_1^T	X_2^T	
1	0	$-E_r^T$	$-E_s^T$	
Y_1	$-E_r$	A_{11}	A_{12}	$-U_1$
Y_2	$-E_t$	A_{21}	A_{22}	$-U_2$
		V_1^T	V_2^T	

		U_1^T	X_2^T	
	d^*	C_1^{*T}	C_2^{*T}	
V_1	B_1^*	A_{11}^*	A_{12}^*	$-X_1$
Y_2	B_2^*	A_{21}^*	A_{22}^*	$-U_2$
		Y_1^T	V_2^T	

Figure 9.13 Illustration of a pivot operation.

so that from (9.17) we obtain

$$v^* = \frac{1}{E_r^T A_{11}^{-1} E_r} \tag{9.22}$$

in other words, v^* *is the reciprocal of the sum of all the components of* A_{11}^{-1}.

Equation (9.18) yields

$$Y_1^* = C_1^* = (A_{11}^{-1})^T E_r,$$

which states that the basic part of Y^* is obtained by summing the row vectors of A_{11}^{-1}. Of course, the nonbasic part of Y^*, that is, Y_2^*, is zero. Since $v^*Y^* = P^*$ we have

$$P^* = \begin{bmatrix} \dfrac{(A_{11}^{-1})^T E_r}{E_r^T A_{11}^{-1} E_r} \\ 0 \end{bmatrix}; \tag{9.23}$$

hence the jth entry of P^*, $(1 \leq j \leq r)$ is obtained from A_{11}^{-1} by summing the components of the jth column of A_{11}^{-1} and normalizing by dividing by the sum of all of the entries of A_{11}^{-1}.

In a similar manner, (9.20) yields

$$Q^* = \begin{bmatrix} \dfrac{A_{11}^{-1} E_r}{E_r^T A_{11}^{-1} E_r} \\ 0 \end{bmatrix} \tag{9.24}$$

which states that the jth, $(1 \leq j \leq r)$ component of Q^* is obtained by summing the components of the jth row of A_{11}^{-1} and normalizing. The existence of a submatrix A_{11} for which (9.22), (9.23), (9.24) hold is the major conclusion of the Shapley-Snow Theorem.

We now wish to give a geometrical meaning to the Equations (9.19) and (9.21). Consider the hyperplane

$$(E_r^T A_{11}^{-1})X = 1,$$

and note that each column of A_{11} lies on this hyperplane since $A_{11}^{-1} A_{11} = I$. But since $C_2^{*T} \geq 0$ we see that

$$E_s^T \leq E_r^T A_{11}^{-1} A_{12}, \tag{9.25}$$

which implies that each column of A_{12} lies above the hyperplane $(E_r^T A_{11}^{-1})X = 1$, that is, each column of A_{12} lies on the opposite side of $(E_r^T A_{11}^{-1})X = 1$ from the origin. Moreover, the vector v^*E_r lies on the hyperplane since

$$(E_r^T A_{11}^{-1})v^*E_r = v^*\left(\frac{1}{v^*}\right) = 1 \tag{9.26}$$

so that the ray τE_r, $(\tau \geq 0)$ pierces the hyperplane $(E_r^T A_{11}^{-1})X = 1$ when $\tau = v^*$. See Figure 9.14.

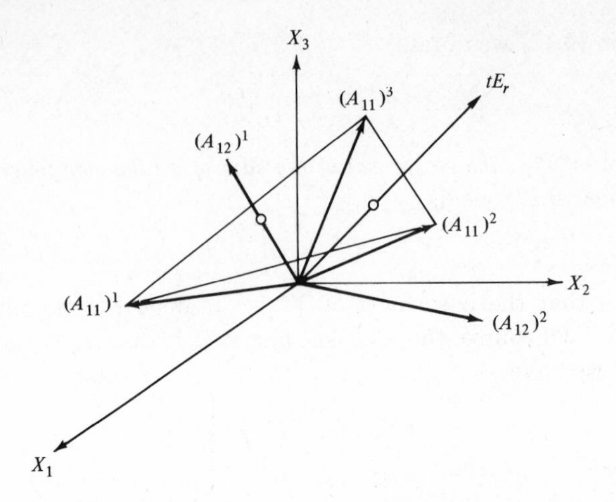

Figure 9.14 Illustration of Equation (9.26). Here $(A_{11})^j$ is the jth column of A_{11}.

In a similar manner, $B_2^* \leq 0$ and Equation (9.21) imply that the rows of A_{21} lie *below* the hyperplane $Y^T A_{11}^{-1} E_r = 1$ generated by the *rows* of A_{11}, and again this hyperplane intersects the ray to E_r when $\tau = v^*$.

For a game with a saddle point one considers three things: the entry in the saddle point position, the remaining components in the row of the saddle point, and the remaining components in the column of the saddle point. The saddle point is characterized by the fact that the entry in the saddle point position is less than or equal to all other components in the row and is greater than or equal to all other components in the column. For a game whose kernel is A_{11} we consider the following three analogous things: (1) the kernel itself, A_{11}, (2) the submatrix A_{12} of A whose rows are the same as those of the kernel but whose columns are all columns not in the kernel, (3) the submatrix A_{21} of A whose columns are the same as those of A_{11} but whose rows are all rows not in the kernel. The analogue of the saddle point property is the statement that each column of A_{12} lies above the hyperplane determined by the columns of A_{11} whereas the rows of A_{21} lie below the hyperplane generated by the rows of A_{11}. Finally, the value of the game is determined by the fact that $v^* E_r$ lies on both of these hyperplanes. Indeed, a kernel of size one is precisely an ordinary saddle point.

It is easy to see that the kernel of a game need not be unique; in fact, a game may have kernels of different sizes. Let us partition an optimal schema as in Figure 9.15. Clearly one can pivot on *any* nonzero element in the center of the dotted cross and still maintain row and column feasibility. Pivots in the extremities will be allowed if they preserve an optimal schema. A kernel obtained *after* a pivot transformation has been carried out may differ in degree from the previous ones. Also the solution pair may change.

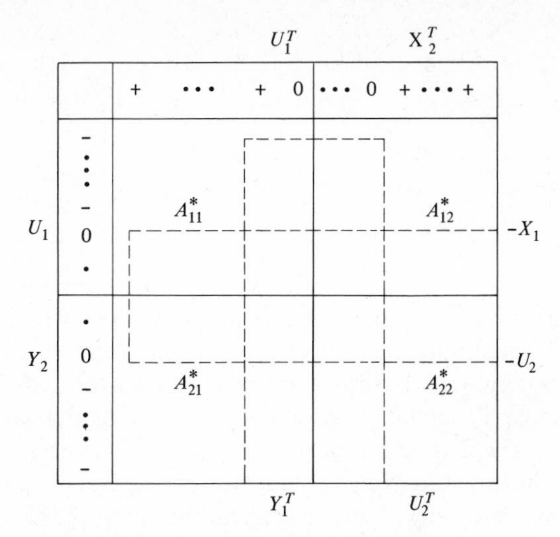

Figure 9.15 A partitioning of schema.

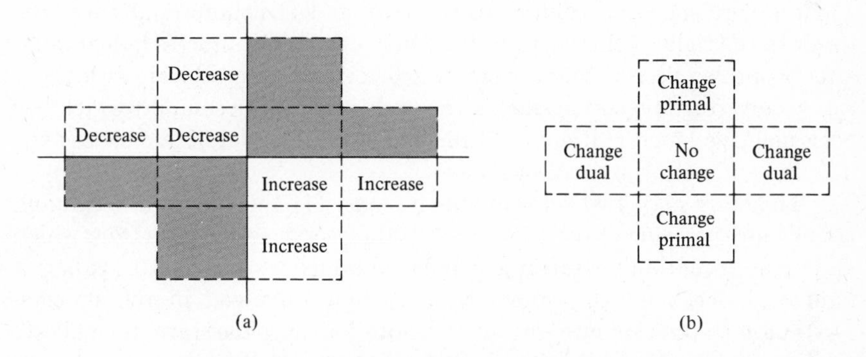

Figure 9.16 Effects of pivoting. (a) Effect of a pivot operation on the kernel degree; (b) Effect on solution of a pivot operation.

Figure 9.16 illustrates the effect of pivoting in the various regions. Note that one can never pivot and change both the primal and dual optimal solutions simultaneously.

9.6. ITERATIVE APPROXIMATION SOLUTION FOR GAMES.

A method for solving two-person, zero-sum games which is often useful is the iterative or successive approximation method. This is sometimes called the method of fictitious play. Consider the following example where the game matrix A has 3 rows and 4 columns:

$$\begin{array}{l} \begin{bmatrix} 2 & 3 & 0 & 5 \\ 3 & 1 & 4 & 2 \\ 4 & 5 & 3 & 2 \end{bmatrix} \begin{matrix} \\ \\ * \end{matrix} \begin{matrix} 5^* & 5 & 5 & 8 & 13 \\ 2 & 6^* & 10^* & 11 & 13 \\ 2 & 5 & 8 & 13^* & 15^* \end{matrix} \\ \begin{matrix} \;\;4 & 5 & 3 & 2^* \\ \;\;6 & 8 & 3^* & 7 \\ \;\;9 & 9 & 7^* & 9 \\ 12 & 10^* & 11 & 11 \\ 16 & 15 & 14 & 13^* \end{matrix} \end{array}$$

First find the maximin for pure strategies and place a star in a new column just opposite the maximin row (here row 3) and copy this row directly underneath the game matrix. Since the player who controls the columns wishes to minimize the payoff we next mark with a star the *smallest* number in this row (here the 2 in column 4) and copy the corresponding column just beyond the initial star. Now place a star on the largest number in this column and enter the sum of the corresponding row of the game matrix (here row 1) and the current bottom row in the tableau just below that row. Once again star the smallest element (here the 3 in the third column) and add the corresponding column to the current last column and enter the result to the right of this column. We can continue this process indefinitely, alternating between row and column extensions of the tableau. In case of ties at any stage place the star on the tied line (row or column) which has been inactive longest; if any tied lines have never been active choose the inactive one with the smallest index.

The stars play two roles in this process: (1) they serve in obtaining bounds for the value of the game and (2) they serve as counters from which approximate optimal strategies can be obtained as well as providing a numerical check on the numerical operations involved in the process.

For each positive integer t let v_t denote $1/t$ times the starred (smallest) element in the tth added row and let w_t denote $1/t$ times the starred (largest) element in the tth added column. Then

$$v_t \leq v \leq w_s, \qquad s, t = 1, 2, \cdots. \tag{9.27}$$

Hence, after z complete steps we have

$$\max_{t \leq z} v_t \leq v \leq \min_{s \leq z} w_s \tag{9.28}$$

as bounds for v. The process terminates when we are willing to accept these bounds as estimates for v, in other words, when the difference $\min w_s - \max v_t$ is considered small enough to be neglected.

Thus in our illustrative example

$$v_1 = 2, \qquad v_2 = 1\frac{1}{2}, \qquad v_3 = 2\frac{1}{3}, \qquad v_4 = 2\frac{1}{2}, \qquad v_5 = 2\frac{3}{5}$$

and

$$w_1 = 5, \qquad w_2 = 3, \qquad w_3 = 3\frac{1}{3}, \qquad w_4 = 3\frac{1}{4}, \qquad w_5 = 3.$$

Thus our best bounds for $z = 5$ steps are

$$2\frac{3}{5} = v_5 \leq v \leq w_2 = w_5 = 3. \tag{9.29}$$

Suppose that the game matrix A has n rows and m columns. Let $a_i(t)$ be the number among the first t column stars which lie in row i, $(i = 1, \cdots, n)$. Here we count the unattached star and those in added columns $1, 2, \ldots, t-1$ but not the star in added column t. Then the vector P_t having $p_{ti} = 1/t\, a_i(t)$ is a mixed strategy for the row player which guarantees an expected value at least v_t for him; indeed, if R_t is the tth added row then

$$P_t^T A = \frac{1}{t} R_t. \tag{9.30}$$

Similarly let $b_j(s)$ denote the number among the first s row stars which lie in column j, $(j = 1, \cdots, m)$. Then the vector Q_s having $q_{sj} = b_j(s)/s$ is a mixed strategy for the column player which guarantees an expected value at most $v_s(B)$ for him; that is,

$$AQ_s = \frac{1}{s} C_s \tag{9.31}$$

where C_s is the sth added column. The results (9.30) and (9.31) give checks on numerical accuracy at any stage. Thus in our example

$$Q_5 = \frac{1}{5}[0 \quad 1 \quad 2 \quad 2]^T \qquad \text{and} \qquad P_5 = \frac{1}{5}[1 \quad 2 \quad 2]^T \tag{9.32}$$

are the best approximate strategies available after $z = 5$ steps; they correspond to (9.29).

It is convenient to add an initial row and column to record the v_t and w_s. Thus after ten steps we have:

v_t				w_s										
					5	3	3 1/3	3 1/4	3	3	2 6/7	2 7/8	2 7/9	2 8/10
	2	3	0	5	5*	5	5	5	13	18*	18	18	23	28*
	3	1	4	2	2	6*	10*	11	13	15	19	23*	25	27
	4	5	3	2 *	2	5	8	13*	15*	17	20*	23	25*	27
2	4	5	3	2*										
1 1/2	6	8	3*	7										
2 1/3	9	9	7*	9										
2 1/2	12	10*	11	11										
2 3/5	16	19	14	13*										
2 1/2	20	20	17	15*										
2 3/7	22	23	17*	20										
2 1/2	26	28	20*	22										
2 2/3	29	29	24	24*										
2 3/5	33	34	27	26*										

Note the use of the tie-breaking rule in assigning stars in added columns 8, 9 and row 10. The best bounds are now given by

$$2\frac{2}{3} = v_9 \leq v \leq w_9 = 2\frac{7}{9}. \tag{9.33}$$

The actual optimal value and strategies are

$$v = 2\frac{21}{29}, \qquad P = \frac{1}{29}[7 \quad 13 \quad 9]^T, \qquad Q = \frac{1}{29}[0 \quad 3 \quad 12 \quad 14]^T. \tag{9.34}$$

We caution the reader against concluding from (9.33) that the best values for s and t are necessarily equal. Indeed, if the number of stages is expanded to 15 the reader will find that the best bounds are given by

$$2\frac{9}{13} = v_{13} \leq v \leq w_{14} = 2\frac{11}{14}$$

with corresponding best approximate strategies

$$P = \frac{1}{13}[3 \quad 5 \quad 5]^T, \qquad Q = \frac{1}{14}[0 \quad 1 \quad 6 \quad 7]^T.$$

9.7. GAMES AGAINST NATURE.

Consider a situation in which a decision maker must select one of p alternatives $R_1, \cdots, R_p$. We suppose that there are n states of nature $S_1, \cdots, S_n$ and consider the matrix $A = [a_{ij}]$ in which a_{ij} is the valuation placed by the decision maker on having selected alternative R_i in case nature turns out to be in state S_j. We assume that the decision maker has no information regarding which state of nature will obtain.

The decision maker could take the point of view that A is the matrix of a two-person, zero-sum game in which he is the row player and nature is the column player. Then he can employ a maximin strategy to obtain a security level. The maximin can be calculated either for pure strategies or for mixed strategies in accordance with which model best fits his environment.

We consider an example. Two youths plan to set up as vendors at the baseball stadium of their high school. They can sell ginger ale or hamburgers, but to avoid a high city license fee vendors must limit their daily purchases to 10 dollars. Ginger ale costs 5 cents and sells at 10 cents per cup. Hamburgers cost 10 cents and sell at 25 cents each. On a warm day they can sell unlimited ginger ale and 40 hamburgers; on a cool day they can sell unlimited hamburgers and 100 cups of ginger ale. Anything not sold is a total loss. The situation is summarized in Table 9.2. The entries are profits in dollars.

Table 9.2. THE YOUNG VENDORS.

	States of nature	
	Warm day	*Cool day*
Program 1 200 cups of ginger ale	10	0
Program 2 100 hamburgers	0	15

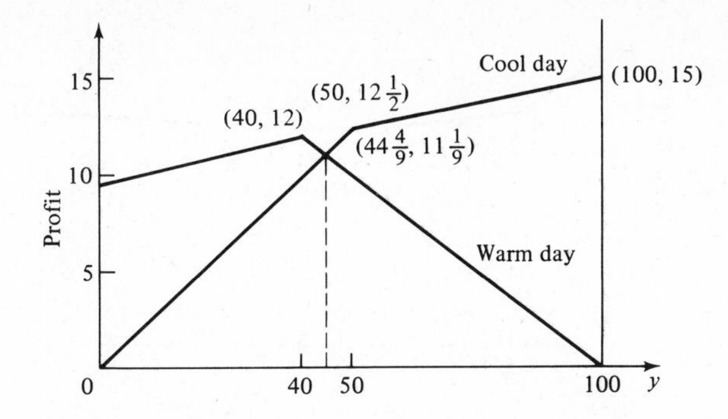

Figure 9.17 The young vendors with physical mixtures.

The maximin strategy is Program 1 with probability $\frac{3}{5}$, Program 2 with probability $\frac{2}{5}$, and the expected profit then is 6 dollars.

If physical mixtures are permitted, then

$$5x + 10y = 1000$$

where x = the number of cups of ginger ale and y = the number of hamburgers. The profit on a warm day is then

$$\frac{x}{10} + \frac{1}{4}\min\{40, y\} - 10 = 10 - \frac{y}{5} + \frac{1}{4}\min\{40, y\},$$

and on a cool day is

$$\frac{1}{10}\min\{x, 100\} + \frac{y}{4} - 10 = \frac{1}{10}\min\{200 - 2y, 100\} + \frac{y}{4} - 10.$$

The situation for physical mixtures is illustrated in Figure 9.17. The mixed program $[111\frac{1}{9} \quad 44\frac{4}{9}]^T$ gives a profit of $11\frac{1}{9}$ dollars regardless of the weather. (The nearest integer values are $[112 \quad 44]^T$ and $[110 \quad 45]^T$ both with guaranteed profit of 11 dollars.) Note that the physical mixture represents a considerable improvement over the minimax mixture of pure strategies; this should be expected, since the number of alternatives is expanded from 2 to infinity.

If the probability p of a warm day is known, the expected return can be somewhat improved as is shown in Figure 9.18.

If the conditions are changed so that there is no market for hamburgers on a hot day or for ginger ale on a cool day, then the minimax solution and expected return are unchanged, but the physical mixture optimal program (assuming no knowledge of state of nature) becomes $[120 \quad 40]^T$ with certain daily profit of 6; i.e., in this case, the only improvement over the minimax solution is the change from expected profit of 6 to a certain profit of 6.

The maximin strategy is not the only possibility for use in games against nature. Indeed, it may represent much too pessimistic an approach.

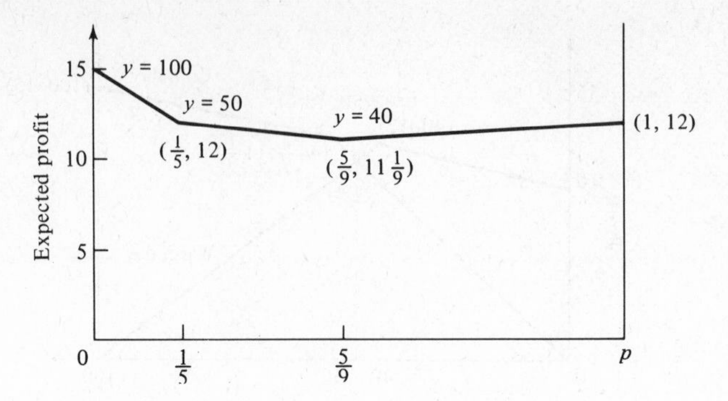

Figure 9.18 Maximum expected return with known probability.

Chapter 13 of Reference 8 summarizes and discusses a number of alternate decision criteria among which we consider here only a subset treated by John Milnor (Reference 34).

Laplace. If the probabilities of the different possible states of nature are unknown, we should assume that they are all equal. Thus if the player chooses the ith row, his expectation is given by the average $(a_{i1} + \cdots + a_{in})/n$, and he should choose a row for which this average is maximized.

Wald (Maximin principle). If the player chooses the ith row, then his payoff will certainly be at least $\min_j a_{ij}$. The safest possible course of action is therefore to choose a row for which $\min_j a_{ij}$ is maximized. This corresponds to the pessimistic hypothesis of expecting the worst.

If mixed strategies for the player are also allowed, then this criterion should be formulated as follows. Choose a probability mixture $(y_1, \cdots, y_m)$ of the rows so that the quantity $\min_j (y_1 a_{1j} + \cdots + y_m a_{mj})$ is maximized. In other words, play as if nature were the opposing player in a zero-sum game.

Hurwicz. Select a constant $0 \leq \alpha \leq 1$ which measures the player's optimism. For each row (or probability mixture of rows), let a denote the smallest component and b the largest. Choose a row (or probability mixture of rows) for which $\alpha b + (1 - \alpha)a$ is maximized. For $\alpha = 0$ this reduces to the Wald Criterion.

Savage (Minimax regret). Define the (negative) regret matrix $[r_{ij}]$ by $r_{ij} = a_{ij} - \max_k a_{kj}$. Thus r_{ij} measures the difference between the payoff which actually is obtained and the payoff which could have been obtained if the true state of nature had been known. Now apply the Wald Criterion to the matrix $[r_{ij}]$. That is, choose a row (or mixture of rows) for which $\min_j r_{ij}$ (or $\min_j (y_1 r_{1j} + \cdots + y_m r_{mj})$) is maximized.

These four criteria are clearly different. This is illustrated by the following example, where the preferred row under each criterion is indicated.

	Laplace	*Wald*	*Hurwicz*	*Savage*
1. Ordering	⊗	⊗	⊗	⊗
2. Symmetry	⊗	⊗	⊗	⊗
3. Strong domination	⊗	⊗	⊗	⊗
4. Continuity	×	⊗	⊗	⊗
5. Linearity	×	×	⊗	×
6. Row adjunction	⊗	⊗	⊗	
7. Column linearity	⊗			⊗
8. Column duplication		⊗	⊗	⊗
9. Convexity	×	⊗		⊗
10. Special row adjunction	×	×	×	⊗

× = Compatibility. Each criterion is characterized by axioms marked ⊗

Figure 9.19 Axioms for criteria.

$$\begin{bmatrix} 2 & 2 & 0 & 1 \\ 1 & 1 & 1 & 1 \\ 0 & 4 & 0 & 0 \\ 1 & 3 & 0 & 0 \end{bmatrix} \begin{matrix} \text{Laplace} \\ \text{Wald} \\ \text{Hurwicz } \left(\text{for } \alpha > \frac{1}{4}\right) \\ \text{Savage} \end{matrix}$$

In an attempt to throw further light on these and other decision procedures, Milnor has considered certain properties or axioms which one might wish to have any procedure satisfy. He determined those axioms which characterize each of the four given criteria and showed that there can be no criterion which satisfies all of his axioms. His results point to the necessity for closer scrutiny of the axioms and further progress is still awaiting the results of such scrutiny (see Figure 9.19).

PROBLEMS

9.1. Given the game defined by the following matrix,

$$\begin{bmatrix} 3 & 2 & 3 \\ 2 & 3 & 4 \\ 5 & 4 & 2 \end{bmatrix},$$

find the value of the game by using the simplex method. What are the optimal strategies?

9.2. Suppose the following matrix is the matrix of a game:

$$\begin{bmatrix} -3 & -2 & 6 \\ 2 & 0 & 2 \\ 5 & -2 & -4 \end{bmatrix}.$$

(a) Is there a saddle point for the game? If so, what is it?
(b) What is an optimal strategy for each of the players?

9.3. Show by using a block pivot that the 3 by 4 game represented in the tableau below has kernel size 2 and that the kernel consists of the 2 by 2 matrix in the upper left-hand corner of the game matrix. Also find the value of the game and the optimal strategies for the players.

0	−1	−1	−1	−1
−1	4	2	5	4
−1	2	4	3	4
−1	2	3	6	4

9.4. Suppose

$$A = \begin{bmatrix} 1 & 5/2 & 3 \\ 2 & 3/2 & 1 \end{bmatrix}$$

is the matrix for a game. Find optimal strategies and the value of the game.

9.5. If G is a positive vector and if A is a p by n matrix with at least one positive column prove that the problem

$$\text{minimize } z = D^T X$$

subject to

$$-AX + I_p X' = -G,$$
$$X \geq 0, \qquad X' \geq 0,$$

can be put into canonical form by means of one pivot transformation. Show how this result applies to the row player's problem as given by (9.9).

9.6. Given the following game matrix

$$A = \begin{bmatrix} 2 & 3 & 7 \\ 6 & 4 & 2 \end{bmatrix},$$

solve this game using (9.10). Also solve this game using (9.9) and then using the schema form (9.12) ff. Finally, solve this game using the approximation method.

9.7. In the following matrix games find the saddle points, if any.

(a) $\begin{bmatrix} 3 & 2 & -2 \\ 1 & -3 & -4 \\ 0 & 1 & -3 \end{bmatrix}$ (b) $\begin{bmatrix} 1 & -3 \\ 4 & 0 \\ 3 & 1 \end{bmatrix}$

(c) $\begin{bmatrix} -7 & 7 & 8 \\ -4 & -3 & -2 \end{bmatrix}$ (d) $\begin{bmatrix} 4 & -2 & 3 \\ 0 & 5 & 6 \end{bmatrix}$

(e) $\begin{bmatrix} 0 & 2 \\ -3 & -6 \\ -4 & -3 \end{bmatrix}$ (f) $\begin{bmatrix} 1 & 2 \\ 3 & 4 \\ 5 & -6 \end{bmatrix}$

(g) $\begin{bmatrix} 1 & 7 & 6 \\ -4 & 3 & -5 \\ 0 & 2 & 7 \end{bmatrix}$ (h) $\begin{bmatrix} 1 & 6 & 8 \\ 2 & 7 & -9 \\ -4 & 9 & 12 \end{bmatrix}$

9.8. Solve the following game by the simplex and approximation methods:

$$\begin{bmatrix} -1 & -6 & 3 & 1 \\ -7 & -4 & 2 & 0 \end{bmatrix}.$$

9.9. Let the following be a matrix representing a game against nature:

$$\begin{bmatrix} 2 & 2 & 0 & 1 \\ 1 & 1 & 1 & 1 \\ 0 & 4 & 0 & 0 \\ 1 & 3 & 0 & 0 \end{bmatrix}.$$

Choose an optimal row according to each of the following criteria:
(a) Wald maximin principle,
(b) Savage minimax regret,
(c) Laplace criterion,
(d) Hurwicz criterion.

9.10. Suppose the following tableaus represent, respectively, the initial and optimal tableaus of a game solved by the simplex algorithm:

0	-1	-1	-1	0	0	0
1	3	2	2	1	0	0
1	2	3	1	0	1	0
1	1	2	3	0	0	1
6/13	0	0	0	3/13	1/13	2/13
3/13	0	1	0	$-5/13$	7/13	1/13
2/13	0	0	1	1/13	$-4/13$	5/13
1/13	1	0	0	7/13	$-2/13$	$-4/13$

.

Find (a) the value of the game, (b) optimal strategies for the players.

9.11. Solve the following game once by the schema method and again by means of the Dantzig formulation:

$$A = \begin{bmatrix} -2 & 4 & 0 \\ 3 & 1 & -1 \\ -3 & 5 & 2 \\ 5 & -2 & -4 \end{bmatrix}.$$

9.12. Given the following game matrices, find optimal mixed strategies for each player and the value of the game by the graphical procedures of Section 9.3.

(a) $\begin{bmatrix} 2 & 0 & 1 \\ 1 & 4 & 3 \end{bmatrix}$ (b) $\begin{bmatrix} -2 & 3 & -3 & 5 \\ 0 & -1 & 2 & -4 \end{bmatrix}$

(c) $\begin{bmatrix} 2 & 3 \\ 5 & 2 \end{bmatrix}$ (d) $\begin{bmatrix} 1 & 2 \\ 3 & 4 \end{bmatrix}$

9.13. Two players play the following game: player 1 chooses a number x from the set $\{1, 2, 3\}$ then player 2, without knowing the first player's choice, chooses a number y from the same set. If $x \geq y$, player 2 pays $x + y$ to player 1; if $x < y$, player 2 pays $x - y$ to player 1.

(a) Draw the tree for this game, labeling nodes and arcs and indicating payoffs.

(b) Characterize this game in normal form by giving the payoff matrix.

(c) What is a maximin strategy for player 1? What is a minimax strategy for player 2? Is there a saddle point for this game?

9.14. The game of two-finger morra is a two person game played as follows. Each player shows one or two fingers, simultaneously calling out his guess as to how many fingers his opponent will show. If only one player guesses correctly, he wins as many pennies as the total number of fingers shown by both players. Otherwise no money changes hands. How many pure strategies does each player have? What is the payoff matrix for this game? Find the value of the game and optimal strategies for the players.

9.15. Consider a matrix game Γ defined by an n by n skew symmetric matrix A. Show that the value of Γ is zero and that if a strategy P is optimal for the first player then it is also optimal for the second player (hint: first show that $X^T A X = 0$ for any X and use this, together with the inequalities of Figure 9.10, to conclude that $\min v^0 \geq 0$ and $\max v_0 \leq 0$). Such a game is said to be *symmetric;* Example 4 of Section 9.3 is an example of a symmetric game.

Special Topics: Decomposition and Upper-bound Constraints

10

10.1. INTRODUCTION.

Suppose a linear programming problem has the following form:

$$\text{minimize } z = c_0 + \sum_{j=1}^{r} C_j^T X_j \tag{10.1}$$

subject to

$$\sum_{j=1}^{r} A_{0j} X_j = B_0, \tag{10.2}$$

$$A_j X_j = B_j, \qquad (j = 1, \cdots, r), \tag{10.3}$$

$$X_j \geq 0, \qquad (j = 1, \cdots, r), \tag{10.4}$$

where for each $j = 1, 2, \cdots, r$

X_j is n_j by 1,
A_j is p_j by n_j,
A_{0j} is p_0 by n_j,
B_j is p_j by 1,
C_j is n_j by 1,
c_0 is a scalar,

and where

$$B_0 \text{ is } p_0 \text{ by } 1,$$

$$X = \begin{bmatrix} X_1 \\ \cdot \\ \cdot \\ \cdot \\ X_r \end{bmatrix}.$$

The special structure of this problem is emphasized in Figure 10.1;

$-c_0$	C_1^T	C_2^T	$\cdots$	C_i^T	$\cdots$	C_r^T
B_0	A_{01}	A_{02}	$\cdots$	A_{0i}	$\cdots$	A_{0r}
B_1	A_1					
B_2		A_2				
$\cdot$			$\cdot$			
$\cdot$			$\cdot$			
$\cdot$			$\cdot$			
B_i				A_i		
$\cdot$					$\cdot$	
$\cdot$					$\cdot$	
$\cdot$					$\cdot$	
B_r						A_r

Figure 10.1 Tableau for the problem stated in Equations (10.1) through (10.4).

the constraints can be partitioned into $r + 1$ subsets such that r subsets [corresponding to (10.3)] are independent of each other in the sense that each subset constrains different unknowns, and one subset, corresponding to (10.2), containing what is called *linking constraints*, constrains all the variables in the problem.

Problems exhibiting this structure arise naturally in a variety of contexts. One, already mentioned in Chapter 5, occurs in a decentralized organization. For example, one in which there are r largely autonomous divisions whose individual operations can be modeled by (10.3) and in which the totality of operations are subject to linking constraints represented by (10.2).

The decomposition procedure we now introduce takes advantage of the special structure of this problem by replacing it with $r + 1$ problems (the r independent subproblems and the linking subproblem) which can be more efficiently solved than the original problem. The procedure itself, which can be used on any linear programming problem since the conditions are always met if we take $r = 1$, was developed by Dantzig and Wolfe (Reference 16).[1] This procedure is especially efficient if the r independent subproblems have a simple structure.

10.2. THE DECOMPOSITION PRINCIPLE.

Let L_j denote the system

$$\begin{aligned} A_jX_j &= B_j, \\ X_j &\geq 0 \end{aligned} \tag{10.5}$$

and let T_j denote the feasible region defined by L_j. The algorithm which we will soon introduce tests each system L_j for feasibility and redundancy. If any L_j is infeasible, of course, the entire problem is infeasible. If L_j has redundant constraints we discard them as detected. Accordingly, for the argument which follows we may and will assume that L_j does not contain redundant constraints and that T_j is not empty. Furthermore, the algorithm does not require that any T_j be bounded.

Now according to Theorem 5 of Section 3.10 we know that there are sequences

$$K_j = (X_j^1, \cdots, X_j^{k_j})$$

and

$$H_j = (X_j^{k_j+1}, \cdots, X_j^{k_j+h_j})$$

with the property that T_j consists of all vectors X_j of the form

$$X_j = \sum_{i=1}^{k_j+h_j} \lambda_j^i X_j^i, \tag{10.6}$$

where $\lambda_j^1, \cdots, \lambda_j^{k_j+h_j}$ are nonnegative scalars such that

$$\sum_{i=1}^{k_j} \lambda_j^i = 1.$$

With this notation the linking constraints (10.2) of the original problem can be written

$$\begin{aligned} &\sum_{j=1}^{r} A_{0j}\left(\sum_{i=1}^{k_j+h_j} \lambda_j^i X_j^i\right) = B_0, \\ &\sum_{i=1}^{k_j} \lambda_j^i = 1, \qquad (j = 1, \cdots, r), \\ &\lambda_j^i \geq 0, \qquad (i = 1, \cdots, k_j + h_j; j = 1, \cdots, r). \end{aligned} \tag{10.7}$$

[1]Dantzig and Wolfe have indicated that their work was inspired by ideas of Ford and Fulkerson (Reference 18). Dantzig (Reference 13) has also indicated that a similar approach appears in Jewell (Reference 25).

Note that for each j we have from (10.6) a $(k_j + h_j)$ by 1 vector

$$\Lambda_j = \begin{bmatrix} \lambda_j^1 \\ \lambda_j^2 \\ \cdot \\ \cdot \\ \cdot \\ \lambda_j^{k_j+h_j} \end{bmatrix}.$$

We also define

$$\Lambda = \begin{bmatrix} \Lambda_1 \\ \Lambda_2 \\ \cdot \\ \cdot \\ \cdot \\ \Lambda_r \end{bmatrix}$$

and observe that if Λ is a feasible program for (10.7), then X defined using (10.6) is a feasible program for (10.2), (10.3), and (10.4). Conversely, if X is feasible for (10.2), (10.3), and (10.4), then there exists at least one vector Λ satisfying (10.6) and (10.7).

Now let $S_j^i = A_{0j}X_j^i$; then (10.7) can be written as

$$\begin{aligned}
&\sum_{i=1}^{k_1+h_1} \lambda_1^i S_1^i + \sum_{i=1}^{k_2+h_2} \lambda_2^i S_2^i + \cdots + \sum_{i=1}^{k_r+h_r} \lambda_r^i S_r^i = B_0, \\
&\sum_{i=1}^{k_1} \lambda_1^i = 1, \\
&\sum_{i=1}^{k_2} \lambda_2^i = 1, \\
&\cdot \quad \cdot \quad \cdot \quad \cdot \quad \cdot \quad \cdot \quad \cdot \quad \cdot \quad \cdot \quad \cdot \quad \cdot \quad \cdot \quad \cdot \quad \cdot \quad \cdot \\
&\sum_{i=1}^{k_r} \lambda_r^i = 1, \\
&\Lambda_j \geq 0, \qquad (j = 1, \cdots, r).
\end{aligned} \tag{10.8}$$

The objective function can be rewritten as follows

$$z = c_0 + \sum_{j=1}^{r} C_j^T X_j = c_0 + \sum_{j=1}^{r} C_j^T \left(\sum_{i=1}^{k_j} \lambda_j^i X_j^i \right) = c_0 + \sum_{=1}^{r} \sum_{i=1}^{k_j} \lambda_j^i (C_j^T X_j^i).$$

Let $c_0 = d_0$ and $d_j^i = C_j^T X_j^i$; then set

$$D_j = \begin{bmatrix} d_j^1 \\ \cdot \\ \cdot \\ \cdot \\ d_j^{k_j+h_j} \end{bmatrix}$$

and the objective function becomes

$$z = d_0 + \sum_{j=1}^{r} D_j^T \Lambda_j. \tag{10.9}$$

We define as the *master problem* that of minimizing (10.9) subject to the constraints (10.8).

Each feasible program for the master problem is a feasible program for the original problem (10.1) through (10.4) and if $X^T = [X_1^T \, X_2^T \cdots X_r^T]$ is a feasible program for the original problem, then there is at least one vector Λ which satisfies (10.6) and (10.7). Hence a feasible solution (Λ, z') for the master problem also yields a feasible solution (X, z) to the original problem with $z = z'$.

It should be noted that a given feasible program X for the original problem may correspond to more than one feasible vector Λ for (10.7). This is true because a feasible program for the jth subproblem may be represented in more than one way in the form (10.6).

The master problem is written in tableau form in Figure 10.2. We denote its matrix by G. There are $m = p_0 + r$ constraints in the problem and $N = \sum_{j=1}^{r} (k_j + h_j)$ variables. The integers k_j, h_j, and N are usually unknown and each can be extremely large. For example, suppose we have a problem in which $r = 10$, $p_0 = 50$, $n_1, \cdots, n_{10}$ are each equal to 200 and $p_1, \cdots, p_{10}$ are each equal to 100. Then $p = 1{,}050$ and $n = 2{,}000$, whereas $m = p_0 + r = 60$ and N is a very large number. Since the speed of solution and machine capacity each depend to an important extent on the number of rows in the constraint matrix of a problem, the decomposition procedure can be described in words as transforming a problem with many rows into one with a much smaller number of rows but a vastly increased number of columns. If one had to deal with the master problem explicitly this approach would be of little practical value. However, the usefulness of the decomposition algorithm rests on the key fact that we are not required to load the master problem into a computer or even to calculate most of its columns. One starts from an initial basic feasible solution and generates only those columns S_j^i which the simplex algorithm brings into the basis at each iteration. This benefit is not without a cost, however, for as we shall soon see, we might have to optimize as many as r subproblems at each iteration.

10.3. AN ITERATIVE ALGORITHM.

Our goal is to adapt the revised simplex algorithm so as to solve the master problem efficiently. We made two assumptions initially: (1) that we have a basis for the master problem, and (2) that the master

$$
G = \left[\begin{array}{c|cccccccccccccccccc}
-d_0 & d_1^1 & \cdots & d_1^{k_1} & d_1^{k_1+1} & \cdots & d_1^{k_1+h_1} & \cdots & d_q^1 & \cdots & d_q^{k_q} & \cdots & d_q^{k_q+h_q} & \cdots & d_r^1 & \cdots & d_r^{k_r} & d_r^{k_r+1} & \cdots & d_r^{k_r+h_r} \\
\hline
B_0 & S_1^1 & \cdots & S_1^{k_1} & S_1^{k_1+1} & \cdots & S_1^{k_1+h_1} & \cdots & & & & & & \cdots & S_r^1 & \cdots & S_r^{k_r} & S_r^{k_r+1} & \cdots & S_r^{k_r+h_r} \\
\cdots & \cdots & \cdots & \cdots & \cdots & \cdots & \cdots & \cdots & \cdots & \cdots & \cdots & \cdots & \cdots & \cdots & \cdots & \cdots & \cdots & \cdots & \cdots & \cdots \\
1 & 1 & \cdots & 1 & 0 & \cdots & 0 & & & & & & & & & & & & & \\
\vdots & \cdots & \cdots & \cdots & \cdots & \cdots & \cdots & \cdots & \cdots & \cdots & \cdots & \cdots & \cdots & \cdots & \cdots & \cdots & \cdots & \cdots & \cdots & \cdots \\
1 & & & & & & & & & & & & & & 1 & \cdots & 1 & 0 & \cdots & 0
\end{array}\right]
$$

$$
G = \begin{bmatrix} -d_0 & D^T \\ A_0 & A \end{bmatrix}
$$

Figure 10.2 Tableau for the master problem stated in Equations (10.8) and (10.9).

problem is not degenerate. We will discuss later the problem of obtaining an initial basic feasible program for the master problem. Degeneracy can be handled by using the random-choice rule introduced earlier in the development of the simplex algorithm and will not be discussed further here.

Consider the basic feasible program for the master problem. Associated with this program are $m = p_0 + r$ basic columns of the tableau in Figure 10.2. An additional notational problem now arises: if we tried to use a single index to designate columns of the matrix G we would have to know the total number N of columns of G. We circumvent this difficulty by using a double index (j, i) which denotes the ith column in the jth block. To put it another way, a basic sequence will now be identified by a sequence of ordered pairs rather than a sequence of integers. A basic sequence thus has the form

$$S = ((j_1, i_1), (j_2, i_2), \cdots, (j_m, i_m)),$$

where $m = p_0 + r$. It then becomes possible to assign the indices i for the columns in block j in the order in which they are generated, that is, as they are needed and used. We also observe that the equation $\sum_{i=1}^{k_j} \lambda_j^i = 1$ guarantees that in any basic feasible program Λ no Λ_j is zero and moreover that at least one column from the first part of each of the r blocks is in the corresponding basis.

By suitable enumeration of the linking constraints and of the vectors X_j^i we may suppose that the basic sequence is[2]

$$S = ((1, 1), (1, 2), \cdots, (1, \mu_1), (1, k_1 + 1), \cdots, (1, k_1 + \nu_1), \\ \cdots, (r, 1), \cdots, (r, \mu_r), (r, k_r + 1), \cdots, (r, k_r + \nu_r)).$$

Here $\sum_{j=1}^{r} (\mu_j + \nu_j) = p_0 + r = m$ and for each j we have $\mu_j > 0$, $\nu_j \geq 0$.

If we were to use the revised simplex method on the master problem we recall that at the beginning of each iteration we would have the following information:

(1) $p_0 + r$ basic columns of G each of the form

$$G_j^i = \begin{bmatrix} d_j^i \\ S_j^i \\ T_j^i \end{bmatrix} = \begin{bmatrix} d_j^i \\ A_j^i \end{bmatrix},$$

[2]Since the integers k_j ordinarily are not known, this procedure would have to be modified for computer programming. For each block the basic program and generators of extreme rays would have to be labeled in separate series each beginning with 1. However, for theoretical exposition the notation we have adopted is logically acceptable and much more convenient.

where G_j^i denotes the ith column in the jth block and where T_j^i has r rows and is the jth unit vector if $i \leq k_j$ and is zero if $i > k_j$. For later use we write

$$T_j^i = \begin{bmatrix} 0 \\ \cdot \\ \cdot \\ \cdot \\ 0 \\ t_j^i \\ 0 \\ \cdot \\ \cdot \\ \cdot \\ 0 \end{bmatrix} \leftarrow \text{row } j$$

where $t_j^i = 0$ if $i > k_j$ and $t_j^i = 1$ if $i \leq k_j$.

(2) A matrix Q which can be written

$$Q = \begin{bmatrix} 1 & [\pi^T \quad \theta^T] \\ 0 & R \end{bmatrix},$$

where, as usual, the matrix Q is the inverse of G_S, R is the inverse of

$$A_S = [A_{j_1}^{i_1} \quad \cdots \quad A_{jm}^{im}] = \begin{bmatrix} S_{j_1}^{i_1} & \cdots & S_{jm}^{im} \\ T_{j_1}^{i_1} & \cdots & T_{jm}^{im} \end{bmatrix}$$

where $m = p_0 + r$, π is m by 1 and θ is r by 1.

Using Q we obtain the current tableau matrix G^* after an iteration of the revised simplex algorithm as

$$G^* = QG.$$

The cost elements d_j^{*i} in G^* are given by

$$d_j^{*i} = d_j^i + \pi^T S_j^i + \theta_j t_j^i$$

and since the revised simplex preserves the simplex procedure we know that π and θ satisfy

$$d_j^{*i} = d_j^i + \pi^T S_j^i + \theta_j t_j^i = 0$$

for each basic index (j, i).

Three possibilities then exist.

Case I: $d_j^i + \pi^T S_j^i + \theta_j t_j^i \geq 0$ for every j and i. An optimal program for the master problem has been found.

Case II: for some index (j, i) we have

$$d_j^i + \pi^T S_j^i + \theta_j t_j^i < 0$$

and $G_j^{*i} = QG_j^i \leq 0$. The objective function is unbounded so we terminate the procedure.

Case III: for some index (j, i) we have

$$d_j^i + \pi^T S_j^i + \theta_j t_j^i < 0$$

and some component of QG_j^i is positive, and, as usual, a pivot operation is indicated.

The key issue, then, is the determination of negative cost elements d_j^{*i} in the desired tableau matrix G^*. The following procedure allows us to check for negative cost elements by studying the subsystems L_j of the initial problem. We have

$$\begin{aligned} d_j^{*i} &= d_j^i + \pi^T S_j^i + \theta_j t_j^i = C_j^T X_j^i + \pi^T (A_{0j} X_j^i) + \theta_j t_j^i \\ &= (C_j^T + \pi^T A_{0j}) X_j^i + \theta_j t_j^i, \end{aligned}$$

where X_j^i is a basic feasible program for $A_j X_j = B_j$. Clearly $d_j^{*i} < 0$ if and only if the problem L_j^*

$$\text{minimize } z_j = (C_j^T + \pi^T A_{0j}) X_j + \theta_j t_j^i \tag{10.10}$$

subject to

$$\begin{aligned} A_j X_j &= B_j, \\ X_j &\geq 0, \end{aligned} \tag{10.11}$$

is unbounded or has an optimal program with negative value z_j.

If for every j minimum $z_j = 0$, then Λ is an optimal program for the master problem and we are in Case I above (the minimum z_j cannot be positive, for if λ_j^i is in a basic feasible program at a positive level, then $(C_j^T + \pi^T A_{0j}) X_j^i + \theta_j t_j^i = 0$, and we have seen that for each j there is an i with $i \leq k_j$ such that $\lambda_j^i > 0$).

Suppose to the contrary that for some j_0 and i_0 we have a program $X_{j_0}^{i_0}$ for which the corresponding z_{j_0} is negative and such that either (a) $i_0 \leq k_{j_0}$ and $X_{j_0}^{i_0}$ is optimal for $L_{j_0}^*$ or (b) $i_0 > k_{j_0}$ and $X_{j_0}^{i_0}$ generates an extreme ray along which z_{j_0} is unbounded. Whichever of (a) or (b) holds we determine whether we are in Case II or Case III. If the former, we stop: the original problem is unbounded. If the latter, we introduce (j_0, i_0) into the basic sequence, determine the outgoing index (j_ρ, i_ρ) in the usual way and pivot in row ρ and column (j_0, i_0) of G^* to obtain a new improved tableau G^{**}. Let F be the matrix which describes this pivot process (F is an identity matrix except for the ρth column, which is as shown below). Here $a_j^i(h)$ is the element in row h of column (j, i) of A (see Section 4.1). Let $Q^* = FQ$; then $G^{**} = FG^* = Q^*G$.

$$F = \begin{bmatrix} & -d_j^i/a_j^i(\rho) & \\ & -a_j^i(1)/a_j^i(\rho) & \\ & \cdot & \\ & \cdot & \\ & \cdot & \\ \cdots & 1/a_j^i(\rho) & \cdots \\ & \cdot & \\ & \cdot & \\ & \cdot & \\ & -a_j^i(m)/a_j^i(\rho) & \end{bmatrix} \rho$$

(column ρ)

We now have:

1. a new basic sequence S^* for the master problem which is obtained from S by replacing $s_\rho = (j_\rho, i_\rho)$ by $s_\rho^* = (j_0, i_0)$,
2. a new matrix multiplier

$$Q^* = \begin{bmatrix} 1 & [\pi^{*T} & \theta^{*T}] \\ 0 & & R^* \end{bmatrix}.$$

This completes an iteration. The proof of convergence is the same as that for the simplex algorithm.

For a numerical example, suppose that in solving a master problem the cost function for the first block [compare (10.10)] is

$$(C_1^T + \pi^T A_{01})X_1 + \theta_1 t_1^i = -x_1 - x_2 - x_3 + 6$$

and that the system L_1 [compare (10.11)] is

$$A_1 = \begin{bmatrix} 1 & 0 & -1 \\ 0 & 1 & -1 \end{bmatrix}, \quad B_1 = \begin{bmatrix} 1 \\ 1 \end{bmatrix}, \quad X_1 = \begin{bmatrix} x_1 \\ x_2 \\ x_3 \end{bmatrix}.$$

Then the subproblem is

$$\text{minimize } -x_1 - x_2 - x_3 + 6$$

subject to

$$\begin{bmatrix} 1 & 0 & -1 \\ 0 & 1 & -1 \end{bmatrix} \begin{bmatrix} x_1 \\ x_2 \\ x_3 \end{bmatrix} = \begin{bmatrix} 1 \\ 1 \end{bmatrix}$$

$$X_1 \geq 0.$$

Solving by means of tableaus gives:

−6	−1	−1	−1
1	1	0	−1
1	0	1	−1
−4	0	0	−3
1	1	0	−1
1	0	1	−1

In the second tableau the third column has a negative cost element and each entry in the third column is negative. T_1 is unbounded and the vector X_1 where $x_1 = -(-1)$, $x_2 = -(-1)$, $x_3 = 1$ generates an extreme ray for the cone defined by $A_1X_1 = 0$, $X_1 \geq 0$. In general, if the columns corresponding to $x_1, \cdots, x_m$ are basic and if corresponding to x_h there is a negative cost element and a nonpositive column, then

$$[-a_{1h} - a_{2h} \cdots -a_{m_1h} \quad 0 \cdots 1 \cdots 0]^T,$$

where 1 appears in the hth component, generates an extreme ray.

We now take this vector to be $X_1^{i_1}$, where $i_1 > k_1$, and introduce $(1, i_1)$ into the basic sequence by listing the column

$$G_1^{*i_1} = QG_1^{i_1} = Q\begin{bmatrix} C_1^T X_1^{i_1} \\ A_{01}X_1^{i_1} \\ 0 \\ \cdot \\ \cdot \\ \cdot \\ 0 \end{bmatrix}$$

in G^* and proceeding as in Case II or Case III.

If, in contrast to this, $X_1^{i_1}$ had been an extreme vector for T_1 (that is, $i_1 \leq k_1$), then we would have inserted the column

$$G_1^{*i_1} = QG_1^{i_1} = Q\begin{bmatrix} C_1^T X_1^{i_1} \\ A_{01}X_1^{i_1} \\ 1 \\ 0 \\ \cdot \\ \cdot \\ \cdot \\ 0 \end{bmatrix}$$

into G^*.

10.4. COMMENTS ON COMPUTATION.

It is usually not necessary at each iteration to optimize each of the subproblems L_j^*, since we need only determine a basic feasible program $X_{j_0}^{i_0}$ for which $z_{j_0} < 0$. We can therefore modify the algorithm and reduce considerably the number of computational steps in each iteration. For example, we might reduce computation time by terminating the solution of subproblems once a basic program $X_{j_0}^{i_0}$ has been found for which the value of the objective function z_{j_0} is negative even though it is not optimal.

It should be observed that each subproblem has a new objective function at each iteration. However, the system of constraints does not change, so one can store a basic sequence and corresponding matrix multiplier from the previous iteration and use it to generate an initial basic feasible program for the current iteration. This means that we need to employ Phase 1 procedures (see Section 4.2) at most once for the constraining system of each subproblem.

10.5. DETERMINING AN INITIAL BASIC FEASIBLE PROGRAM.

We use artificial variables w_i to find an initial basic feasible program for the master problem which we now write in the form

$$\text{minimize } z = d_0 + \sum_{j=1}^{r} D_j^T \Lambda_j \tag{10.12}$$

subject to

$$\begin{aligned} A\Lambda &= A_0, \\ \Lambda &\geq 0, \end{aligned} \tag{10.13}$$

where

$$A_0 = \begin{bmatrix} B_0 \\ 1 \\ \cdot \\ \cdot \\ \cdot \\ 1 \end{bmatrix}$$

and A is the submatrix of G given in Figure 10.2. We can assume that $B_0 \geq 0$ in the original problem since, if necessary, we can multiply through by -1. Now insert artificial variables and obtain the following Phase 1 problem:

$$\text{minimize } w_0 = \sum_{i=1}^{p_0+r} w_i \tag{10.14}$$

<table>
<tr><td>0</td><td>0</td><td>1</td><td>⋯</td><td>1</td></tr>
<tr><td rowspan="2" colspan="2">G</td><td>0</td><td>⋯</td><td>0</td></tr>
<tr><td colspan="3">I_{p_0+r}</td></tr>
</table>

Figure 10.3 Tableau for phase (1).

subject to

$$\begin{aligned} A\Lambda + I_m W &= A_0, \\ \Lambda &\geq 0, \\ W &\geq 0. \end{aligned}$$

This problem is displayed in tableau form in Figure 10.3. The problem can be solved as a new decomposed problem with the original constraints on X. An initial feasible basis consists, as usual, of artificial vectors. If at the end of Phase 1 the minimum value of (10.14) is positive, then no feasible program for the original problem exists. If the minimum value of (10.14) is zero, a basic feasible program for the original problem is determined. In this case either artificial vectors are in the basis at a zero level or no artificial vectors are in the basis. In the latter case we discard the artificial columns and proceed to the algorithm developed above. In the former case we retain the basic artificial columns, assign a value of zero to the corresponding variables in the function (10.12) and proceed to the iterative algorithm. If at any subsequent iteration an incoming column corresponding to λ_j^i has a nonzero entry in row h in which an artificial variable appears, we pivot in the position (h, j). This removes the corresponding artificial variable from the program and we can then ignore the corresponding column (the pivot element itself can be positive or negative since an artificial variable is in a Phase 2 basis at zero level).

For an example, suppose Phase 1 has been completed and that one artificial variable, w_1, is basic with a value of zero. Suppose also that the remaining basic columns correspond to λ_1^1 and λ_2^1 and that the master problem, after premultiplication by the matrix Q, appears in the following tableau form:

		λ_1^1 ⋯	λ_2^1 ⋯	w_1	
	$-d_0'$	0	0	0	
	4	1	0	0	$p_0 = 1$
$G =$	3	0	1	0	$r = 2$
	0	0	0	1	$h = 3$

Suppose the first block has an optimal program with a negative value so that a new column corresponding to λ_1^2 is to be introduced into the basis and suppose that transforming this column by Q gives the column $[-2 \quad 3 \quad 2 \quad -4]^T$. This is shown in the following more complete representation of G:

		λ_1^1	λ_1^2	$\cdots$	λ_2^1	$\cdots$	w_1
	$-d_0'$	0	-2		0		0
	4	1	3		0		0
$G =$	3	0	2		1		0
	0	0	-4		0		1

The entry -4 is in a row containing an artificial variable, so we can pivot at position (3, 2) (note that the entry for this row in the A_0 column is zero). We get the following tableau:

		λ_1^1	λ_1^2	$\cdots$	λ_2^1	$\cdots$	w_1
	$-d_0'$	0	0		0		$-1/2$
	4	1	0		0		3/4
$G^* =$	3	0	0		1		1/2
	0	0	1		0		$-1/4$

The artificial variable w_1 is no longer basic so the corresponding column can be discarded in subsequent iterations.

Note that if the incoming column (after transformation by Q) had had a zero entry in row 3, no pivot operation would make w_1 nonbasic; therefore it is possible in the case of degeneracy or redundancy to obtain an optimal program having an artificial variable at a zero level. We could, nevertheless, determine an optimal Λ and corresponding program X.

10.6. UPPER-BOUND CONSTRAINTS.

A variable in a linear programming problem can be in any one of the following four classes: (1) it can be a free variable, (2) it can have a lower bound but no upper bound, ($x \geq l$), (3) it can have an upper bound but no lower bound ($x \leq k$), (4) it can have both a lower and an upper bound ($l \leq x \leq k$). By some simple algebraic manipulations these four classes reduce to three: free variables, nonnegative variables, and nonnegative variables having an upper bound (finite). For example, suppose $x \geq l$; then we can define a new variable $x' = x - l$ and have $x' \geq 0$. Also, if $x \leq k$ we introduce $x' = k - x$ and again we have $x' \geq 0$. Finally, if $l \leq x \leq k$, then we can take $x' = x - l$ and have $0 \leq x' \leq k - l$. Moreover, since free variables can be treated by the procedures introduced in

Chapter 3, variables with upper and lower bounds can be handled in full generality in terms of the following formulation:

$$\text{minimize } z = d + C_1^T X_1 + C_2^T X_2 \tag{10.15}$$

subject to

$$A_1 X_1 + A_2 X_2 = B, \tag{10.16}$$

$$X_1 \geq 0, \tag{10.17}$$

$$X_2 \geq 0, \tag{10.18}$$

$$X_1 \leq K, \tag{10.19}$$

where $K > 0$, A_1 is p by n_1, A_2 is p by n_2, $n = n_1 + n_2$. We assume that slack variables have been inserted as needed to obtain (10.16). This problem can be solved, at least theoretically, by the simplex algorithm. One would insert an additional slack variable for each component of the vector X_1 and proceed. However, this can enlarge the problem greatly and lengthen considerably the computing time required for solution. Indeed, since X_1 has n_1 components, this would increase the matrix of constraints by n_1 rows and the number of variables by n_1.

We now introduce a procedure developed by Dantzig (Reference 12) which is a modification of the simplex algorithm and which solves this problem efficiently. We will make extensive use of substitutions of the form $x_i' = k_i - x_i$, which for obvious geometrical reasons are called *reflections*. Since X_1 is of degree n_1 there are 2^{n_1} substitutions each involving reflections $x_i' = k_i - x_i$ for certain of the n_1 indices i. To be explicit, consider an n_1 vector

$$I = \begin{bmatrix} i_1 \\ \cdot \\ \cdot \\ \cdot \\ i_{n_1} \end{bmatrix},$$

where each i_j is zero or one. We call I an *indicator vector* and associate with it the transformation $T(I)$ which leaves x_j unchanged if $i_j = 0$ and replaces x_j by $x_j' = k_j - x_j$ if $i_j = 1$. In particular, the indicator vector

$$I_0 = \begin{bmatrix} 0 \\ \cdot \\ \cdot \\ \cdot \\ 0 \end{bmatrix}$$

yields the identity transformation which leaves each of the x_j unchanged. Denote the original problem (10.15) through (10.19) by H and define $H(I)$ to be the problem obtained by applying the transformation $T(I)$ to H. In particular $H = H(I_0)$. The 2^{n_1} problems so obtained are equivalent.

We may associate with the problem H a new problem H_R defined by (10.15) through (10.18), that is, a problem in which the upper-bound constraints have been removed. Correspondingly, with each equivalent problem $H(I)$ we associate the problem $H(I)_R$ obtained by removing its upper-bound constraints. We call the totality of 2^{n_1} problems $H(I)_R$ the *set of problems related to* H.

For example, suppose H is the problem

$$\text{minimize } -x_1 - x_2 + 5$$

subject to

$$\begin{aligned} x_1 + 3x_2 + x_3 &= 2, \\ 2x_1 - x_2 + x_4 &= 5, \\ x_j &\geq 0, \qquad (j = 1, \cdots, 4) \\ x_1 &\leq 10, \\ x_2 &\leq 20, \end{aligned}$$

in which $n_1 = 2$, $n_2 = 2$, $n = 4$. Let $I = \begin{bmatrix} 0 \\ 1 \end{bmatrix}$; then $T(I)$ replaces x_2 by $20 - x_2'$ and the new equivalent problem $H(I)$ is

$$\text{minimize } - x_1 + x_2' - 15$$

subject to

$$\begin{aligned} x_1 - 3x_2' + x_3 &= -58, \\ 2x_1 + x_2' + x_4 &= 25, \\ x_1 &\geq 0, \\ x_2' &\geq 0, \\ x_3 &\geq 0, \\ x_4 &\geq 0, \\ x_1 &\leq 10, \\ x_2' &\leq 20. \end{aligned}$$

If $\begin{bmatrix} X_1 \\ X_2 \end{bmatrix}$ is an optimal feasible program for H we can reflect each x_i which is at its capacity k_i and obtain an equivalent problem $H(I)$ with an optimal feasible program in which every nonzero variable x_i is below its capacity and hence is also an optimal feasible program for the corresponding related problem $H(I)_R$.

Our strategy consists of the following steps.

Step 1. Begin with a basic feasible program for some related problem $H(I)_R$ which is also feasible for $H(I)$ and therefore for H. If this program is optimal for $H(I)_R$, then we are done.

Step 2. If this program is not optimal, then the simplex algorithm will indicate some subscript q and corresponding variable x_q as a candidate for entering the basis in the problem $H(I)_R$. As usual, we would like to make x_q as large as possible but we must now consider three possible types of bounds for x_q: Case (a), if x_q is a bounded variable we cannot exceed its bound; Case (b), as x_q increases some basic variables may decrease, leading to the usual simplex bound for x_q; Case (c), some of the basic variables may increase with x_q and none of these can be allowed to exceed its bound. According as Case (a), Case (b), or Case (c) yields the smallest bound we reflect on x_q, we carry out the usual pivot operation, or we reflect on x_i and enter q in place of i in the basic sequence following an appropriate pivot.

A sequence of pivots and reflections will enable us to reach an optimal program if one exists.

Step 3. To determine if a basic feasible program for a related problem exists, we pivot in a basis without regard to satisfying (10.17), (10.18), or (10.19). Then if in the corresponding basic solution any variables exceed their upper bounds we proceed as indicated below to obtain a new related problem in which the only departure from canonical form is that certain of the constants are negative. At this stage we consider a feasible subproblem of maximal size and apply the ideas of Chapter 3 to either demonstrate infeasibility or obtain a basic feasible solution for problem H. In solving the subproblem we use Steps 1 and 2 above.

Suppose then that we have obtained a feasible basic solution to problem H_R and that for some h, $b_h > k_{s_h}$. We reflect on x_{s_h} and then pivot in position (h, s_h) to obtain a new problem with $b_h^* = k_{s_h} - b_h < 0$ and $b_i^* = b_i$ for all $i \neq h$. (The pivot operation actually requires only changing signs in row h.) This process is applied until we have $B \leq K_S$.

Before developing the algorithm we introduce the following notation: we write the related problem H_R as in Figure 10.4 where two additional rows are appended at the top of the tableau to remind us that $H = H(I)$ for

$$I = \begin{bmatrix} 0 \\ \cdot \\ \cdot \\ \cdot \\ 0 \end{bmatrix}$$

and that in the upper-bound problem the vector X_1 has the bounds indicated (these rows will not enter into the calculations explicitly). The superscripts zero identify the original data given in (10.15) through (10.19) and are used so that the symbols without numerical superscripts can be used to indicate a general stage in the problem (compare Figure 10.5).

	K^T	
	$0 \cdots 0$	
$-d^0$	C_1^{0T}	C_2^{0T}
B^0	A_1^0	A_2^0

Figure 10.4 Tableau for the problem H_R.

Using the example introduced above, the tableau for H_R is:

	10	20		
	0	0		
−5	−1	−1	0	0
2	1	3	1	0
5	2	−1	0	1

More generally, Figure 10.5 indicates a tableau for the related problem $H(I)_R$. Note that the indicator row tells which reflections have been made and that the top row does not change. The body of the augmented tableau is, of course, changed in accordance with whatever reflections and pivots have been made.

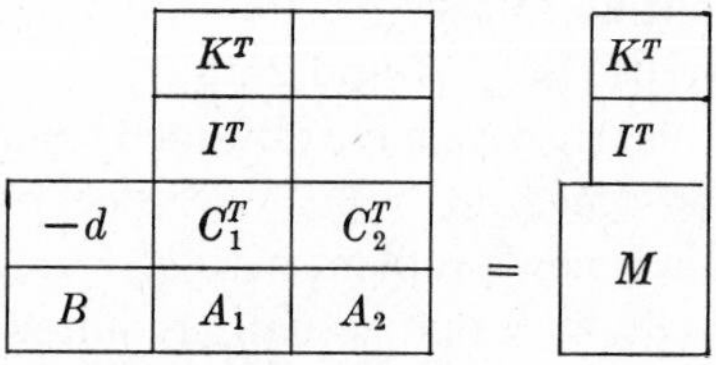

	K^T			K^T
	I^T			I^T
$-d$	C_1^T	C_2^T	$=$	M
B	A_1	A_2		

Figure 10.5 Tableau for the problem $H(I)_R$.

Following the strategy set forth above we begin with the assumption that some related problem $H(I)_R$ is in canonical form with respect to a basic sequence $S = (s_1, \cdots, s_p)$ and that the corresponding basic vector X^S has $X_1^S \leq K$ (X^S is also feasible for H). If now $C \geq 0$, we have an optimal solution. Otherwise, there exists an index q such that $c_q < 0$. Let

$$y_1 = \begin{cases} k_q & \text{if } q \leq n_1 \\ \infty & \text{if } q > n_1, \end{cases}$$

$$y_2 = \min\left\{\frac{b_i}{a_{iq}} \mid a_{iq} > 0\right\},$$

$$y_3 = \min\left\{\frac{b_i - k_{s_i}}{a_{iq}} \mid a_{iq} < 0,\ s_i \leq n_1\right\},$$

with the understanding that in case there is no positive a_{iq} we take $y_2 = \infty$ and similarly if there is no negative a_{iq} we take $y_3 = \infty$.

Let $y = \text{minimum } \{y_1, y_2, y_3\}$. If $y = \infty$, there is no lower bound for the objective function and we stop. Otherwise, Step 2 of the strategy separates into three cases according as $y = y_1$ or $y = y_2$ or $y = y_3$. We consider case (1), $y = y_1$; we carry out the reflection

$$x_q = k_q - x_q'$$

and consider the new augmented tableau (Figure 10.6).

K^T
I^{*T}
M^*

Figure 10.6 Subsequent tableau for a problem $H(I)_R$.

It differs from that in Figure 10.5 only in columns 0 and q. For column 0 we have

$$M_0^* = M_0 - k_q M_q$$

and for column q we have

$$M_q^* = -M_q \qquad \text{and} \qquad i_q^* = 1 - i_q.$$

The change from M to M^* can be described in terms of matrix multiplication as

$$M^* = MG_q$$

where G_q is given as

$$G_q = \begin{bmatrix} 1 & & & & & & \\ & \cdot & & & & & \\ & & \cdot & & & & \\ & & & \cdot & & & \\ -k_q & & & -1 & & & \\ & & & & \cdot & & \\ & & & & & \cdot & \\ & & & & & & \cdot \\ & & & & & & 1 \end{bmatrix} \tag{10.20}$$

where rows $0, 1, \cdots, q-1, q+1, \cdots, n$ of G_q are equal to the corresponding rows of the identity matrix I_{n+1} and row q of G_q has $g_{q_1} = -k_q$, $g_{qq} = -1$ and all other elements zero. G_q is called a *simple reflection matrix* (we say simple because only one variable is reflected). Note that the basis is not changed although the indicator vector is changed. We now return to Step 1 with the new augmented tableau of Figure 10.6.

If $y = y_2 = \dfrac{b_n}{a_{nq}}$ [Case (b)], we pivot on position (h, k), obtain a new basis S^*, retain the same indicator vector I and return to Step 1 with the resulting new augmented tableau.

If $y = y_3 = (b_n - k_{s_n})/a_{hq}$ [Case (c)], then first reflect on x_{s_q} and second pivot at position (h, q). Replace s_h by q in the basic sequence, replace i_{s_h} by $1 - i_{s_h}$ in the indicator vector, and return to Step 1. (Note that both basic sequence and indicator vector are changed.)

Except for degenerate problems there is strict improvement in the value of the objective function in all three cases. Degeneracy could be handled using a lexicographic argument but for practical purposes random breaking of ties in the choice of h will suffice to assure convergence of the algorithm.

A flow chart implementing the strategy would begin with Step 3 to construct an initial basic feasible solution and proceed to Steps 1 and 2.

10.7. AN ILLUSTRATIVE EXAMPLE.

Suppose we have the following problem, where the upper bound vector K^T appears at the top of the tableau,

	1	4	3	1	5
	0	0	0	0	0
3	0	0	−3	2	−2
5	1	0	1	2	0
2	0	1	1	1	−1

The corresponding basic program is infeasible since $x_1 = 5$ and the upper bound on x_1 is $k_1 = 1$. We reflect on x_1,

$$x_1' = k_1 - x_1 = 1 - x_1,$$

and pivot on position (1, 1), obtaining the following tableau (the vector K^T, since it is unchanged, is not reproduced further in this illustration)

	1	0	0	0	0
3	0	0	−3	2	−2
−4	1	0	−1	−2	0
2	0	1	1	1	−1

We have $k_4 = 1 < 2$, so we use row 1 of the matrix M as a temporary objective function and reflect on x_4, getting

	1	0	0	1	0
1	0	0	−3	−2	−2
−2	1	0	−1	2	0
1	0	1	1	−1	−1

Now $q = 3$ and we pivot on position (2, 3); the result is

	1	0	0	1	0
4	0	3	0	−5	−5
−1	1	1	0	1	−1
1	0	1	1	−1	−1

We have an unbounded subproblem so we pivot at position (1, 5). The resulting tableau provides a first feasible solution

	1	0	0	1	0
9	−5	−2	0	−10	0
1	−1	−1	0	−1	1
2	−1	0	1	−2	0

Here $q = 4$, and we are in Case (c), $y = y_3$. Hence we reflect on x_3 and then pivot at position (2, 4). The corresponding results are as follows, the first tableau indicating the results of the reflection and the second those of the pivot operation,

	1	0	1	1	0
9	−5	−2	0	−10	0
1	−1	−1	0	−1	1
−1	−1	0	−1	−2	0

,

	1	0	1	1	0
14	0	−2	5	0	0
3/2	−1/2	−1	1/2	0	1
1/2	1/2	0	1/2	1	0

.

Now $q = 2$; we see that $x_5 = 3/2 + x_2$, $k_2 < 3/2 + k_4$; hence we reflect on x_5 and pivot on position (1, 2):

	1	0	1	1	1
14	0	−2	5	0	0
−7/2	−1/2	−1	1/2	0	1
1/2	1/2	0	1/2	1	0

,

	1	0	1	1	1
21	1	0	4	0	2
7/2	1/2	1	−1/2	0	−1
1/2	1/2	0	1/2	1	0

.

The second of these tableaus provides an optimal solution

$$X = \begin{bmatrix} 1 \\ 7/2 \\ 3 \\ 1/2 \\ 5 \end{bmatrix}$$

for which $z = -21$.

10.8. THE REVISED MODIFIED ALGORITHM.

We have discussed an algorithm for solving any linear programming problem with upper and lower bounds on the activities involved without introducing the activity bounds as constraint inequalities in the tableau. The purpose for developing this algorithm was to increase the efficiency of computer solutions for such problems. However, the algorithm as presented above needs further modification before it is efficient for computer calculations. These modifications resemble those used in developing the revised simplex method of Chapter 4 from the simplex algorithm of Chapter 3. Thus, beginning with the problem H_R we introduce a full set of artificial variables and carry out a Phase 1 procedure which either provides a basic feasible solution for H_R or demonstrates its infeasibility (and therefore also that of H). After a basic feasible program is attained which contains no artificial variables, we can proceed to determine an optimal program by the procedures discussed above. At any intermediate point we need only store the matrix of transition from the initial tableau (including the artificial columns) to the current tableau.

One matter that still requires discussion is the relationship between pivots and reflections. Let M be the full tableau matrix at the beginning of Phase (2) of the solution procedure and let M^0 be the initial tableau matrix prior to addition of artificial variables. Then $M = Q^0M^0$ for some nonsingular matrix Q^0 (which may be specified in product form with pivot matrices as its factors). We know that a pivot operation replaces M by FM, where F is a pivot matrix and a reflection on x_q replaces M by MG_q where G_q is defined in (10.20). Since $(FM)G_q = F(MG_q)$ we see that the order of carrying out a pivot and a reflection is immaterial. Moreover, the reflections generate a commutative group of order 2^{n_1} whose elements are in one-to-one correspondence with the indicator vectors. Indeed, if I is an indicator vector, then the matrix G_I defined by

row zero $= [1 \quad 0 \quad \cdots \quad 0]$;
if $i \in I$, row $i = [-k_i \quad 0 \quad \cdots \quad 0 \quad -1 \quad \cdots \quad 0]$,
with the -1 in position (i, i);
if $i \notin I$, row $i = [0 \quad \cdots \quad 1 \quad \cdots \quad 0]$,
with the 1 in position (i, i);

has the property that

$$MG_I$$

is the tableau obtained from M by carrying out all of the reflections designated by I. It follows that if M^* is obtained from M by a sequence of pivots and reflections then we may write

$$M^* = QMG_I$$

where I is the current indicator vector and Q is the product of all of the pivot matrices in the sequence. Thus we see that the algebraic effect of reflections can readily be incorporated into the revised simplex method either in full inverse form or in product form. Of course, a flow chart for the algorithm must take account of the situations which involve reflections rather than just pivots.

PROBLEMS

10.1. Solve the following problem:

$$\text{maximize } f = x_1 + x_2 - x_3 + 2x_4$$

subject to

$$\begin{aligned} x_1 + x_2 + 2x_3 - x_4 &\leq 9, \\ x_1 + 2x_2 &\leq 8, \\ x_1 + x_2 &\leq 5, \\ x_3 + 3x_4 &\leq 5, \\ 3x_3 + 2x_4 &\leq 6, \\ x_j &\geq 0, \qquad (j = 1, \ldots, 4). \end{aligned}$$

10.2. Solve the following problem:

$$\text{maximize } g = 2x_1 - x_2 - x_3 + x_4$$

subject to

$$\begin{aligned} 3x_1 + x_2 + 2x_3 + x_4 &\leq 10 \\ -2x_1 + x_2 &\leq 3 \\ x_1 - x_2 &\leq 4 \\ x_3 + x_4 &\leq 7 \\ 9x_3 + 6x_4 &\leq 54 \\ x_j &\geq 0 \qquad (j = 1, \ldots, 4). \end{aligned}$$

10.3. Solve the following problem:

$$\text{maximize } z = 100x_1 - 75x_2 + 80x_3$$

subject to

$$\begin{aligned} 2x_1 + 2x_2 + 3x_3 &\leq 40, \\ 4x_1 + x_2 + 2x_3 &\leq 40, \\ 3x_1 + x_2 + 4x_3 &\leq 40, \\ x_1 &\geq 0, \\ x_2 &\geq 0, \\ x_3 &\geq 0, \\ x_1 &\leq 6, \\ x_2 &\leq 5. \end{aligned}$$

10.4. Solve the following problem:

$$\text{maximize } z = 60x_1 + 300x_2$$

subject to

$$\begin{aligned} x_1 + 8x_2 &\leq 120, \\ 5x_1 + 15x_2 &\leq 300, \\ x_1 &\geq 0, \\ x_2 &\geq 0, \\ x_1 &\leq 15, \\ x_2 &\leq 10. \end{aligned}$$

10.5. Minimize $z = x_1 + 2x_2 - 3x_3 + 5x_4 + 7x_5$
subject to

$$\begin{aligned} x_1 + x_2 - 5x_3 - 7x_4 + 10x_5 &\leq 10, \\ 3x_1 - 5x_2 - 8x_3 + x_4 - 2x_5 &\leq 6, \\ 2x_1 + 3x_2 + x_3 + 5x_4 + x_5 &\leq 9, \\ x_j &\geq 0, \qquad (j = 1, \ldots, 5), \\ x_3 &\leq 5. \end{aligned}$$

10.6. Minimize $z = -10x_1 - 12x_2$
subject to

$$\begin{aligned} 3x_1 + 4x_2 &\leq 23, \\ 5x_1 + 2x_2 &\leq 20, \\ x_1 &\geq 0, \\ x_2 &\geq 0, \\ x_1 &\leq 2, \\ x_2 &\leq 4. \end{aligned}$$

10.7. Minimize $z = -2x_1 - 5x_2$
subject to

$$\begin{aligned} 3x_1 + x_2 + x_3 &= 21, \\ x_1 + x_2 + x_4 &= 9, \\ x_1 + 4x_2 &\leq 24, \\ x_j &\geq 0, \qquad (j = 1, \ldots, 4), \\ x_1 &\leq 3, \\ x_2 &\leq 2. \end{aligned}$$

Foundations, Sets, and Functions

Appendix A

A.1. SETS.

In developing the mathematical tools used in model building we could begin with topics such as matrices, inequalities, boolean algebra, differential and difference equations, computer programming, probability, and statistics. We choose here, however, to begin with set theory for several reasons.

First, on logical grounds, we must recognize that mathematics is a language and as such it must logically begin with certain undefined terms and concepts. Although logicians and specialists in foundations may prefer to dig deeper, it is customary for mathematicians to take the concepts of "set" and "member of a set" as undefined primitives and to build from there by logical argument and construction.

Second, on systematic grounds, a close examination of the model building process discloses that set theory is actually the most universally applicable topic in mathematics. Thus to bypass set theory would naturally lead to a dangerously incomplete concept of model building. Moreover, we may regard set theory as setting up the basic part of the "language" we shall use in building mathematical models of real-world situations which we wish to study. There are, indeed, many situations in which the more common language aspects of model building provide the main contribution. This is especially true in situations where abstract mathematical formulation brings to light significant analogies between apparently unrelated problems and thus enables us to apply already developed techniques to new problems.

Third, on pedagogical grounds, the use of set concepts makes the model-building process structurally more precise and teachable. We illustrate this by reference to a common early experience of all students in model building, to wit, handling "word problems." Clearly, a word problem is a two-part exercise consisting of a model formulation stage followed by a solution stage. The difficulty commonly attributed to word problems is concentrated in the first stage, and this difficulty may well spring from the absence of explicit use of set concepts in the format. The customary admonition to the student to use "common sense" in interpreting a word problem (for example avoidance of negative mass) usually conceals an implicit instruction to determine carefully for each variable involved its revelent set of values. Thus, teaching model building without set concepts is about on a par pedagogically with teaching arithmetic using roman numerals.

We begin our formal development of mathematics with *set* and *member of a set* as primitive, undefined terms. This does not mean, however, that these terms are meaningless nor that they cannot be illustrated. For example, the employees of the Ford Motor Company form a set, as do the members of the United States House of Representatives. Moreover, any individual either is or is not a Ford employee (we set aside such cases as an ostensible Ford employee who is actually a spy in the pay of General Motors). A Ford employee might also belong to the set of all Republicans. A firm has a set of inputs (raw materials, capital, and labor) and a set of outputs. A business is owned by a set of stockholders and is controlled by a set called the Board of Directors. We can speak of the set of grains of sand on a beach or of the set of hairs on a man's head.

Thus, intuitively, a *set* is a collection of objects of any sort. The words *collection, class, family, ensemble, aggregate* are synonyms for set. We will ordinarily designate sets by capital letters. Each object in a set is called a *member* (*element*) of the set. We ordinarily use lower case letters to designate elements of sets.

If S is a set and b is an element of S then we write

$$b \in S;$$

this is read, "b belongs to S," or "b is an element of S." If b does not belong to (lie in) S we write

$$b \notin S.$$

From the logical point of view, the terms "set" and "element of a set" are undefined; we do take as the fundamental axiom (postulate or assumption) of set theory that for any set S and element b exactly one of the statements "$b \in S$" or "$b \notin S$" is true. In a logical development of mathematics from set theory, one might proceed first to study sets and then to build up the number system. Since our purpose here is not to give a complete logical treatment, we bypass most of this development, retaining only those parts which seem essential for understanding the material in the book.

Thus, we shall assume general familiarity with numbers and use certain collections of numbers as further illustrations of sets. These sets occur so frequently that we shall assign certain letters to designate them throughout Appendix A. They are

- $\boldsymbol{Z}$: the set of *counting numbers* or *positive integers* z ($\boldsymbol{Z}$ consists of the numbers $1, 2, 3, \cdots$);
- $\boldsymbol{J}$: the set of all *integers* j; this set includes all of the elements of $\boldsymbol{Z}$ together with zero and the negative integers;
- $\boldsymbol{Q}$: the set of *rational numbers* $q = j/z$, where j is an integer and z is a positive integer;
- $\boldsymbol{R}$: the set of all *real numbers* r;
- $\boldsymbol{C}$: the set of all *complex numbers* $c = x + yi$ where x and y are real and $i^2 = -1$.

Each of these sets contains every element of the preceding one if we identify the real number r with the complex number $r + 0i$, and the fraction $j/1$ with the integer j. The real numbers can be characterized as unending or infinite decimals, and the rational numbers then appear as those decimals which are eventually periodic, that is, repeat in blocks. For example, $1/7 = .142857142857\cdots$, $50/3 = 16.6666\cdots$, and $1/81 = .01234567890123456789\cdots$ are rational numbers. Another way in which we frequently picture real numbers is in terms of points on a line. We mark an arbitrary point on a line to be the origin, select an arbitrary unit of measurement, and then relate a positive number r to the point at distance r to the right of the origin and the negative number $-r$ to the point at distance r to the left of the origin. All of the elements of the sets $\boldsymbol{R}$, $\boldsymbol{Q}$, $\boldsymbol{J}$, and $\boldsymbol{Z}$ are thereby related to points on the line. To interpret elements of $\boldsymbol{C}$ geometrically, two dimensions are needed.

SET NOTATION.

A convenient and generally accepted method for designating a set containing only a few elements is to list its elements between braces. Thus

$$T = \{\text{black, blue, brown, purple}\}$$

is the set whose elements are the four colors listed; and

$$S = \left\{3, \frac{2}{5}, 0, -17\right\}$$

is the set whose elements are the numbers 3, 2/5, 0, and -17.

For infinite sets, we obviously cannot list all of the members. We still use braces, however, and include within them the condition or conditions of set membership; this device is also used for finite sets when individual listing is inconvenient—especially for sets with many elements. For example,

$$M = \{m \mid m \text{ is a Mason}\}$$

indicates the set of all Masons (this is finite but large). Again

$$\boldsymbol{Z} = \{z \mid z \text{ is a positive integer}\};$$

this set is also designated (less precisely) as

$$\boldsymbol{Z} = \{1, 2, 3, \cdots, n, \cdots\}.$$

We write

$$S = \{s \mid s \text{ has property } P,\ s \text{ has property } Q, \cdots\}$$

to indicate that S consists of precisely those elements which possess all of the properties listed. For example, one could think of S as being a club whose members are individuals s who meet the membership conditions "s is over 21," "s has been nominated and elected," "s has paid his dues," and "s has not violated any of the rules which would call for expulsion." What we are saying is that a set is completely determined by its membership conditions.

The collection of symbols $\{s \mid \cdots \text{properties} \cdots\}$ is called the *set builder*.

By the *order* of a set we mean the number of elements in it; we denote the order of S by the symbol $O(S)$. For sets such as $\boldsymbol{Z}$, $\boldsymbol{J}$, $\boldsymbol{Q}$, $\boldsymbol{R}$, and $\boldsymbol{C}$ the order is infinite. The sets T and S listed above each have order 4, and the set M of Masons has finite order (although it is unlikely that anyone knows exactly what this order is).

QUANTIFIERS.

There are two important classes of statements concerning sets for which we introduce a special notation. Consider the equation $x^2 + 3x - 1 = 0$. The statement, "there exists at least one real root of this equation," is an example of an *existential* statement. On the other hand the statement, "for every real number x, $x^2 - 1 = (x - 1)(x + 1)$," is a *universal* statement.

For an arbitrary set S we introduce the symbol $\exists$ and by the statement,

$$\exists\, x \in S, \qquad x \text{ has property } P,$$

we mean, "there exists an element x in S such that x has property P." We also introduce the symbol $\forall$ and by the statement,

$$\forall\, x \in S, \qquad x \text{ has property } P,$$

we mean, "for every $x \in S$, x has property P." We call $\exists$ the *existential quantifier*, and call $\forall$ the *universal quantifier*. Note that $\exists\, x$ calls for at least one x and does not deny the possibility of several elements with the stated property.

Several quantifiers may appear in a single sentence; when this happens the sequence of occurrence is important. For example, the statement

$$\forall\, x \in \boldsymbol{R} \quad \exists\, y \in \boldsymbol{R}, \qquad x + y = 10$$

means that for every real number x there is at least one real number y (that is, $10 - x$) such that $x + y = 10$; this is a valid statement. On the other hand the statement

$$\exists\, x \in \boldsymbol{R} \quad \forall\, y \in \boldsymbol{R}, \qquad x + y = 10$$

is false since it would require a fixed value of x for which $x + y = 10$ would hold for all real numbers y.

The negation of an existential statement is a universal statement and vice versa. Thus the negation of the statement, "there exists a cow which is purple," is the statement, "for every cow its color is not purple," or more simply, "no cow is purple." Similarly, the negation of the universal statement, "all cows are red," is the statement, "there exists at least one cow which is not red."

Confusion in the use of quantifiers is rife in every-day discourse as well as in that of beginning mathematics students. Careful use of the quantifier notation can be an aid in avoiding such confusion and marks a beginning in precise mathematical discourse.

A.2. SUBSETS.

Consider the sets of numbers

$$A = \{1, 3, 7, 9\}, B = \{1, 3, 7, 9, 11\}, C = \{1, 3, 9, 7, 3, 1\}.$$

Every element of A is also an element of B; thus we say that A is a subset of B. More generally, if two sets S and T have the property that every member s of S is also a member of T, then we say that S is a *subset* of T and write

$$S \subset T \qquad \text{or} \qquad T \supset S.$$

These are read, respectively, "S is contained in T," and "T contains S."

If both $S \subset T$ and $T \subset S$, then we say that S and T are *equal*, and write

$$S = T.$$

In other words, two sets are equal if every element of each is an element of the other, in other words, if they consist of exactly the same elements. For example, in the sets listed above, $A = C$. Note that neither the order in which the elements are written down nor the number of times an element is repeated within the braces is significant for set equality.

If S is not equal to T, we write $S \neq T$. For two sets to be unequal it is necessary and sufficient that at least one of them should contain an element which does not belong to the other.

If S is a subset of T and $S \neq T$, then we say that S is a *proper subset* of T, and write $S < T$ or $T > S$. These are read, respectively, "S is properly contained in T," and "T contains S properly." Note that for S to be a proper subset of T, every element of S must also be an element of T, and in addition, there must be at least one element of T which is not in S. For example, in the sets above $A < B$, since $A \subset B$ and 11 is in B but is not in A.

Consider the set

$$S = \{s \mid s \text{ is a counting number and } s \text{ is divisible by } 3\}.$$

To belong to S, an element s must satisfy two conditions, the first of which guarantees that $s \in \boldsymbol{Z}$; thus $S \subset \boldsymbol{Z}$. Since there are many counting numbers that are not divisible by 3 (for example, 2) we can write $S < \boldsymbol{Z}$. One could also write

$$S = \{s \mid s \in \boldsymbol{Z} \text{ and } s \text{ is divisible by } 3\},$$

and this form displays the fact that S is a subset of $\boldsymbol{Z}$.

More generally, if S is any set one writes

$$T = \{t \mid t \in S \text{ and } t \text{ has properties } P, Q, \cdots\}$$

to indicate that T is the subset of S consisting of those elements which have the further properties $P, Q, \cdots$.

For another example, consider the set S of items s for sale in Woolworth's. Then one might be interested in the subset

$$T = \{t \mid t \in S \text{ and } t \text{ was a profit-making item in each of the past five years}\}.$$

One could also write

$$W = \{w \mid w \in S \text{ and was a profit-maker last year, and } w \text{ was a loss maker last year}\}.$$

It is obvious in this case that W cannot have any members at all. We still call it a set: the *empty set* or *void set*. The existence of the empty set is a great convenience in making statements about sets since it protects one from having to know in advance whether or not there are any elements which possess some string of properties. In our example above, the set T might be empty if we included a depression year, but this fact would not always be evident in advance. On the other hand, to be able to say that some set is not empty is sometimes very important. Thus, one might be interested in the set of profitable products to manufacture and to know that this set is not empty is an encouragement to potential investors.

We use the symbol ϕ to designate the empty set. There is only one empty set; it makes no difference whether we are considering the set of no potatoes, or the set of no numbers, or the set of no people; by our definition of equality for sets, these are all just ϕ. Moreover, ϕ is a proper subset of every nonempty set S.

The set $A = \{a, b\}$ has $4 = 2^2$ subsets, $\{a\}$, $\{b\}$, ϕ, A itself and no others. It is not difficult to show that a set S having n elements has exactly 2^n subsets. This fact motivates the notation 2^S for the collection of all subsets of a given set S. The collection 2^S is called the *power set* of S (we use the word "collection" here to avoid the phrase "set of subsets," although that phrase is also correct). In dealing with collections whose elements are themselves sets, one is embarking upon the development of a heirarchy of sets which is the source of certain well-known paradoxes in the foundations of mathematics. To avoid some of these paradoxes we must distinguish between an element a and the set $\{a\}$. The need for this distinction is also demonstrated by the examples of ϕ and the set $\{\phi\} = 2^\phi$; the former has no elements and the latter is a collection containing exactly one element, namely the set ϕ.

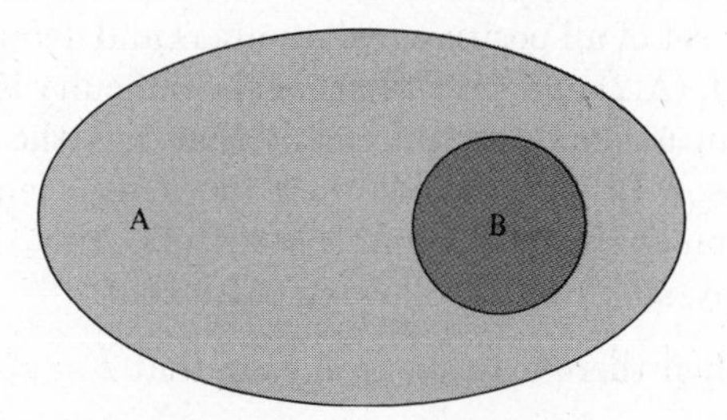

Figure A.1 Set and subset.

An important source of sets is provided by geometry, where we encounter various sets of points beginning with lines and circles and leading up to more complicated curves and then to regions. Geometric diagrams (called Venn diagrams) can be helpful in reasoning about sets even in nongeometric problems. Thus Figure A.1, a Venn diagram, can be interpreted as a region A and a subregion B in the plane, or it can also be used to talk about an arbitrary set A with a given subset B. In using such geometric aids, one must be careful to avoid drawing conclusions from geometric intuition without validating them on purely logical grounds.

A.3. MATHEMATICAL INDUCTION.

Although we forego a detailed development of the number system, some remarks on the set $\boldsymbol{Z}$ of counting numbers seem appropriate since the entire tower of number systems rests on the structure and properties of this system. Our first task is to translate our intuitive understanding of counting numbers into a set of axioms which characterize the system.

Intuitively, the counting numbers consist of what one can obtain by counting, beginning with 1 and proceeding indefinitely. If we picture the counting numbers arranged in a long string, each number has a unique successor and each number except 1 has also a unique predecessor. For each integer z we may designate its successor by z^+. Actually $z^+ = z + 1$, but addition will be a defined concept in our axiomatic system and hence should not be used in our axioms. The successor operation has the following properties.

For $z \in \boldsymbol{Z}$ there exists a unique $z^+ \in \boldsymbol{Z}$. (A.1)

There exists an element $1 \in \boldsymbol{Z}$ such that there is no $z \in \boldsymbol{Z}$ with

$$z^+ = 1. \tag{A.2}$$

$$\text{If } z_1^+ = z_2^+ \text{ then } z_1 = z_2. \tag{A.3}$$

These properties alone are not enough to characterize the integers. For example, consider the set of all positive real numbers and define $r^+ = r + 1$. This set satisfies (A.1), (A.2) and (A.3). Part of the difficulty is that we have not yet introduced an axiomatic requirement that 1 is the only element which is not a successor. In our example there are many elements without predecessors; for example, any real number r with $0 < r \leq 1$. We can remove this difficulty by making the following statement.

$$\text{If } z \neq 1 \text{ then there exists } z_1 \in \boldsymbol{Z} \text{ such that } z = z_1^+. \tag{A.4}$$

Unfortunately, this is still not enough, for consider the set $S = \{\boldsymbol{Z}, J'\}$ where $J' = \{j + \frac{1}{2} \mid j \in \boldsymbol{J}\}$; that is, J' is the set of all halves of odd numbers (positive or negative). Then define $s^+ = s + 1$ for all $s \in S$ and (A.1), (A.2), (A.3), (A.4) are all satisfied.

We need to introduce the property that every number in $\boldsymbol{Z}$ can be reached by counting beginning with 1. This is done by the axiom of induction:

If S is a subset of $\boldsymbol{Z}$ for which

(a) $1 \in S$

and

(b) If $k \in S$ then $k^+ \in S$,

then $S = \boldsymbol{Z}$. (A.5)

Intuitively, this axiom seems acceptable since by (a) $1 \in S$ and then by (b) $1^+ = 2 \in S$. Again by (b) $2^+ = 3 \in S$, and proceeding in this manner we can reach any counting number.

It can be proved that these five axioms are categorical; in other words, there is at most one system which satisfies them. However, it cannot be proved that any system exists which does satisfy them; stated differently there is at most one system of counting numbers but we cannot *prove* that there is at least one such system. Thus almost all of mathematics rests on an act of faith, namely the assumption that the set $\boldsymbol{Z}$ exists.

Once the positive integers are available the extensions of $\boldsymbol{Z}$ to $\boldsymbol{J}$, $\boldsymbol{Q}$, $\boldsymbol{R}$, and $\boldsymbol{C}$ present no logical difficulties although the step from $\boldsymbol{Q}$ to $\boldsymbol{R}$ is not without mathematical complexities.

Starting with the five axioms one can define addition and multiplication and derive their usual properties. One can then define $a < b$ to mean that there exists $c \in \boldsymbol{Z}$ for which $b = a + c$, and in terms of $<$ there are two useful alternate formulations of the induction principle.

If S is a subset of $\boldsymbol{Z}$ for which

(a) $1 \in S$,

and

(b) If $k \in S$ for all $k < n$ then $n \in S$,

then $S = \boldsymbol{Z}$. (A.6)

If S is a nonempty subset of $\boldsymbol{Z}$ then S has a least or smallest number. (A.7)

The property (A.7) is sometimes described by the statement that $\boldsymbol{Z}$ is a *well-ordered set.*

One important application of the induction axiom is in proofs of theorems. For example, let $n \in \boldsymbol{Z}$ and consider the proposition

$$P(n) : 1^2 + 2^2 + \cdots + n^2 = \frac{n(n+1)(2n+1)}{6}.$$

Let S be its *truth set:*

$$S = \{n \mid n \in \boldsymbol{Z},\ P(n) \text{ is true}\}.$$

Clearly $1 \in S$, since $1^2 = 1(1+1)(2+1)/6$; thus S satisfies part (a) of (A.6).

Next, suppose that for some n, S contains all $k < n$; in particular $(n-1) < n$ and hence $P(n-1)$ is true. Now

$$\begin{aligned} 1^2 + 2^2 + \cdots + n^2 &= [1^2 + 2^2 + \cdots + (n-1)^2] + n^2 \\ &= \frac{(n-1)(n)(2n-1)}{6} + n^2 \qquad (\text{since } P(n-1) \text{ is true}) \\ &= \frac{n[(n-1)(2n-1) + 6n]}{6} \\ &= \frac{n(n+1)(2n+1)}{6} \end{aligned}$$

and hence the truth of $P(n)$ follows from the truth of $P(k)$ for all $k \in n$; this establishes part (b) of (A.6) and we conclude that $S = \boldsymbol{Z}$, or in other words $P(n)$ is true $\forall\, n \in \boldsymbol{Z}$.

Remark 1. The proof as it stands could be readily adapted to fit the original induction axiom (A.5) and might seem simpler in that form. However, there are many cases where establishment of P_n depends on the truth of P_k for several values of k and not just for $k = n - 1$; for such cases proofs cannot be based directly on (A.5). On the other hand (A.6) can be used for any case which can be handled using (A.5); hence it seems pedagogically more efficient to concentrate on learning to use the more general formulation.

Remark 2. Note that proof by induction applies to theorems which can be broken down into individual statements, one for each $n \in \boldsymbol{Z}$. However, if in (A.6) we replace (a) by (a′) $m \in S$, then (a′) and (b) are sufficient to establish that S contains all integers $n \geq m$ and this holds even for m a negative integer or zero.

A second important application of the induction axiom is in definition of functions with domain $\boldsymbol{Z}$ (see Section A.8 below for the definitions of *function* and *domain*). For example, consider the factorial function $n \to n!$ We recall that

$$1! = 1,\ 2! = 1 \cdot 2,\ 3! = 1 \cdot 2 \cdot 3, \cdots.$$

An inductive definition of $n!$ is provided by the statements,

(a) $1! = 1$;
(b) $\forall\, n > 1, \quad n! = (n - 1)! \cdot n$.

Remark 3. Strictly speaking, any symbol such as $1^2 + 2^2 + \cdots + n^2$ is meaningful only via inductive definition: $1^2 + 2^2 + \cdots + (n - 1)^2 + n^2 = [1^2 + 2^2 + \cdots + (n - 1)^2] + n^2$. In such straightforward situations it is our custom to use the three dots $\cdots$ to indicate both the ellipsis and the implied inductive definition.

A third important application of induction is in the logical construction of computer programs involving *loops*. The introduction of loops enables the program writer to utilize one instruction (or one collection of instructions) many times. Typically one will carry out an instruction for $i = s$ and then go on to $i = s + 1$. In this application the induction axiom guarantees that the program will eventually reach some terminal value $i = n$. Pedagogically, the writing of computer programs is a useful aid in understanding induction.

A.4. OPERATIONS ON SETS.

Let S and T be two sets. Then the *union* of S and T is the set of all elements which lie in at least one of S and T. The union is designated $S \cup T$ and we can write formally

$$S \cup T = \{u \mid u \in S \quad \text{or} \quad u \in T\}.$$

Synonyms for union are *logical sum* and *join*. The *intersection* of S and T, designated by $S \cap T$, is the set of elements that lie in both S and T; formally,

$$S \cap T = \{u \mid u \in S \quad \text{and} \quad u \in T\}.$$

Synonyms for intersection are *logical product* and *meet*. The symbols $\cup$ and $\cap$ are called, respectively, *cup* and *cap*.

Note that in the definition of union—and generally in mathematics—the word "or" is used in the inclusive sense and thus an element which lies in both S and T also lies in their union $S \cup T$. The exclusive "or", meaning one or the other but not both, is given a logical position in the introduction of the symmetric difference to be defined below; however, we shall always use the word or itself in the inclusive sense.

The exclusive or is illustrated by the following sentence: "I have room in my car for one more person; if either John or Mary wishes to come I will be delighted." The inclusive or occurs in the following sentence: "If it rains or snows we will call the picnic off." Clearly, it is not likely that a picnic will be held if it both rains and snows.

The *difference* of S and T, designated by $S - T$, is the set of elements that lie in S but not in T; formally

$$S - T = \{u \mid u \in S \quad \text{and} \quad u \notin T\}. \tag{A.8}$$

If T is a subset of S, then $S - T$ is called the *relative complement* of T in S and is denoted by $C_S T$; formally, $C_S T$ is defined by the right-hand side of Equation (A.8). In case all sets under consideration are subsets of some set U we call U the *universal set*. Then we shorten $C_U T$ to $C(T)$ and call it the *complement* of T. Then we may shorten the previous expression to

$$C(T) = \{u \mid u \notin T\}.$$

The *symmetric difference* of S and T, designated by $S \oplus T$, is the set of elements that lie in exactly one of S and T; formally

$$S \oplus T = \{u \mid (u \in S \quad \text{and} \quad u \notin T) \quad \text{or} \quad (u \in T \quad \text{and} \quad u \notin S)\}$$

or equivalently,

$$S \oplus T = (S - T) \cup (T - S).$$

We observe that the only way in which the symmetric difference can be the empty set is to have $S = T$. Note also that $S - T = \phi$ is equivalent to the statement "S is a subset of T."

We illustrate union, intersection, difference, symmetric difference, and complement in Figure A.2, and also by the following example; let

$$A = \{2, 4, 6, 8\}, B = \{3, 6, 9\}, D = \{2, 4\}.$$

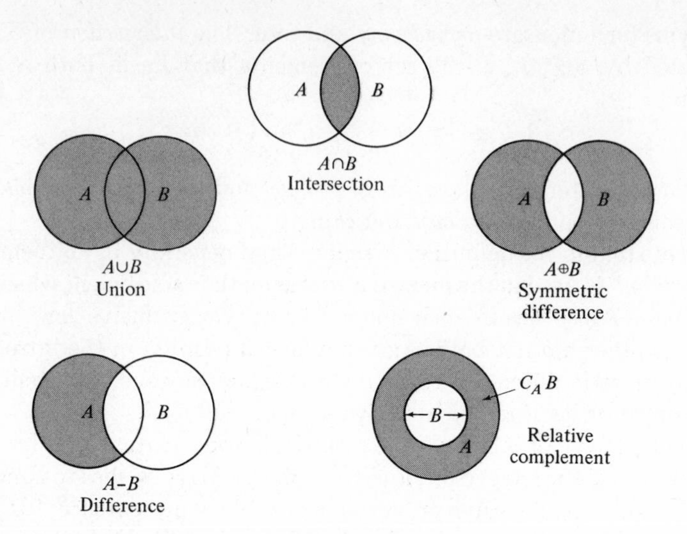

Figure A.2 Set operations.

Then

$$
\begin{aligned}
A \cup B &= \{2, 3, 4, 6, 8, 9\}, \\
A \cap B &= \{6\}, \\
A - B &= \{2, 4, 8\}, \\
A \oplus B &= \{2, 3, 4, 8, 9\},
\end{aligned}
$$

and

$$C_A D = \{6, 8\}.$$

The union of a finite collection $T = \{S_1, S_2, \cdots, S_n\}$ of sets is defined to be the set consisting of all elements which belong to at least one of the sets in the collection. In symbols we write

$$\bigcup_{\alpha=1}^{n} S_\alpha = \{t \mid \text{there is at least one } S_\alpha \text{ for which } t \in S_\alpha\}.$$

If we denote by $A = \{1, 2, \cdots, n\}$ the set over which the index varies, we can also denote this union by

$$\bigcup_{\alpha \in A} S_\alpha; \tag{A.9}$$

we also sometimes denote it by

$$\bigcup_{S \in T} S. \tag{A.10}$$

We do not require that notation for a set be limited to one label. For example, let A be the set of children in an elementary school and let T be the set of rooms. Then S_α means α's room. For two students α and β the statement $S_\alpha = S_\beta$ means that α and β are in the same room.

These concepts and notations can be readily extended to the case of arbitrary collections of sets. Thus if

$$T = \{S \mid S \text{ has certain properties}\}$$

we may wish to introduce an index set $A = \{\alpha\}$ for use as labels for the elements of T and then write

$$T = \{S_\alpha \mid \alpha \in A\}.$$

Then the two notations (A.9) and (A.10) for union can be employed and both mean the set of all elements t which lie in at least one set in the collection T.

Similarly, if A (and therefore also T) is not empty we write

$$\bigcap_{\alpha \in A} S_\alpha \qquad \text{or} \qquad \bigcap_{S \in T} S$$

for the set of all elements t which lie in each member of the collection T; this new set is called the intersection of the sets in T.

If for a given argument, we restrict ourselves to a given universe U, then there is an important *duality* between union and intersection:

$$C(S \cup T) = C(S) \cap C(T) \qquad \text{and} \qquad C(S \cap T) = C(S) \cup C(T).$$

These are known as De Morgan's Rules (see Figure A.3). According to them we see that to obtain the complement of any combination involving unions and intersections of sets, one replaces each set by its complement, all unions by intersections, and all intersections by unions.

As another illustration of De Morgan's Rules, consider the following case: $U = \{1, 2, 3, 4, 5, 6\}$, $S = \{1, 2, 3\}$, $T = \{3, 4, 5\}$. Then $S \cup T =$

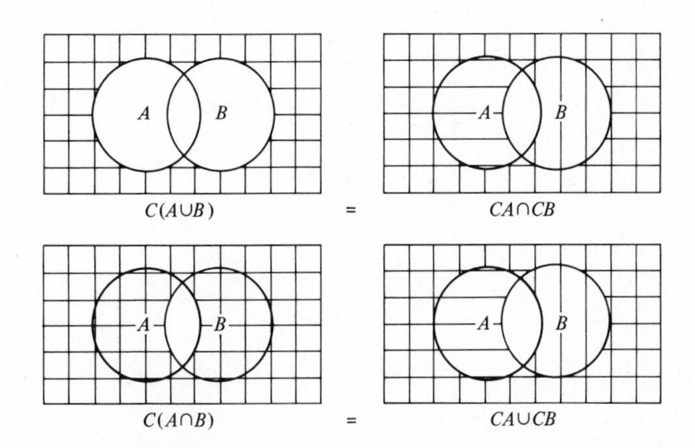

Figure A.3 De Morgan's rules.

$\{1, 2, 3, 4, 5\}$, $S \cap T = \{3\}$, $C(S) = \{4, 5, 6\}$, and $C(T) = \{1, 2, 6\}$. Now $C(S \cup T) = \{6\} = C(S) \cap C(T)$ and $C(S \cap T) = \{1, 2, 4, 5, 6\}$ $= C(S) \cup C(T)$.

From the way in which complements are defined, it is clear that for all subsets S of U

$$C(C(S)) = S$$

and

$$S \cup C(S) = U, S \cap C(S) = \phi.$$

A.5. CARTESIAN PRODUCTS.

In analytic geometry we encounter the idea of describing points by assigning them numbers called coordinates. Thus (5, 1) stands for the point whose abscissa (x coordinate) is 5 and whose ordinate (y coordinate) is 1 (see Figure A.4). This point is not the same as (1, 5); in contrast with listing set elements, the order in which these numbers are written has significance. Repetitions such as (3, 3) are also significant and refer to points on the line $y = x$.

More generally, if n is any positive integer, we use the notation

$$(a_1, a_2, \cdots, a_n)$$

to designate an *ordered set* of any n numbers (repetitions permitted), the terms *n-tuple* and *vector of degree n* are also used. (In Appendix B and in the text we have used column matrices rather than n-tuples to designate vectors. A reason for this is the greater convenience of column matrices in dealing with systems of linear equations.) Two n-tuples $(a_1, a_2, \cdots, a_n)$ and $(b_1, b_2, \cdots, b_n)$ are said to be *equal* if and only if $a_1 = b_1$, $a_2 = b_2, \cdots, a_n = b_n$. The case $n = 3$ is familiar from the geometry of three dimensions, although

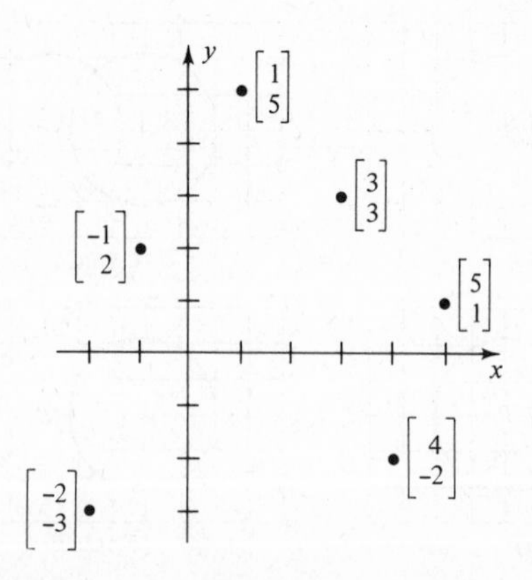

Figure A.4 Cartesian coordinates.

here, to designate a point, we frequently write (x, y, z) instead of (a_1, a_2, a_3). The development of the theory and properties of numerical n-tuples belongs to linear algebra; our interest here is with more general types of n-tuples.

If A and B are any two sets we call the set of all ordered pairs (a, b) with $a \in A$ and $b \in B$ the *Cartesian product* of A and B and designate it by $A \times B$; formally,

$$A \times B = \{(a, b) \mid a \in A \quad \text{and} \quad b \in B\}. \tag{A.11}$$

Note, in particular, that for the set $\boldsymbol{R}$ of real numbers, $\boldsymbol{R} \times \boldsymbol{R}$ is the set of all points in the plane. The name Cartesian is in honor of René Descartes, the founder of analytic geometry.

As a further example let $A = \{a, b\}$ and $B = \{1, 2, 3\}$. Then $A \times B$ consists of the six ordered pairs

$$(a, 1), (a, 2), (a, 3), (b, 1), (b, 2), (b, 3).$$

In general, as in this example, for finite sets A and B we have

$$O(A \times B) = O(A) \cdot O(B); \tag{A.12}$$

the order or number of elements in $A \times B$ is equal to the product of the orders of A and B. This equality is a reason for the word "product" in the term Cartesian product.

Suppose that $A = \{a_1, a_2, a_3\}$ and $B = \{b_1, b_2, b_3, b_4, b_5\}$. Then we can represent $A \times B$ in (the tabular form) given by Figure A.5. This rectangular display is convenient for many purposes and can even be used symbolically when A or B are of infinite order. The relation (A.12) holds also for infinite sets provided multiplication is suitably interpreted.

In higher-dimensional geometry and in vector problems we need more than two coordinates and thus are led to consider Cartesian products of more than two factors. The Cartesian product of the sets $S_1, S_2, \cdots, S_n$ in the order given is the set

$$S_1 \times S_2 \times \cdots \times S_n = \mathop{\times}_{i=1}^{n} S_i$$
$$= \{(s_1, s_2, \cdots, s_n) \mid s_1 \in S_1, s_2 \in S_2, \cdots, s_n \in S_n\}.$$

The special case $\boldsymbol{R} \times \boldsymbol{R} \times \cdots \times \boldsymbol{R}$ (n factors) in which each S_i is the real field $\boldsymbol{R}$ is the set V_n of all vectors of degree n.

	b_1	b_2	b_3	b_4	b_5
a_1	(a_1, b_1)	(a_1, b_2)	(a_1, b_3)	(a_1, b_4)	(a_1, b_5)
a_2	(a_2, b_1)	(a_2, b_2)	(a_2, b_3)	(a_2, b_4)	(a_2, b_5)
a_3	(a_3, b_1)	(a_3, b_2)	(a_3, b_3)	(a_3, b_4)	(a_3, b_5)

Figure A.5 Rectangular display of Cartesian product.

If $S_1, S_2, \cdots, S_n, \cdots$ is an infinite sequence of sets, we denote the Cartesian product by

$$\mathop{\times}_{i=1}^{\infty} S_i;$$

this is the set of all infinite sequences $(s_1, \cdots, s_n, \cdots)$ where $s_i \in S_i$ $(i = 1, \cdots, n, \cdots)$. Still more generally if $P = \{p, \cdots\}$ is any index set and $T = \{S_p \mid p \in P\}$ is any collection of sets we denote by

$$\mathop{\times}_{p \epsilon P} S_p$$

the set C of all collections $(\cdots, S_p, \cdots)$ of one element s_p from each set S_p and again call this set C a Cartesian product.

The critical reader may object that our definition of ordered pair (a, b) is not itself entirely in set language since the concepts of "first" and "second" have not been defined but were taken from intuition. It can be shown, however, that the concept of ordered pair does not require the use of concepts outside of set theory. First we remark that our definition must meet the qualification that $(a, b) = (c, d)$ if and only if $a = c$ and $b = d$. Now, for $a \in A$ and $b \in B$ we define

$$(a, b) = \{\{a\}, \{a, b\}\}. \tag{A.13}$$

Now suppose that $(a, b) = (c, d)$. If $a \neq b$ then (a, b) consists of two elements, one a set which properly contains the other. Then clearly $(a, b) = (c, d)$ requires $\{a\} = \{c\}$, $\{a, b\} = \{c, d\}$, hence $\{b\} = \{a, b\} - \{a\} = \{c, d\} - \{c\} = \{d\}$, and so $a = c$, $b = d$. If $a = b$ then $(a, b) = \{\{a\}\}$ and so (c, d) must also reduce to a single one element set $\{c\}$ and $a = c$. Thus in all cases $(a, b) = (c, d)$ requires $a = c$, $b = d$.

Once an ordered pair has been defined the concept of ordered triple is easy, for example, $(a, b, c) = ((a, b), c)$, and the ordered n-tuple $(a_1, \cdots, a_n)$ can be defined inductively as $((a_1, \cdots, a_{n-1}), a_n)$. Now that we have shown that the concept of ordered pair (and ordered n-tuple) can be based on purely set-theoretic grounds we never refer again to (A.13), but use (A.11).

A.6. RELATIONS.

The concept of relation is central (at least implicitly) in almost every application of mathematics to the real world. One observes certain connections or relationships between persons, or prices, or colors, or events or experiments in a laboratory and then may attempt to capture the essence of these relations by a mathematical model. We now give a mathematical definition of relation and introduce some of the important properties of relations.

Let A and B be two sets. A subset R of $A \times B$ is said to be a relation in $A \times B$.

This formal definition should be supplemented by a more intuitive one. A relation R between the elements of A and B picks our certain ordered pairs (a, b) which are related by R whereas all other ordered pairs are not so related.

For example, let $A = \{1, 2, 4, 6, 8\}$ and $B = \{1, 2, 3, 4\}$ and let the relation R pick out those pairs (a, b) for which $a = 2b$; then R consists of the following four (out of the total set of 20) elements in $A \times B$:

$$R = \{(2, 1), (4, 2), (6, 3), (8, 4)\}.$$

For another example, consider the used car dealer whose stock consists of the set

$A = \{$Ford, Chevrolet, Plymouth, Rambler, Buick, Chrysler, Lincoln, Volkswagen, and Cadillac$\}$.

Under current marketing practices advertised prices usually end in the digits 99; thus we take for the set of possible car prices (in dollars) the set

$$B = \{99, 199, 299, \cdots, 4999\}.$$

In this case the Cartesian product $A \times B$ represents the set of all possible car-price pairings. However, the relation

$R = \{$(Ford, 399), (Chevrolet, 499), (Plymouth, 399), (Rambler, 699), (Buick, 299), (Chrysler, 399), (Lincoln, 1399), (Volkswagen, 699), (Cadillac, 3999)$\}$

represents an assignment by the dealer of a price to each of his cars. The relation R in this case then completely describes the dealer's selling prices.

The usual set membership notation applies for relations. Thus for the first example above we write

$$(2, 1) \in R \qquad \text{and} \qquad (2, 2) \notin R$$

to indicate that the members of the first pair are related and the second are not. We also write

$$2\,R\,1 \qquad \text{and} \qquad 2 \not{R}\, 2$$

and read these respectively, "2 is in the relation R to 1," and "2 is not in the relation R to 2." This phrasing is more in accord with everyday experience with relations. In the used car example, the phrase "Ford R 399" can be read "the Ford sells for 399 dollars." Similarly, the sentences "John loves Mary," "Peter is the son of Mark," "$7 < 9$," "Beethoven is better known than Pergolesi," "brown is a more intense color than chartreuse," "line a is parallel to line b," all are in the form $a\,R\,b$ where a and b are nouns and the relation R is indicated by a verbal phrase. Later we will encounter other verbal variants for the expression of relations.

To summarize, in order to define a relation R we must first specify two sets A and B and then indicate which ordered pairs (a, b) are included in R and which are excluded from R.

Since relations are sets, equality of relations is defined as set equality. Thus if R and S are two relations in $A \times B$ we say that $R = S$ if for every pair $(a, b) \in A \times B$, $a\, R\, b$ implies $a\, S\, b$ and $a\, S\, b$ implies $a\, R\, b$. Similarly, we say that $R \subset S$ if $a\, R\, b$ implies $a\, S\, b$. Thus the smaller the set R the stricter (or stronger) the relation is. In this sense the universal relation $U = A \times B$ is the weakest of all relations in $A \times B$, whereas the empty relation ϕ is the strongest. We have along with the relations R and S also their union $R \cup S$ and their intersection $R \cap S$ as relations.

For example, suppose that $A = B = \boldsymbol{J}$; let $a\, R\, b$ mean that $a - b$ is divisible by 3, and let $a\, S\, b$ mean that $a - b$ is divisible by 7. Then $a(R \cap S)\, b$ means that $a - b$ is divisible by both 3 and 7 (and hence by 21), and a $(R \cup S)\, b$ means that $a - b$ is divisible by at least one of 3 and 7. Note in this case that $R \cup S > R > R \cap S$.

We come now to a rather tricky point. We defined equality of two relations R and S in $A \times B$ to be set equality, but what do we say when R and S are equal as sets but R is in $A \times B$ and S is in $A^* \times B^*$ where $A^* \times B^*$ is not equal to $A \times B$? There is no universally accepted answer to this question. In this book we will take the point of view that equality of R and S as relations means equality as sets without reference to their containing sets $A \times B$ and $A^* \times B^*$. In other words, we regard a relation as consisting of a set of ordered pairs; thus the statement "R is a relation in $A \times B$" merely specifies that A and B are large enough so that $A \times B$ contains R. Making either of them still larger does not change R.

For example, let $A = B = \boldsymbol{R}$, let $A^* = B^* = \{x \mid x \text{ is real and } -1 \leq x \leq 1\}$, let $R = \{(x, y) \mid x^2 + y^2 = 1,\ (x, y) \in A \times B\}$, and let $S = \{(x, y) \mid x^2 + y^2 = 1,\ (x, y) \in A^* \times B^*\}$. Then according to our convention $R = S$. For another example, let $A = B = \boldsymbol{Q}$, let $A^* = B^* = \boldsymbol{Z}$, let $R = \{x \mid x \in \boldsymbol{Q} \text{ and } x^2 - 3x + 2 = 0\}$, and let $S = \{x \mid x \in \boldsymbol{Z} \text{ and } x^2 - 3x + 2 = 0\}$. Then $R = \{1, 2\} = S$. The concepts of "range" and "domain" introduced below will provide the smallest possible sets B and A for a given R.

Suppose that A, A^*, B, B^* are sets with $A^* \subset A$ and $B^* \subset B$. Then $A^* \times B^*$ is a relation in $A \times B$. Now let R be any relation in $A \times B$ and define the relation R^* by

$$R^* = R \cap (A^* \times B^*),$$

that is, R^* is the set of all ordered pairs (a, b) for which $a\, R\, b$, $a \in A^*$, and $b \in B^*$. According to our convention R^* can be regarded as being a relation in $A \times B$ or in $A^* \times B^*$. We call R^* the *restriction* of R to $A^* \times B^*$

and call R an *extension* of R^* to $A \times B$. Note that whereas given A, B, A^*, B^*, a given relation R in $A \times B$ can have only one restriction to $A^* \times B^*$, an R^* in $A^* \times B^*$ can have many extensions R in $A \times B$. In the two examples above $S = R^*$ and also $R = R^*$, that is, R is equal to its restriction. For a more typical example, let $A = B = \boldsymbol{Q}$, let $A^* = B^* = \boldsymbol{Z}$, and let R be $<$. Then $R^* = R \cap (A^* \times B^*)$ consists of all ordered pairs (a^*, b^*) of positive integers for which $a^* < b^*$.

We now introduce the concept of "transpose" which helps give the first and second members of a Cartesian product equal status. For an example, let A be a set of women and B a set of men. The relation $a\ R\ b$ in $A \times B$ which states, "a is the wife of b" is important in our society. If we wish to put the husband first and say equivalently "b is the husband of a," we then have a relation in $B \times A$. This new relation we denote by R^T and call it the transpose of R. Note that R^T is merely R read in reversed or transposed order, hence the name. Moreover, R and R^T state exactly the same basic facts, namely which couples are husband and wife and which are not.

More generally, if A and B are any sets and R is a relation in $A \times B$ then we define the *transpose* R^T of R to be the relation defined by

$$R^T = \{(b, a) \mid a\ R\ b\}.$$

Some authors use the term "inverse" instead of transpose. We prefer to restrict the term inverse for the special case of the transpose of a one-to-one correspondence (see Section A.8).

We list some further examples of transposes:

1. $A = B = \boldsymbol{Z}$, R is the relation "less than" $<$, then R^T is the relation $>$. Thus $2\ R\ 3$ and $3\ R^T\ 2$, or, as we usually write it, $2 < 3$ and $3 > 2$.
2. A and B are sets of men, R is the relation "is the father of," then R^T is the relation "is a son of."
3. $A = B =$ a set of people, R is the relation "is a cousin of," then R^T is the same as R.
4. It can be shown (see Section B.4) that the transpose of a matrix is a special case of the transpose of a relation (this is one reason for our preference for the term "transpose" as opposed to "inverse").

Note that in two of the examples illustrating the concept of transpose we had $A = B$. This is an important special case in the theory of relations. When $A = B$ we abbreviate the expression "R is a relation in $A \times A$" to "*R is a relation in A*."

If A, B, C are three sets, a subset of $A \times B \times C$ is called a *relation in* $A \times B \times C$; this is described as a *ternary relation* because there are three factors in the Cartesian product. Similarly, one can speak of m-ary relations in Cartesian products having m factors. Our initial case of relations in

$A \times B$ comprises *binary relations*. Ordinarily, it is clear from the context how many factors are involved in the Cartesian product, but in the case of equal factors encountered above the expression, "R is a relation in A," might leave some question as to whether R is a subset of $A \times A$ or of $A \times A \times A$ and so we sometimes say more precisely, "R is a binary relation in A," or, "R is a ternary relation in A," to indicate the number of factors.

Let us return to the "is the wife of" relation introduced above. In the set B there may be some men who are married to women in A and others who are not, that is, who are bachelors, are widowers, or are married to women not in A. The subset of men married to women in A is given the name *range* of the relation R and can be defined formally as

$$\{b \mid b \in B \text{ and there exists } a \in A \text{ such that } a\, R\, b\}.$$

In general, let A and B be any two sets and let R be a relation in $A \times B$. For each $a \in A$ we denote by $R(a)$ the set of all $b \in B$ for which $a\, R\, b$; $R(a)$ is called the *image* of a under R. Thus in a monogamous society the image $R(a)$ of any woman a is either a one-man set or the empty set; however in a polyandrous society $R(a)$ could be larger.

For any subset S of A we denote by $R(S)$ the set of all $b \in B$ for which $a\, R\, b$ for at least one $a \in S$; we call $R(S)$ the image of S under R. In particular, the image of A under R, the set $R(A)$, is called the *range* of R.

For example, if R is the relation, "is the wife of," and S is a set of women, then $R(S)$ is the set of husbands of women in S.

Formally, we have

$$\begin{aligned} R(a) &= \{b \mid b \in B \text{ and } a\, R\, b\}, \\ R(S) &= \{b \mid b \in B \text{ and there exists } a \in S \text{ such that } a\, R\, b\} \\ &= \bigcup_{a \in S} R(a). \end{aligned}$$

Similarly, the *counterimage* of an element b under R is the set of all $a \in A$ for which $a\, R\, b$; this is the same as the set $R^T(b)$ and will be so designated. The counterimage of a subset V of B is the set of all $a \in A$ such that $a\, R\, b$ for at least one $b \in V$; this is the same as the set $R^T(V)$. In particular, the set $R^T(B)$ is called the *domain* of the relation R. Thus the domain of the relation R, "is a wife of," is the set of women in A who are married to men in B. A relation is sometimes called a *correspondence* between its domain and its range.

For example, let A and B be the sets shown in Figure A.6 and let the relation R be identified by the arrows between elements in A and B. Then $R = \{(a_1, b_1), (a_2, b_2), (a_2, b_4), (a_3, b_3), (a_4, b_6), (a_5, b_7)\}$. The range of the relation is the subset of B consisting of those elements of B that are related to elements of A under R, the set $\{b_1, b_2, b_3, b_4, b_6, b_7\}$. Also, consider the ele-

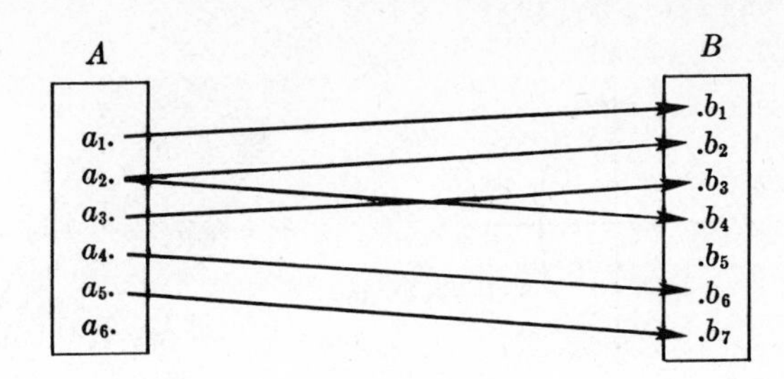

Figure A.6 A relation.

ment $a_2 \in A$; the *image* of a_2 under R is the set $\{b_2, b_4\}$ and the image of a_5 under R is the set $\{b_7\}$. The domain of the relation is $\{a_1, a_2, a_3, a_4, a_5\}$, a subset of A. The counterimage of $b_2 \in B$ is the set $\{a_2\}$.

If S is a relation in $\boldsymbol{R} \times \boldsymbol{R}$ we consider the *graph* of S to be the set of all points in the Cartesian plane whose coordinates (as ordered pairs) belong to S. Examples of relations in $\boldsymbol{R} \times \boldsymbol{R}$ are as follows. Let $x\,S\,y$ mean $\{(x, y) \mid 4x^2 + 9y^2 = 36\}$, $x\,Q\,y$ mean $\{(x, y) \mid 4x^2 + 9y^2 \leq 36\}$ and $x\,P\,y$ mean $\{(x, y) \mid 2x + 5y \leq 6\}$. Then

$$S(2) = \left\{\frac{\sqrt{20}}{3}, -\frac{\sqrt{20}}{3}\right\}, Q(2) = \left\{y \mid -\frac{\sqrt{20}}{3} \leq y \leq \frac{\sqrt{20}}{3}\right\},$$

$$\text{domain } S = \text{domain } Q = \{x \mid -3 \leq x \leq 3\},$$
$$\text{range } S = \text{range } Q = \{y \mid -2 \leq y \leq 2\},$$
$$\text{domain } P = \boldsymbol{R}, \text{ range } P = \boldsymbol{R}.$$

Also, S^T is obtained from S by interchanging the roles of x and y, so that $S^T = \{(x, y) \mid 9x^2 + 4y^2 = 36\}$. Likewise, $P^T = \{(x, y) \mid 5x + 2y \leq 6\}$. Geometrically speaking, the graph of the transpose of a relation is the reflection (mirror image) about the line $y = x$ of the graph of the relation, as is seen by comparing the graphs of S and P in Figure A.7 with those of S^T

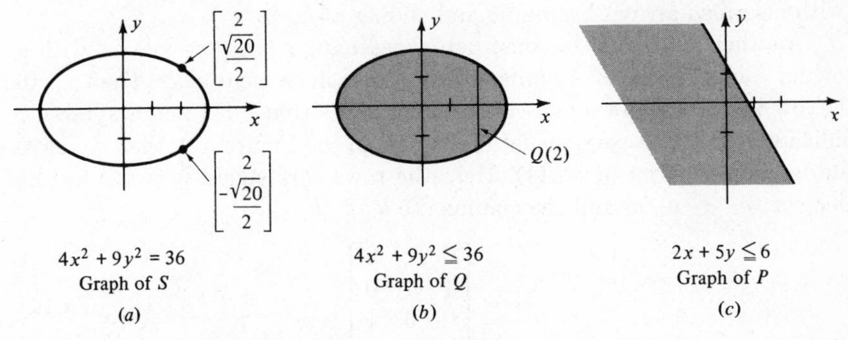

Figure A.7 Graphs of relations.

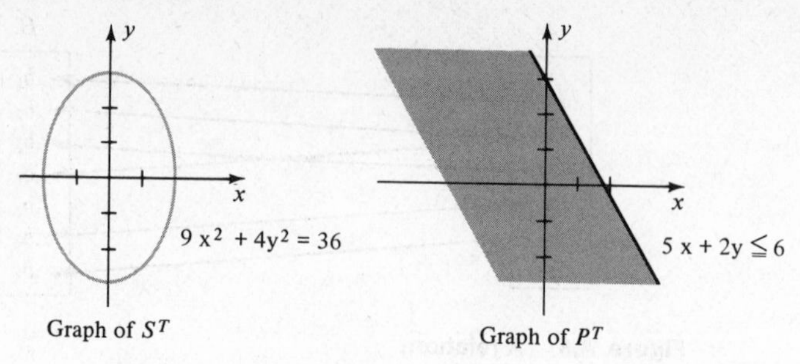

Figure A8 Graphs of relations.

and P^T in Figure A.8, respectively. The relations P and P^T are examples of linear relations.

Observe that for relations in $\boldsymbol{R} \times \boldsymbol{R}$ the range is the projection of the graph onto the y axis and the domain is the projection onto the x axis.

We now see that if R is a relation in $A \times B$ and if A^*, B^* are any sets for which $A \supset A^* \supset$ (domain R) and $B \supset B^* \supset$ (range R) then $R = R^*$ since

$$R^* = R \cap A^* \times B^* \supset R \cap (\text{domain } R \times \text{range } R) = R,$$

which, together with the universal inequality $R \supset R^*$ for restrictions of relations, yields $R = R^*$.

If $A = \{a_1, \cdots, a_s\}$ and $B = \{b_1, \cdots, b_t\}$ are finite sets then a relation R in $A \times B$ can be described by certain graphical and tabular devices. One method is to list the elements of A in one column and those of B in a parallel column and then join a_i to b_j if and only if $a_i \, R \, b_j$. This method is illustrated in Figures A.6, A.9, and A.11. If A and B are the same set then it is sometimes desirable to list each element only once, in polygonal form, and indicate by an arrow on the joining line which element is first and which second in the relation. Thus Figure A.10 illustrates the relation $R = \{(b_1, b_3), (b_1, b_4), (b_3, b_3), (b_4, b_1)\}$ on the set $B = \{b_1, b_2, b_3, b_4\}$.

Note that a segment joining a and b with arrows in each direction indicates that both $a \, R \, b$ and $b \, R \, a$; note also that $a \, R \, a$ is indicated by a loop (with inscribed arrow) beginning and ending at a.

Another method is to construct a rectangular array or matrix with a row for each element of A and a column for each element of B. Then a 1 in the intersection of row a and column b indicates that $a \, R \, b$ and a symbol 0 indicates $a \not R \, b$. For example, let R be given in the Figure A.9; then we have the matrix M given in (A.14). Here the rows correspond in order to the elements a_1, a_2, a_3, a_4 and the columns to b_1, b_2, b_3:

$$M = \begin{bmatrix} 0 & 1 & 0 \\ 1 & 1 & 0 \\ 0 & 0 & 1 \\ 1 & 0 & 1 \end{bmatrix}. \tag{A.14}$$

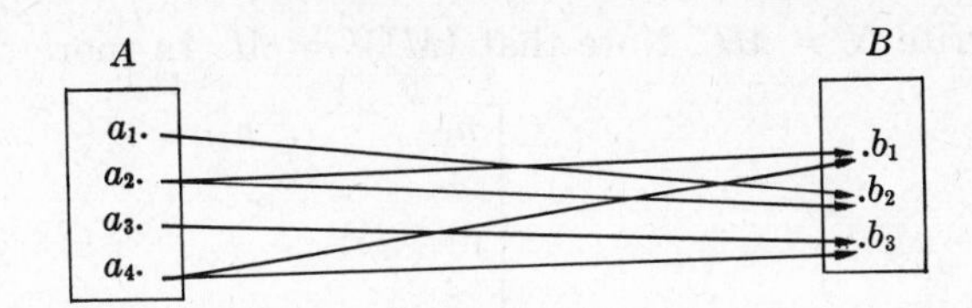

Figure A.9 Illustrating a relation.

Note that M^T is the matrix for the relation R^T (see Section B.4 for a definition of the transpose of a matrix).

Let $A = \{a_1, \cdots, a_p\}$ and $B = \{b_1, \cdots, b_n\}$ be finite ordered sets, let R be a relation in $A \times B$, and let M be the matrix for R. Then

$$M = \begin{bmatrix} m_{11} & \cdots & m_{1j} & \cdots & m_{1n} \\ \cdot & & \cdot & & \cdot \\ \cdot & \cdots & \cdot & \cdots & \cdot \\ \cdot & & \cdot & & \cdot \\ m_{i1} & \cdots & m_{ij} & \cdots & m_{in} \\ \cdot & & \cdot & & \cdot \\ \cdot & \cdots & \cdot & \cdots & \cdot \\ \cdot & & \cdot & & \cdot \\ m_{p1} & \cdots & m_{pj} & \cdots & m_{pn} \end{bmatrix} \tag{A.15}$$

where $m_{ij} = 1$ if $a_i \, R \, b_j$ and $m_{ij} = 0$ if $a_i \not R \, b_j$. We call the place in M where the ith row intersects the jth column *position* (i, j); thus m_{ij} is the *entry* or *element* in position (i, j).

Let $N = [n_{ij}]$ be the matrix for R^T. Then N has n rows and p columns and $\forall_{ij}$, $n_{ij} = m_{ji}$, that is, N is obtained from M by making rows into columns and columns into rows. In matrix algebra (see Appendix B) the term "transpose" is used to describe the relation between M and N and

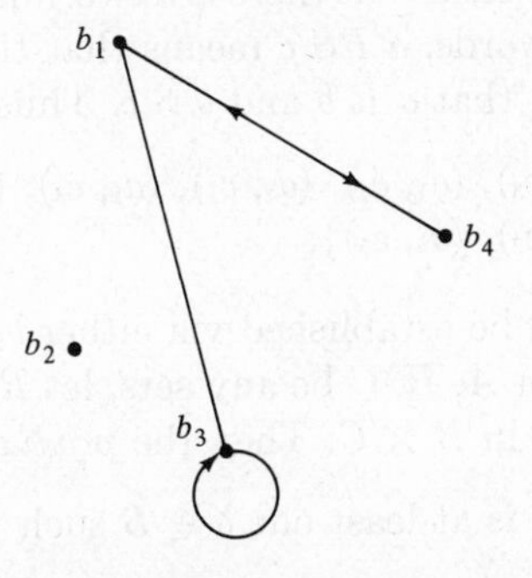

Figure A.10 Graphical presentation of a relation.

we write $N = M^T$. Note that $(M^T)^T = M$. In more detail we have

$$M^T = N = \begin{bmatrix} n_{11} & \cdots & n_{1j} & \cdots & n_{1p} \\ \cdot & & \cdot & & \cdot \\ \cdot & \cdots & \cdot & \cdots & \cdot \\ \cdot & & \cdot & & \cdot \\ n_{i1} & \cdots & n_{ij} & \cdots & n_{ip} \\ \cdot & & \cdot & & \cdot \\ \cdot & \cdots & \cdot & \cdots & \cdot \\ \cdot & & \cdot & & \cdot \\ n_{n1} & \cdots & n_{nj} & \cdots & n_{np} \end{bmatrix}$$

$$= \begin{bmatrix} m_{11} & \cdots & m_{j1} & \cdots & m_{p1} \\ \cdot & & \cdot & & \cdot \\ \cdot & \cdots & \cdot & \cdots & \cdot \\ \cdot & & \cdot & & \cdot \\ m_{1i} & \cdots & m_{ji} & \cdots & m_{pi} \\ \cdot & & \cdot & & \cdot \\ \cdot & \cdots & \cdot & \cdots & \cdot \\ \cdot & & \cdot & & \cdot \\ m_{1n} & \cdots & m_{jn} & \cdots & m_{pn} \end{bmatrix}. \qquad \text{(A.16)}$$

A.7. PRODUCTS OF RELATIONS.

We introduce the concept of product of relations with the following example. Consider a communication network consisting of a set A of four initiating stations, a set B of three relay stations, and a set C of five receiving stations (Figure A.11). Let R be the relation $\{(a_1, b_2), (a_2, b_1), (a_2, b_2), (a_3, b_3), (a_4, b_1), (a_4, b_3)\}$ in $A \times B$, and let S be the relation $\{(b_1, c_1), (b_2, c_1), (b_2, c_3), (b_2, c_4), (b_3, c_2), (b_3, c_5)\}$ in $B \times C$. Here $a\ R\ b$ means that there is a communication channel linking a to b, and $b\ S\ c$ has a similar meaning (see Figure A.11). By RS we mean the relation in $A \times C$ consisting of all pairs (a, c) such that there is a two-link communication channel from a to c; in other words, $a\ RS\ c$ means that there is at least one intermediate station b such that $a\ R\ b$ and $b\ S\ c$. Thus

$$RS = \{(a_1, c_1), (a_1, c_3), (a_1, c_4), (a_2, c_1), (a_2, c_3), (a_2, c_4), (a_3, c_2), (a_3, c_5), (a_4, c_1), (a_4, c_2), (a_4, c_5)\}.$$

Note that $a_2\ RS\ c_1$ can be established via either b_1 or b_2.

More generally, let A, B, C be any sets, let R be a relation in $A \times B$, and let S be a relation in $B \times C$. Then the *product* RS is defined by:

$$RS = \{(a, c) \mid \text{there is at least one } b \in B \text{ such that } a\ R\ b \text{ and } b\ S\ c\}.$$

The domain of RS is part of the domain of R and the range of RS is part of

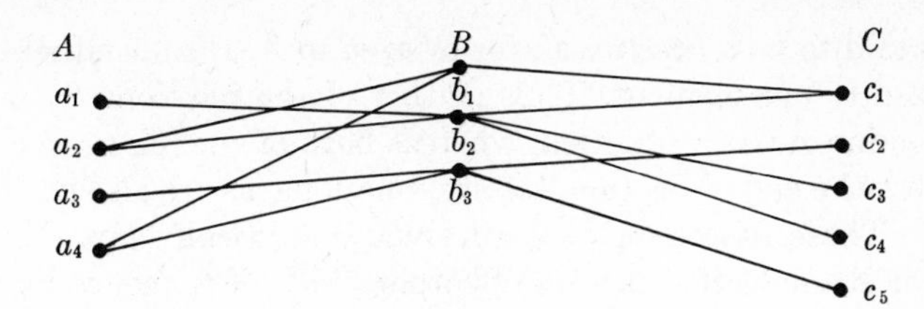

Figure A.11 A communication network.

the range of S. The product of two relations is not empty if and only if the intersection of the range of the first factor and the domain of the second is not empty.

The product of relations used here is sometimes called *iterative product* to indicate the sequential nature of the process involved in first applying R and then S; it is also called *relative product* by some authors. A familiar example of product is given by the expression $y = \log \sin x$. Here one has $R = \sin$ and $S = \log$, for $y = \log \sin x$ if and only if there exists z such that $z = \sin x$ (i.e., $x\ R\ z$) and $y = \log z$ (i.e., $z\ S\ y$).

The special case $A = B = C$, $R = S$ is of considerable importance and then we ordinarily abbreviate RR to R^2. Similarly R^3 denotes RRR and R^n means a product of n equal factors R. For example, if $a\ R\ b$ means a is a child of b then $a\ R^2\ c$ means a is a grandchild of c.

For another example, let S, Q be relations in $R \times R$ defined by

$$x\ S\ y \text{ means } y = x^3,$$
$$x\ Q\ y \text{ means } y = 2x,$$

then SQ, QS, and S^2, respectively, are defined by

$$x\ SQ\ y \text{ means } y = 2x^3,$$
$$x\ QS\ y \text{ means } y = 8x^3,$$
$$x\ S^2\ y \text{ means } y = x^9.$$

To check the second of these note that $x\ QS\ y$ means there exists z such that $z = 2x$ and $y = z^3$.

Products of more than two relations satisfy the associative law

$$(PQ)S = P(QS);$$

as a consequence of this it can be proved that parentheses are unnecessary in products of any length.

A.8. FUNCTIONS.

Although historically the concepts of relation and function were developed more or less independently, it is now customary and instructive to regard functions as special kinds of relations. A relation F in $A \times B$

is said to be a *function* if, for every a in A, $F(a)$ is either empty or contains exactly one element. Thus neither of the relations S and Q introduced in Section A.6 is a function, whereas both of the relations S and Q introduced near the end of Section A.7 are functions, as are also their products SQ, QS, S^2. These functions were all given analytically but this is not necessary. For example, if A is a set of corporations, B is a set of businessmen, and the relation F is defined by the statement, "$a\ F\ b$ means that the president of a is b," then F is a function in $A \times B$. If corporation a_1 has no president then $F(a_1)$ is empty; it is also possiblef or one man b to be president of more than one corporation; however, the essence of the situation is that no corporation can have more than one president.

The case of functions in $\boldsymbol{R} \times \boldsymbol{R}$ has an interesting geometric explanation. A relation S in $\boldsymbol{R} \times \boldsymbol{R}$ is a function if and only if no vertical line meets the graph of S in more than one point. Thus in Figure A.12 examples (a) and (b) are functions, whereas (c) is not a function. In many older texts (c) would be called a "two-valued function" to account for the fact that vertical lines in the domain cut the graph in two points. We prefer in this exposition to limit use of the word function strictly in accordance with the definition above.

An example of a function in $\boldsymbol{R} \times \boldsymbol{R}$ that is useful in many applications is the absolute-value function. Let $x \in \boldsymbol{R}$; then this function is defined by

$$F(x) = \begin{cases} x \text{ if } x \geq 0, \\ -x \text{ if } x < 0. \end{cases} \tag{A.17}$$

It is conventional to denote the value of the function for any x by the symbols $|x|$ instead of $F(x)$. A graph of this function is shown in Figure A.13; the domain is the set $\boldsymbol{R}$ and the range of F is the set of all nonnegative real numbers.

Still another example of a function in $\boldsymbol{R} \times \boldsymbol{R}$—and one which emphasizes that functions need not be defined by equations—is the postage

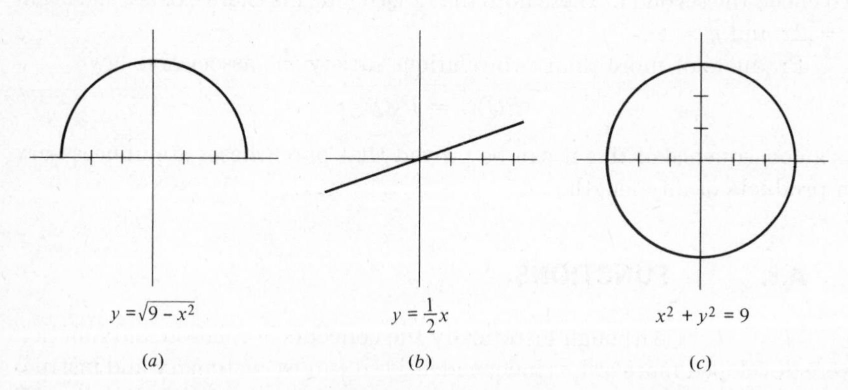

Figure A.12 Graphical interpretation of a function.

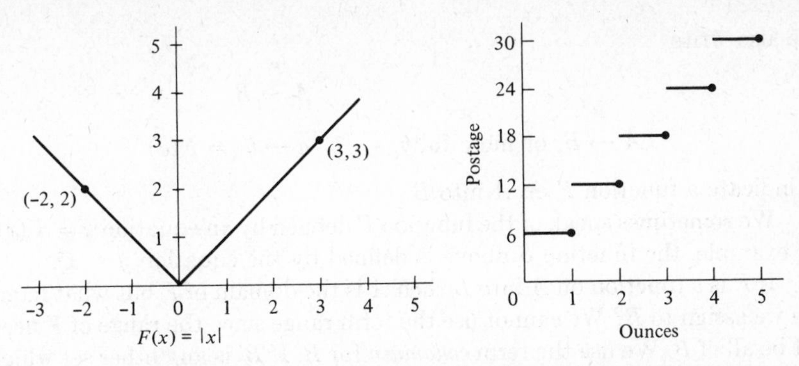

Figure A.13 Absolute value function and postage function.

function. The rate of postage on first-class letters in the United States (at this writing) is 6 cents per ounce or fraction thereof. If we let x denote weight and $P(x)$ denote postage, then when $0 < x \leq 1$, $P(x) = 6$ cents; when $1 < x \leq 2$, $P(x) = 12$ cents; $\cdots$. A graph of this function (an example of a step function) appears in Figure A.13. We can let the domain be the set of all positive real numbers; the range $\{6, 12, 18, \cdots\}$ then is the set of all possible amounts of first-class postage. Alternatively, if we let the domain be the set of all nonnegative real numbers, then the range is $\{0, 6, 12, 18, \cdots\}$. The function in this case indicates that no postage is necessary when no letters are mailed.

If F is a function in $A \times B$ and F has domain A then we say that F is a *function on A into B*. When the term function is used without qualification it is this special case which is meant. For this case we also use the terms *mapping* and *transformation* and write $b = F(a)$ instead of $a\ F\ b$. Note that this is a change from our general convention of using $F(a)$ to designate the set consisting of all elements b for which $a\ F\ b$. Since F is a function we would have under the old notation $F(a) = \{b\}$ rather than $F(a) = b$. We are not saying that $\{b\}$ and b are the same thing, but rather we are altering our notation and using $F(a)$ to mean the unique element in the image of a under F rather than the image considered as a set. We also sometimes call b the image of a under F. This is one of many places where mathematical notation is not strictly logical; the justification is that the convenience of using one symbol for several things outweighs the logical malpractice, especially if the situation is one where confusion is not likely to result.

We sometimes write

$$F: a \to b = F(a)$$

to emphasize that F is a mapping. The notation may be extended still further to indicate the connection in more detail. For example, the function S of Section A.7 could be defined by

$$S: x \to y = S(x) = x^3.$$

We also write

$$A \xrightarrow{F} B, \text{ or more fully, } \qquad \begin{array}{c} A \xrightarrow{F} B \\ a \to b = F(a) \end{array}$$

to indicate a function F on A into B.

We sometimes speak of the function F defined by an equation $y = F(x)$; for example, the function S above is defined by the equation $y = x^3$.

If F is a function on A into B then A is the domain of F but what name can we assign to B? We cannot use the term range since the range of F need not be all of B. We use the term *codomain* for B. If B' is any other set which contains the range of F it would also serve as a codomain of F. In other words a function has many codomains, but only one domain and one range.

If F is a function whose domain is A and whose range is B we say that F is *a function on* A *onto* B, and write

$$A \overset{F}{\twoheadrightarrow} B.$$

Referring once again to Figure A.12, let A and B, respectively, be the intervals $\{x \mid -3 \leq x \leq 3\}$ and $\{y \mid 0 \leq y \leq 3\}$. Then Figure A.12($a$) is (1) a function in $\boldsymbol{R} \times \boldsymbol{R}$, (2) a function on A into $\boldsymbol{R}$, and (3) a function on A onto B. In all three cases the domain is A and the range is B. In Case (2) $\boldsymbol{R}$ serves as codomain and in Case (3) B is codomain as well as range. In Figure A.12(b) the function is on $\boldsymbol{R}$ onto $\boldsymbol{R}$.

The real need for the concept of codomain is in instances where the range is not obvious. For example, let A be the set of positive real numbers and let F be the function

$$F: x \to y = F(x) = \frac{12}{x} + 2x + x^5$$

on A into A (see Figure A.14). It is clear that the range of this function has

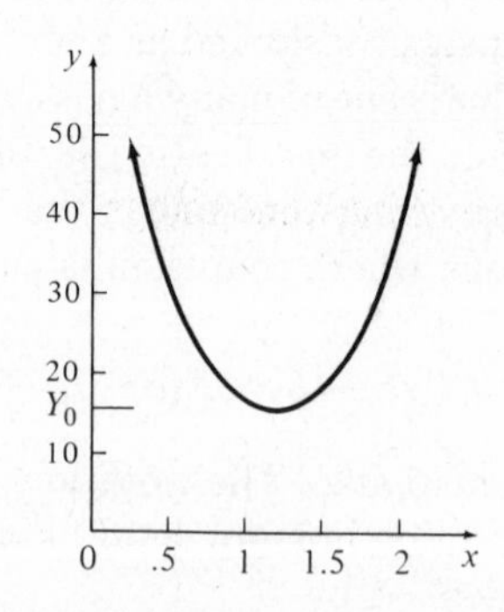

Figure A.14 Function whose range is not obvious: $F(x) = \dfrac{12}{x} + 2x + x^5$.

the form $\{y \mid y \geq y_0\}$, but it is not easy to find the numerical value of y_0. It is also clear that the properties of this function are not altered by enlarging the codomain from A to $\boldsymbol{R}$.

As a consequence of our definition of equality for relations we see that two functions F and G are equal if and only if

$$\text{(1)} \quad \text{domain } F = \text{domain } G,$$

and

$$\text{(2)} \quad F(a) = G(a) \text{ for all } a \text{ in this domain.}$$

This suggests an alternate direct procedure for defining the concept of function. *A function F on A into B is a correspondence which assigns to each element a in A a unique image $b = F(a)$ in B.*

An extreme point of view in discussing equality of functions is to require also (3) codomain F = codomain G, but there is little, if anything to be gained by such insistence on the uniqueness of codomain. Hence, we follow here for functions the convention already established in Section A.6 for equality of relations.

In mathematical literature the distinction between the function F and its value $F(x)$ for the element x is frequently not made, with the symbol $F(x)$ playing both roles. This is almost universally true in older elementary textbooks.

We observe that the product of two functions (considered as relations) is again a function. For if F is a function on A into B, G is a function on B into C, and $H = FG$, then $a\, H\, c$ means that there exists $b \in B$ for which $a\, F\, b$ and $b\, G\, c$. But the only b for which $a\, F\, b$ is $b = F(a)$ and the only c for which $b\, G\, c$ is then $c = G(b) = G(F(a))$. Hence FG is a function on A into C. Thus

$$FG\colon a \rightarrow c = G(b) = G(F(a));$$

note that FG means first F then G and this results in a reversal of order when the function notation is applied. This reversal applies for products with any number of factors. Thus

$$FG \cdots H\colon a \rightarrow c = (FG \cdots H)(a) = H(\cdots G(F(a)) \cdots).$$

Our definitions of restriction and extension for relations hold also for functions. The restriction of a function is always a function; however, an extension of a function need not be a function. For example, the function whose graph appears in Figure A.12(a) is a restriction of the relation in Figure A.12(c). It is frequently useful to construct functions which are restrictions of relations. For example, in trigonometry one restricts the arcsine relation[1] to the codomain $-\pi/2 \leq y \leq \pi/2$ and obtains the Arcsine function.

[1]Here arcsine designates the transpose of sine.

Many problems in mathematics take the form of finding an extension of a function which has certain desired properties. For example, fitting a curve to a set of empirically defined points can be regarded as the problem of finding an extension of a function whose graph is the given set of points to a continuous function whose domain is an interval of real numbers. Again, in an inventory model one may be interested in an optimal number n of items to order and the profit will be given by a real-valued function F defined on the set $\boldsymbol{Z}$ of positive integers. By extending the domain of F to the positive reals one may be able to apply the tools of calculus in maximizing F.

If the domain of a function F is the set of all positive integers $\boldsymbol{Z}$, we call F an *infinite sequence*. Instead of denoting values of this function by $F(1), F(2), \cdots, F(n), \cdots$, we sometimes write $F_1, F_2, \cdots, F_n, \cdots$; that is, we use subscripts rather than numbers within parentheses to indicate the first element of each ordered pair. A typical ordered pair in F can then be written (j, F_j) for $j \in \boldsymbol{Z}$. We also make extensive use of finite sequences which are defined as follows. For any $n \in \boldsymbol{Z}$ we define $\boldsymbol{Z}_n$ to be the set consisting of the integers less than or equal to n. For example, $\boldsymbol{Z}_4 = \{1, 2, 3, 4\}$. A function whose domain is $\boldsymbol{Z}_n$ is called a *finite sequence*. Such a function can be completely described by writing down the n-tuple $(F_1, \cdots, F_n)$, which lists the function values in order. In this way we can develop a close association between the concepts of finite sequence and n-tuple.

We have given special attention to relations F in $\boldsymbol{R} \times \boldsymbol{R}$ whose graphs are cut by each vertical line in at most one point. It is natural to ask if the same property for horizontal lines is of interest. Since, in general, transposition interchanges the roles of domain and range it will in the case of relations in $\boldsymbol{R} \times \boldsymbol{R}$ interchange horizontal and vertical (see Figure A.15 for examples of relations and their transposes) and amount to a reflection of the graph in the line $y = x$. Thus F^T is a function in $\boldsymbol{R} \times \boldsymbol{R}$ if and only if each horizontal line cuts the graph of F in at most one point. A case of special interest is when both F and F^T are functions [see for example, Figure A.15 (*c*) and (*d*)]. Then knowledge of either coordinate of a point on the graph is sufficient to determine the other coordinate uniquely. This one-to-one rela-

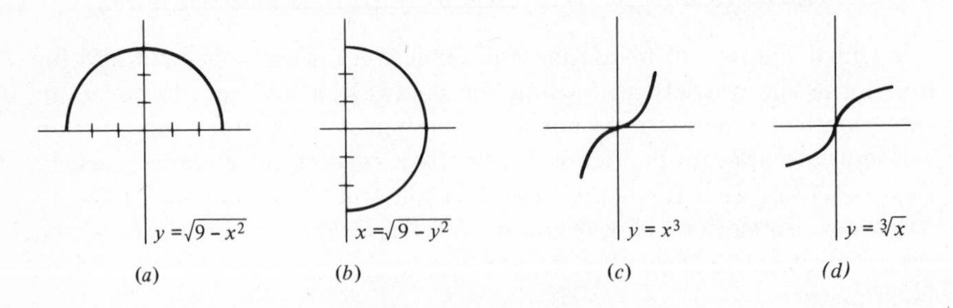

Figure A.15 Geometric meaning of transposition.

tionship between coordinates is of considerable importance and leads to the following definition of one-to-one function.

If both F and F^T are functions we say that F is a *one-to-one function.* A one-to-one function on A onto B is called a *one-to-one correspondence* between A and B. In this case we call the transpose of F the *inverse* of F and denote it by F^{-1}. Then we have

$$\forall\, a \in A, \quad (FF^{-1})(a) = F^{-1}(F(a)) = a$$

and

$$\forall\, b \in B, \quad (F^{-1}F)(b) = F(F^{-1}(b)) = b;$$

we summarize these relations by introducing the concept of *identity function* on a set. For any set A we introduce the function

$$\forall\, a \in A, \quad E_A{:}a \rightarrow a = E_A(a);$$

E_A is a function on A onto A and is called the *identity function* (or *identity relation*) on A. In the case of the reals the graph of E_A is the 45 degree line $y = x$. For any function F with an inverse we have

$$FF^{-1} = E_A \qquad \text{and} \qquad F^{-1}F = E_B. \tag{A.18}$$

Note that the relationship between function and inverse is reciprocal; if $G = F^{-1}$ then $F = G^{-1}$; or alternatively, if F^{-1} exists then $(F^{-1})^{-1} = F$.

The functions of Figure A.15(c) and (d) are inverses of each other. The natural logarithm and exponential functions are also inverses of each other (although not with the same domain).

In general, suppose that F is a relation in $A \times B$ with domain A and range B. Then we have

$$FF^T \supset E_A \qquad \text{and} \qquad F^TF \supset E_B.$$

If, in addition, F is a function we also have

$$F^TF = E_B.$$

For, suppose that bF^TFc holds. Then by the definition of product $\exists a \in A$, bF^Ta and aFc. But bF^Ta implies aFb, and now, since F is a function, $b = c$ and hence $F^TF \subset E_B$; we conclude from this and the universal relation $F^TF \supset E_B$ that $F^TF = E_B$.

As an illustration let $A = \boldsymbol{R}$, let $B = \{y \mid y \in \boldsymbol{R},\ -1 \leq y \leq 1\}$, and

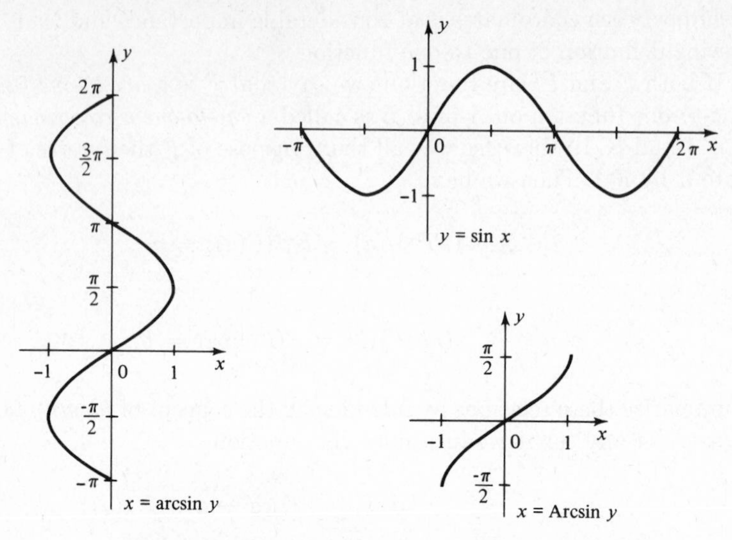

Figure A.16 Graphs of the sine, arcsine, and Arcsine.

let F = sin. Then F^T is arcsin (see Figure A.16) and $F^T F = E_B$. Note that $FF^T > E_A$ and has as its graph two infinite sets of parallel lines (see Figure A.17). If we restrict the range of arcsin to the interval $A' = \{y \mid -\pi/2 \leq y \leq \pi/2\}$ we obtain a function G commonly called the Arcsine function (referred to in calculus texts as the principle value of the arcsine). For this function we have

$$\text{Arcsin} \cdot \sin = E_B,$$

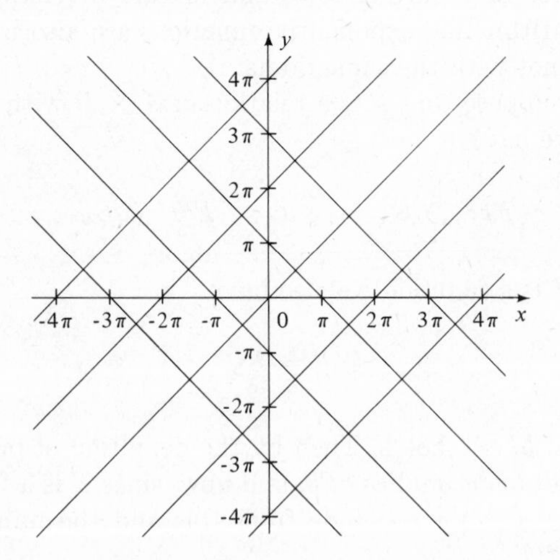

Figure A.17 Graph of the (sin·arcsin) relation.

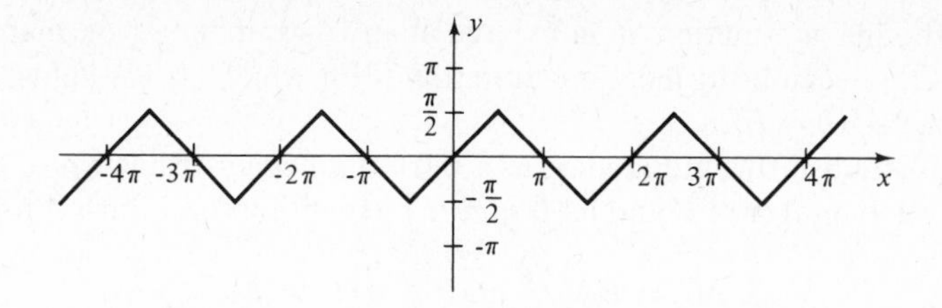

Figure A.18 Graph of the (sin · Arcsin) function.

but not sin · Arcsin $= E_A$ (see Figure A.18). If we next restrict the domain of sin to A' and denote the resulting function by Sin then we have $\mathrm{Sin}^T = \mathrm{Arcsin}$ and these functions are inverses.

Suppose that F is a function on A into B, that G is a function on B into A, and that

$$FG = E_A.$$

We then say that F is a *left inverse* for G, and that G is a *right inverse* for F. Then, since the range of a product is contained in the range of the second factor we must have range $G = A$; in other words, *a right inverse must be a function onto* (the domain of its left inverse).

Next, suppose that $F(x) = F(x')$ for $x, x' \in A$. Then

$$x = E_A x = (FG)(x) = G(F(x)) = G(F(x')) = E_A x' = x'$$

and hence *a left inverse must be one to one.*

Let F be any one-to-one function on A into B. Then F^T is a function in $B \times A$. Let G^* be any function on (B − range F) into A. Then $G = F^T \cup G^*$ is a function on B into A which is a right inverse of F. Thus we have a simple construction which shows that every *one-to-one function has a right inverse.*

The corresponding result that *every function which is onto has a left inverse* is true but its proof lies deeper. Let G be a function on B onto A. Suppose that B^* is a subset of B such that the restriction G^* of G to B^* is a one-to-one function on B onto A; then $F = G^{*T}$ is a left inverse for G. The difficulty lies in the selection of B^*. For each $a \in A$ we must have $B^* \cap G^T(a)$ a one-element set $\{b^*\}$ and B^* must be the union of all such one-element sets. In other words, for each $a \in A$ we must select exactly one element $b^* \in G^T(a)$ and then let B^* be the set of all such selected b^*.

For a relatively simple function G such as the one which maps $B = \boldsymbol{R}$ onto the nonnegative reals A according to the formula $y \rightarrow x = y^2$ we can select $B^* = A$. But if the sets A and $G^T(a)$ are more complex no such simple construction may present itself and the selection of B^* may involve an infinite number of arbitrary choices. This requires the introduction of the axiom of choice or some equivalent axiom; discussion of this axiom is beyond the scope of this book.

Although a function F can have many left inverses or many right inverses there can be at most one function G for which (A.18) holds, namely its transpose $G = F^T$.

We establish this uniqueness as a part of an even stronger result. Let F be a function on A into B and let G and H be functions on B into A for which

$$FG = E_A \qquad \text{and} \qquad HF = E_B.$$

Then

$$H = HE_A = H(FG) = (HF)G = E_BG = G.$$

Now, since F has a right inverse it is one to one and, since it has a left inverse it is onto; hence, the inverse $F^{-1} = F^T$ of F exists and we conclude by taking in turn $H = F^{-1}$ and $G = F^{-1}$ in the above argument that $G = H = F^{-1}$ holds for any left inverse H and right inverse G. Thus a function can have at most one inverse and if it has both a left and a right inverse they are both equal to F^T.

To summarize, we see that (1) a function has a right inverse if and only if it is one to one, (2) a function has a left inverse if and only if it is onto, and (3) a function has an inverse if and only if it is both one to one and onto. Moreover, if a function has an inverse then F^T is its only left or right inverse.

A.9. FUNCTIONS IN $\boldsymbol{R} \times \boldsymbol{R}$.

Since the study of functions in $\boldsymbol{R} \times \boldsymbol{R}$ has its roots in the early history of mathematics, it should not be surprising to find that the heavy hand of tradition has saddled us with many notational defects both of commission and omission. The parentheses in the symbolism $F(x)$ are not really needed and are indeed omitted in many advanced texts and research papers where one finds simply Fx (or even xF which has the advantage of not requiring a reversal of order for products, that is, $x(FG) = (xF)G$. However, the parentheses cause no logical difficulty and we shall retain them here.

One of the logical flaws has been the use of $F(x)$ to indicate both the function F and the image or *value* of F for a given x. We shall avoid this dual usage: here $F(x)$ will always mean the image of x under the function F. Thus such phrases as, "the function $y = x$," or, "the function $y = 2 + 3x + x^2$," or, "the function x^2," or "the function $\sin x$," are avoided in this book. Traditional usage is not entirely consistent. Such phrases as, "the sine function," "the logarithm function," and "the exponential function" are never replaced by "the $\sin x$ function," "the $\log x$ function," and "the $\exp x$ function."

Although we will avoid such phrases as, "the function $y = x \sin x$," it is nevertheless desirable to have some means of relating functions to formulas. If it is understood that we are discussing only functions in $\boldsymbol{R} \times \boldsymbol{R}$ we will use the phrase, "the function f defined by $f(x)$ = some formula," to mean the mapping $f{:}x \rightarrow y = f(x)$ whose domain is the set of all x for which the formula is meaningful and where y is then given by the formula. For example, the function defined by the formula $\sqrt{9 - x^2}$ has domain $\{x \mid -3 \leq x \leq 3\}$ and the function defined by the formula $x/(x^2 + x - 2)$ has as domain all of $\boldsymbol{R}$ except $x = 1$, and $x = -2$.

A major defect in traditional notation is the failure to use a symbol for the identity function E_R (which we shall shorten to E). Since iterates of E are all just E we will not use exponents on E to mean iteration but rather will denote by E^n for any positive integer n the function

$$E^n: x \longrightarrow E^n(x) = x^n \tag{A.19}$$

with domain $\boldsymbol{R}$.

As an illustration the product E^2 sin is the function

$$x \longrightarrow (E^2 \sin)(x) = \sin (E^2x) = \sin (x^2),$$

with domain $\boldsymbol{R}$.

We introduce the symbol E^0 for the function which maps each x in $\boldsymbol{R}$ into the number 1. For any k in $\boldsymbol{R}$ we denote by kE^0 the function which maps each x in $\boldsymbol{R}$ into k.

We use the Formula (A.19) to define E^n for negative integers n on the domain $\boldsymbol{R} - \{0\}$.

The operations of addition and multiplication in $\boldsymbol{R}$ can be used to introduce addition and a new multiplication of functions. Let F and G be functions in $\boldsymbol{R} \times \boldsymbol{R}$. Then we define a function H with

$$\text{domain } H = \text{domain } F \cap \text{domain } G$$

by

$$H: x \rightarrow y = H(x) = F(x) + G(x);$$

we call H the *sum* of F and G, written

$$H = F + G.$$

Similarly, we define a function K with the same domain as H by

$$K: x \rightarrow y = k(x) = F(x) \circ G(x);$$

we call K the *product* of F and G, written

$$K = F \circ G.$$

The symbol $\circ$ is important here since otherwise there might be confusion with the iterative product FG.

For any $k \in \boldsymbol{R}$ we define kF to be $kE^0 \circ F$, that is,

$$kF: x \to kF(x).$$

To illustrate these concepts we give several examples. Let $F_1 = E^2$, $F_2 = \sin$, $F_3 = E$, $F_4 = E^0$. Then we have

$$\begin{aligned} &F_1 \circ F_2: x \to x^2 \sin x = (F_1 \circ F_2)(x), \\ &F_1 + F_2: x \to x^2 + \sin x = (F_1 + F_2)(x), \\ &F_1F_2: x \to \sin x^2, \\ &2F_4 + 3F_3 + F_1: x \to 2 + 3x + x^2, \\ &(F_1 + F_1 \circ F_1)F_2: x \to \sin (x^2 + x^4), \\ &F_1 \circ (F_1F_2): x \to x^2 \sin x^2, \end{aligned}$$

and

$$F_2 \circ F_2 = F_2F_1: x \to (\sin x)^2.$$

This last term is sometimes written $\sin^2 x$, but we refrain from this since in our notation $\sin^2 x$ means $\sin (\sin x)$.

Note in particular the contrast between EF, FE, and $E \circ F$ for any function F in $\boldsymbol{R} \times \boldsymbol{R}$; we have

$$EF = FE = F$$

whereas $E \circ F$ is the function

$$E \circ F: x \to xF(x).$$

A.10. BINARY RELATIONS ON A SET.

In Section A.6 we introduced the concept of a relation in $A \times B$. The special case where A and B are the same set is of considerable importance; we use the terminology, "*R is a relation on A*," as an abbreviation for, "*R is a relation in $A \times A$*." We introduce and assemble in one place for easy reference some important classes of relations on a set. Each of these is defined first in set language and then in terms of individual elements.

A relation R on a set A is said to be:

Identical	if $R = E_A$,	if $a\ R\ b$ is equivalent to $a = b$;
Reflexive	if $R \supset E_A$,	if for every a in A we have $a\ R\ a$;
Irreflexive	if $R \cap E_A = \phi$,	if for no a in A is $a\ R\ a$;
Symmetric	if $R = R^T$,	if $a\ R\ b$ implies $b\ R\ a$;

Antisymmetric if $R \cap R^T \subset E_A$, if $a\,R\,b$ and $b\,R\,a$ imply $a = b$;

Asymmetric if $R \cap R^T = \phi$, if for no a, b do we have both $a\ R\ b$ and $b\ R\ a$;

Transitive if $R^2 \subset R$, if $a\,R\,b$ and $b\,R\,c$ imply $a\ R\ c$;

Acyclic if $R^n \cap E_A = \phi$ for all n in Z, if $a_1\ R\ a_2$, $a_2\ R\ a_3$, $\cdots$, $a_{n-1}\,R\,a_n$ imply $a_1 \neq a_n$;

Complete if $R \cup R^T = A \times A$, if for each pair a, b either $a\,R\,b$ or $b\ R\ a$;

Trichotomous if $R \cup R^T \cup E_A = A \times A$ and $R \cap R^T = \phi$, if for each pair a, b exactly one of the relations $a\ R\ b$, $b\ R$ a, $a = b$ holds.

These relations are of importance individually but are even more important taken in groups. We shall consider some of the important groupings in later sections.

We have already discussed the identity relation. One of its important properties is that given any additional sets B_1 and B_2 and any relations S_1 in $B_1 \times A$ and S_2 in $A \times B_2$ we have $S_1E_A = S_1$ and $E_AS_2 = S_2$; that is, E_A behaves under (iterative) multiplication much like the number 1 in numerical multiplication. If A is a finite sequence $(a_1, \cdots, a_p)$ the matrix $M(E_A)$ (see Section A.6) for E_A is the identity matrix. This is illustrated for a set A having four elements.

$$M(E_A) = \begin{bmatrix} 1 & 0 & 0 & 0 \\ 0 & 1 & 0 & 0 \\ 0 & 0 & 1 & 0 \\ 0 & 0 & 0 & 1 \end{bmatrix}.$$

The positions (1, 1), (2, 2), $\cdots$, (p, p), where $M(E_A)$ has entries of one, constitute the *diagonal* of a square matrix. Thus E_A is characterized by the property that its matrix has ones on the diagonal and zeros elsewhere.

A reflexive relation on a finite set A is characterized by the property that the diagonal entries in its matrix are all one; for an irreflexive relation the diagonal entries are all zero. For example,

$$\begin{bmatrix} 1 & 0 & 1 & 0 \\ 1 & 1 & 1 & 0 \\ 0 & 0 & 1 & 1 \\ 0 & 1 & 1 & 1 \end{bmatrix}$$

is the matrix of a reflexive relation, and

$$\begin{bmatrix} 0 & 1 & 1 & 1 \\ 1 & 0 & 0 & 1 \\ 0 & 1 & 0 & 0 \\ 1 & 0 & 1 & 0 \end{bmatrix}$$

is the matrix of an irreflexive relation.

A square matrix $M = [m_{ij}]$ is said to be *symmetric* if for every i and j we have $m_{ij} = m_{ji}$; that is, if the entry in position (i, j) is the same as that in position (j, i). This can be abbreviated to $M = M^T$. A symmetric relation is characterized by the property that its matrix is symmetric. For example

$$\begin{bmatrix} 0 & 1 & 1 & 0 \\ 1 & 1 & 0 & 1 \\ 1 & 0 & 1 & 1 \\ 0 & 1 & 1 & 0 \end{bmatrix}$$

is the matrix of a symmetric relation.

The concepts of antisymmetric and asymmetric differ only in that the second is stricter. Both require that for two distinct elements a, b at most one of the relations $a\ R\ b$, $b\ R\ a$ holds; in the asymmetric case $a\ R\ a$ is not permitted. Thus a relation is asymmetric if and only if it is both antisymmetric and irreflexive. In matrix language antisymmetry requires that we have at least one of m_{ij} and m_{ji} equal to zero if $i \neq j$ and for asymmetry we also require $m_{ii} = 0$. For example,

$$\begin{bmatrix} 0 & 1 & 1 & 1 \\ 0 & 1 & 0 & 1 \\ 0 & 0 & 1 & 0 \\ 0 & 0 & 1 & 0 \end{bmatrix}$$

is the matrix of an antisymmetric relation, and

$$\begin{bmatrix} 0 & 1 & 1 & 1 \\ 0 & 0 & 0 & 1 \\ 0 & 0 & 0 & 0 \\ 0 & 0 & 1 & 0 \end{bmatrix}$$

is the matrix of an asymmetric relation.

Let A be any set and let R be a relation on A. A sequence $a_1, \cdots, a_n$ of distinct elements in A is said to be a *chain of length n relative to R* if $a_1\ R\ a_2$, $a_2\ R\ a_3, \cdots, a_{n-1} R a_n$. If also $a_n\ R\ a_1$ the sequence is called a *cycle of length n relative to R.*

The relation R is transitive if whenever a_1, a_2, a_3 is a chain of length 3 relative to R then $a_1 R a_3$. But if $a_1, a_2, \cdots, a_n$ is a chain relative to R, so is any subset $a_i, \cdots, a_j$ for $j > i$. Hence if R is transitive and $a_1, \cdots, a_n$ is any chain of length n relative to R we can shorten the chain one link at a time, dropping out first a_2, then a_3, and continuing until we get $a_1 R a_n$. Thus a relation R is transitive if and only if for every chain $a_1, \cdots, a_n$ relative to R we have $a_1 R a_n$. Thus the hypothesis that chains of length three can be shortened to length two has as a consequence that chains of any length can be shortened to length two.

A relation R is acyclic if there are no cycles of any length $n = 1, 2, \cdots$ relative to it. Since $a R a$ is a cycle of length one we see that the acyclic property implies the irreflexive property.

The term trichotomous refers to the fact that such a relation subdivides the set $A \times A$ into three nonoverlapping subsets. The term complete refers to the fact that every element in A belongs to either the domain or to the range of R.

We give a few illustrative examples. The relation, "is a brother of," is symmetric and irreflexive, but it is not transitive, acyclic, or complete. The relation $<$ on the real numbers is asymmetric, transitive, trichotomous, and acyclic, whereas the relation $\leq$ is antisymmetric, transitive, and complete. The relation, "is at least as hard as," is reflexive, transitive, but not antisymmetric if there are two distinct objects of equal hardness. The relation R, "is the son of," on the set of all men is irreflexive. Then R^2 is the relation, "is the grandson of" and $S = \bigcup_{n=1}^{\infty} R^n$ is the relation, "is a direct descendent of." S is transitive as well as irreflexive; moreover S is the smallest transitive relation that contains R. The relation, "has the same color as," is reflexive, symmetric, and transitive. Consider a basketball league with a round-robin schedule in which all tie games are played off in overtime. Then the relation "beat" on the set of teams is irreflexive, asymmetric, and complete, but is not transitive or acyclic. Indeed it is the existence of chains such as, "a beats b, b beats c, and c beats a," which add to the excitement of the game.

We next consider connections between some of these classes of binary relations and also some ways in which relations can be constructed which belong to one or more of these classes.

Let A be any set and let R be a relation on A. Then $S = \bigcup_{n=1}^{\infty} R^n$ is transitive and is the smallest transitive relation which is an extension of R. Similarly, $R \cup R^T$ is symmetric and is the smallest symmetric extension of R; the intersection $R \cap R^T$ and the products RR^T and R^TR are also symmetric. If the domain of R is A, then RR^T is reflexive.

If R is asymmetric then $S = R \cup E_A$ is antisymmetric and if S is antisymmetric then $S - E_A$ is asymmetric. If R is trichotomous then $S = R \cup E_A$ is complete and if S is complete then $S - E_A$ is trichotomous.

A.11. EQUIVALENCE RELATIONS AND PARTITIONS.

We are all familiar with the concept of a will which partitions an estate. A will is not satisfactory unless it has two properties: each item in the estate must be allocated, and no item can be allocated more than once. These properties, suitably abstracted, define the mathematical concept of partition.

Let A be a set. A *partition* $P = \{M\}$ of A is a collection of subsets M of A such that

P1. Each element a of A lies in at least one M in P,

and

P2. If a is in both M_1 and M_2 then $M_1 = M_2$.

These two properties can be paraphrased, "each element a in A lies in exactly one subset M in P." The terms "exhaustive" and "exclusive" are frequently used to describe properties P1 and P2, respectively.

Another familiar concept is that of grouping together things which although distinct can be regarded as equivalent for some purpose. Grouping of people may be defined by considerations such as sex, color, height, occupation, and income. The concept of indifference as applied to commodity bundles is another example of such equivalence.

A relation R on a set A is said to be an *equivalence relation* if it is reflexive, symmetric, and transitive. If R is an equivalence relation on A and $a \in A$ call the image set $R(a) = \{b \mid a\, R\, b\}$ an *equivalence class* relative to R.

For example, let A be the set $\boldsymbol{J}$ of all integers and let R_3 be the relation defined by "$a\, R_3\, b$" means "$a - b$ is a multiple of 3." Now 0 is a multiple of 3; since $a - b = 3k$ implies $b - a = 3(-k)$ and since $a - b = 3k$, $b - c = 3m$ imply $a - c = 3(k - m)$, we see that R_3 is an equivalence relation on $\boldsymbol{J}$. For any $a \in \boldsymbol{J}$ we have

$$R_3(a) = \{a + 3k \mid \forall\, k \in \boldsymbol{J}\},$$

so that $R_3(a)$ is an equivalence class relative to R_3.

In particular, let $a = 1$, then $R_3(1) = \{1 + 3k \mid \forall\, k \in \boldsymbol{J}\}$,

$$R_3(1) = \{1, 4, 7, 10, 13, \cdots\}.$$

For $a = 2$ we have $R_3(2) = \{2, 5, 8, 11, 14, \cdots\}$. Moreover, $R_3(2) = R_3(5) = \cdots = R_3(2 + 3k)$; and each $R_3(a)$ is identical with one of the three equivalence classes $R_3(0)$, $R_3(1)$, $R_3(2)$, and no pair of these has a member in common. Hence $P = \{R_3(0), R_3(1), R_3(2)\}$ is a partition of A. We also say that any two members of an equivalence class are equivalent relative to the relation; for example, the elements 5 and 14 are equivalent relative to R_3 (R_3 has the technical name "congruent modulo 3").

This case illustrates the following general connection between partitions and relations. For any equivalence relation R the collection P of all equivalence classes relative to R is a partition of A. Conversely, let P be a partition of A and let R be the relation defined by the statement $a\,R\,b$ holds when and only when a and b lie in a common subset $M \in P$; then R is an equivalence relation on A.

To prove these assertions we assume first that R is an equivalence relation on A and let P be the set of all equivalence classes relative to R. To establish P1 we observe that if $a \in A$ it follows from the reflexivity of R that $a\,R\,a$ and hence $a \in R(a)$. Next, suppose that $c \in R(a)$; then $R(a) \supset R(c)$. For, by the transitivity of R, $a\,R\,c$ and $c\,R\,d$ implies $a\,R\,d$; hence if $d \in R(c)$ we have also $d \in R(a)$. But by the symmetry of R if $c \in R(a)$ then $a \in R(c)$; hence $R(a) = R(c)$. It follows that if two equivalence classes $R(a)$ and $R(b)$ have an element c in common they are both equal to $R(c)$ and hence to each other; thus $P2$ is established.

Conversely, suppose that P is a partition of A. We define a relation R by the statement $a\,R\,b$ if and only if there exists $M \in P$ for which $a \in M$ and $b \in M$. By the symmetry of a and b in this definition R is symmetric. The reflexivity of R follows at once from P1. Finally, if the relations $a\,R\,b$ and $b\,R\,c$ are established from $a \in M$, and $b \in M$ and $b \in M'$, $c \in M'$ then $b \in (M \cap M')$ and hence by P2, $M = M'$ and now $a\,R\,c$ follows.

We denote by A/R (read A modulo R) the partition P defined by an equivalence relation R and we denote by $R(P)$ the equivalence relation defined by a partition P. It is easy to verify that

$$R(A/R) = R \text{ and } P = A/R(P);$$

in other words, if R defines A/R then R is in turn the equivalence relation which A/R defines, and if P defines $R(P)$ then P is in turn the partition defined by the equivalence classes of $R(P)$.

Equivalence is the mathematical and logical basis of classification and thus equivalence relations will appear in many applications of mathematics to real-world problems. Because of the correspondence between partitions and equivalence relations we may find applications worded in terms of either of the two concepts. We give a few examples.

The sorting of nuts and bolts is based on the equivalence relation, "—has the same size and shape as—." The partition of elementary school students into classes is based on the equivalence relation, "—has the same teacher as—." This example illustrates the desirability of permitting more than one label for a set in a partition. For, two mothers attending a PTA meeting may wish to know if their children are in the same schoolroom. The mothers will identify the class in terms of their respective children as "Mary's room," and "Johnny's room." The label on the door will identify the class as "Miss Jones' room." However, if one mother has not yet learned the teacher's name the two parents may not yet know whether their children are in the same room. But then Mary's mother may recall that Peter is a tall blond boy who sits behind Mary and pulls her pigtails and then when Johnny's mother recognizes Peter as the big bully who has been teasing Johnny they finally conclude that

$$\text{Mary's room} = \text{Johnny's room.}$$

A bowl of beads can be partitioned on the basis of the equivalence relation, "—has the same color as—."

The set of all formal quotients p/q where p, q are positive integers can be classified by the equivalence relation of $=$, defined by $p/q = r/s$ if and only if $ps = qr$. The equivalence classes under this relation form the positive rational numbers. Similar equivalence classes are used throughout the extension of the number system from the positive integers to the complex numbers.

Let $m \in \boldsymbol{J}$, then the relation R_m defined by "$a \, R_m \, b$" means "$a - b$ is a multiple of m" is an equivalence relation whose m distinct equivalence classes are sometimes called *congruence classes modulo m*.

In geometry we classify triangles according to the equivalence relations of congruence and similarity.

Much of the content of linear algebra can be expressed in terms of various equivalence relations on matrices. *Row equivalence* is related to the solution of systems of linear equations; *equivalence* and *similarity* appear in the general study of linear transformations; *congruence, conjunction,* and *unitary equivalence* are key concepts in the treatment of length and angle in real and complex vector spaces (applications of these include least-squares curve fitting, resonance phenomena in vibrating systems, analysis of finite-state Markov processes, and time series analysis). These equivalence relations on matrices are discussed in Appendix B.

In the body of this text we discuss a number of equivalences between linear programming problems. Two problems in the canonical form of Section 3.2 are said to be equivalent if there is a sequence of pivot operations leading from one to the other. This is a genuine equivalence relation. However, we sometimes employed the term "equivalent" in a less technical sense as in the statement, "two linear programming problems are considered to be equivalent if the optimal solutions of each can be obtained (in some simple manner) from those of the other." For example, replacing a free variable x by a difference $x = x' - x''$ of two nonnegative variables leads to a new problem which is equivalent to the former in this broader sense.

Let R and S be two equivalence relations on A. If $R \subset S$ we say that R is *stronger* than S. This terminology follows from the fact that $R \subset S$

means that $a\,R\,b$ implies $a\,S\,b$ whereas we may have $a\,S\,b$ and $a \not R\, b$. For example, in geometry congruence is a stronger equivalence relation than similarity, since any two congruent triangles are similar.

It is easy to verify that $R \cap S$ is again an equivalence relation, but in general $R \cup S$ is not. For example, $R_3 \cup R_7$ is not an equivalence relation on $\boldsymbol{J}$.

Let S be any relation on A, and let $W(S)$ denote the set of all equivalence relations T(on A) that contain S. Since $A \times A$ is one such equivalence relation the formula

$$R = \bigcap_{T \in W(S)} T \tag{A.20}$$

defines a valid intersection. Since each T is an equivalence relation, (1) R contains E_A, (2) R is symmetric, and (3) R is transitive. Hence R is an equivalence relation. Since each $T \supset S$, so does R; clearly R is the strongest equivalence relation which contains S, and we call R the *equivalence relation defined by* S. It is not difficult to give a more direct construction for R. Let

$$S_1 = \bigcup_{n=1}^{\infty} S^n;$$

then S_1 is the smallest transitive relation which contains S. Next, consider

$$R = E_A \cup S_1 \cup S_1^T; \tag{A.21}$$

clearly $R \supset S$ and R is obviously reflexive and symmetric. An easy argument shows that R is also transitive; hence $R \in W(S)$. Now, if $T \in W(S)$ it must be transitive and contain S, therefore $T \supset S_1$. But T is symmetric and reflexive and so T also contains S_1^T and E_A. Hence, if $T \in W(S)$, $R \subset T$. It follows that R is equal to the right-hand side of Equation (A.20) as well as that of (A.21).

A.12. INVARIANTS AND CANONICAL FORMS.

When an equivalence relation is given we frequently wish to find some effective means of determining whether or not two elements are related. This is the origin of the concept of invariant in mathematics. A property shared by all elements of each equivalence class is called an invariant. For example, the length of the longest side is an invariant for congruence of triangles. If two triangles have unequal longest sides they cannot possibly be congruent. On the other hand, two triangles need not be congruent just because their longest sides are equal; to guarantee congruence we must have all three corresponding sides equal. Thus for an equivalence relation we are interested in determining a set of invariants large enough so that equality for them will guarantee equivalence.

More formally, let R be an equivalence relation on a set B. A function F on B is said to be an *invariant relative to* R if $a\ R\ b$ implies $F(a) = F(b)$. A set of invariants $\{F, G, \cdots\}$ relative to an equivalence relation R is said to be *complete* if

$$F(a) = F(b),\ G(a) = G(b), \cdots,$$

together imply $a\ R\ b$.

Viewed in this light much of elementary geometry can be recognized as a search for invariants or for complete sets of invariants for equivalence relations on various classes of geometric objects. The common phraseology, "a necessary and sufficient condition for K is that $P_1, P_2, \cdots$ all hold," is one guise in which we frequently find the property that $P_1, P_2, \cdots$ constitute a complete set of invariants for an equivalence relation associated with K. Equivalence relations and complete sets of invariants play a fundamental role in the theory of matrices and vector spaces.

Note that in defining invariant we specified only the domain and said nothing about the codomain. This was deliberate since there is no simple qualification for the range of an invariant. The value $F(a)$ may be for example, a number, a geometric object, a color, or a matrix.

We are frequently interested in selecting from each equivalence class one or more representatives from which some invariants can be easily calculated. Such representatives are said to be in *normal form* or *standard form*. For example, in analytic geometry the equation

$$(x - 3)^2 + (y + 2)^2 = 25$$

identifies the same circle as does

$$x^2 + y^2 - 6x + 4y - 12 = 0,$$

but the former indicates the center and radius which constitute a complete set of invariants for circles under the relation of equality. We call $(x - h)^2 + (y - k)^2 = a^2$ a standard form for the equation of a circle. We used the concept of standard form for a linear program in Section 3.2 above.

The general concept of *normal form* or *standard form* does not have a precise mathematical definition although in specific instances it is precisely defined. On the more technical level we introduce the concept of *canonical form*. Let R be an equivalence relation on a set B. A function which assigns to each equivalence class $R(a)$ one of its members is called a *canonical form* relative to R. The representative thus selected is called the *canonical member* of the equivalence class and is said *to be in canonical form*. Thus in matrix theory there are canonical forms for row equivalence, equivalence, congruence, similarity, and orthogonal congruence, and in analytic geometry we use translation and rotation of axes to reduce the equations of the conics to canonical forms. In the case of canonical forms it is customary, though not logically necessary, to select a representative which displays some complete set of invariants; thus, a canonical form is usually also a normal form. In many cases the canonical form is accompanied by a process or algorithm by means of which one can pass step by step from any member of an equivalence class to the canonical member.

For example, in Section B.12 we provide an algorithm (in flow-chart form) for reduction of a p by n matrix to echelon form (the canonical form for row equivalence of matrices).

The simplex algorithm as developed in Chapter 3 utilizes the concept, "canonical form with respect to a basic sequence $S = (s_1, \cdots, s_p)$," and a given linear programming problem in standard form therefore is equivalent to as many canonical forms as there are acceptable basic sequences. Of course, the central feature of the simplex algorithm is successive selection of basic sequences S until one is found which solves the problem. In a sense the need to use a collection of canonical forms, rather than just one, measures the complexity of linear programming in contrast to solution of linear equations without optimization or sign restrictions on the variables. The uniqueness property of canonical forms plays a central role in several places: in establishing convergence of the simplex algorithm, in the computational considerations of Chapter 4, in our proof of the Farkas Lemma, and in establishing the duality theorems.

Indeed, the use of canonical forms and equivalence permeates our treatment of linear programming so thoroughly that it is hard to find any major result that does not depend on these concepts. More generally, expository treatments of algorithms which do not explicitly recognize the roles of the concepts of equivalence and canonical form tend to be either, on the one hand, obscure and hard to read or, on the other hand, reduce to sets of instructions which do not provide insights into what is really happening.

Remark 1. In practice we sometimes use the term canonical form without insisting on absolute uniqueness. For example, in the case of similarity of matrices with distinct eigenvalues the usual canonical form is a diagonal matrix whose entries are the eigenvalues. Thus if the matrix has degree n there are $n!$ possible diagonal matrices having the given eigenvalues. To meet the full logical requirements for a canonical form one could order the field of scalars and then select as the canonical member that matrix whose eigenvalues are in order from top to bottom. In cases such as this we rarely invoke the full uniqueness requirement and are content with a nonunique form and the knowledge that uniqueness would be easily obtained if really needed.

Remark 2. The concept of invariant is sometimes associated with that of a group of transformations on a set. The reader who is familiar with the concept of group will be able to verify that a transformation group on a set defines an equivalence relation on that set and that an invariant for the group is the same as an invariant relative to this equivalence relation.

A.13. ORDER RELATIONS.

The concept of "order" is an essential part of the theory of measurement. In its most primitive form measurement consists of statements such as, "picture A is preferred to picture B," "light A is brighter than light B," "Cheerios are no better than Wheaties," "football team A can beat football team B," "John loves Mary more than he loves Sally." At the opposite extreme measurement consists of assigning real numbers to describe an item or situation, for instance, temperature, length, weight, cost are described by real numbers which can then be used in turn to generate comparative statements between objects without having to examine the objects in a common environment.

We will begin our formal discussion of order relations by giving a string of definitions based on the properties of binary relations introduced in Section A.10. In each case we are considering a binary relation on a set A. The symbols employed for these relations are the ones most frequently encountered in the literature, although variations also exist.

Preorder, $\precsim$, reflexive and transitive.
Partial order, $\leq$, reflexive, antisymmetric, and transitive.
Weak order, $\lesssim$, reflexive, transitive, and complete.
Chain order, $\leq$, reflexive, antisymmetric, transitive, and complete.

The transposes of $\precsim, \lesssim, \leq$, are designated respectively by $\succsim, \gtrsim \geq$. For example, if $\precsim$ is, "is no better than," then $b \succsim a$ might be read, "b is at least as good as a."

Chain order is illustrated by the real numbers with $\leq$ given its usual meaning. A somewhat more complex example of a chain order is the lexicographic order for vectors introduced in Section 3.4. A vector

$$D = \begin{bmatrix} d_1 \\ \cdot \\ \cdot \\ \cdot \\ d_p \end{bmatrix}$$

in V_p is *lexicographically positive*, abbreviated lex pos and written[2] $D \succ 0$, if for some h with $1 \leq h \leq p$ we have $d_h > 0$ but $d_i = 0$ for $i = 1, \cdots, h-1$, that is, if the first nonzero coordinate of D is a positive number.

[2]The use of $\succ$ (and $\succeq$) for lexicographic order is conventional in the literature, and therefore is used here despite the danger of confusion with the concept of preorder.

Then for two vectors D_1 and D_2 in V_p, we say that $D_1 \succ D_2$ (D_1 is *lexicographically greater* than D_2) if $D_1 - D_2 \succ 0$. The binary relation $\succ$ is transitive and asymmetric; the corresponding relation $\succeq$ is reflexive, antisymmetric, and transitive.

This relation $\succ$ is also trichotomous. Furthermore, when lex pos vectors in V_p are added, the resulting vector is lex pos: if $X \succ 0$, $Y \succ 0$, then $X + Y \succ 0$. Lex pos is preserved under positive scalar multiplication in V_p: if $k > 0$, $X \succ 0$, then $kX \succ 0$. The second of these properties means that all lex comparisons between pairs of vectors can be reduced to comparisons with the zero vector. It follows that $\succeq$ is complete and hence that it is a chain order on V_p.

The concept of *weak order* arises frequently in economics when studying valuation of various commodity bundles. Suppose we are considering two commodities, say potatoes and steak, and wish to compare different bundles of these commodities. For example, we might ask an individual, "would six units of potatoes and one of steak be at least as good as two units of potatoes and two units of steak?" This problem can be given a geometric frame of reference by letting the bundle consisting of x units of potatoes and y units of steak correspond to the point (x, y) in the first quadrant of the ordinary Cartesian plane (see Figure A.19). Two bundles are said to be *indifferent* or *equivalent* if each is judged at least as good as the other.

This indifference relation is trivially reflexive and the symmetry of its formulation makes it symmetric. It is customary to assume that it is also transitive and consequently a genuine equivalence relation. In the literature of economics one frequently encounters the phrase, "a rational decision

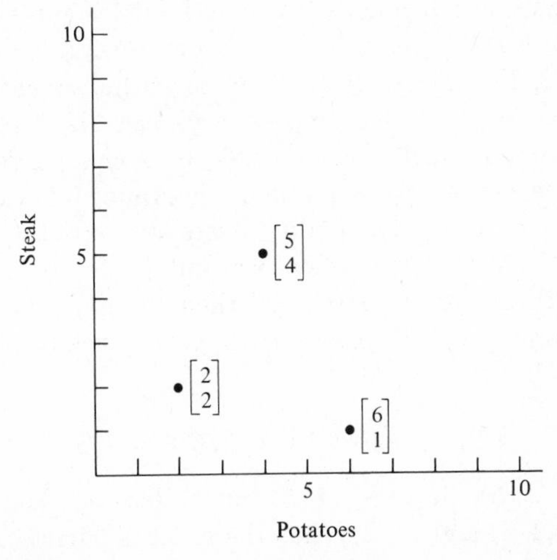

Figure A.19 Graphical representation of commodity bundles.

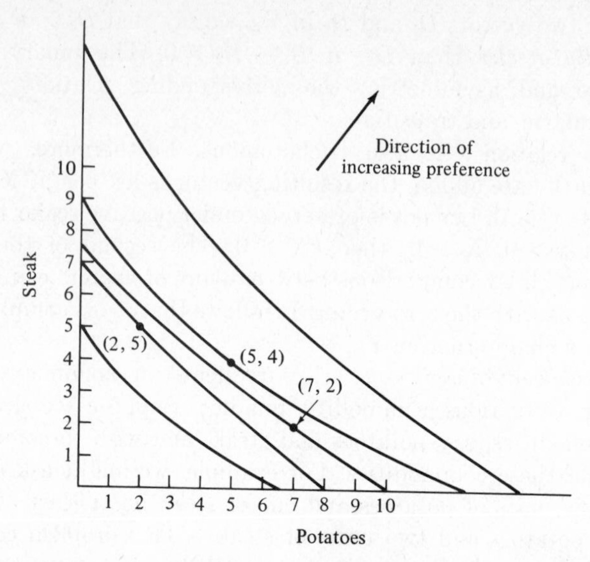

Figure A.20 Indifference curves for commodity bundles.

maker;" in this and in similar phrases the word "rational" is ordinarily interpreted to imply transitivity. Equivalence classes under this relation are called *indifference classes.*

Two cases are distinguished for the set of all commodity bundles: discrete and continuous. In the discrete case only integer values are allowed for x and y; in the continuous x and y can take arbitrary real values (subject perhaps to bounds on size). In the continuous case we ordinarily assume that the graph of each equivalence class, called an indifference curve, is continuous (this continuity is another tacit property of rationality). In Figure A.20 we picture several indifference curves. The bundles (2, 5) and (7, 2) are seen to be indifferent, whereas (5, 4) is on a higher curve. In this example the phrase, "is at least as good as," defines a weak order which is clearly not also a chain order.

Indifference curves are sometimes determined purely by cost considerations. Then if all commodity costs are strictly proportional to numbers of units considered, they turn out to be straight lines. If unit prices vary with quantities purchased then the linearity is lost.

We may also illustrate *partial order* geometrically. For points (x, y) in the first quadrant we write

$$(x, y) \leq (x', y') \text{ if } x \leq x' \text{ and } y \leq y'.$$

Thus $(2, 3) \leq (2, 5)$, $(3, 2) \leq (4, 5)$ but $\leq$ does not hold in either order for the points (2, 3) and (3, 2); thus the order is partial rather than complete. In this case there are no indifference curves, and for no two distinct points (x, y) and (x', y') can we have both $(x, y) \leq (x', y')$ and $(x', y') \leq (x, y)$ as is possible for weak order. This example can clearly be extended to n-tuples (see Section B.3).

The phrase, "at least as good as," captures the essence of the concept of preorder. Indifference is permitted and completeness is not required. An example might be provided by a collection of pictures in a gallery which an expert is asked to rate. There could be some cases of indifference, or ties, and there could be other cases where he would balk at any comparison. For example, he might refuse to make any preference judgment at all between a classical portrait and a modern abstraction.

Returning now to formal development, let $\precsim$ be a preorder on a set B. Then we define a new relation $\sim$ by the definition

$$a \sim b \text{ means } a \precsim b \text{ and } b \precsim a.$$

This new relation $\sim$ is an equivalence relation (the symbol $\sim$ [tilde] is ordinarily reserved to designate equivalence relations). To see this we observe that the definition of $\sim$ is symmetric; the reflexivity and transitivity of $\sim$ follow at once from the same properties of $\precsim$.

Let C be the set $B/\sim$ of equivalence classes $[a] = \sim(a)$ relative to $\sim$. Then *the relation* $\leq$ *defined by*

$$[a] \leq [b] \text{ means } a \precsim b$$

is a partial order on C.

To establish this statement we must first verify that $\leq$ is well defined, for as it stands the definition of $\leq$ apparently depends on the choice of representatives a and b from $[a]$ and $[b]$ rather than on $[a]$ and $[b]$ themselves. To verify that $\leq$ is well defined we need to show that if $a \sim a'$, $b \sim b'$ then $a' \precsim b'$ if and only if $a \precsim b$. This follows easily from the definition of $\sim$ and the transitivity of $\precsim$.

Clearly, the reflexivity and transitivity of $\leq$ follow from the same properties for $\precsim$. Next, suppose that $[a] \leq [b]$ and $[b] \leq [a]$; then $a \precsim b$ and $b \precsim a$; hence $a \sim b$ and $[a] = [b]$. This shows that $\leq$ is a partial order.

These results can be summarized by the statement that *any preorder can be decomposed into an equivalence relation and a partial order*. If $\precsim$ is also complete then so is the related $\leq$; hence *a weak order can be decomposed into an equivalence relation and a chain order*, or, alternatively stated, the equivalence classes relative to a weak order form a chain order.

Next, we introduce a new relation $\prec$ by the definition

$$a \prec b \text{ means } a \precsim b \text{ but } b \not\precsim a.$$

Observe that $\prec$ is transitive and asymmetric. A verbal translation of $\prec$ might be, "is not as good as." Similarly, if $\leq$ is a partial order relation then

$$a < b \text{ means } a \leq b \text{ but } b \not\leq a.$$

Note that $<$ is also transitive and asymmetric, moreover, $\leq$ is the union of $<$ and $=$ and $<$ is the difference of $\leq$ and $=$.

Sometimes the theory of partial order is developed starting with a relation $<$ which is transitive and asymmetric. This is justified on the ground that the corresponding relation $\leq$ is reflexive, transitive, and antisymmetric. The relation $<$ is for this reason sometimes called a partial order. Similarly, the theory of chain orders is sometimes developed with a relation such as $\preceq$ or, alternatively, with the relation $\prec$, which is transitive, asymmetric, and complete. These variations exist because elements that are not equal are ordered the same way whether one uses the relation $\prec$ or $\preceq$.

Suppose that $\leq$ is a partial order on a set B. An element b of B is said to be an *upper bound* for a subset C of B if $c \leq b$ for all c in C; if also $b < e$ for every upper bound e of C then b is said to be a *least upper bound* (l.u.b.) for C. A set C can have at most one least upper bound. An element d in C is said to be *maximal* in C if $d \leq c$ for c in C requires $d = c$. Note that (1) the least upper bound of C may or may not be in C, (2) a set C may have many maximal elements, (3) if the least upper bound c lies in C then it is maximal in C and is the only maximal element in C.

If in these definitions we replace $\leq$ by $\geq$, the resulting concepts are called, respectively, *lower bound*, *greatest lower bound* (g.l.b.) and *minimal element.*

Example 1. Let B be the set of real numbers. Then B has no g.l.b. or l.u.b., and no maximal, or minimal elements. The subset $C = \{x \mid 0 \leq x < 1\}$ has 0 as its g.l.b. and unique minimal element, has 1 as its l.u.b., and has no maximal element.

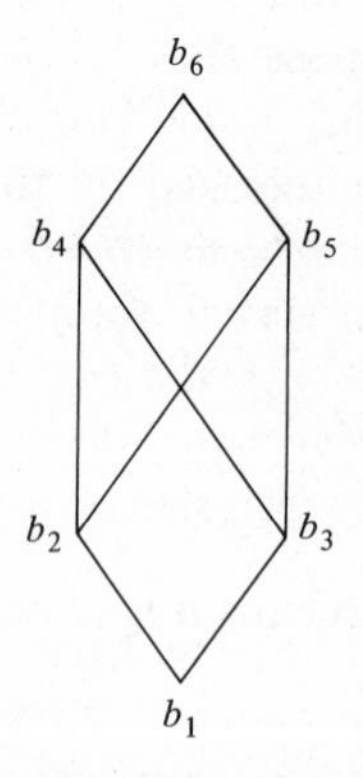

Figure A.21 Graph of a partial order.

Example 2. Let B consist of the six elements b_1, b_2, b_3, b_4, b_5, b_6, and suppose that $\leq$ is given by (A.22); thus $b_i \leq b_j$ if and only if $m_{ij} = 1$ (compare A.15)

$$M = [m_{ij}] = \begin{bmatrix} 1 & 1 & 1 & 1 & 1 & 1 \\ 0 & 1 & 0 & 1 & 1 & 1 \\ 0 & 0 & 1 & 1 & 1 & 1 \\ 0 & 0 & 0 & 1 & 0 & 1 \\ 0 & 0 & 0 & 0 & 1 & 1 \\ 0 & 0 & 0 & 0 & 0 & 1 \end{bmatrix}. \tag{A.22}$$

Then b_6 is the l.u.b. and b_1 is the g.l.b. of B. The subset $C = \{b_2, b_3\}$ has no l.u.b. and has b_1 as its g.l.b. The subset $D = \{b_2, b_3, b_4, b_5\}$ has b_4 and b_5 as maximal elements and b_2 and b_3 as minimal elements.

Let $\leq$ be a partial order relation on a set B. If for two elements b, c in B with $b < c$ the relation

$$b \leq x \leq c$$

requires $x = b$ or $x = c$, then we say that c *covers* b. If B is a finite set and $b \leq c$ then there exist elements $b_0 = b, b_1, \cdots, b_r = c$ such that b_i covers b_{i-1}, $i = 1, \cdots, r$. Thus every relation $\leq$ on B is completely specified by indicating only the cases where covering is involved. This fact provides the basis for a graphical representation of a finite partial order.

The graph has a set of nodes, one for each element of B, arranged so that if $b < c$ then the node for b is lower than the node for c. Then the node for b is joined to the node for c whenever c covers b. For example, the relation of the matrix M above is given graphically in Figure A.21. Further examples of partial orders are given graphically in Figure A.22.

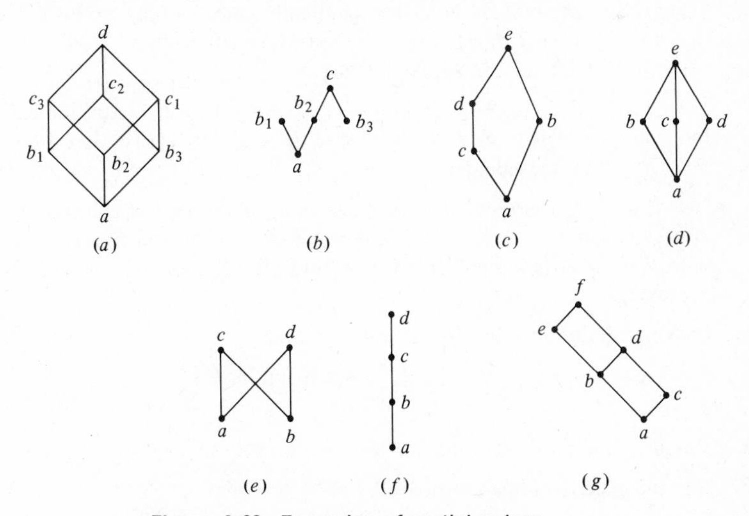

Figure A.22 Examples of partial orders.

PROBLEMS

A.1. Given the sets $A = \{3, 9, -5, 6\}$, $B = \{-5, y, d\}$, $C = \{13, p, q, 100\}$, determine each of the following sets: $A \cup B$, $B \cup C$, $A \cap B$, $A \cap C$, $A \cap B \cap C$, $(A \cup B) \cap C$, $A \cap (B \cup C)$, $A - B$, $B - A$, $A \oplus B$.

A.2. Which of the following are correct and why? (a) $1 = \{1\}$, (b) $1 \in \{1\}$, (c) $\{1\} \in \{\{1\}\}$, (d) $\{1\} \subset \{\{1\}\}$, (e) $\{1\} \subset \{1, \{1\}\}$, (f) $0 = \phi$, (g) $0 \in \phi$.

A.3. Let $S = [-2, 3, 1\}$, $T = \{5, -2, 7, 3, 9, 1\}$, $W = \{-2, 5, 3, 7, 9, 13, 1, 3, 5\}$. What are the sets $C_T S$, $C_W S$? What is the set $C_W T$?

A.4. Let $S = \{1, 3, 5, 7, 9\}$, $T = \{2, 4, 6, 8\}$, $V = \{1, 2, 3, 4\}$, $Z = \{3, 4, 5\}$, $U = \{1, 2, 3, 4, 5, 6, 7, 8, 9\}$, with U considered as the universe. (a) Determine the following sets: $S \cap T$, $S \cap V$, $S \cap V \cap Z$. (b) Show that $S \cup (V \cap Z) = (S \cup V) \cap (S \cup Z)$ and that $T \cap (V \cup Z) = (T \cap V) \cup (T \cap Z)$. (c) If C represents complement relative to the universal set U, show that $C(S \cap V) = C(S) \cup C(V)$ and that $C(S \cup V) = C(S) \cap C(V)$. (d) Determine the power set of V and the power set of S. (e) Recall that the symbolism $O(A)$ denotes the order or number of elements in the set A; show that $O(S) + O(C(S)) = O(U)$. Also show that $O(S) + O(V) = O(S \cap V) + O(S \cup V)$ and that $O(S - V) = O(S) - O(S \cap V)$.

A.5. Prove by induction that for all positive integers

$$1 + 2 + \cdots + n = \frac{n(n+1)}{2}.$$

A.6. Prove by induction that for all positive integers

$$1^3 + 2^3 + 3^3 + \cdots + n^3 = (1 + 2 + \cdots + n)^2.$$

A.7. Using the set builder notation, indicate each of the following sets: (a) $A = \{1, 3, 5, 7, 9, 11, 13\}$, (b) $B = \{100, 101, 102, \cdots, 199\}$, (c) $C = \{20, 22, 24, 26, 28, 30, 32, 34\}$.

A.8. Let $R = \{1, 2, 3, 4\}$, $S = \{1, 3, 9, -1\}$, $T = \{-1, -2, 7\}$; (a) indicate the following sets: $R \times S$, $R \times T$, $S \times R$, $R \times R$; (b) verify that $O(R \times S) = O(R) \cdot O(S)$; (c) calculate $R \times S \times T$.

A.9. Let $A = \{h, t\}$ represent the outcomes on a single toss of a coin. Determine $A \times A$, $A \times A \times A$. Is $A \times A$ a representation of the outcomes associated with two tosses of a coin? What is an interpretation of $A \times A \times A$ in this context?

A.10. Prove by induction that for all positive integers

$$1^2 + 2^2 + \cdots + n^2 = \frac{n(n+1)(2n+1)}{6}.$$

A.11. Prove by induction that for all positive integers $n^3 + 2n$ is divisible by 3.

A.12. For any positive integer n prove by induction that

$$3 + 6 + 9 + \cdots + 3n = \frac{3}{2}\, n(n+1).$$

A.13. Let $A = \{1, 2, 3, 4, 5\}$ and $B = \{6, 9, 12, 14, 18\}$; construct two different relations in $A \times B$.

A.14. Let $R = \{2, 4, 5, 9\}$, $S = \{1, 3, 6, 7, 8\}$, and let a relation Q in $R \times S$ be given as

$$Q = \{(2, 3), (4, 7), (4, 8), (5, 3), (9, 6)\}.$$

What is the domain of Q? What is the range of Q? What is the image of the element 4 in R? What is the counterimage of the element 3 in S? What is Q^T?

A.15. Given the sets $A = \{a_1, a_2, a_3\}$, $B = \{b_1, b_2, b_3, b_4\}$, and $C = \{c_1, c_2, c_3\}$, construct two different binary relations and two different ternary relations.

A.16. Let M and S denote the relations developed in Problem A.13. Determine a matrix for M and a matrix for S; determine the relations M^T and S^T; determine matrices for M^T and S^T. Show that M^T is a relation in $B \times A$.

A.17. Let A and B be finite sets; prove that $A \times B = B \times A$ if and only if $A = B$.

A.18. Prove that $2^n > n$ for every positive integer n.

A.19. Let $S = \{(x, y) \mid x + y \leq 4\}$ and $T = \{(x, y) \mid 2x - y \leq 5\}$; graph the relations $S \cap T$ and $S \cup T$.

A.20. Let $A_1 = \{(x, y) \mid x + y \geq 5\}$, $A_2 = \{(x, y) \mid x + 3y \leq 18\}$, $A_3 = \{(x, y) \mid x \geq 0\}$ and $A_4 = \{(x, y) \mid y \geq 0\}$. Graph the relation $A_1 \cap A_2 \cap A_3 \cap A_4$.

A.21. Sketch a graph of the relations A_1^T and A_2^T, where A_1 and A_2 are given in Problem A.20. Also sketch a graph of $A_1^T \cap A_2^T$.

A.22. Sketch a graph of the relations defined by each of the following and indicate which are functions. (a) $x \; F \; 6x$, domain is the set of all positive integers; (b) $x \; G \; \frac{3}{4}$ for all $x \leq 0$, $\frac{1}{2} \; G \; 1$, $1 \; G \; 1$, $2 \; G \; \frac{3}{4}$, $2 \; G - 3$, domain of G is the set $\{-1, -\frac{1}{2}, -\frac{1}{4}, 0, \frac{1}{2}, 1, 2\}$; (c) $x \; H \; x^3$, domain is the set of all real numbers.

A.23. Which of the following equations define functions from $\boldsymbol{R}$ into $\boldsymbol{R}$? (a) $f(x) = x^2 - x + 1$, (b) $g(x) = x - 2$, (c) $F(x) = y^2 - y + 1$, (d) $G(x) = 3$, (e) $H(x) = 2x^2 - x$.

A.24. Which of the equations in Problem A.23 define one to one functions from $\boldsymbol{R}$ into $\boldsymbol{R}$?

A.25. Let the functions F and G be defined as follows:

$$F: x \to y = F(x) = \frac{x^2 - 1}{x - 1}, \quad x \in \boldsymbol{R}, x \neq 1;$$

$$G: x \to y = G(x) = x + 1, \quad x \in \boldsymbol{R}.$$

Is $F = G$? Explain.

A.26. Let the functions G and H be defined as follows:

$$G: x \to y = G(x) = \frac{1}{x - 1}, \qquad H: x \to y = H(x) = x + 1.$$

Chose domains for these functions and form the product functions GH and HG. State domains for each of the product functions.

A.27. Let F be the function $F:x \to y = F(x) = 2x^2 - 5x$. Specify a domain of this function. What is F^T? Specify domain of F^T.

A.28. Let F and G be functions defined as follows:

$$F:x \to y = F(x) = 5x + 9, \quad x \in \boldsymbol{R},$$
$$G:x \to y = G(x) = 3x^2 - 2x + 1, \quad x \in \boldsymbol{R}.$$

What are the functions FG and GF^2? Indicate domains of each.

A.29. Given F and G in Problem A.28, what are the functions F^T, G^T, $F + G$, $F - G$, kF, kG? State domains of each.

A.30. Given F, G as in Problem A.28, let H be the function $H:x \to y = H(x) = x - 7$. What are the functions HGF and FGH?

A.31. Given the functions F and G in Problem A.28 and H in Problem A.30, which functions have inverses?

A.32. Let $F = \{(4, 3), (5, 7), (3, 2)\}$ and $G = \{(7, 1), (2, 9), (3, 13), (8, 4)\}$. Find $GF(4)$, $GF(3)$, $FG(2)$, $FG(7)$.

A.33. Let $F = \{(1, 6), (2, 4), (3, 8), (5, 8)\}$ and $G = \{(2, 9), (6, 8), (4, 10), (8, 9), (10, 11)\}$. What are the functions $F + G$, $F - G$, $2(F + 3G)$?

A.34. Let $F:x \to y = F(x) = 4 - x^2$, and let the domain be $\{x \mid -2 \leq x \leq 2\}$. Graph this function and the function F^T. Is it true that $F^T = F^{-1}$?

A.35. Let $G:x \to y = G(x) = |x - 3|$; graph this function. What is the relation G^T? Is $G^T = G^{-1}$?

A.36. Let $F:x \to y = F(x) = x - |x|$; graph this function and the relation F^T.

A.37. Let $G:x \to y = 2x + 1$, domain $\{x \mid 0 \leq x \leq 5\}$. What is G^{-1}? Graph these functions.

A.38. Let R be the relation "less than or equal to," $\boldsymbol{J}$ (see Section A.1) be the set R is defined on, and consider the ten classes of relations defined in Section A.10. Indicate which of these classes R belongs to.

A.39. Do Problem A.38 when R is replaced by the relation, "is taller than," and $\boldsymbol{J}$ is replaced by the set of all people.

A.40. Do Problem A.38 when R is replaced by the relation, is a subset of, and $\boldsymbol{J}$ is replaced by the collection of all subsets of $\boldsymbol{J}$.

A.41. Prove that if S and T are equivalence relations on a set M, then $S \cap T$ is an equivalence relation on M.

A.42. Do Problem A.38 when R is replaced by the relation, "is a brother of," and $\boldsymbol{J}$ is replaced by the set of all people.

A.43. Consider the relation, "has the same color hair as," defined on the set of all girls. Verify that this is an equivalence relation.

A.44. Let the relation, "is the same age as," be defined on the set of all baseball players in the American League. Is this an equivalence relation?

A.45. Consider the relation of, "lexicographically greater than," defined on the set of all vectors in V_2. Verify that this relation is transitive, asymmetric, and trichotomous. Is this relation a chain order on V_2?

A.46. For the equivalence relation in Problem A.43, what are the equivalence classes?

A.47. Let R be a relation on a set A and let R be transitive and asymmetric; prove that R is also irreflexive.

A.48. Let S be any binary relation in a set A. Prove that $S \cap S^T$ and $S \cup S^T$ are symmetric.

A.49. Suppose that the binary relation R is transitive and reflexive; prove that $R \cup R^T$ is an equivalence relation.

A.50. Let an infinite sequence $a_1, a_2, \cdots, a_n, \cdots$ be defined by $a_1 = 1$, $a_2 = 2, \cdots$, $a_n = a_{n-1} + (n-1)^2 a_{n-2}$, for all integers $n > 2$. Prove by induction that $a_n = n!$

A.51. Show that Figure A.22(f) defines a chain order.

A.52. In Figure A.22(d) show that l.u.b. $\{b$, g.l.b.$\{b, d\}\} \neq$ g.l.b. $\{$l.u.b.$\{b, c\}$, l.u.b.$\{b, d\}\}$.

A.53. Show that under lexicographic (dictionary) order the set of all English words is a chain order.

Linear Algebra

Appendix B

B.1. VECTORS.

Vectors arise in many contexts and at various levels of abstraction. One important and readily comprehended example is the use of a vector to describe and deal with a set of related numbers. For example the shopping list

2 loaves of bread,
10 pounds of sugar,
3 steaks,
1 head of lettuce,

can be described by a column of numbers as follows,

$$\begin{bmatrix} 2 \\ 10 \\ 3 \\ 1 \end{bmatrix}. \tag{B.1}$$

Here we understand that the first "component," 2, refers to bread, the second, 10, refers to sugar, and so on. We call this column of numbers a "column vector" and designate it by a single symbol, say, X. When the term vector is used without qualification we shall mean column vector (we will introduce "row vectors" later in connection with the concept of "transpose"). We call X a *column vector of degree* n if it describes n related numbers.

If in addition to the shopping list

$$X = \begin{bmatrix} 2 \\ 10 \\ 3 \\ 1 \end{bmatrix},$$

we have also a second shopping list containing different amounts of the same items

$$Y = \begin{bmatrix} 2 \\ 0 \\ 4 \\ 3 \end{bmatrix} \tag{B.2}$$

we may consider the composite of the two shopping lists which we call the sum, written $X + Y$. We have, by adding corresponding components of the vectors X and Y

$$X + Y = \begin{bmatrix} 2 \\ 10 \\ 3 \\ 1 \end{bmatrix} + \begin{bmatrix} 2 \\ 0 \\ 4 \\ 3 \end{bmatrix} = \begin{bmatrix} 2+2 \\ 10+0 \\ 3+4 \\ 1+3 \end{bmatrix} = \begin{bmatrix} 4 \\ 10 \\ 7 \\ 4 \end{bmatrix}. \tag{B.3}$$

We can regard (B.3) as indicating the total purchases (item by item) made by an agent of two housewives who have needs represented respectively by X and Y. Note that even though the second list contained no sugar, it was still necessary to use four components, for otherwise the accounting would become confused.

Historically, vectors were regarded as generalizations of numbers and the term *scalar* was used to designate a vector with one component. Thus "scalar" is just a synonym for "number," and we use the term in this sense.

More generally, if X and Y are vectors of degree n, say

$$X = \begin{bmatrix} x_1 \\ \cdot \\ \cdot \\ \cdot \\ x_n \end{bmatrix}, \qquad Y = \begin{bmatrix} y_1 \\ \cdot \\ \cdot \\ \cdot \\ y_n \end{bmatrix},$$

then the vector $X + Y$ called the *sum* of X and Y is

$$X + Y = \begin{bmatrix} x_1 + y_1 \\ \cdot \\ \cdot \\ \cdot \\ x_n + y_n \end{bmatrix}. \tag{B.4}$$

This operation has many of the familiar properties of addition of scalars. For example, *associativity*,

$$(X + Y) + Z = X + (Y + Z), \tag{B.5}$$

and *commutativity*

$$X + Y = Y + X; \tag{B.6}$$

these relations hold for arbitrary vectors X, Y, Z, each having the same number of components. Moreover, there is a *zero vector*

$$0 = 0_n = \begin{bmatrix} 0 \\ 0 \\ \cdot \\ \cdot \\ \cdot \\ 0 \end{bmatrix} \tag{B.7}$$

for which

$$0 + X = X + 0 = X, \tag{B.8}$$

and for each vector X there is a *negative*, designated $-X$ and given by

$$-X = \begin{bmatrix} -x_1 \\ -x_2 \\ \cdot \\ \cdot \\ \cdot \\ -x_n \end{bmatrix}$$

for which

$$X + (-X) = -X + X = 0. \tag{B.9}$$

Subtraction of vectors is analogous to subtraction of scalars; by $X - Y$ we mean a vector Z which, when added to Y, gives X. Using the concept of negative of a vector we can evaluate Z as $X - Y = X + (-Y)$. For example,

$$\begin{bmatrix} 2 \\ 1 \\ 3 \end{bmatrix} - \begin{bmatrix} 4 \\ -2 \\ 1 \end{bmatrix} = \begin{bmatrix} 2 \\ 1 \\ 3 \end{bmatrix} + \begin{bmatrix} -4 \\ 2 \\ -1 \end{bmatrix} = \begin{bmatrix} 2 \\ 3 \\ 2 \end{bmatrix}.$$

Returning to our shopping list (B.1), we might wish to double it to provide for twice as long a period. Thus

$$2\begin{bmatrix} 2 \\ 10 \\ 3 \\ 1 \end{bmatrix} = \begin{bmatrix} 4 \\ 20 \\ 6 \\ 2 \end{bmatrix}$$

is the vector representing the doubled list. This leads to the definition of a second operation: multiplication of a vector by a scalar,

$$cX = c\begin{bmatrix} x_1 \\ x_2 \\ \cdot \\ \cdot \\ \cdot \\ x_n \end{bmatrix} = \begin{bmatrix} cx_1 \\ cx_2 \\ \cdot \\ \cdot \\ \cdot \\ cx_n \end{bmatrix}. \tag{B.10}$$

This operation has the following properties: for all scalars c, c_1, c_2 and vectors X, X_1, X_2 of degree n

$$(c_1 + c_2)X = c_1X + c_2X, \tag{B.11}$$

$$c(X_1 + X_2) = cX_1 + cX_2, \tag{B.12}$$

$$c_1(c_2X) = (c_1c_2)X, \tag{B.13}$$

and

$$1X = X. \tag{B.14}$$

The scalars used in this book will ordinarily be real numbers, although for some purposes they may be further restricted by the nature of an application, or by conditions requiring them to be positive, or nonnegative, or integral, or some combination of these. Occasionally, we may even wish to employ complex scalars, and in higher mathematics, still more general classes of scalars find use. The set of all scalars that are available for use in a given situation is called the *scalar domain.*

Let $V = \{X, Y, \cdots\}$ be a set and let $F = \{c, \cdots\}$ be a scalar domain. Suppose that V is closed under an addition operation $X + Y$ and under a multiplication, cX, of elements X of V by scalar c (that is, if X, Y are any elements in V and c is any scalar in F, then $X + Y$ and cX are in V). Then if these operations also satisfy (B.5), (B.6), (B.8), (B.9), (B.11), (B.12), (B.13), and (B.14), we call V *a vector space over* F and give the name *vector* to elements of V.

It can be shown that under a finiteness assumption which will be discussed below as the assumption of "finite dimension" any vector space V can be described analytically by column vectors of some degree n. In the language of algebra we say that V is *isomorphic* to the set of all column vectors of degree n where n is an integer called the *dimension* of V. We denote by $V_n(F)$ the set of all column vectors of degree n whose components lie in F. When there is no doubt about the identity of the scalar domain, we abbreviate this to V_n.

Returning once more to (B.1), we introduce a *cost* vector

$$C = \begin{bmatrix} 20 \\ 6 \\ 90 \\ 15 \end{bmatrix}$$

whose components are, in order, the costs per unit (in cents) of the commodities in the shopping list. Then the cost of the entire list is given by

$$20\cdot 2 + 6\cdot 10 + 90\cdot 3 + 15\cdot 1 = 385.$$

In the following section we shall use the concept of combining a cost vector with a requirement vector (shopping list) to provide a natural interpretation for matrix multiplication.

It is also convenient to represent vectors in two- and three-dimensional coordinate space geometrically (the concept of dimension will be defined carefully in Section B.7). This can be done by regarding any point in the space as the end point of a line segment whose initial point is the origin and whose end point is the given point. A vector is thus determined by its end point and the corresponding line segment is the geometric representation of the vector. Examples are shown in Figure B.1. Also, given $X = \begin{bmatrix} x_1 \\ x_2 \end{bmatrix}$ and $Y = \begin{bmatrix} y_1 \\ y_2 \end{bmatrix}$, the vector $X + Y$ has for its geometric representation the diagonal of the parallelogram in Figure B.1.

The operation of multiplication of a vector by a scalar also has a simple geometric interpretation. If $X = \begin{bmatrix} x_1 \\ x_2 \end{bmatrix}$, then cX can be regarded as a

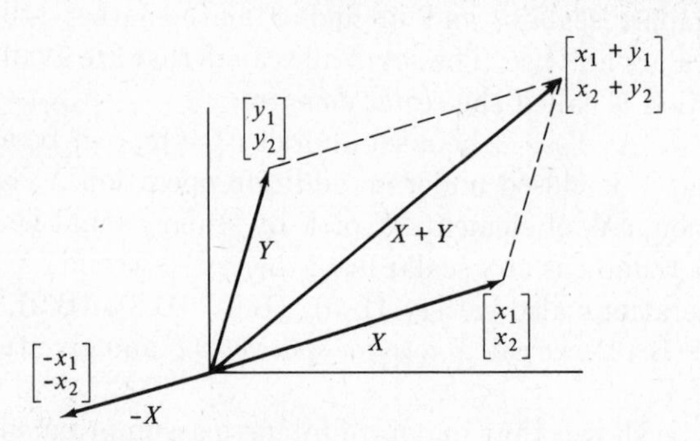

Figure B.1 Geometric representation of vectors.

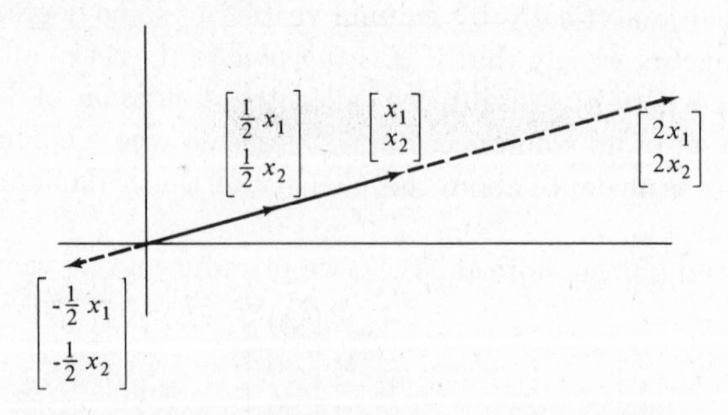

Figure B.2 Multiplication of a vector by a scalar.

"stretching" of the vector if $c > 1$ and a "contraction" if $0 \leq c < 1$. If $c < 0$, we have a stretching or a contraction followed by a rotation or "flip over." These possibilities are illustrated in Figure B.2.

B.2. INPUT-OUTPUT MATRICES.

We continue our use of everyday situations to motivate the introduction of the basic concepts of linear algebra. Having begun with the concept of vector we now continue by considering matrices.

To motivate the use of matrices and to develop some of the operations of matrices let us consider a simple example in an economic context. Suppose there is a factory which manufactures three products as *outputs:* leather easy chairs, wooden desks, and conference chairs with leather trimming. We denote these outputs by G_1, G_2, and G_3, respectively.

In order to produce these items various materials are needed. To simplify the problem, let us suppose that all that is needed is high-grade lumber, low-grade lumber, leather, and labor. These are *inputs* and we denote high-grade lumber by F_1, low-grade lumber by F_2, leather by F_3, and labor by F_4.

If we know how much of each input is required for each unit of output, we can tabulate these requirements as follows:

Input \ *Output*	G_1	G_2	G_3
F_1	1	6	2
F_2	3	1	1
F_3	4	0	2
F_4	3	5	2

(B.15)

Thus to make one G_3 we need 2 units of high-grade lumber, 1 unit of low-grade lumber, 2 units of leather, and 2 units of labor. Suppose the factory receives an order for 6 easy chairs, 4 desks, and 22 conference chairs. We shall then be interested in determining how much of each input is needed for the required output. The required output can be interpreted as being a vector X which we write as

$$X = \begin{bmatrix} 6 \\ 4 \\ 22 \end{bmatrix}. \tag{B.16}$$

If we leave off the labels in (B.15) and consider only the rectangular array of numbers in the table, we have what is called a matrix. We call this matrix a *technology matrix*; we denote it by the symbol A and thus write

$$A = \begin{bmatrix} 1 & 6 & 2 \\ 3 & 1 & 1 \\ 4 & 0 & 2 \\ 3 & 5 & 2 \end{bmatrix}. \tag{B.17}$$

To determine the required input we introduce a new vector

$$Y = \begin{bmatrix} y_1 \\ y_2 \\ y_3 \\ y_4 \end{bmatrix} \tag{B.18}$$

called the *input vector*. Here the component y_2, for example, is the number of units of low-grade lumber necessary to fill the order. To find y_2 we note by (B.15) that to make 6 units of G_1 we need $3 \cdot 6$ units of F_2; to make 4 units of G_2 we need $1 \cdot 4$ units of F_2; and to make 22 units of G_3 we need $1 \cdot 22$ units of F_2. Thus the total requirement of F_2 is

$$y_2 = 3 \cdot 6 + 1 \cdot 4 + 1 \cdot 22 = 44.$$

Similarly, we find

$$\begin{aligned} y_1 &= 1 \cdot 6 + 6 \cdot 4 + 2 \cdot 22 = 74, \\ y_3 &= 4 \cdot 6 + 0 \cdot 4 + 2 \cdot 22 = 68, \\ y_4 &= 3 \cdot 6 + 5 \cdot 4 + 2 \cdot 22 = 82. \end{aligned}$$

Here Y is determined by A and X, and we use this relationship to introduce matrix multiplication. Thus we write

$$Y = AX \tag{B.19}$$

or in full detail

$$\begin{bmatrix} y_1 \\ y_2 \\ y_3 \\ y_4 \end{bmatrix} = \begin{bmatrix} 1 & 6 & 2 \\ 3 & 1 & 1 \\ 4 & 0 & 2 \\ 3 & 5 & 2 \end{bmatrix} \cdot \begin{bmatrix} 6 \\ 4 \\ 22 \end{bmatrix} = \begin{bmatrix} 74 \\ 44 \\ 68 \\ 82 \end{bmatrix}. \tag{B.19$'$}$$

Matrix operations are discussed formally in the following sections. However, the idea is that each row of A combines with X to give the corresponding row of $Y = AX$. The method of combination is called *row-by-column multiplication* and its form is determined by the logic of the problem. The product of a row

$$[a_1 \quad a_2 \quad a_3]$$

and a column

$$\begin{bmatrix} x_1 \\ x_2 \\ x_3 \end{bmatrix}$$

is then defined by

$$[a_1 \quad a_2 \quad a_3] \begin{bmatrix} x_1 \\ x_2 \\ x_3 \end{bmatrix} = a_1x_1 + a_2x_2 + a_3x_3. \qquad \text{(B.20)}$$

Note that this product is a scalar, not a vector.

In the discussion of our problem we have not yet considered the costs of the materials involved in the production. We will suppose that the unit costs (in dollars) of high-grade lumber, low-grade lumber, leather, and labor are, respectively, 4, 2, 5, 9. We call

$$D = \begin{bmatrix} d_1 \\ d_2 \\ d_3 \\ d_4 \end{bmatrix} = \begin{bmatrix} 4 \\ 2 \\ 5 \\ 9 \end{bmatrix} \qquad \text{(B.21)}$$

the *cost vector*. Then the total cost of the order will be, given Y as in (B.19′),

$$74d_1 + 44d_2 + 68d_3 + 82d_4$$
$$= 74 \cdot 4 + 44 \cdot 2 + 68 \cdot 5 + 82 \cdot 9 = 1{,}462. \qquad \text{(B.22)}$$

This can be determined by means of matrix multiplication if we introduce the concept of *transpose* of a vector. The transpose of the column vector D is the row vector

$$[d_1 \quad d_2 \quad d_3 \quad d_4]$$

and is denoted by D^T. Then (B.22) can be written in vector notation as

$$D^TY = [d_1 \quad d_2 \quad d_3 \quad d_4] \cdot \begin{bmatrix} y_1 \\ y_2 \\ y_3 \\ y_4 \end{bmatrix} = [4 \quad 2 \quad 5 \quad 9] \cdot \begin{bmatrix} 74 \\ 44 \\ 68 \\ 82 \end{bmatrix} = 1{,}462. \qquad \text{(B.23)}$$

Now if $c_1 = 59$, $c_2 = 85$, and $c_3 = 40$ are, respectively, the market values per unit of outputs G_1, G_2, and G_3, then the total return realized by the manufacturer from the sale of the order is

$$6c_1 + 4c_2 + 22c_3 = 6 \cdot 59 + 4 \cdot 85 + 22 \cdot 40 = 1{,}574 \qquad \text{(B.24)}$$

or in matrix notation, if we let $C = \begin{bmatrix} c_1 \\ c_2 \\ c_3 \end{bmatrix}$ we can write

$$C^TX \qquad \text{(B.25)}$$

for this value. The profit will be the difference between total returns and total costs,

$$P = C^TX - D^TY = 1{,}574 - 1{,}462 = 112. \qquad \text{(B.26)}$$

But we know that $Y = AX$, hence

$$C^TX - D^TY = C^TX - D^T(AX) = C^TX - (D^TA)X$$
$$= (C^T - D^TA)X. \qquad (B.27)$$

(For justification of these manipulations with the matrices see the associative and distributive laws in Section B.4.) Now let $H^T = C^T - D^TA$; then

$$H^T = [59 \quad 85 \quad 40] - [4 \quad 2 \quad 5 \quad 9]\begin{bmatrix} 1 & 6 & 2 \\ 3 & 1 & 1 \\ 4 & 0 & 2 \\ 3 & 5 & 2 \end{bmatrix}$$

$$= [59 \quad 85 \quad 40] - [57 \quad 71 \quad 38]$$
$$= [2 \quad 14 \quad 2],$$

or

$$H = \begin{bmatrix} 2 \\ 14 \\ 2 \end{bmatrix}.$$

The components of the vector H are the unit profits for the three products. Thus to get the total profit P we form

$$P = h_1x_1 + h_2x_2 + h_3x_3 = 2\cdot 6 + 14\cdot 4 + 2\cdot 22 = 112, \qquad (B.28)$$

as in (B.26). In matrix form

$$P = H^TX = 2x_1 + 14x_2 + 2x_3; \qquad (B.29)$$

this P gives the profit for an arbitrary order X.

The unit profit vector H clearly provides useful information to management. In actual practice the accounting policies involved in assessing indirect costs or "loads" may make determination of H difficult. Indeed, in some cases the assessment of loads is so arbitrary as to make H meaningless.

Next, suppose that a closer look at the factory discloses that it has two divisions: the first for woodworking and the second for assembly. Each division has its own technology matrix. Thus, the woodworking might be described by the matrix

$$A_1 = \begin{bmatrix} 1 & 6 & 2 \\ 3 & 1 & 1 \\ 0 & 0 & 0 \\ 1 & 3 & 1 \end{bmatrix} \qquad (B.30)$$

and the assembly by the matrix

$$A_2 = \begin{bmatrix} 0 & 0 & 0 \\ 0 & 0 & 0 \\ 4 & 0 & 2 \\ 2 & 2 & 1 \end{bmatrix}. \qquad (B.31)$$

The zeros in the third row of A_1 indicate that no leather is used in the woodworking; similarly, the zeros in the first two rows of A_2 indicate that no wood is used in assembly except for that which has already been processed in the woodworking division. Note that the available units of labor have been allocated to the two divisions. It is natural to call A the *sum* of A_1 and A_2; that is,

$$\begin{bmatrix} 1 & 6 & 2 \\ 3 & 1 & 1 \\ 0 & 0 & 0 \\ 1 & 3 & 1 \end{bmatrix} + \begin{bmatrix} 0 & 0 & 0 \\ 0 & 0 & 0 \\ 4 & 0 & 2 \\ 2 & 2 & 1 \end{bmatrix} = \begin{bmatrix} 1 & 6 & 2 \\ 3 & 1 & 1 \\ 4 & 0 & 2 \\ 3 & 5 & 2 \end{bmatrix}. \tag{B.32}$$

In general, matrix addition resembles vector addition in that the sum is obtained by adding corresponding entries.

It may be instructive to note that the zero rows in A_1 and A_2 serve as "place markers". If one were concerned only with the woodworking division the third row would not have been used; similarly, the assembly division would by itself require only two rows. The use of the rows of zeros makes it possible to combine the two individual technology matrices into a single one. This is one reason that it is desirable to allow some room for expansion in fixing the number of components or in *dimensioning* a problem. This is especially important when writing computer programs.

Industrial processes which involve first making subassemblies and then putting these together in various combinations to construct final products provide a natural introduction to multiplication of matrices. For example, a factory whose final outputs are 12 models of radios, 7 models of television sets, 5 models of phonographs, 3 radio-phonograph combinations, and 4 radio-phonograph-television combinations has 31 final outputs. These are constructed using perhaps several hundred different components such as tubes, resistors, cabinets, turntables, motors, control knobs. The manufacture of these components involves the use of suitable amounts of some collection of raw materials such as various types of metals, grades of glass, types of woods. If there are, say, 200 different components then the assembly is described by a technology matrix C having 200 rows and 31 columns.

The preliminary processes in which the various raw materials, say 80 in all, are made into components is described by a technology matrix B having 80 rows and 200 columns.

Clearly, if one traces through the combined process he can construct a technology matrix A having 80 rows and 31 columns which will give the amount of each raw material needed for each final product. It is natural to ask if there is any simple connection between the three technology matrices A, B, C. The answer is, yes; it turns out that we define matrix multiplication in such a way that $A = BC$. This means that matrix multiplication can be regarded as providing a systematic way for organizing and manipulating the data involved in tracing through the combined process.

The numbers of rows and columns involved in our hypothetical illustrative example are quite small as compared with technology matrices being used routinely in today's industries. However, for further illustration we introduce an example with smaller size. Suppose that 4 raw materials F_1, F_2, F_3, and F_4 are used to construct 5 components G_1, G_2, G_3, G_4, and G_5, and that the components $G_1, \cdots, G_5$ are used as inputs in the second process to produce 3 final products H_1, H_2, and H_3. The input-output tables for the construction of components and final assembly processes are, respectively

Input \ Output	G_1	G_2	G_3	G_4	G_5
F_1	4	3	2	1	2
F_2	1	2	0	2	0
F_3	0	1	2	1	1
F_4	0	0	3	1	2

Input \ Output	H_1	H_2	H_3
G_1	2	4	1
G_2	1	3	0
G_3	0	1	2
G_4	1	0	2
G_5	2	1	1

from which the corresponding technology matrices are

$$B = \begin{bmatrix} 4 & 3 & 2 & 1 & 2 \\ 1 & 2 & 0 & 2 & 0 \\ 0 & 1 & 2 & 1 & 1 \\ 0 & 0 & 3 & 1 & 2 \end{bmatrix}, \quad C = \begin{bmatrix} 2 & 4 & 1 \\ 1 & 3 & 0 \\ 0 & 1 & 2 \\ 1 & 0 & 2 \\ 2 & 1 & 1 \end{bmatrix}.$$

Now let A be the technology matrix relating the raw materials F_1, F_2, F_3, and F_4 to the final products H_1, H_2, and H_3; clearly A must have 4 rows and 3 columns,

$$A = \begin{bmatrix} a_{11} & a_{12} & a_{13} \\ a_{21} & a_{22} & a_{23} \\ a_{31} & a_{32} & a_{33} \\ a_{41} & a_{42} & a_{43} \end{bmatrix};$$

we seek to determine the 12 entries a_{ij} of the matrix A. Consider a_{11}; it represents the amount of F_1 needed to produce one unit of H_1. According to the matrix C one unit of H_1 requires 2 units of G_1, 1 unit of G_2, 0 units of G_3, 1 unit of G_4, and 2 units of G_5. Moreover, according to the matrix B, the number of units of F_1 needed per unit of G_1, G_2, G_3, G_4, and G_5 are, respectively, 4, 3, 2, 1, and 2. We conclude that

$$\begin{aligned} a_{11} &= 4\cdot 2 + 3\cdot 1 + 2\cdot 0 + 1\cdot 1 + 2\cdot 2 \\ &= 8 + 3 + 0 + 1 + 4 = 16. \end{aligned}$$

We observe that a_{11} is obtained by our row-by-column rule for multiplication using the first row of B and the first column of C.

Further reflection leads us to the conclusion that a_{12} will be the product of the first row of B and the second column of C,

$$a_{12} = 4 \cdot 4 + 3 \cdot 3 + 2 \cdot 1 + 1 \cdot 0 + 2 \cdot 1 = 29,$$

and more generally, for each i and j, a_{ij} is the product of the ith row of B with the jth column of C.

This reasoning motivates the formal definition of matrix multiplication which is given in Section B.4. In the present case we get

$$A = \begin{bmatrix} 16 & 29 & 12 \\ 6 & 10 & 5 \\ 4 & 6 & 7 \\ 5 & 5 & 10 \end{bmatrix}.$$

One might ask if the two processes in the wood products factory should not be handled by (matrix) multiplication instead of by addition. The answer is that both ways are available for that situation. We may proceed as follows. Consider that the purpose of the first processing is to produce finished kits of wooden parts, one for each final product, and denoted W_1, W_2, and W_3. We also consider leather and labor as outputs of the first process. We regard leather as an output whose only input is leather and with technology coefficient 1; that is, the row and column for leather have zero entries except for a one where they intersect. This is one way of stating that leather is not really used in the first process. Then we regard labor as an output of the first process whose only input is labor; this really means that some of our labor potential is reserved for the second process.

For example, the input-output table for the first process is as shown below. The intermediate products (W_1, W_2, W_3, F_1, and F_2) are now inputs

Output / *Input*	W_1	W_2	W_3	*Leather* F_3	*Labor* F_4
F_1	1	6	2	0	0
F_2	3	1	1	0	0
F_3	0	0	0	1	0
F_4	1	3	1	0	1

to the second process, whose input-output table is as follows.

Output / *Input*	G_1	G_2	G_3
W_1	1	0	0
W_2	0	1	0
W_3	0	0	1
F_3	4	0	2
F_4	2	2	1

We now have related the original 4 inputs, the original 3 outputs, and the 5 intermediate products (the three kits of wooden parts plus leather and labor). The matrices B and C for the two processes are, respectively,

$$B = \begin{bmatrix} 1 & 6 & 2 & 0 & 0 \\ 3 & 1 & 1 & 0 & 0 \\ 0 & 0 & 0 & 1 & 0 \\ 1 & 3 & 1 & 0 & 1 \end{bmatrix}, \quad C = \begin{bmatrix} 1 & 0 & 0 \\ 0 & 1 & 0 \\ 0 & 0 & 1 \\ 4 & 0 & 2 \\ 2 & 2 & 1 \end{bmatrix}.$$

One can easily verify that the original technology matrix A is equal to the product BC, as required. Note that the final two columns of the matrix B account for leather and labor held over for the second process and that the columns of C state that each final product is made from a corresponding kit together with appropriate leather and labor.

B.3. MATRICES.

In this section and the three which follow we give formal definitions of matrices and matrix operations, and state (without proof) certain important properties of matrix algebra.

A matrix is a rectangular array of numbers (or scalars) arranged in rows and columns. For example,

$$A = \begin{bmatrix} 1 & 5 & 2 \\ 4 & 1 & 1 \\ 3 & 0 & 1 \\ 4 & 5 & 2 \end{bmatrix}, \quad B = \begin{bmatrix} 3 & 4 & 1 \\ 2 & 3 & 0 \\ 1 & 1 & 1 \\ 2 & 3 & 6 \end{bmatrix}, \quad C = \begin{bmatrix} 1 \\ 3 \\ 7 \end{bmatrix},$$

$$D = \begin{bmatrix} 4 & -2 & 5 \\ 6 & 4 & 0 \\ 3 & 1 & 7 \end{bmatrix}, \quad E = \begin{bmatrix} 1 & 1 & 1 & 1 \\ 1 & 1 & 1 & 1 \\ 1 & 1 & 1 & 1 \end{bmatrix}, \quad F = [f_1 \; f_2 \; f_3 \; f_4],$$

$$G = \begin{bmatrix} g_{11} & g_{12} & g_{13} \\ g_{21} & g_{22} & g_{23} \end{bmatrix}, \quad H = \begin{bmatrix} 1 & 4 & 3 & 4 \\ 5 & 1 & 0 & 5 \\ 2 & 1 & 1 & 2 \end{bmatrix}, \quad X = \begin{bmatrix} x_1 \\ x_2 \\ x_3 \end{bmatrix}$$

are matrices. In general, we shall use capital letters to represent matrices, and we shall use small letters with two subscripts for the scalar *entries* or *elements* of the matrix. The first subscript indicates the row and the second subscript the column in which the entry lies. Thus, g_{12} is the entry in the first row and the second column of G; and, similarly, $a_{12} = 5$, $b_{23} = 0$, $c_{31} = 7$, and $e_{ij} = 1$ for all i and j. If a matrix has only one row or only one column then just one subscript need be used, as is the case with the vectors F and X above.

Other symbols used to denote a matrix are large parentheses and parellel bars; thus

$$\begin{bmatrix} 2 & 1 \\ 3 & 0 \end{bmatrix}, \quad \left\|\begin{matrix} 2 & 1 \\ 3 & 0 \end{matrix}\right\|, \quad \begin{pmatrix} 2 & 1 \\ 3 & 0 \end{pmatrix}$$

are three ways of writing the same matrix. The first and second are best adapted to typesetting whereas the third is convenient for handscript. Single bars have been assigned by tradition to denote determinant.

We sometimes use the abbreviated notation

$$G = [g_{ij}]. \tag{B.33}$$

The shape of a matrix is determined by two integers called its *row degree* and *column degree* which are, respectively, the number of rows and columns in the matrix. For example, the matrix G above has row degree 2 and column degree 3. We also say that G is a 2 by 3 matrix; similarly A is a 4 by 3 and C is a 3 by 1 matrix. If, as with D, the number of rows is the same as the number of columns, we say that the matrix is *square* and use the term *degree* to indicate this common number. Thus D has degree 3.

A matrix such as C, whose column degree is one, is called a *column vector;* similarly if, as in F, the row degree is one we speak of a *row vector*. The entries of a vector are also called its *components*; for example, the third component of C is 7.

Two matrices, P and Q, are said to be *equal*, written $P = Q$, if they are identical, that is, if they have the same shape and if the entries in corresponding positions are the same. For example, $C = X$ if and only if $1 = x_1, 3 = x_2, 7 = x_3$.

Let $P = [p_{ij}]$ and $Q = [q_{ij}]$ be matrices with the same shape. We define $P < Q$ to mean that

$$p_{ij} < q_{ij} \qquad \text{for all } i \text{ and } j. \tag{B.34}$$

Similarly, we define $P \leq Q$ to mean that

$$p_{ij} \leq q_{ij} \qquad \text{for all } i \text{ and } j, \tag{B.35}$$

and we define $P \lesssim Q$ to mean that $P \leq Q$ but $P \neq Q$, that is (B.35) holds but with at least one strict inequality.

For example, let $A_1 = [1 \quad 2 \quad 4]$, $A_2 = [1 \quad 3 \quad 7]$, $A_3 = [2 \quad 4 \quad 6]$ then

$$A_1 \leq A_1,\ A_1 \lesssim A_2,\ A_1 < A_3 \tag{B.36}$$

but none of the relations holds between A_2 and A_3 since $1 < 2$ and $7 > 6$.

A matrix (vector) is said to be *nonnegative* if each entry (component) is nonnegative. Thus, we write

$$X \geq 0 \tag{B.37}$$

to indicate that each component x_i of X is a nonnegative real number. Such vectors arise naturally in activity analysis, since a negative level of operation of a manufacturing process is meaningless.

A *positive* matrix (vector) is one each of whose entries (components) is positive. Thus, we write

$$X > 0 \tag{B.38}$$

to indicate that each component x_i of X is a positive real number. For example, if X is a program vector then (B.38) indicates that every activity is being engaged in at a positive level.

B.4. MATRIX OPERATIONS.

ADDITION OF MATRICES.

We can add two matrices if and only if they have the same shape. To add matrices we add their corresponding entries. For example,

$$\begin{bmatrix} a_{11} & a_{12} \\ a_{21} & a_{22} \\ a_{31} & a_{32} \end{bmatrix} + \begin{bmatrix} b_{11} & b_{12} \\ b_{21} & b_{22} \\ b_{31} & b_{32} \end{bmatrix} = \begin{bmatrix} a_{11} + b_{11} & a_{12} + b_{12} \\ a_{21} + b_{21} & a_{22} + b_{22} \\ a_{31} + b_{31} & a_{32} + b_{32} \end{bmatrix},$$

or, we can write $[a_{ij}] + [b_{ij}] = [a_{ij} + b_{ij}]$. Another example is

$$\begin{bmatrix} 3 & 2 \\ 7 & 3 \\ 4 & 4 \end{bmatrix} + \begin{bmatrix} 4 & 7 \\ 2 & 5 \\ 5 & 6 \end{bmatrix} = \begin{bmatrix} 3+4 & 2+7 \\ 7+2 & 3+5 \\ 4+5 & 4+6 \end{bmatrix} = \begin{bmatrix} 7 & 9 \\ 9 & 8 \\ 9 & 10 \end{bmatrix}.$$

More generally, for any two m by n matrices P and Q, we have $P + Q = R$ where $R = [r_{ij}]$ is again an m by n matrix and has

$$r_{ij} = p_{ij} + q_{ij}, \qquad (i = 1, \cdots, m; j = 1, \cdots, n). \tag{B.39}$$

MULTIPLICATION OF MATRICES.

Before giving a general definition of matrix multiplication, we first consider the product of a row vector $P = [p_j]$ and a column vector $Q = [q_i]$ each having four components. As was indicated in Section B.2,

$$PQ = [p_1 \quad p_2 \quad p_3 \quad p_4] \begin{bmatrix} q_1 \\ q_2 \\ q_3 \\ q_4 \end{bmatrix} = p_1q_1 + p_2q_2 + p_3q_3 + p_4q_4.$$

If the number of components is n then

$$PQ = p_1q_1 + p_2q_2 + \cdots + p_nq_n = \sum_{j=1}^{n} p_jq_j. \tag{B.40}$$

For example,

$$[1 \quad 3 \quad -1] \begin{bmatrix} 4 \\ 0 \\ 2 \end{bmatrix} = 1 \cdot 4 + 3 \cdot 0 + (-1) \cdot 2 = 2.$$

Note that the product PQ of two vectors P and Q is not defined unless

(a) P is a row vector and Q is a column vector, and
(b) Column degree of P equals row degree of Q. (B.41)

Moreover, note that the product is a scalar.

More generally, the product of two matrices D and E exists if and only if the column degree of the first equals the row degree of the second. This product, if it exists, will be a matrix F. Suppose D is an m by n matrix and E is an n by p matrix. The product DE has the row degree of D and the column degree of E and is an m by p matrix. Schematically the operation can be illustrated as follows:

$$\overset{n}{m\begin{bmatrix} o & o & o & o \\ o & o & o & o \\ o & o & o & o \end{bmatrix}} \; \overset{p}{n\begin{bmatrix} o & o & o & o & o \\ o & o & o & o & o \\ o & o & o & o & o \\ o & o & o & o & o \end{bmatrix}} = \overset{p}{m\begin{bmatrix} o & o & o & o & o \\ o & o & o & o & o \\ o & o & o & o & o \end{bmatrix}}.$$

$$D \qquad\qquad\qquad E \qquad\qquad\qquad F$$

In words, the row degree of the product of two matrices is equal to the row degree of the left factor and the column degree of the product is equal to the column degree of the right factor.

To get the entry f_{11} (the entry in the first row and first column) of F, multiply the first row of D by the first column of E, considering each as vectors. In general, to get the entry f_{ij} of F, multiply the ith row of D by the jth column of E considered as vectors.

Example:

$$\underset{A}{\begin{bmatrix} 1 & 0 & 1 \\ 3 & 1 & 2 \\ 2 & -4 & 1 \end{bmatrix}} \underset{B}{\begin{bmatrix} 6 & 2 & 1 \\ -3 & 4 & 0 \\ 1 & 2 & -6 \end{bmatrix}} = \begin{bmatrix} 1\cdot 6 + 0\cdot(-3) + 1\cdot 1 & 1\cdot 2 + 0\cdot 4 + 1\cdot 2 & 1\cdot 1 + 0\cdot 0 + 1\cdot(-6) \\ 3\cdot 6 + 1\cdot(-3) + 2\cdot 1 & 3\cdot 2 + 1\cdot 4 + 2\cdot 2 & 3\cdot 1 + 1\cdot 0 + 2\cdot(-6) \\ 2\cdot 6 + (-4)\cdot(-3) + 1\cdot 1 & 2\cdot 2 + (-4)\cdot 4 + 1\cdot 2 & 2\cdot 1 + (-4)\cdot 0 + 1\cdot(-6) \end{bmatrix},$$

$$= \begin{bmatrix} 7 & 4 & -5 \\ 17 & 14 & -9 \\ 25 & -10 & -4 \end{bmatrix}.$$

Note that the product

$$\underset{B}{\begin{bmatrix} 6 & 2 & 1 \\ -3 & 4 & 0 \\ 1 & 2 & -6 \end{bmatrix}} \underset{A}{\begin{bmatrix} 1 & 0 & 1 \\ 3 & 1 & 2 \\ 2 & -4 & 1 \end{bmatrix}} = \begin{bmatrix} 14 & -2 & 11 \\ 9 & 4 & 5 \\ -5 & 26 & -1 \end{bmatrix}$$

and thus, a product AB is not necessarily equal to BA. In fact, it is possible that the product of two matrices L and R may exist if L is the left factor and R is the right factor, but may not exist if L is the right factor and R the left factor. For example, if

$$L = \begin{bmatrix} 1 & 0 & 2 \\ -1 & 1 & 3 \end{bmatrix} \quad \text{and} \quad R = \begin{bmatrix} -5 \\ 1 \\ 6 \end{bmatrix},$$

then, $LR = \begin{bmatrix} 1 & 0 & 2 \\ -1 & 1 & 3 \end{bmatrix} \begin{bmatrix} -5 \\ 1 \\ 6 \end{bmatrix} = \begin{bmatrix} 7 \\ 24 \end{bmatrix}$,

but RL is not defined.

In general, let $D = [d_{ij}]$ and $E = [e_{ij}]$ be matrices with shapes m by n and n by p, respectively. Then the product

$$DE = F = [f_{ij}] \tag{B.42}$$

is an m by p matrix and

$$\begin{aligned} f_{ij} &= d_{i1}e_{1j} + d_{i2}e_{2j} + \cdots + d_{in}e_{nj} \\ &= \sum_{k=1}^{n} d_{ik}e_{kj}, \quad (i = 1, \cdots, m; j = 1, \cdots, p). \end{aligned} \tag{B.43}$$

Note that if P^i is the ith row of D and Q_j is the jth column of E then

$$f_{ij} = P^iQ_j. \tag{B.44}$$

MULTIPLICATION OF A MATRIX BY A SCALAR.

A scalar, as has been noted before, is merely a number. To multiply a scalar and a matrix we multiply each entry in the matrix by the scalar. For example,

$$2\begin{bmatrix} 6 & 2 & 1 \\ -3 & 4 & 0 \\ 1 & 2 & -6 \end{bmatrix} = \begin{bmatrix} 12 & 4 & 2 \\ -6 & 8 & 0 \\ 2 & 4 & -12 \end{bmatrix}.$$

If $P = [p_{ij}]$ is any matrix and k is any scalar then

$$kP = [kp_{ij}] \tag{B.45}$$

and has the same shape as P.

PROPERTIES OF MATRIX OPERATIONS.

Almost all of the rules of elementary algebra hold for matrices; these are:

1. The sum of two matrices (if it exists) is the same in whatever order the matrices are added:

$$A + B = B + A \tag{B.46}$$

This is called the *commutative law* for addition.

2. The sum of three or more matrices is the same in whatever way the matrices are grouped:

$$(A + B) + C = A + (B + C) = A + B + C. \quad \text{(B.47)}$$

   This is called the *associative law* for addition.
3. Recall that the commutative law for multiplication of matrices does not hold; thus $AB \neq BA$ in general, and also, even though AB exists, BA may not.
4. The product of three or more matrices (if it exists) is the same in whatever way the matrices are grouped:

$$(AB)C = A(BC) = ABC. \quad \text{(B.48)}$$

   This is the *associative law* for multiplication.
5. The product of a matrix and the sum of other matrices (when it exists) is the same as the sum of the products obtained by multiplying each of the matrices of the sum by that matrix:

$$A(B + C) = AB + AC, \quad \text{(B.49)}$$

$$(A + B)C = AC + BC. \quad \text{(B.50)}$$

   These are the *distributive laws.*

Note: In ordinary arithmetic, there is only one distributive law, but since the commutative law for multiplication does not hold, we need two distributive laws for matrices.

TRANSPOSITION OF MATRICES.

The *transpose* P^T of a p by q matrix $P = [p_{ij}]$ is the q by p matrix $Q = [q_{ij}]$ obtained by interchanging the rows and columns of P, thus

$$P^T = Q = [q_{ij}] \quad \text{(B.51)}$$

where

$$q_{ij} = p_{ji}, \qquad (i = 1, \cdots, q; j = 1, \cdots, p). \quad \text{(B.52)}$$

For example, if

$$P = \begin{bmatrix} 1 & -1 & 3 & 2 \\ 3 & 7 & -4 & 5 \end{bmatrix},$$

then

$$P^T = \begin{bmatrix} 1 & 3 \\ -1 & 7 \\ 3 & -4 \\ 2 & 5 \end{bmatrix},$$

and if

$$X = \begin{bmatrix} 5 \\ -1 \\ 0 \\ 4 \end{bmatrix}$$

then

$$X^T = [5 \quad -1 \quad 0 \quad 4].$$

We note several important properties of transposes. First, *the transpose of the transpose is the original matrix;* for every matrix P we have

$$(P^T)^T = P. \tag{B.53}$$

Second, *the transpose of a sum is the sum of the transposes*

$$(P + Q)^T = P^T + Q^T; \tag{B.54}$$

and this can be extended to any number of summands,

$$(P_1 + P_2 + \cdots + P_s)^T = P_1^T + P_2^T + \cdots + P_s^T.$$

For example, if

$$A = \begin{bmatrix} 1 & 2 \\ 3 & 4 \end{bmatrix}, \qquad B = \begin{bmatrix} 2 & -1 \\ 1 & 3 \end{bmatrix}$$

then

$$A + B = \begin{bmatrix} 3 & 1 \\ 4 & 7 \end{bmatrix}$$

and

$$(A + B)^T = \begin{bmatrix} 3 & 4 \\ 1 & 7 \end{bmatrix} = \begin{bmatrix} 1 & 3 \\ 2 & 4 \end{bmatrix} + \begin{bmatrix} 2 & 1 \\ -1 & 3 \end{bmatrix}$$
$$= A^T + B^T.$$

Third, *the transpose of a scalar times a matrix is the same scalar times the transpose of that matrix,*

$$(kP)^T = kP^T. \tag{B.55}$$

Fourth, *the transpose of a product is the product of the transposes in reverse order,*

$$(PQ)^T = Q^T P^T; \tag{B.56}$$

this can also be extended to any number of factors,

$$(P_1 P_2 \cdots P_s)^T = P_s^T \cdots P_2^T P_1^T. \tag{B.57}$$

For example, with A and B as above

$$AB = \begin{bmatrix} 4 & 5 \\ 10 & 9 \end{bmatrix},$$

and

$$(AB)^T = \begin{bmatrix} 4 & 10 \\ 5 & 9 \end{bmatrix} = \begin{bmatrix} 2 & 1 \\ -1 & 3 \end{bmatrix} \begin{bmatrix} 1 & 3 \\ 2 & 4 \end{bmatrix} = B^T A^T.$$

If we regard a scalar as being a 1 by 1 matrix, then we note that for any scalar b we have

$$b = b^T. \tag{B.58}$$

Next, let X and Y be two column vectors each of row degree n. Then X^TY is a scalar; hence

$$X^TY = (X^TY)^T = Y^T(X^T)^T = Y^TX. \tag{B.59}$$

The product X^TY of two column vectors is sometimes called their *inner product* (or *dot product* or *scalar product*). It follows from (B.59) that the value of an inner product does not depend on the order of its factors; in more technical language, we say that the inner product is a *symmetric function* of two vectors.

B.5. SUBMATRICES, PARTITIONING.

Consider the two matrices

$$A = \begin{bmatrix} 2 & 1 & 3 & 7 \\ 1 & 2 & -6 & 4 \\ 3 & 0 & 1 & 3 \end{bmatrix}, \qquad B = \begin{bmatrix} -6 & 4 \\ 1 & 3 \end{bmatrix}. \tag{B.60}$$

B consists of the part of A that is left after discarding its first row and its first two columns; we call B a *submatrix* of A and designate it by

$$B = A(1 \mid 1, 2). \tag{B.61}$$

More generally, we denote by

$$A(i_1, i_2, \cdots, i_r \mid j_1, j_2, \cdots, j_s) \tag{B.62}$$

the submatrix obtained from A by deleting the rows $i_1, i_2, \cdots, i_r$ and the columns $j_1, j_2, \cdots, j_s$.

We also write

$$B = A[2, 3 \mid 3, 4] \tag{B.63}$$

to indicate that B is made up of the elements in A where the rows 2 and 3 intersect the columns 3 and 4. More generally,

$$A[i_1, i_2, \cdots, i_r \mid j_1, j_2, \cdots, j_s] \tag{B.64}$$

is the r by s submatrix of A consisting of the elements where the rows $i_1, i_2, \cdots, i_r$ meet the columns $j_1, j_2, \cdots, j_s$.

The rows and columns selected for consideration need not be adjacent; thus if A is the matrix in (B.60)

$$A(2 \mid 1, 3, 4) = \begin{bmatrix} 1 \\ 0 \end{bmatrix}, \quad A[2 \mid 1, 3, 4] = [1 \quad -6 \quad 4]. \tag{B.65}$$

The extreme case of partitioning is subdivision into 1 by 1 submatrices.

The two submatrices given by (B.62) and (B.64) are called *complementary* and an example of complementary submatrices is given in (B.65).

Let $C = A(1 \mid 3, 4)$, $D = A(2, 3 \mid 1, 2)$ and $F = A(2, 3 \mid 3, 4)$. Then

$$C = A(1 \mid 3, 4) = \begin{bmatrix} 1 & 2 \\ 3 & 0 \end{bmatrix}, \qquad D = A(2, 3 \mid 1, 2) = [3 \quad 7],$$
$$F = A(2, 3 \mid 3, 4) = [2 \quad 1].$$

Now let B be the submatrix defined in (B.60); we can write

$$A = \begin{bmatrix} F & D \\ C & B \end{bmatrix} = \begin{bmatrix} [2 \quad 1] & [3 \quad 7] \\ \begin{bmatrix} 1 & 2 \\ 3 & 0 \end{bmatrix} & \begin{bmatrix} -6 & 4 \\ 1 & 3 \end{bmatrix} \end{bmatrix}. \tag{B.66}$$

We say that A has been *partitioned* into the submatrices F, D, C, and B.

The notation on the right in (B.66) suggests that A can be regarded as a matrix made up of matrices. This in turn leads to the observation that matrix operations can be carried out on partitioned matrices, treating the submatrices like scalars provided that (1) in all products, the left-right order is strictly maintained and (2) the submatrices are of the proper shapes for the operations being performed.

In more precise language: the sum of two identically partitioned matrices is another matrix with the same partitioning and whose submatrices are the sums of the corresponding submatrices of the summands,

$$[A_{ij}] + [B_{ij}] = [A_{ij} + B_{ij}]. \tag{B.67}$$

For multiplication of two partitioned matrices A and B, the column partitioning of the first factor must coincide with the row partitioning of the second factor. Then the product will be a partitioned matrix with its rows subdivided like those of the first factor and with its columns subdivided like those of the second factor. The submatrices of the product are formed from those of the factors according to the usual row-by-column multiplication rule:

$$[A_{ij}] \quad [B_{ij}] = [\sum_k A_{ik}B_{kj}], \tag{B.68}$$

where k is summed over the number of subdivisions in the columns of A. For example, let

$$A = \left[\begin{array}{c|cc} 3 & 1 & 2 \\ 0 & -1 & 2 \end{array}\right], \qquad B = \left[\begin{array}{cccc} 2 & -6 & -6 & 2 \\ \hline -1 & 1 & 2 & 0 \\ 0 & 2 & 3 & 0 \end{array}\right]$$

be partitioned as is suggested by the dashed lines, and let

$$A_{11} = \begin{bmatrix} 3 \\ 0 \end{bmatrix}, \quad A_{12} = \begin{bmatrix} 1 & 2 \\ -1 & 2 \end{bmatrix}, \quad B_{11} = [2 \quad -6 \quad -6 \quad 2],$$
$$B_{21} = \begin{bmatrix} -1 & 1 & 2 & 0 \\ 0 & 2 & 3 & 0 \end{bmatrix}.$$

We can write

$$A = [A_{11} \quad A_{12}], \qquad B = \begin{bmatrix} B_{11} \\ B_{21} \end{bmatrix}.$$

Moreover, the column partitioning of A is the same as the row partitioning of B and

$$AB = [A_{11}B_{11} + A_{12}B_{21}].$$

Note that A_{11} is 2 by 1, B_{11} is 1 by 2, so $A_{11}B_{11}$ is defined, and that A_{12} is 2 by 2 and B_{21} is 2 by 4, so that $A_{12}B_{21}$ is likewise defined. We write out the operations in detail:

$$AB = \left[\begin{bmatrix} 3 \\ 0 \end{bmatrix}[2 \quad -6 \quad -6 \quad 2]\right.$$
$$\left. + \begin{bmatrix} 1 & 2 \\ -1 & 2 \end{bmatrix}\begin{bmatrix} -1 & 1 & 2 & 0 \\ 0 & 2 & 3 & 0 \end{bmatrix}\right],$$
$$= \left[\begin{bmatrix} 6 & -18 & -18 & 6 \\ 0 & 0 & 0 & 0 \end{bmatrix} + \begin{bmatrix} -1 & 5 & 8 & 0 \\ 1 & 3 & 4 & 0 \end{bmatrix}\right]$$
$$= \begin{bmatrix} 5 & -13 & -10 & 6 \\ 1 & 3 & 4 & 0 \end{bmatrix}.$$

For another example, let

$$A = \begin{bmatrix} A_{11} & A_{12} & A_{13} & A_{14} \\ A_{21} & A_{22} & A_{23} & A_{24} \\ A_{31} & A_{32} & A_{33} & A_{34} \end{bmatrix}, \qquad B = \begin{bmatrix} B_{11} & B_{12} \\ B_{21} & B_{22} \\ B_{31} & B_{32} \\ B_{41} & B_{42} \end{bmatrix}$$

be two partitioned matrices in which the column degree of A_{ih} equals the row degree of B_{hj} for all i and j; then

$$AB =$$
$$\begin{bmatrix} A_{11}B_{11} + A_{12}B_{21} + A_{13}B_{31} + A_{14}B_{41} & A_{11}B_{12} + A_{12}B_{22} + A_{13}B_{32} + A_{14}B_{42} \\ A_{21}B_{11} + A_{22}B_{21} + A_{23}B_{31} + A_{24}B_{41} & A_{21}B_{12} + A_{22}B_{22} + A_{23}B_{32} + A_{24}B_{42} \\ A_{31}B_{11} + A_{32}B_{21} + A_{33}B_{31} + A_{34}B_{41} & A_{31}B_{12} + A_{32}B_{22} + A_{33}B_{32} + A_{34}B_{42} \end{bmatrix}.$$

Note that the row partitioning in the product matrix is that of the first factor and the column partitioning is that of the second factor.

Partitioned matrices are more useful for describing properties of matrices than in performing calculations, but their usefulness in computer computations is considerable (as we will indicate below when we consider the inverse of a square matrix in terms of a partitioned matrix). One of the most important examples of partitioning is provided by subdivision of a matrix into columns or rows. Let A be an m by n matrix, we can then write

$$A = [A_1 \quad \cdots \quad A_n] = \begin{bmatrix} A^1 \\ \cdot \\ \cdot \\ \cdot \\ A^m \end{bmatrix} \tag{B.69}$$

to indicate the partitioning of A into its columns and rows.

If B is an n by p matrix and $C = AB$ then the product formulation

$$AB = \begin{bmatrix} A^1 \\ A^2 \\ \cdot \\ \cdot \\ \cdot \\ A^m \end{bmatrix} [B_1 \quad B_2 \quad \cdots \quad B_p] = \begin{bmatrix} A^1B_1 & A^1B_2 & \cdots & A^1B_p \\ A^2B_1 & A^2B_2 & \cdots & A^2B_p \\ & \cdots & & \\ A^mB_1 & A^mB_2 & \cdots & A^mB_p \end{bmatrix} \tag{B.70}$$

illustrates via partitioning the fact that $c_{ij} = A^iB_j$ is the product of the ith row of A and the jth column of B. To illustrate this notation let A be given by (B.60); then

$$A^3 = [3 \quad 0 \quad 1 \quad 3] \qquad \text{and} \qquad A_2 = \begin{bmatrix} 1 \\ 2 \\ 0 \end{bmatrix}.$$

In order to avoid confusion between the product $A^2 = AA$ and the second row of A we sometimes write $A^{(2)}$ for the latter.

B.6. SOME SPECIAL MATRICES.

We will denote by E_{mn} the m by n matrix each of whose entries is 1. The special case of E_{m1} is abbreviated to E_m; for example,

$$E_{23} = \begin{bmatrix} 1 & 1 & 1 \\ 1 & 1 & 1 \end{bmatrix}, \qquad E_3 = \begin{bmatrix} 1 \\ 1 \\ 1 \end{bmatrix}, \qquad E_4^T = [1 \quad 1 \quad 1 \quad 1]. \tag{B.71}$$

E_m is sometimes called the *summation vector* since E_m^TX is the sum of the components of X.

Let C be a column vector with m components. If $E_m^TC = 1$ and if the components of C are nonnegative, then C is called a *probability vector.* For example,

$$\begin{bmatrix} 1 \\ 0 \\ 0 \end{bmatrix}, \quad \begin{bmatrix} 1/2 \\ 1/3 \\ 1/6 \end{bmatrix}, \quad \begin{bmatrix} .25 \\ .13 \\ .34 \\ .28 \end{bmatrix},$$

are probability vectors.

A square matrix A is said to be *symmetric* if

$$A^T = A; \tag{B.72}$$

and is said to be *skew symmetric* if

$$A^T = -A. \tag{B.73}$$

(Examples of these and other special matrices are given at the end of this section.)

Let B be any square matrix and k a scalar; consider the matrices $k(B + B^T)$ and $k(B - B^T)$. Now

$$[k(B + B^T)]^T = k(B + B^T)^T = k(B^T + (B^T)^T) = k(B^T + B) = k(B + B^T)$$

and

$$[k(B - B^T)]^T = k(B - B^T)^T = k(B^T - B) = -k(B - B^T).$$

This means that $k(B + B^T)$ is symmetric and $k(B - B^T)$ is skew symmetric. If B is any square matrix, the following holds,

$$B = \frac{1}{2}(B + B^T) + \frac{1}{2}(B - B^T). \tag{B.74}$$

Hence, any square matrix can be expressed as the sum of two matrices, one symmetric and one skew symmetric. It is easy to show that there is only one such decomposition (see exercise B.19).

If the scalars are complex numbers, we denote by $\bar{A}$ the matrix whose elements are the complex conjugates of those of A (a complex number can be written as $a + bi$, where a and b are real numbers and $i = \sqrt{-1}$; if $a + bi$ is a complex number, then $a - bi$ is the complex conjugate of $a + bi$). A square matrix A is said to be *Hermitian* if

$$\bar{A}^T = A, \tag{B.75}$$

and is said to be *skew Hermitian* if

$$\bar{A}^T = -A. \tag{B.76}$$

We frequently write A^* for $\bar{A}^T$. Thus a matrix is Hermitian if $A^* = A$ and is skew Hermitian if $A^* = -A$.

Let A be a square matrix of degree n; then the elements $a_{11}, a_{22}, \cdots, a_{nn}$ are said to constitute the *diagonal* of A. A square matrix

$$D = \begin{bmatrix} d_{11} & & & & \\ & d_{22} & & & \\ & & \cdot & & \\ & & & \cdot & \\ & & & & \cdot \\ & & & & & d_{nn} \end{bmatrix} \tag{B.77}$$

is said to be *a diagonal matrix* if all of its nondiagonal entries are zero, in other words, if $d_{ij} = 0$ for all $i \neq j$. If the nonzero entries in a diagonal matrix are all equal, the matrix is said to be a *scalar matrix*,

$$C = \begin{bmatrix} c & & & & & \\ & c & & & & \\ & & \cdot & & & \\ & & & \cdot & & \\ & & & & \cdot & \\ & & & & & c \end{bmatrix}. \tag{B.78}$$

If $c = 1$, we have as a special case the *identity matrix*

$$I_n = \begin{bmatrix} 1 & & & & & \\ & 1 & & & & \\ & & \cdot & & & \\ & & & \cdot & & \\ & & & & \cdot & \\ & & & & & 1 \end{bmatrix}. \tag{B.79}$$

If every entry above the diagonal in a square matrix A is zero, then A is said to be *(lower) triangular*.

The identity matrix has the multiplicative properties resembling those of the scalar 1; if A has n rows and B has n columns

$$I_nA = A, \qquad BI_n = B. \tag{B.80}$$

Note that for C as in (B.78) we have

$$C = cI_n \tag{B.81}$$

and also

$$CA = cA, \qquad BC = cB. \tag{B.82}$$

The p by n matrix all of whose entries are zero is called a *zero matrix*. It has additive and multiplicative properties resembling those of the zero scalar. For example,

$$\begin{bmatrix} 0 & 0 & 0 \\ 0 & 0 & 0 \end{bmatrix} + \begin{bmatrix} 2 & 1 & 3 \\ -1 & 2 & 4 \end{bmatrix} = \begin{bmatrix} 2 & 1 & 3 \\ -1 & 2 & 4 \end{bmatrix}$$

and

$$\begin{bmatrix} 0 & 0 & 0 \\ 0 & 0 & 0 \end{bmatrix} \begin{bmatrix} x_1 \\ x_2 \\ x_3 \end{bmatrix} = \begin{bmatrix} 0 \\ 0 \end{bmatrix}.$$

When desirable we use the symbol 0_{pn} to denote the p by n zero matrix.

Examples of special matrices are as follows:

Symmetric $\begin{bmatrix} 2 & 3 \\ 3 & -1 \end{bmatrix}$,

Skew symmetric $\begin{bmatrix} 0 & 4 & -1 \\ -4 & 0 & 2 \\ 1 & -2 & 0 \end{bmatrix}$,

Hermitian $\begin{bmatrix} 2 & 1+i \\ 1-i & 3 \end{bmatrix}$,

Skew Hermitian $\begin{bmatrix} 3i & 4+i \\ -4+i & 2i \end{bmatrix}$,

Diagonal $\begin{bmatrix} 2 & 0 & 0 \\ 0 & -4 & 0 \\ 0 & 0 & 6 \end{bmatrix}$, $\begin{bmatrix} 2 & 0 \\ 0 & 0 \end{bmatrix}$,

Scalar $\begin{bmatrix} 3 & 0 & 0 \\ 0 & 3 & 0 \\ 0 & 0 & 3 \end{bmatrix}$,

Identity $I_3 = \begin{bmatrix} 1 & 0 & 0 \\ 0 & 1 & 0 \\ 0 & 0 & 1 \end{bmatrix}$,

Lower triangular $\begin{bmatrix} 3 & 0 & 0 \\ 4 & 1 & 0 \\ 1 & -1 & 2 \end{bmatrix}$, $\begin{bmatrix} 0 & 0 & 0 \\ 4 & 0 & 0 \\ 2 & 1 & 0 \end{bmatrix}$,

Upper triangular $\begin{bmatrix} 3 & 1 & 2 \\ 0 & 4 & 5 \\ 0 & 0 & 6 \end{bmatrix}$,

Zero $0_{23} = \begin{bmatrix} 0 & 0 & 0 \\ 0 & 0 & 0 \end{bmatrix}$, $0_{33} = \begin{bmatrix} 0 & 0 & 0 \\ 0 & 0 & 0 \\ 0 & 0 & 0 \end{bmatrix}$.

B.7. DIMENSION AND BASIS.

Let V be a vector space over a scalar domain F and let $S = (X_1, \cdots, X_s)$ be a sequence of vectors in V. We say that S is *dependent* (linearly dependent) if there exist scalars $c_1, \cdots, c_s$ not all zero such that

$$c_1X_1 + \cdots + c_sX_s = 0. \tag{B.83}$$

If the only solution of (B.83) is the trivial one, $c_1 = c_2 = \cdots = c_s = 0$, we say that S is an *independent* sequence.

For example, consider the sequences $S_1 = (X_1, X_2, X_3)$, $S_2 = (X_1, X_2)$, $S_3 = (Y_1, Y_2, Y_3)$, where

$$X_1 = \begin{bmatrix} 1 \\ 0 \end{bmatrix}, X_2 = \begin{bmatrix} 0 \\ 1 \end{bmatrix}, X_3 = \begin{bmatrix} 1 \\ -2 \end{bmatrix}, Y_1 = \begin{bmatrix} 2 \\ 1 \\ 3 \end{bmatrix}, Y_2 = \begin{bmatrix} 3 \\ 2 \\ 4 \end{bmatrix}, Y_3 = \begin{bmatrix} 2 \\ 2 \\ 2 \end{bmatrix}.$$

Then the relations $-X_1 + 2X_2 + X_3 = 0$, and $-2Y_1 + 2Y_2 - Y_3 = 0$ show that S_1 and S_3 are dependent. On the other hand, $c_1X_1 + c_2X_2 = \begin{bmatrix} c_1 \\ c_2 \end{bmatrix}$ and this is zero if and only if $c_1 = c_2 = 0$; hence S_2 is independent.

We introduce the Kronecker delta function δ whose domain is the set of all ordered pairs of positive integers and which is defined by

$$\begin{cases} \delta_{ij} = 0, & i \neq j \\ \delta_{ii} = 1, & i = 1, 2, \cdots. \end{cases} \tag{B.84}$$

For example, we can represent compactly an identity matrix by means of this symbol,

$$I_n = [\delta_{ij}];$$

when $j = i$ we get the diagonal elements ($\delta_{ii} = 1$, $i = 1, \cdots, n$) and when $i \neq j$ we get the off-diagonal elements ($\delta_{ij} = 0$, $i \neq j$).

In the vector space V_n the vectors $U_1, \cdots, U_n$ defined by

$$U_j = [\delta_{ij}] = \begin{bmatrix} 0 \\ \cdot \\ \cdot \\ \cdot \\ 1 \\ 0 \\ \cdot \\ \cdot \\ \cdot \\ 0 \end{bmatrix} \qquad (j = 1, \cdots, n) \tag{B.85}$$

are called the *natural basis vectors*. The sequence $N = (U_1, \cdots, U_n)$ *is* independent. For example, when $n = 3$

$$U_1 = \begin{bmatrix} 1 \\ 0 \\ 0 \end{bmatrix}, \qquad U_2 = \begin{bmatrix} 0 \\ 1 \\ 0 \end{bmatrix}, \qquad U_3 = \begin{bmatrix} 0 \\ 0 \\ 1 \end{bmatrix};$$

then

$$c_1U_1 + c_2U_2 + c_3U_3 = c_1 \begin{bmatrix} 1 \\ 0 \\ 0 \end{bmatrix} + c_2 \begin{bmatrix} 0 \\ 1 \\ 0 \end{bmatrix} + c_3 \begin{bmatrix} 0 \\ 0 \\ 1 \end{bmatrix} = \begin{bmatrix} c_1 \\ c_2 \\ c_3 \end{bmatrix}. \tag{B.86}$$

This is equal to the zero vector if and only if $c_1 = c_2 = c_3 = 0$. Hence (U_1, U_2, U_3) is independent.

Let $S = (X_1, \cdots, X_s)$ be any sequence of vectors in V. *A vector* Y *is* said to be *dependent on* S if there exist scalars $c_1, \cdots, c_s$ such that

$$Y = c_1X_1 + \cdots + c_sX_s; \tag{B.87}$$

Y is then called a *linear combination* of the vectors in S. A sequence $T = (Y_1, \cdots, Y_t)$ is said to be *dependent on* S if each vector in T is dependent on S.

For example, if $Y = \begin{bmatrix} y_1 \\ \cdot \\ \cdot \\ \cdot \\ y_n \end{bmatrix}$ is any vector in V_n then the equation

$$Y = y_1U_1 + \cdots + y_nU_n \tag{B.88}$$

shows that Y is dependent on N, the sequence of natural basis vectors in V_n. Moreover, since this is true for every vector in V_n it follows that V_n is dependent on N.

A sequence S of V is said to *span* V if every vector in V is dependent on S. We then write

$$V = \langle S \rangle. \tag{B.89}$$

For example, let $X_1 = \begin{bmatrix} 2 \\ 5 \end{bmatrix}$, $X_2 = \begin{bmatrix} 6 \\ 1 \end{bmatrix}$, $X_3 = \begin{bmatrix} -2 \\ 3 \end{bmatrix}$, $Y = \begin{bmatrix} 8 \\ 4 \end{bmatrix}$ and let $S = (X_1, X_2, X_3)$. From Figure B.3 we see that there are several ways of expressing Y as a linear combination of the vectors in S. One way is $Y = 4/7X_1 + 8/7\, X_2 + 0X_3$, and the geometry indicates that Y is the sum of $4/7\, X_1$, $8/7\, X_2$, and $0X_3$,

$$\frac{4}{7}\begin{bmatrix} 2 \\ 5 \end{bmatrix} + \frac{8}{7}\begin{bmatrix} 6 \\ 1 \end{bmatrix} + 0\begin{bmatrix} -2 \\ 3 \end{bmatrix} = \begin{bmatrix} 8/7 \\ 20/7 \end{bmatrix} + \begin{bmatrix} 48/7 \\ 8/7 \end{bmatrix} = \begin{bmatrix} 56/7 \\ 28/7 \end{bmatrix} = \begin{bmatrix} 8 \\ 4 \end{bmatrix}.$$

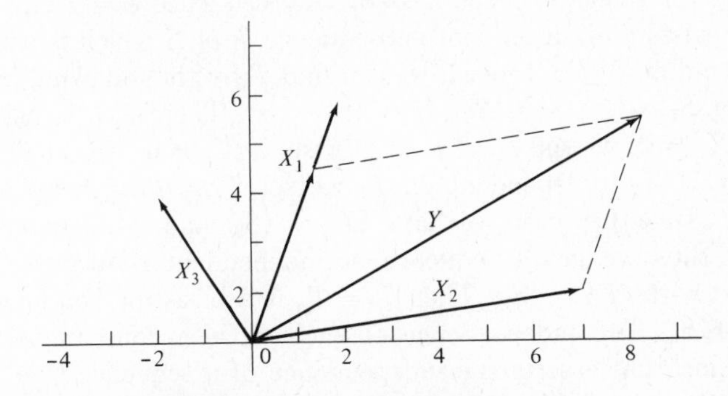

Figure B.3 Addition of vectors.

Another way of expressing Y as a linear combination of the vectors in S is

$$Y = -\frac{1}{7} X_1 + \frac{12}{7} X_2 + X_3.$$

Thus Y is dependent on the sequence S and it can be shown that every vector in V_2 depends on S, which means that V_2 is dependent on S. The sequence S spans V_2 (or is a *spanning sequence* for V_2) and we can write $V_2 = \langle S \rangle$ or $V_2 = \langle X_1, X_2, X_3 \rangle$.

If S is an arbitrary subset of V we define $\langle S \rangle$ to consist of all vectors in V that are dependent on any sequence of elements in S.

A sequence $S = (X_1, \cdots, X_s)$ of V is said to be a *basis* for a vector space V if

$$\begin{aligned} &\text{(a) } S \text{ is independent,} \\ &\text{(b) } S \text{ spans } V. \end{aligned} \tag{B.90}$$

Thus a basis is an independent spanning sequence. Referring to the example above, the sequence $S = (X_1, X_2, X_3)$ is a dependent spanning sequence because the vector equation

$$c_1X_1 + c_2X_2 + c_3X_3 = 0$$

is satisfied for some c_i not all zero. Hence this sequence is not a basis. However, any subsequence of two distinct elements of S has the property that it is independent and it can be shown that each such subsequence spans V_2 (V_2 is dependent on each such subsequence) and is therefore a basis of V_2, for example,

$$S' = (X_1, X_3)$$

is a basis for V_2.

It can be shown that if $T = (Y_1, \cdots, Y_t)$ is also a basis for V then $s = t$ and this common number is called the *dimension* of V, written $s = \dim V$. For example, $N = (U_1, \cdots, U_n)$ where U_j is defined in (B.85) is a basis for V_n and hence V_n has dimension n.

If a vector space V is spanned by any sequence $S = (X_1, \cdots, X_s)$ then there exists an independent subsequence T of S which is a basis for V and hence $\dim V \leq s$. Indeed, we can find T by the following inductive process. Let $S_i = (X_1, \cdots, X_i)$, $(i = 1, \cdots, s)$. Then if S_1 is dependent, that is, if $X_1 = 0$, we set $T_1 = \phi$; otherwise, if $X_1 \neq 0$, we set $T_1 = (X_1) = S_1$. Next, if X_2 is dependent on T_1 we set $T_2 = T_1$, otherwise set $T_2 = (T_1, X_2)$. In either case we have $\langle T_2 \rangle = \langle S_2 \rangle$ and T_2 is independent. Proceeding thus we get a sequence of independent sequences T_i with $T_i \subset S_i$ and with $\langle T_i \rangle = \langle S_i \rangle$. Then $T = T_n$ is a basis for V and is a subsequence of S_n. The independence of each T_i is a consequence of the general lemma that asserts the independence of a sequence $Q = (Q', X)$ on the hypotheses that Q' is independent and that X is not dependent on Q'.

A vector space V is said to have *infinite dimension* if it is not spanned by any finite sequence. For example, if V is the set of all polynomials $a_0 + a_1x + \cdots + a_rx^r$ in one scalar variable x then V is a vector space of infinite dimension. For another example, the real numbers R may be regarded as an infinite-dimensional vector space over the scalar domain Q of the rational numbers. In what follows we shall be concerned exclusively with finite-dimensional vector spaces.

B.8. TESTS FOR DEPENDENCE.

Let $S = (A_1, \cdots, A_s)$ be a sequence of vectors in V_n. In the preceding section we saw the importance of knowing whether S is dependent or independent, in other words, whether or not the vector equation

$$x_1A_1 + \cdots + x_sA_s = 0 \tag{B.91}$$

has a solution other than the trivial one $x_1 = \cdots = x_s = 0$. Let $A_j = [a_{ij}]$, $(j = 1, \cdots, s)$; then (B.91) is equivalent to the system of n scalar equations

$$\begin{aligned}
a_{11}x_1 + a_{12}x_2 + \cdots + a_{1s}x_s &= 0,\\
a_{21}x_1 + a_{22}x_2 + \cdots + a_{2s}x_s &= 0,\\
\cdot \qquad\qquad \cdot\ &\ \ \cdot\\
\cdot \qquad\qquad \cdot\ &\ \ \cdot\\
\cdot \qquad\qquad \cdot\ &\ \ \cdot\\
a_{n1}x_1 + a_{n2}x_2 + \cdots + a_{ns}x_s &= 0.
\end{aligned} \tag{B.92}$$

This is called a homogeneous linear system because each variable enters with degree one and the constant terms on the right are all zero. For example, if $S = (A_1, A_2, A_3)$ where

$$A_1 = \begin{bmatrix} 3\\ 1\\ 4\\ 6 \end{bmatrix}, \qquad A_2 = \begin{bmatrix} 2\\ 4\\ -7\\ -3 \end{bmatrix}, \qquad A_3 = \begin{bmatrix} 8\\ 6\\ 1\\ 9 \end{bmatrix},$$

then (B.91) becomes

$$\begin{aligned}
3x_1 + 2x_2 + 8x_3 &= 0\\
x_1 + 4x_2 + 6x_3 &= 0\\
4x_1 - 7x_2 + x_3 &= 0\\
6x_1 - 3x_2 + 9x_3 &= 0.
\end{aligned}$$

In this case $X = \begin{bmatrix} x_1\\ x_2\\ x_3 \end{bmatrix} = \begin{bmatrix} 2\\ 1\\ -1 \end{bmatrix}$ is seen to be a (nontrivial) solution, so that S is dependent.

We discuss solutions of systems of linear equations in later sections; our present discussion has as its main objective to show that *actual determination of dependence or independence of a set of vectors leads to solution of a system of homogeneous linear equations.*

Now let $K = [k_i]$ be a vector in V_n and ask if K is dependent on S, that is, if the vector equation

$$x_1A_1 + \cdots + x_sA_s = K \tag{B.93}$$

has any solution (whether trivial or nontrivial makes no difference this time). Once again the scalar formulation leads to a system of n linear equations, this time called *nonhomogeneous* because of the constants k_i on the right-hand sides. We have

$$\begin{aligned} a_{11}x_1 + a_{12}x_2 + \cdots + a_{1s}x_s &= k_1 \\ a_{21}x_1 + a_{22}x_2 + \cdots + a_{2s}x_s &= k_2 \\ \cdot \qquad\qquad\qquad \cdot\ & \quad \cdot \\ \cdot \qquad\qquad\qquad \cdot\ & \quad \cdot \\ \cdot \qquad\qquad\qquad \cdot\ & \quad \cdot \\ a_{n1}x_1 + a_{n2}x_2 + \cdots + a_{ns}x_s &= k_n. \end{aligned} \tag{B.94}$$

For example, with S as before, let $K = [13 \quad 11 \quad -2 \quad 12]^T$. Then (B.94) becomes

$$\begin{aligned} 3x_1 + 2x_2 + 8x_3 &= 13 \\ x_1 + 4x_2 + 6x_3 &= 11 \\ 4x_1 - 7x_2 + x_3 &= -2 \\ 6x_1 - 3x_3 + 9x_3 &= 12. \end{aligned}$$

The solution $X = [1 \quad 1 \quad 1]^T$ shows that K is dependent on S (the reader might wish to find if there is a solution X having $x_3 = 0$).

B.9. SUBSPACES OF A VECTOR SPACE, SUM AND INTERSECTION.

Let V be a vector space. A subset W of V is called a *subspace* if it is closed under addition and multiplication by scalars or equivalently if every linear combination

$$Y = c_1Y_1 + \cdots + c_tY_t \tag{B.95}$$

of vectors in W is again in W. A subspace W of V is a vector space in its own right and it can be shown that

$$\dim W \leq \dim V \tag{B.96}$$

and hence that *any subspace of a finite-dimensional space is finite dimensional.*

If $V = V_3$ is ordinary real 3-space, then the subspaces of V, aside from $\{0\}$ and V itself, are the lines and planes through the origin. For example, let $Z = \begin{bmatrix} 1 \\ 1 \\ 1 \end{bmatrix}$; then $W_1 = \langle Z \rangle$ is the line $x_1 = x_2 = x_3$. Let $X_1 = \begin{bmatrix} 1 \\ 0 \\ 0 \end{bmatrix}$ and $X_2 = \begin{bmatrix} 0 \\ 1 \\ 0 \end{bmatrix}$; then $W_2 = \langle X_1, X_2 \rangle$ is the x_1, x_2 plane. Let $X = \begin{bmatrix} x_1 \\ x_2 \\ x_3 \end{bmatrix}$, where $x_1 + x_2 - x_3 = 0$; then $W_3 = \langle X \rangle$ is the plane through the origin whose traces in the x_1, x_3 and x_2, x_3 planes are, respectively, the 45 degree lines $x_1 = x_3$ and $x_2 = x_3$.

If S is any sequence of vectors in V or from a subset of V, then $\langle S \rangle$ can be shown to be a *subspace* of V. We sometimes call this subspace the *span* of the sequence S. For example, if $Y = \begin{bmatrix} 2 \\ 5 \end{bmatrix}$, then the span of Y, $\langle Y \rangle$, is the line through the origin of V_2 containing the end point of the vector Y, and the span of the sequence (U_1, U_2, U_3), where

$$U_1 = \begin{bmatrix} 1 \\ 0 \\ 0 \end{bmatrix}, \qquad U_2 = \begin{bmatrix} 0 \\ 1 \\ 0 \end{bmatrix}, \qquad U_3 = \begin{bmatrix} 0 \\ 0 \\ 1 \end{bmatrix},$$

is V_3.

If U and W are subspaces of V then their intersection

$$U \cap W \tag{B.97}$$

is again a subspace; it may consist of the zero vector alone. If U is the x_1, x_2 plane in three space and W is the x_3 axis, then $U \cap W$ is the origin. On the other hand if U is the x_1, x_2 plane and W' is taken to be the x_2, x_3 plane, then $U \cap W'$ is the x_2 axis. In either case the intersection is a subspace; the set consisting of the zero vector alone is a subspace since it satisfies (trivially) the definition of a subspace.

If S and T are subsets of V then their *sum* $S + T$ is defined as the set of all sums $X + Y$ for X in S and Y in T, formally

$$S + T = \{X + Y \mid X \in S, Y \in T\}. \tag{B.98}$$

For example, if S is the line $y = x$ in V_2 and $T = \left\{\begin{bmatrix} 1 \\ 2 \end{bmatrix}\right\}$ then $S + T$ is the line $y = x + 1$ parallel to $y = x$ (see Figure B.4). The sum

$$U + W \tag{B.99}$$

of two subspaces is again a subspace of V. The dimensions of these subspaces are connected by the important formula

$$\dim (U + W) = \dim U + \dim W - \dim (U \cap W), \tag{B.100}$$

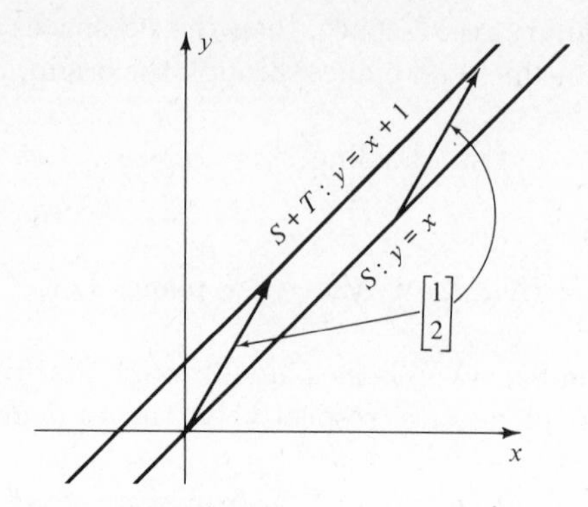

Figure B.4 Addition of sets.

or, in words, *the dimension of a sum equals the sum of the dimensions of the summands minus the dimension of the intersection.*

For example, in V_3 let U be the x_1, x_2 plane and let W be the x_2, x_3 plane. Then $U + W$ is V_3 and $U \cap W$ is the x_2 axis, as noted earlier. Observe that the dimension of $U + W$ satisfies (B.100).

B.10. SYSTEMS OF LINEAR EQUATIONS.

As we saw in Section B.8, questions of dependence and independence of sequences and of vectors on sequences of vectors lead to systems of linear equations. In the language of an application, if the technology matrix and an input vector for a factory are given there will be a feasible program using up all of each input if and only if the input vector is dependent on the columns of the technology matrix with nonnegative coefficients in the dependence relation; each nonnegative solution of the corresponding system of nonhomogeneous linear equations defines a feasible program. This is the way linear equations arise in linear programming.

For dimensions less than four one can draw figures to illustrate the solutions of a linear system (see Figure B.4). When the dimension of the underlying space is more than three, geometric visualization is no longer available although geometric language is still useful. The basic algebraic problem in any case is the solution of systems of linear equations. This topic is, moreover, of central importance for many applications of mathematics. In the present section we give an intuitive approach and in the sections immediately following we give a more comprehensive discussion.

Example. Consider the equations

$$\begin{aligned}
&\text{(a)} \quad x + 2y - 2z = -1,\\
&\text{(b)} \quad 2x - 3y + z = -1,\\
&\text{(c)} \quad 4x + y - 3z = -3,\\
&\text{(d)} \quad x - 5y + 3z = 0.
\end{aligned} \tag{B.101}$$

We can use equation (a) to eliminate the variable x from the other three equations by adding -2 times equation (a) to equation (b), and so on, and getting

$$\begin{array}{ll} (\text{a}') & x + 2y - 2z = -1, \\ (\text{b}') & \quad - 7y + 5z = 1, \\ (\text{c}') & \quad - 7y + 5z = 1, \\ (\text{d}') & \quad - 7y + 5z = 1. \end{array} \tag{B.102}$$

The solutions of the system (B.102) are exactly the same as those of (B.101). Note however that three of the equations in (B.102) are the same. Hence we need only consider the system

$$\begin{array}{ll} (\text{a}') & x + 2y - 2z = -1 \\ (\text{b}') & \quad - 7y + 5z = 1. \end{array} \tag{B.103}$$

Solving for x and y in terms of z we have

$$\begin{array}{l} (\text{a}'') \ y = \frac{5}{7}z - \frac{1}{7} \\ \qquad x = -2y + 2z - 1 = -2\left(\frac{5}{7}z - \frac{1}{7}\right) + 2z - 1 \text{ or} \\ (\text{b}'') \ x = \frac{4}{7}z - \frac{5}{7}. \end{array} \tag{B.104}$$

Thus for any value given to z we get a value for x and y. That we get an infinite number of solutions for x and y should not be surprising for the intersection of the two planes in (B.103) will be a line in V_3.

Again the use of matrices can simplify the computation. A linear equation in x, y, and z is completely determined once we know what the coefficients of the variables are, and what the constant terms are. Each equation in the system can be regarded as an inner product C^TX set equal to a scalar. Thus Equation (B.101 (a)) is $C^TX = -1$ where $C = \begin{bmatrix} 1 \\ 2 \\ -2 \end{bmatrix}$ is made up of the coefficients and $X = \begin{bmatrix} x \\ y \\ z \end{bmatrix}$. The entire system in matrix form can be written

$$AX = K \tag{B.105}$$

where

$$A = \begin{bmatrix} 1 & 2 & -2 \\ 2 & -3 & 1 \\ 4 & 1 & -3 \\ 1 & -5 & 3 \end{bmatrix}, \quad X = \begin{bmatrix} x \\ y \\ z \end{bmatrix}, \quad K = \begin{bmatrix} -1 \\ -1 \\ -3 \\ 0 \end{bmatrix}.$$

Here A is called the *coefficient matrix* of the system.

We frequently combine A and K into the *augmented matrix* of the linear equation system,

$$B = [A \quad K] = \begin{bmatrix} 1 & 2 & -2 & -1 \\ 2 & -3 & 1 & -1 \\ 4 & 1 & -3 & -3 \\ 1 & -5 & 3 & 0 \end{bmatrix}. \tag{B.106}$$

The process of elimination used to get (B.102) can be achieved, using matrices, in several ways. One method would be to add -2 times row 1 of B to row 2 of B, then add -4 times row 1 to row 3, and so on. Multiplying B on the left by the vector $[-2 \quad 1 \quad 0 \quad 0]$ in effect adds -2 times row 1 to row 2, this is incorporated as row 2 in the following product,

$$\begin{bmatrix} 1 & 0 & 0 & 0 \\ -2 & 1 & 0 & 0 \\ 0 & 0 & 1 & 0 \\ 0 & 0 & 0 & 1 \end{bmatrix} \begin{bmatrix} 1 & 2 & -2 & -1 \\ 2 & -3 & 1 & -1 \\ 4 & 1 & -3 & -3 \\ 1 & -5 & 3 & 0 \end{bmatrix} = \begin{bmatrix} 1 & 2 & -2 & -1 \\ 0 & -7 & 5 & 1 \\ 4 & 1 & -3 & -3 \\ 1 & -5 & 3 & 0 \end{bmatrix}.$$

Similarly, multiplying B on the left by $[-4 \quad 0 \quad 1 \quad 0]$ essentially adds -4 times row 1 to row 3 and multiplying by $[-1 \quad 0 \quad 0 \quad 1]$ on the left adds -1 times row 1 to row 4. All this can be accomplished in one operation by multiplying B on the left by the matrix

$$Q_1 = \begin{bmatrix} 1 & 0 & 0 & 0 \\ -2 & 1 & 0 & 0 \\ -4 & 0 & 1 & 0 \\ -1 & 0 & 0 & 1 \end{bmatrix},$$

for

$$Q_1B = \begin{bmatrix} 1 & 0 & 0 & 0 \\ -2 & 1 & 0 & 0 \\ -4 & 0 & 1 & 0 \\ -1 & 0 & 0 & 1 \end{bmatrix} \begin{bmatrix} 1 & 2 & -2 & -1 \\ 2 & -3 & 1 & -1 \\ 4 & 1 & -3 & -3 \\ 1 & -5 & 3 & 0 \end{bmatrix}$$

$$= \begin{bmatrix} 1 & 2 & -2 & -1 \\ 0 & -7 & 5 & 1 \\ 0 & -7 & 5 & 1 \\ 0 & -7 & 5 & 1 \end{bmatrix} = C,$$

which is the augmented matrix of the system (B.102). The zeros in the first column of the matrix C indicate that we have eliminated x from the second, third, and fourth equations.

We can now add -1 times row 2 of C to rows 3 and 4, or multiply C on the left by

$$\begin{bmatrix} 1 & 0 & 0 & 0 \\ 0 & 1 & 0 & 0 \\ 0 & -1 & 1 & 0 \\ 0 & -1 & 0 & 1 \end{bmatrix}$$

and get

$$C' = \begin{bmatrix} 1 & 0 & 0 & 0 \\ 0 & 1 & 0 & 0 \\ 0 & -1 & 1 & 0 \\ 0 & -1 & 0 & 1 \end{bmatrix} \begin{bmatrix} 1 & 2 & -2 & -1 \\ 0 & -7 & 5 & 1 \\ 0 & -7 & 5 & 1 \\ 0 & -7 & 5 & 1 \end{bmatrix}$$

$$= \begin{bmatrix} 1 & 2 & -2 & -1 \\ 0 & -7 & 5 & 1 \\ 0 & 0 & 0 & 0 \\ 0 & 0 & 0 & 0 \end{bmatrix},$$

which corresponds to (B.103).

The second row of C' can be multiplied by $-1/7$. This is equivalent to multiplying Equation (B.103(b′)) by $-1/7$ and clearly this does not affect the solution. Thus solutions of a system of equations whose augmented matrix is

$$C'' = \begin{bmatrix} 1 & 2 & -2 & 1 \\ 0 & 1 & -5/7 & -1/7 \\ 0 & 0 & 0 & 0 \\ 0 & 0 & 0 & 0 \end{bmatrix}$$

are the same as those of the system whose augmented matrix is C'. Finally we eliminate the variable y from the first equation by adding (-2) times row 2 to row 1. This yields

$$D = \begin{bmatrix} 1 & 0 & -4/7 & -5/7 \\ 0 & 1 & -5/7 & -1/7 \\ 0 & 0 & 0 & 0 \\ 0 & 0 & 0 & 0 \end{bmatrix}$$

One can go directly from C to D by means of premultiplying C by a matrix Q_2, where

$$Q_2 = \begin{bmatrix} 1 & 2/7 & 0 & 0 \\ 0 & -1/7 & 0 & 0 \\ 0 & -1 & 1 & 0 \\ 0 & -1 & 0 & 1 \end{bmatrix}.$$

That is,

$$Q_2 C = D,$$

as the reader can verify. The matrices Q_1 and Q_2 are examples of what we call Gaussian pivot matrices, or more briefly, pivot matrices. These matrices will be discussed more fully below.

Translating the matrix D back to equations we have (after transposing the z terms),

$$x = \frac{4}{7}z - \frac{5}{7},$$
$$y = \frac{5}{7}z - \frac{1}{7},$$

which is the system (B.104).

B.11. ELEMENTARY OPERATIONS; THE PIVOT PROCESS.

We have seen that certain operations on the augmented matrix of a system of linear equations do not alter the solutions of the system. There are three such operations, called *elementary row operations.*

E1. We can interchange any two rows. This is equivalent to writing the equations in a different order and obviously this does not affect the solutions.

E2. We can multiply a row by a nonzero constant. Multiplying an equation by a nonzero constant does not alter the solution of the system.

E3. We can replace a row by the sum of that row and a multiple of any other row.

In two dimensions we can interpret the third operation to mean that we replace one of two intersecting lines by another line passing through the point of intersection of the original pair. If a solution exists for the system all the other lines must pass through the intersection of the original pair. In higher dimensions the locus of a linear equation is called a hyperplane. The solution of a system consists of the intersection of the corresponding hyperplanes. An operation of type E3 involves two hyperplanes and replaces one of them by a new one which contains the intersection of the first two.

We next introduce the concept of an *echelon matrix* as one having the following properties.

(a) The first nonzero entry in each nonzero row is a 1, and the nonzero rows precede any zero rows.

(b) This 1 is the only nonzero element in its column.

(c) In each row after the first the number of zeros preceding this entry 1 is larger than the corresponding number of zeros in the preceding row.

Example. If the augmented matrix of a system of linear equations in the variables x_1, x_2, x_3, x_4, x_5, x_6, x_7, x_8, is

$$B = \begin{bmatrix} 0 & 2 & -4 & 0 & 0 & 3 & -7 & 7 & 16 \\ 0 & 1 & -2 & 1 & 0 & 1 & 0 & 2 & 5 \\ 0 & -1 & 2 & 2 & 0 & 1 & 0 & -3 & 0 \\ 0 & 0 & 0 & 1 & 0 & -1 & 5 & -2 & -5 \end{bmatrix}, \quad (B.107)$$

then by a succession of elementary row operations we can obtain the echelon matrix

$$D = \begin{bmatrix} 0 & 1 & -2 & 0 & 0 & 0 & 1 & 2 & 2 \\ 0 & 0 & 0 & 1 & 0 & 0 & 2 & -1 & -1 \\ 0 & 0 & 0 & 0 & 0 & 1 & -3 & 1 & 4 \\ 0 & 0 & 0 & 0 & 0 & 0 & 0 & 0 & 0 \end{bmatrix}. \quad (B.108)$$

In equation form this gives

$$\begin{aligned} x_2 - 2x_3 \qquad\qquad + x_7 + 2x_8 &= 2 \\ x_4 \quad + 2x_7 - x_8 &= -1 \\ x_6 - 3x_7 + x_8 &= 4. \end{aligned} \quad (B.109)$$

We will regard this as the solved form of the original system since these equations determine x_2, x_4, x_6 in terms of the remaining variables. For certain purposes we may prefer the equivalent form

$$\begin{aligned} x_2 &= 2x_3 - x_7 - 2x_8 + 2 \\ x_4 &= -2x_7 + x_8 - 1 \\ x_6 &= 3x_7 - x_8 + 4. \end{aligned} \quad (B.110)$$

The form (B.109) is especially useful in linear programming whereas the form (B.110) is better adapted for writing out solutions in vector form as we will soon see.

In the systems above (and in the system in the example of the preceding section) there are infinitely many solutions. Hence the instruction, "find all solutions," or "find the general solution," cannot mean, "write down all solutions." We study the question of the nature of the set of solutions in later sections. However, we can give one interpretation here. Solving a system can be regarded as a separation of the variables into two classes: *basic* (or dependent) and *nonbasic* (or independent). In the solved form (B.109) each basic variable appears in exactly one equation and with coefficient one. The nonbasic variables can be regarded as parameters to which arbitrary numerical values can be assigned. Every solution can be obtained by assigning suitable values to the nonbasic variables, the values of the basic variables then being determined by the equations (in either form B.109 or B.110).

The process of solution includes selecting which variables are to be basic and which nonbasic. One feature of the echelon matrix is that the basic variables correspond to the columns in which the initial nonzero entries of the rows appear.

We give a convenient way for expressing what we call the general solution in vector form. In the vector X replace each basic variable by its expression in terms of the nonbasic variables and then expand as a constant vector plus linear combination of vectors where the scalar coefficients are the nonbasic variables.

Thus in the example in Section B.10 we have x, y as basic variables and z as the nonbasic variable and then

$$X = \begin{bmatrix} x \\ y \\ z \end{bmatrix} = \begin{bmatrix} 4/7\,z & -5/7 \\ 5/7\,z & -1/7 \\ z & \end{bmatrix} = \begin{bmatrix} -5/7 \\ -1/7 \\ 0 \end{bmatrix} + \begin{bmatrix} 4/7\,z \\ 5/7\,z \\ z \end{bmatrix}$$

$$= \begin{bmatrix} -5/7 \\ -1/7 \\ 0 \end{bmatrix} + z \begin{bmatrix} 4/7 \\ 5/7 \\ 1 \end{bmatrix}. \tag{B.111}$$

Every solution is of this form, and every vector in this form is a solution. In the example of this section we have x_2, x_4, x_6 as basic variables and then

$$X = \begin{bmatrix} x_1 \\ x_2 \\ x_3 \\ x_4 \\ x_5 \\ x_6 \\ x_7 \\ x_8 \end{bmatrix} = \begin{bmatrix} x_1 \\ 2x_3 - x_7 - 2x_8 + 2 \\ x_3 \\ -2x_7 + x_8 - 1 \\ x_5 \\ 3x_7 - x_8 + 4 \\ x_7 \\ x_8 \end{bmatrix}$$

$$= \begin{bmatrix} 0 \\ 2 \\ 0 \\ -1 \\ 0 \\ 4 \\ 0 \\ 0 \end{bmatrix} + \begin{bmatrix} x_1 \\ 0 \\ 0 \\ 0 \\ 0 \\ 0 \\ 0 \\ 0 \end{bmatrix} + \begin{bmatrix} 0 \\ 2x_3 \\ x_3 \\ 0 \\ 0 \\ 0 \\ 0 \\ 0 \end{bmatrix} + \begin{bmatrix} 0 \\ 0 \\ 0 \\ 0 \\ x_5 \\ 0 \\ 0 \\ 0 \end{bmatrix} + \begin{bmatrix} 0 \\ -x_7 \\ 0 \\ -2x_7 \\ 0 \\ 3x_7 \\ x_7 \\ 0 \end{bmatrix} + \begin{bmatrix} 0 \\ -2x_8 \\ 0 \\ x_8 \\ 0 \\ -x_8 \\ 0 \\ x_8 \end{bmatrix}$$

$$= \begin{bmatrix} 0 \\ 2 \\ 0 \\ -1 \\ 0 \\ 4 \\ 0 \\ 0 \end{bmatrix} + x_1 \begin{bmatrix} 1 \\ 0 \\ 0 \\ 0 \\ 0 \\ 0 \\ 0 \\ 0 \end{bmatrix} + x_3 \begin{bmatrix} 0 \\ 2 \\ 1 \\ 0 \\ 0 \\ 0 \\ 0 \\ 0 \end{bmatrix} + x_5 \begin{bmatrix} 0 \\ 0 \\ 0 \\ 0 \\ 1 \\ 0 \\ 0 \\ 0 \end{bmatrix} + x_7 \begin{bmatrix} 0 \\ -1 \\ 0 \\ -2 \\ 0 \\ 3 \\ 1 \\ 0 \end{bmatrix} + x_8 \begin{bmatrix} 0 \\ -2 \\ 0 \\ 1 \\ 0 \\ -1 \\ 0 \\ 1 \end{bmatrix}. \tag{B.112}$$

This final expression is the general solution of this example. The variables x_1, x_3, x_5, x_7, x_8 appearing here are the nonbasic variables.

The *rank* of a matrix can be defined as the number of nonzero rows in the corresponding echelon matrix. We state the following theorem without proof.

A system of nonhomogenous linear equations has a solution if and only if the rank of the coefficient matrix equals the rank of the augmented matrix.

For consistent systems (those which have at least one solution) we have the following situation. The rank is the number of basic variables. The number of unknowns minus the rank, that is, the number of nonbasic variables, can be regarded as the number of degrees of freedom in the general solution.

By imposing additional conditions on the nonbasic variables we get a unique solution. If the rank equals the number of unknowns there is exactly one solution. Note that it is the rank rather than the number of equations that is important in the solution.

The solution of a system of equations can be effected by successive elimination of variables. The basic step in this (Gaussian) elimination process is given the name *pivot operation* and consists of a suitable sequence of elementary operations of type $E2$ and $E3$. Suppose that the system of equations is denoted $AX = K$, where A is a p by n matrix and K is a p by 1 vector. If the coefficient of x_k in equation h, a_{hk}, is not zero, then we first divide equation h by a_{hk} and then for each $i \neq h$ we replace equation i by its sum with the product of equation h by $(-a_{ik}/a_{hk})$. This process leads to a new system equivalent to the original one in which x_k appears only in equation h and there has coefficient 1, that is, the kth column of the new coefficient matrix is the hth unit vector.

This can be stated in matrix language as follows. Let $B = [A \quad K]$ be the augmented matrix of the initial equation system and let $B^* = [A^* \quad K^*]$ be the augmented matrix of the derived system. Then

$$B^* = QB,$$

where Q is the matrix

$$\begin{array}{c} \\ \\ \\ \\ \\ \\ \\ \\ h \\ \\ \\ \\ \\ \end{array}\overset{\displaystyle \qquad\qquad\qquad h}{\begin{bmatrix} 1 & \cdots & -a_{1k}/a_{hk} & \cdots & 0 \\ \cdot & \cdot & \cdot & \cdot & \cdot \\ \cdot & \cdot & \cdot & \cdot & \cdot \\ \cdot & \cdot & \cdot & \cdot & \cdot \\ 0 & \cdots & -a_{ik}/a_{hk} & \cdots & 0 \\ \cdot & \cdot & \cdot & \cdot & \cdot \\ \cdot & \cdot & \cdot & \cdot & \cdot \\ \cdot & \cdot & \cdot & \cdot & \cdot \\ 0 & \cdots & 1/a_{hk} & \cdots & 0 \\ \cdot & \cdot & \cdot & \cdot & \cdot \\ \cdot & \cdot & \cdot & \cdot & \cdot \\ \cdot & \cdot & \cdot & \cdot & \cdot \\ 0 & \cdots & -a_{pk}/a_{hk} & \cdots & 1 \end{bmatrix}},$$

the pivot matrix corresponding to P_{hk}, the pivot operation on the position or element (h, k).

For example, consider the linear equation system

$$\begin{aligned}
-x_1 - x_2 - x_3 - x_4 - x_5 &= 0\\
x_1 + 3x_2 - 2x_3 + \frac{13}{5}x_4 + \frac{10}{3}x_5 &= 1\\
2x_1 + \frac{5}{2}x_3 + x_4 + \frac{1}{2}x_5 &= 1\\
3x_1 + 2x_2 + x_3 + \frac{8}{5}x_4 + x_5 &= 1.
\end{aligned}$$

The augmented matrix of the system is

$$B = \begin{bmatrix} -1 & -1 & -1 & -1 & -1 & 0\\ 1 & 3 & -2 & 13/5 & 10/3 & 1\\ 2 & 0 & 5/2 & 1 & 1/2 & 1\\ 3 & 2 & 1 & 8/5 & 1 & 1 \end{bmatrix} = [A \quad K].$$

We can pivot in position $(h, k) = (4, 1)$ and have as the pivot matrix

$$Q = \begin{bmatrix} 1 & 0 & 0 & 1/3\\ 0 & 1 & 0 & -1/3\\ 0 & 0 & 1 & -2/3\\ 0 & 0 & 0 & 1/3 \end{bmatrix},$$

so that

$$B^* = QB = \begin{bmatrix} 0 & -1/3 & -2/3 & -7/15 & -2/3 & 1/3\\ 0 & 7/3 & 5/3 & 31/15 & 3 & 2/3\\ 0 & -4/3 & 11/6 & -1/15 & -1/6 & 1/3\\ 1 & 2/3 & 1/3 & 8/15 & 1/3 & 1/3 \end{bmatrix} = [A^* \quad K^*].$$

B.12. FLOW CHART FOR ECHELON FORM.

We now wish to introduce the concept of a logical flow chart for the Gaussian elimination process, since the development of a flow chart is the first step in preparing a problem for machine computations. A computing algorithm is a process for step-by-step solution of a numerical problem. The process must be logically complete in the sense that it provides for all alternatives that may arise. It is ordinarily inductive in character, that is, the end of each stage must either complete the solution or prepare the problem for the beginning of the next stage. The stages themselves are ordinarily labeled with one or more integers called indices; these indices are arranged in some increasing order and the conclusion of a stage with one index either prepares for a stage with a greater index or completes the process. Since a good algorithm must be able to deal with a large set of different problems, some short and some long, it is ordinarily not possible to specify in advance just how many stages will be needed for a solution. Upper bounds for this are sometimes available. The algorithm must include explicit "stop" rules which indicate that the process is complete.

The final steps of any stage which prepare for the next stage are sometimes called "closing the loop." The algorithm must specify a first step and a full description of the steps in a "general" stage, including stop rules and rules for transition to a later stage.

We begin with a matrix

$$A = \begin{bmatrix} a_{11} & \cdots & a_{1n} \\ & \cdots & \\ a_{p1} & \cdots & a_{pn} \end{bmatrix}$$

having p rows and n columns. The stage indices for the process of reduction to echelon form will appear as an ordered pair (i, j) of integers with i indicating a row and j a column. A stage (i, j) will follow a stage (h, k) if $i \geq h$ and $j \geq k$ with at least one inequality holding. At the conclusion of stage (i, j) the first j columns of the matrix will be in echelon form, and in these columns all rows below the ith row will be zero. We start with stage $(1, 1)$, that is, with the original matrix as it stands; this can be regarded as the end of stage $(0, 0)$.

The steps in any stage are of two classes: (1) questions which can be answered "yes" or "no" and (2) operations derived from elementary row operations whose legitimacy of application is guaranteed by the questions and answers which precede them in the stage.

These questions and operations are themselves indexed by i or j or both. The questions are:

R_i: Is $i < p$?
C_j: Is $j < n$?
Q_{ij}: Is $a_{hj} = 0$ for all $h \geq i$?
T_{ij}: Is $a_{ij} = 0$?

The operations are:

M_{ij}: Find $h > i$ for which $a_{hj} \neq 0$,
I_{ih}: Interchange row i and row h,
N_{ij}: Divide row i by a_{ij},
S_{ij}: Sweep out column j using a_{ij} as the pivot element,
P_{ij}: The pivot operation at (i, j) which consists of N_{ij} followed by S_{ij}.

To avoid burdensome notation at each stage or substage we use the symbol a_{ij} to mean the element then in row i and column j.

All of the above are self-explanatory, except for S_{ij} which consists of of adding $-a_{hj}/a_{ij}$ times row i to row h for each $h \neq i$.

The legitimacy conditions for our operations are:

P_{ij}: "no" for T_{ij},
M_{ij}: "no" for Q_{ij} and "yes" for T_{ij}.

We describe state (i, j) by the following flow chart.

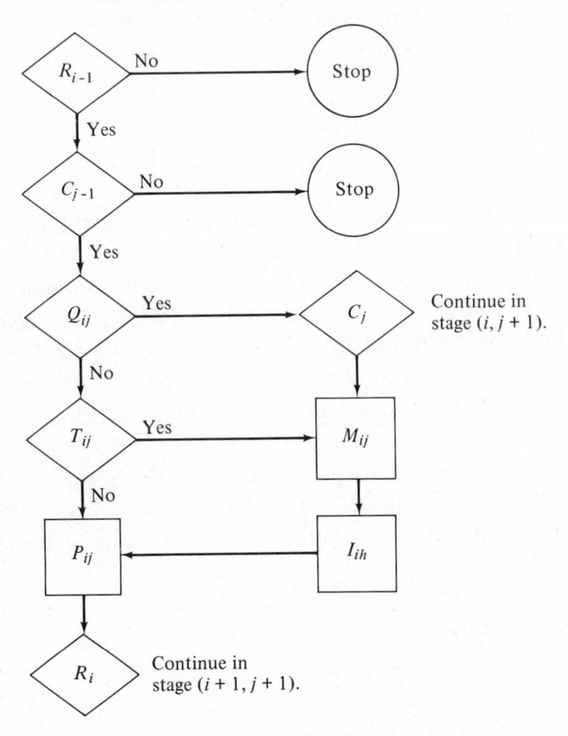

The process begins in stage (1, 1) with question R_0 and continues until a stop is reached. There are four places where stage (i, j) can be completed; the first two of these lead to stops, the third leads to the second step of stage $(i, j + 1)$, and the fourth leads to the first step of stage $(i + 1, j + 1)$. In the case of the third we could have gone to the first step of stage $(i, j + 1)$ but this was not done since the question R_{i-1} had already been answered affirmatively.

In counting steps we consider T_{ij} to be step 4 and P_{ij} to be step 5. We consider M_{ij} to be step 4a and I_{ih} to be step 4b. Step 5 is the final step in stage (i, j). In more complicated flow charts counting of steps is not always convenient; the important consideration is that the step following any given step be well defined in terms of the process.

We give an illustrative example using a 4 by 6 matrix,

$$\begin{bmatrix} -2 & 4 & 1 & -2 & -7 & 0 \\ 2 & -4 & 2 & 1 & 8 & 13 \\ 1 & -2 & 1 & 2 & 7 & 5 \\ -1 & 2 & 1 & 3 & 5 & -2 \end{bmatrix}.$$

The process can be represented as follows.

Stage (1, 1)

$$R_0 \xrightarrow{\text{yes}} C_0 \xrightarrow{\text{yes}} C_{11} \xrightarrow{\text{no}} T_{11} \xrightarrow{\text{no}}$$

$$P_{11} \rightarrow \begin{bmatrix} 1 & -2 & -1/2 & 1 & 7/2 & 0 \\ 0 & 0 & 3 & -1 & 1 & 13 \\ 0 & 0 & 3/2 & 1 & 7/2 & 5 \\ 0 & 0 & 1/2 & 4 & 17/2 & -2 \end{bmatrix}$$

Stage (2, 2) Stage (2, 3)

$$R_1 \xrightarrow{\text{yes}} C_1 \xrightarrow{\text{yes}} Q_{22} \xrightarrow{\text{yes}} C_2 \xrightarrow{\text{yes}} Q_{23} \xrightarrow{\text{no}} T_{23} \xrightarrow{\text{no}}$$

$$P_{23} \rightarrow \begin{bmatrix} 1 & -2 & 0 & 5/6 & 11/3 & 13/6 \\ 0 & 0 & 1 & -1/3 & 1/3 & 13/3 \\ 0 & 0 & 0 & 3/2 & 3 & -3/2 \\ 0 & 0 & 0 & 25/6 & 25/3 & -25/6 \end{bmatrix}$$

Stage (3, 4)

$$R_2 \xrightarrow{\text{yes}} C_3 \xrightarrow{\text{yes}} Q_{34} \xrightarrow{\text{no}} T_{34} \xrightarrow{\text{no}}$$

$$P_{34} \rightarrow \begin{bmatrix} 1 & -2 & 0 & 0 & 2 & 3 \\ 0 & 0 & 1 & 0 & 1 & 4 \\ 0 & 0 & 0 & 1 & 2 & -1 \\ 0 & 0 & 0 & 0 & 0 & 0 \end{bmatrix}$$

Stage (4, 5) Stage (4, 6) Stage (4, 7)

$$R_3 \xrightarrow{\text{yes}} C_4 \xrightarrow{\text{yes}} Q_{45} \xrightarrow{\text{yes}} C_5 \xrightarrow{\text{yes}} Q_{46} \xrightarrow{\text{yes}} C_6 \xrightarrow{\text{no}} \text{Stop.}$$

In hand calculations the questions are rarely written down, and occasional deviations from the process may be found which shorten the calculation.

A slightly altered form of the algorithm for reduction to echelon form can be described more succinctly in the following manner. Stage One: (a) If the first column is not zero permute rows so that the first element a_{11} is not zero; next multiply the first row by $1/a_{11}$; then subtract suitable multiples of this row from later ones so as to make the rest of the first column zero; and finally proceed to Stage Two with the submatrix obtained by deleting the first row and first column of this matrix. (b) If the first column is zero proceed to Stage Two with the submatrix obtained by deleting the first column. Stage Two: Repeat Stage One unless the submatrix under consideration has no nonzero elements; in that case proceed to Stage Three with the entire matrix. After enough steps one always reaches Stage Three. Stage Three: Subtract suitable multiples of each row from higher rows so as to obtain zeros above the leading elements in each nonzero row. The result will be an echelon matrix row equivalent to the original one. The sequence of operations corresponds to successive elimination of unknowns in the equations.

B.13. HOMOGENEOUS LINEAR EQUATIONS.

We frequently have occasion to determine if a sequence of vectors $S = (A_1, \cdots, A_n)$ in V_p is dependent or independent; as observed earlier this in turn reduces to determining whether or not the single vector equation

$$A_1x_1 + \cdots + A_nx_n = 0 \tag{B.113}$$

has a *nontrivial solution* $x_1, \cdots, x_n$ (one in which not every $x_i = 0$). We have seen that the single vector equation (B.113) is equivalent to the p scalar equations, called *homogeneous linear equations*,

$$\begin{aligned} a_{11}x_1 + \cdots + a_{1n}x_n &= 0 \\ \cdots \qquad \cdots & \\ a_{p1}x_1 + \cdots + a_{pn}x_n &= 0 \end{aligned} \tag{B.114}$$

where
$$A_j = \begin{bmatrix} a_{1j} \\ \cdot \\ \cdot \\ \cdot \\ a_{pj} \end{bmatrix} \qquad (j = 1, \cdots, n);$$

and this in turn is equivalent to the single matrix equation

$$AX = 0 \tag{B.115}$$

where $X = \begin{bmatrix} x_1 \\ \cdot \\ \cdot \\ \cdot \\ x_n \end{bmatrix}$ and $A = [a_{ij}]$ is the matrix whose columns are the vectors in the sequence S.

If the only solution of (B.115) is the zero vector we say that the system (B.114) has only *the trivial solution;* this is true, of course, if and only if the sequence S is independent. If (B.114) has a nontrivial solution then the sequence S is dependent.

The problem of solving (B.113), or (B.114), or (B.115) can be resolved into three questions:

Question E. Under what circumstances do (nontrivial) solutions *exist?*
Question N. What is the *nature* of the set of solutions?
Question C. How are the solutions *computed?*

These same questions arise again for nonhomogeneous linear equations, for homogeneous inequalities, and for nonhomogeneous inequalities.

In the homogeneous case the first two questions are answered by the three key theorems which follow (these theorems are stated without proof). The answer to the third question in detail depends on the tools available for solution and is clearly different for pencil and paper solution than when large digital computers are available. In Section B.14 we shall give a general method of solution which is modified in practice to suit the available tools. The idea underlying this method has already been presented in Sections B.11 and B.12.

Theorem B.1. A system of homogeneous linear equations with more unknowns than equations has a nontrivial solution.

This theorem is equivalent to the statement that in V_p any sequence S with more than p vectors is dependent. For example, any four vectors in V_3 are dependent.

Let $L(A)$ denote the subspace of V_p spanned by the columns of A; $L(A)$ is called the *column space* of A. We consider some simple examples. Let

$$A_1 = \begin{bmatrix} 2 & 4 \\ 3 & 1 \end{bmatrix}, \quad A_2 = \begin{bmatrix} 2 & 4 \\ 3 & 6 \end{bmatrix}, \quad A_3 = \begin{bmatrix} 2 & 4 & 5 & 1 \\ 3 & 1 & 2 & 1 \end{bmatrix},$$

$$A_4 = \begin{bmatrix} 1 & 0 \\ 0 & 1 \\ 0 & 0 \end{bmatrix}, \quad A_5 = \begin{bmatrix} 1 & 0 & 0 \\ 0 & 1 & 0 \\ 0 & 0 & 1 \end{bmatrix}.$$

Then, since the columns of A_1 are not proportional they span V_2, that is, $L(A_1) = V_2$. In A_2 the second column is dependent on the first one, hence $L(A_2) = \left\{cB \mid \text{where } c \in R \text{ and } B = \begin{bmatrix} 2 \\ 3 \end{bmatrix}\right\}$ is the line through the origin having slope $\frac{3}{2}$. Since the first two columns of A_3 span V_2 we know without even looking at the others that $L(A_3) = V_2$. $L(A_4)$ is the x_1, x_2 plane in V_3, and $L(A_5) = V_3$. In general, $L(I_p) = V_p$.

Theorem B.2. For any matrix A

$$\dim L(A) = \dim L(A^T). \tag{B.116}$$

This common dimension is called the *rank* of A and is designated $r(A)$ or, for short, just r. This theorem states that for any matrix the maximum number of independent rows is equal to the maximum number of independent columns (in other words, for any matrix, the row rank and the column rank are equal). This definition of rank is equivalent to the one given in Section B.11 as the number of nonzero rows in the corresponding echelon matrix. The advantage of the earlier characterization is its constructive nature.

The set of all solutions of $AX = 0$ is called the *column kernel* of A and is designated $L^*(A)$. We illustrate this concept with some examples. Let

$$A_1 = \begin{bmatrix} 1 & 0 \\ 0 & 1 \end{bmatrix}, \quad A_2 = \begin{bmatrix} 1 & 2 & 0 & 1 \\ 0 & 0 & 1 & -2 \end{bmatrix},$$

$$A_3 = \begin{bmatrix} 1 & 0 & 0 & 2 & 5 \\ 0 & 1 & 0 & 3 & 6 \\ 0 & 0 & 1 & 4 & 7 \end{bmatrix}. \tag{B.117}$$

Then $A_1X = 0$ means $x_1 = 0$ and $x_2 = 0$ and hence $L^*(A_1) = \{0\}$. If $A_2X = 0$ then (by separating into basic and nonbasic variables) $x_1 = -2x_2 - x_4$ and $x_3 = 2x_4$; hence any solution X has the form

$$X = \begin{bmatrix} x_1 \\ x_2 \\ x_3 \\ x_4 \end{bmatrix} = \begin{bmatrix} -2x_2 - x_4 \\ x_2 \\ 2x_4 \\ x_4 \end{bmatrix} = x_2 \begin{bmatrix} -2 \\ 1 \\ 0 \\ 0 \end{bmatrix} + x_4 \begin{bmatrix} -1 \\ 0 \\ 2 \\ 1 \end{bmatrix} \tag{B.118}$$

and $L^*(A_2)$ is the space spanned by the vectors $[-2 \quad 1 \quad 0 \quad 0]^T$ and $[-1 \quad 0 \quad 2 \quad 1]^T$. Similarly, $L^*(A_3)$ is the space spanned by the vectors $[-2 \quad -3 \quad -4 \quad 1 \quad 0]^T$ and $[-5 \quad -6 \quad -7 \quad 0 \quad 1]^T$. In general $L^*(I_p) = \{0\}$.

Theorem B.3. The set of solutions $L^*(A)$ of $AX = 0$ is a vector space whose dimension is the number of columns of A minus the rank of A:

$$\dim L^*(A) = \text{column degree } A - r(A). \tag{B.119}$$

The statement (B.119) has several equivalent formulations:

$$\dim L^*(A) + \dim L(A) = \text{column degree } A; \tag{B.119$'$}$$

The number of independent solutions plus the number of independent equations equals the number of unknowns; (B.119″)

$$\dim L^*(A) = n - r, \text{ where } r \text{ is the rank of } A. \tag{B.119$'''$}$$

Since for $n > r$ the number of solutions of (B.115) is infinite we cannot hope to write them all down. We interpret the request "to find all solutions" or "to solve the system" to mean to find a set of vectors $B_1, \cdots, B_t$ which form a spanning sequence, or better yet, a basis, for $L^*(A)$.

Let $B_1, B_2, \cdots, B_t$ be a spanning sequence for the vector space of solutions $L^*(A)$ and let B be the n by t matrix whose jth column is B_j. Then if A is the p by n matrix in (B.115)

$$AB = A[B_1 \cdots B_t] = [AB_1 \cdots AB_t] = [0 \cdots 0] = 0, \qquad \text{(B.120)}$$

since $B_j \in L^*(A)$; moreover $r(A) + r(B) = r + (n - r) = n$. It should be noted that the number n can be described equivalently as the number of unknowns, or as the column degree of A, or as the row degree of B.

Two matrices A, B satisfying (B.120) and having the property that $r(A) + r(B) = n$, where n is the column degree of A (and row degree of B), are said to be an *apolar pair*. The problem of solving the equation system $AX = 0$ can be restated as that of finding a matrix B apolar to A (the columns of B then span the vector space of solution vectors X).

For examples of apolarity, the matrices, A_1, A_2, A_3 in (B.117) are, respectively, apolar to

$$B_1 = \begin{bmatrix} 0 \\ 0 \end{bmatrix}, \quad B_2 = \begin{bmatrix} -2 & -1 \\ 1 & 0 \\ 0 & 2 \\ 0 & 1 \end{bmatrix}, \quad B_3 = \begin{bmatrix} -2 & -5 \\ -3 & -6 \\ -4 & -7 \\ 1 & 0 \\ 0 & 1 \end{bmatrix} \qquad \text{(B.121)}$$

because $A_iB_i = 0$ for $i = 1, 2, 3$ and $r(A_1) + r(B_1) = 2$, $r(A_2) + r(B_2) = 4$, and $r(A_3) + r(B_3) = 5$.

The concept of apolarity provides a convenient summary of the salient facts concerning the theory of solutions of homogeneous linear equations. The following three conditions are all equivalent to (B.120); that is, each is necessary and sufficient for A and B to be an apolar pair:

$$B^T \text{ and } A^T \text{ are an apolar pair,} \qquad \text{(B.120')}$$

$$L^*(A) = L(B), \qquad \text{(B.120'')}$$

$$L^*(B^T) = L(A^T). \qquad \text{(B.120''')}$$

Note that, in general, apolarity is not symmetric; if A and B are an apolar pair, it does not follow from (B.120) that B and A are an apolar pair; however, it does follow that B^T and A^T are an apolar pair.

B.14. SOLUTION OF HOMOGENEOUS LINEAR EQUATIONS; INVERSES.

The system $AX = 0$ is said to be *equivalent* to a second system

$$A'X = 0 \qquad \text{(B.122)}$$

if every solution of one is a solution of the other, that is, if $L^*(A) = L^*(A')$. There are three basic operations, called elementary row operations (compare Section B.11), which lead from any system to an equivalent system and an appropriate sequence of such operations leads to a system whose solution is obvious. We saw in Section B.12 that every matrix A could be mapped into an echelon matrix D by a sequence of elementary operations. The system

$$DX = 0 \tag{B.123}$$

is equivalent to the system $AX = 0$ and has the same number of equations. Moreover, the solution set $L^*(A) = L^*(D)$ is easily determined from the echelon form (B.123). For an example, consider an equation system whose matrix is that given in (B.107). By means of elementary row transformations this matrix is transformed into the echelon matrix (B.108). The corresponding linear equation system is (B.109); this is simpler than the original system and the equivalent form (B.110), simpler yet, is one from which solutions to the original system—and any system equivalent to it—can be easily obtained.

Two matrices A and A' are said to be *row equivalent* if there is a sequence of elementary row operations leading from one to the other. It is easy to verify that row equivalence is reflexive, symmetric, and transitive; in other words, it is an equivalence relation on the set of all matrices of any fixed shape p by n.

It can be proved that:

Every matrix A is row equivalent to a unique echelon matrix D; D is said to *correspond* to A. (B.124)

Two matrices A and A' are row equivalent to each other if and only if they correspond to the same echelon matrix D. (B.125)

Two matrices A, A' have the same column kernel; that is, $L^*(A) = L^*(A')$, if and only if their corresponding echelon matrices are identical except perhaps for the number of zero rows. (B.126)

Two matrices A, A' have the same row space if and only if their corresponding echelon matrices are identical except perhaps for the number of zero rows. (B.127)

We recall that the rank of a matrix is the number of nonzero rows in the corresponding echelon matrix.

The *spread* of a nonzero row or column vector is the index of its first nonzero entry; thus [0 3 0 1] has spread 2 and [2 1 3] has spread 1.

Let D be an echelon matrix of rank r, and let e_i be the spread of the ith row of D, $(i = 1, \cdots, r)$. Then

$$e_1, e_2, \cdots, e_r \tag{B.128}$$

are called the *echelon indices* of D and also of any matrix A that is row equivalent to D. We observe that

$$1 \leq e_1 < e_2 < \cdots < e_r \leq n. \tag{B.129}$$

For example, the echelon matrix

$$D = \begin{bmatrix} 0 & 1 & 0 & 3 & 0 & 4 \\ 0 & 0 & 1 & -2 & 0 & 7 \\ 0 & 0 & 0 & 0 & 1 & 6 \\ 0 & 0 & 0 & 0 & 0 & 0 \end{bmatrix} \tag{B.130}$$

has echelon indices (2, 3, 5). In the system

$$DX = 0 \tag{B.131}$$

each index variable x_{e_i} appears in only one equation so that the general solution can be obtained by solving for the index variables (basic) in terms of the remaining ones (nonbasic). For example, in the case of (B.131) we would have

$$\begin{aligned} x_2 &= -3x_4 - 4x_6 \\ x_3 &= 2x_4 - 7x_6 \\ x_5 &= - 6x_6 \end{aligned} \tag{B.132}$$

or for the general solution X

$$X = \begin{bmatrix} x_1 \\ x_2 \\ x_3 \\ x_4 \\ x_5 \\ x_6 \end{bmatrix} = \begin{bmatrix} x_1 \\ -3x_4 - 4x_6 \\ 2x_4 - 7x_6 \\ x_4 \\ -6x_6 \\ x_6 \end{bmatrix} = \begin{bmatrix} 1 \\ 0 \\ 0 \\ 0 \\ 0 \\ 0 \end{bmatrix} x_1 + \begin{bmatrix} 0 \\ -3 \\ 2 \\ 1 \\ 0 \\ 0 \end{bmatrix} x_4 + \begin{bmatrix} 0 \\ -4 \\ -7 \\ 0 \\ -6 \\ 1 \end{bmatrix} x_6, \tag{B.133}$$

and here the vectors playing the role of coefficients of x_1, x_4, and x_6 form a basis for the space $L^*(A)$ of all solutions of $AX = 0$, and x_1, x_4, x_6 can be regarded as parameters in the solution.

We can insert and delete zero rows so that from an echelon matrix D we obtain a *square matrix* H, called a Hermite matrix, with the same number of nonzero rows as D but with rows arranged so that each leading nonzero element is on the diagonal. For example, if D is given by (B.130) then the Hermite matrix H is

$$H = \begin{bmatrix} 0 & 0 & 0 & 0 & 0 & 0 \\ 0 & 1 & 0 & 3 & 0 & 4 \\ 0 & 0 & 1 & -2 & 0 & 7 \\ 0 & 0 & 0 & 0 & 0 & 0 \\ 0 & 0 & 0 & 0 & 1 & 6 \\ 0 & 0 & 0 & 0 & 0 & 0 \end{bmatrix}; \tag{B.134}$$

D is 4 by 6 and H is 6 by 6. Note that each diagonal element in H is either 1 or 0 and that when the leading nonzero element in any row is 1 all other elements in the column containing this element are 0.

The Hermite matrix H has an interesting property: $HH = H^2 = H$ (a nonzero matrix that is equal to its square is said to be *idempotent*). A proof of this statement is helpful in understanding more fully the role that Hermite matrices play in solving systems of linear equations. Let $H = [h_{ij}]$ be an n by n Hermite matrix and let $H^2 = [p_{ij}]$; we want to show that $h_{ij} = p_{ij}$ for each i and j. From matrix multiplication we have

$$p_{ij} = \sum_{k=1}^{n} h_{ik}h_{kj}$$

Suppose row i is a zero row of H; then $h_{ik} = 0$ for every k and $\sum_{k=1}^{n} h_{ik}h_{kj} = 0 = p_{ij} = h_{ij}$. Suppose row i is not a zero row of H; then at least one h_{ik} in the products $h_{ik}h_{kj}$ is not zero. Now consider h_{kj}; if row k is a zero row of H, $h_{kj} = 0$. On the other hand, if row k is not a zero row of H there are two possibilities for the products $h_{ik}h_{kj}$: (1) $k = i$ and $h_{ik}h_{kj} = h_{ii}h_{ij} = 1h_{ij} = h_{ij}$, since the diagonal element for each nonzero row is a 1; (2) $k \neq i$, which means that $h_{ik} = 0$, since the first nonzero element in row k is 1 and all other elements in that column must be zero. Therefore the only nonzero product $h_{ik}h_{kj}$ is $h_{ii}h_{ij} = h_{ij}$ and $p_{ij} = h_{ij}$ when row i is a nonzero row of H.

This is illustrated by forming the product of the matrix H in (B.134) with itself. In this matrix product consider the product of row 2 (a nonzero row) with column 6; this gives the element p_{26} in H^2,

$$\begin{aligned} p_{26} &= h_{21}h_{16} + h_{22}h_{26} + h_{23}h_{36} + h_{24}h_{46} + h_{25}h_{56} + h_{26}h_{66} \\ &= (0)(0) + (1)(4) + (0)(7) + (3)(0) + (0)(6) + (4)(0) = 4 = h_{26}. \end{aligned}$$

For $k = i = 2$ we have $h_{22}h_{26} = (1)(4) = 4$, and for $k \neq i$, row k a nonzero row of H, we have, since the nonzero rows of H other than row 2 are rows 3 and 5,

$$\begin{aligned} h_{23}h_{36} &= (0)(7) = 0 \\ h_{25}h_{56} &= (0)(6) = 0. \end{aligned}$$

Note that the first nonzero element in row 3 is $h_{33} = 1$ and $h_{i3} = 0$ for $i \neq k$; an analogous statement can be made concerning the term involving row 5.

Now consider the product

$$H(I - H) = H - HH = H - H = 0,$$

where H is any n by n Hermite matrix having rank $r(H) = r$. It can be shown that $r(I - H) = n - r$; this means that H and $I - H$ are apolar matrices. The relevance of these results to the problem of solving any linear equation system $AX = 0$ is clear from the following solution procedure: (1) transform A into echelon form D; (2) determine the Hermite matrix H

from D; (3) calculate $I - H$; since H and $I - H$ are apolar, the columns of $I - H$ span the set of solutions of the system $AX = 0$; that is, they span $L^*(A) = L^*(D) = L^*(H)$. In particular, the nonzero columns of $I - H$ are a *basis* of $L^*(A)$. For example, if H is given by (B.134), then

$$I - H = \begin{bmatrix} 1 & 0 & 0 & 0 & 0 & 0 \\ 0 & 0 & 0 & -3 & 0 & -4 \\ 0 & 0 & 0 & 2 & 0 & -7 \\ 0 & 0 & 0 & 1 & 0 & 0 \\ 0 & 0 & 0 & 0 & 0 & -6 \\ 0 & 0 & 0 & 0 & 0 & 1 \end{bmatrix} \tag{B.135}$$

and the nonzero columns of this matrix are a basis for $L^*(D) = L^*(H) = L^*(A)$. Finally, these nonzero columns are the vectors appearing as coefficients in the right hand side of (B.133).

The precise computational steps followed in practical solution of systems of equations depend on the actual computational aids available. However, most of the so-called direct methods are based on the ideas discussed in this and the preceding sections.

Viewed in another light the solution of a system of equations involves finding the "inverse" of some submatrix of the coefficient matrix. There is no division operation with matrices and the concept of inverse matrix is a partial replacement for division. We now give a brief discussion of this concept.

If two square matrices A and B, each of degree p, satisfy the equation

$$AB = I_p \tag{B.136}$$

we call B the *inverse* of A, written $B = A^{-1}$. A matrix is said to be *nonsingular* if it has an inverse; otherwise it is called *singular*.

Every nonsquare matrix is singular and a square matrix A of degree p is nonsingular if and only if any one of the following equivalent conditions holds:

$$\text{rank } A = p, \tag{B.137}$$

$$L(A) = V_p, \tag{B.138}$$

$$L^*(A) = \{0\}, \tag{B.139}$$

$$\text{the echelon matrix corresponding to } A \text{ is } I_p. \tag{B.140}$$

It can be proved that a nonsingular matrix A has only one inverse A^{-1}, and moreover, that A^{-1} is also nonsingular with

$$(A^{-1})^{-1} = A. \tag{B.141}$$

For nonsquare matrices (B.136) can hold but then B is not unique and is called a *right inverse* of A and A is called a *left inverse* of B. A matrix A has a right inverse if and only if rank A = row degree A; A has a left inverse if and only if rank A = column degree A.

If a matrix A is nonsingular its inverse can be found in the following way. Let

$$G = [A \quad I_p]; \tag{B.142}$$

then the echelon matrix corresponding to G is

$$D = [I_p \quad A^{-1}]. \tag{B.143}$$

If one does not know whether a square matrix A is singular or nonsingular one can still form G and calculate its corresponding echelon matrix D. If D has the form $[I_p \quad B]$ then A is nonsingular and $B = A^{-1}$. If D does not have the form $[I_p \quad B]$ then A is singular; this will be the case if and only if the pth echelon index of D is greater than p.

For example, let $A = \begin{bmatrix} 3 & 2 \\ 2 & 1 \end{bmatrix}$;
then

$$G = \begin{bmatrix} 3 & 2 & 1 & 0 \\ 2 & 1 & 0 & 1 \end{bmatrix}$$

has echelon form

$$D = \begin{bmatrix} 1 & 0 & -1 & 2 \\ 0 & 1 & 2 & -3 \end{bmatrix},$$

whence

$$A^{-1} = \begin{bmatrix} -1 & 2 \\ 2 & -3 \end{bmatrix}.$$

The inverse of a nonsingular matrix can be calculated in terms of its submatrices. Inversion by submatrices can be desirable from the standpoint of machine computation and has theoretical applications as well. Suppose a matrix A is partitioned into four submatrices A_{ij} so that the submatrices on the diagonal are square. Then if A^{-1} has submatrices denoted D_{ij} we can write

$$\begin{bmatrix} A_{11} & A_{12} \\ A_{21} & A_{22} \end{bmatrix} \begin{bmatrix} D_{11} & D_{12} \\ D_{21} & D_{22} \end{bmatrix} = \begin{bmatrix} I & 0 \\ 0 & I \end{bmatrix},$$

where all matrices are understood to be partitioned in the same manner. Multiplying by means of partitioned matrices gives

$$\begin{aligned} A_{11}D_{11} + A_{12}D_{21} &= I \qquad & A_{11}D_{12} + A_{12}D_{22} &= 0 \\ A_{21}D_{11} + A_{22}D_{21} &= 0 \qquad & A_{21}D_{12} + A_{22}D_{22} &= I. \end{aligned}$$

Now solve for D_{12} and D_{21} in the equations having 0 for right-hand side,

$$A_{22}D_{21} = -A_{21}D_{11}, \qquad A_{11}D_{12} = -A_{12}D_{22}.$$

If A_{22} and A_{11} are nonsingular, we obtain

$$D_{21} = -A_{22}^{-1}(A_{21}D_{11}) \qquad D_{12} = -A_{11}^{-1}(A_{12}D_{22}). \tag{B.144}$$

Substituting in the other two equations above gives

$$[A_{11} - A_{12}A_{22}^{-1}A_{21}]D_{11} = I \qquad [A_{22} - A_{21}A_{11}^{-1}A_{12}]D_{22} = I,$$

which imply that

$$D_{11} = [A_{11} - A_{12}A_{22}^{-1}A_{21}]^{-1} \qquad D_{22} = [A_{22} - A_{21}A_{11}^{-1}A_{12}]^{-1}. \quad \text{(B.145)}$$

For example, consider the matrix in Equation (4.12′),

$$M_{S*} = \begin{bmatrix} 1 & C_{S*}^T \\ 0 & I_p \end{bmatrix}.$$

In this case the partitioning is such that $A_{11} = 1$ and $A_{22} = I_p$ (so that $A_{11}^{-1} = 1$ and $A_{22}^{-1} = I_p$); also $A_{12} = C_{S*}^T$ and $A_{21} = 0$. Also

$$\begin{aligned} D_{11} &= [A_{11} - A_{12}A_{22}^{-1}A_{21}]^{-1} = [1 - C_{S*}^T I_p(0)]^{-1} = 1, \\ D_{22} &= [A_{22} - A_{21}A_{11}^{-1}A_{12}]^{-1} = [I_p - (0)(1)C_{S*}^T]^{-1} = I_p, \\ D_{21} &= -A_{22}^{-1}(A_{21}D_{11}) = -I_p(0 \cdot 1) = 0, \\ D_{12} &= -A_{11}^{-1}(A_{12}D_{22}) = -1(C_{S*}^T I_p) = -C_{S*}^T. \end{aligned}$$

Therefore,

$$M_{S*}^{-1} = \begin{bmatrix} 1 & -C_{S*}^T \\ 0 & I_p \end{bmatrix}.$$

For another example, consider the more general form of M_{S*} given in (4.12″)

$$M_{S*} = \begin{bmatrix} 1 & C_{S*}^T \\ 0 & A_{S*} \end{bmatrix};$$

then

$$\begin{aligned} D_{11} &= [A_{11} - A_{12}A_{22}^{-1}A_{21}]^{-1} = [1 - C_{S*}^T A_{S*}^{-1}(0)]^{-1} = 1, \\ D_{22} &= [A_{22} - A_{21}A_{11}^{-1}A_{12}]^{-1} = [A_{S*} - (0)(1)C_{S*}^T]^{-1} = A_{S*}^{-1}, \\ D_{21} &= -A_{22}^{-1}(A_{21}D_{11}) = -A_{S*}^{-1}(0)(1) = 0, \\ D_{12} &= -A_{11}^{-1}(A_{12}D_{22}) = -1(C_{S*}^T A_{S*}^{-1}) = -C_{S*}^T A_{S*}^{-1}, \end{aligned}$$

and

$$M_{S*}^{-1} = \begin{bmatrix} 1 & -C_{S*}^T A_{S*}^{-1} \\ 0 & A_{S*}^{-1} \end{bmatrix},$$

which is (4.16) of Section 4.1.

B.15. DOUBLE DESCRIPTION AND DUALITY.

Let W be a subspace of V_p. Then there exists a matrix B whose columns span W. Let C be a matrix such that B^T and C are an apolar pair. From the equivalence of the statements (B.120) and (B.120′) we know also that C^T and B are an apolar pair. Let $A = C^T$; then A and

B are an apolar pair and since the columns of B span the subspace W, W must be the space of solutions of the system $AX = 0$;

$$W = L(B) = L^*(A). \tag{B.146}$$

This is known as the theorem of *double description* for subspaces and can be paraphrased, "any subspace can be described by a (finite) spanning set or by a (finite) system of equations." We sometimes call the description by means of a spanning set a *sum description* since if $B = [B_1 \cdots B_t]$ is a partition of B into columns then

$$W = \langle B_1 \rangle + \cdots + \langle B_t \rangle; \tag{B.147}$$

that is, W is the sum of the subspaces spanned by the columns of the matrix B. Similarly, if $A = \begin{bmatrix} A_1 \\ \cdot \\ \cdot \\ \cdot \\ A_p \end{bmatrix}$ is a partition of A into rows then the relation

$$W = L^*(A_1) \cap L^*(A_2) \cap \cdots \cap L^*(A_p) \tag{B.148}$$

expresses the fact that $W = L^*(A)$ is made up of all vectors which satisfy each of the p equations. Thus we speak of a description by means of equations as an *intersection description.*

Earlier we discussed computational methods for finding matrices A and B for which (B.146) holds; our present purpose is to call attention to the concept of double description and its relation to solution of homogeneous linear equations. In practice a subspace W is ordinarily encountered in one of the forms $L(B)$ or $L^*(A)$. We can therefore regard the solution of a system $AX = 0$ as the obtaining of a double description for the space $W = L^*(A)$.

The connection between (B.120″) and (B.120‴) is an example of a general concept called *duality.* Formally, taking transposes in a condition such as (B.120′) gives what we call the *dual* condition. If X and Y are two vectors in V_p the equation

$$X^TY = 0 \tag{B.149}$$

can be regarded in two ways; we can think of $x_1, \cdots, x_p$ as being coefficients of an equation and $y_1, \cdots, y_p$ as forming a solution, or by taking transposes we interchange the roles of X and Y. If W is any subspace of V_p we define a new subspace $W^\perp$ (read, W perpendicular) by

$$W^\perp = \{Y \mid X^TY = 0 \text{ for all } X \in W\}. \qquad \text{1B5(.)0}$$

The geometric meaning of $W^\perp$ is that it contains all vectors that are perpendicular to each vector in W (see Section B.17).

The mapping $\perp$ has several important properties:

$$(W_1 + W_2)^\perp = W_1^\perp \cap W_2^\perp, \tag{B.151}$$

$$(W_1 \cap W_2)^\perp = W_1^\perp + W_2^\perp, \tag{B.152}$$

$$(W^\perp)^\perp = W, \tag{B.153}$$

$$W_1 \subset W_2 \Rightarrow W_1^\perp \supset W_2^\perp. \tag{B.154}$$

Now, if $W = L^*(A)$ then $W^\perp = L(A^T)$, and if $W = L(B)$ then $W^\perp = L^*(B^T)$. This approach may help explain the equivalence of conditions (B.120′) and (B.120‴); it also shows that given either the first or the second member of an apolar pair the problem of finding the other can be reduced to solving a system of homogeneous linear equations. This illustrates the general role of duality in mathematics; duality makes each result do double work. In classical mathematics a major application of duality is in projective geometry where, for example, in two dimensions the dual elements are lines and points and a statement such as, "any two points determine (lie on) a unique line," has as its dual the statement, "any two lines intersect in (lie on) a unique point."

The concepts of double description and duality will play important roles in our discussions of nonhomogeneous equations and of inequalities. It is important to be able to distinguish between them when they arise. Double description always involves a sum-type description and an intersection-type description, whereas in matrix algebra duality can ordinarily be identified by the appearance of transposes.

If complex coefficients are used instead of real ones it is a frequent practice to replace (B.149) by

$$X^*Y = 0 \tag{B.155}$$

where $X^* = \overline{X}^T$ is the conjugate transpose of X. Then if (B.150) is replaced by

$$W^\perp = \{Y \mid X^*Y = 0 \text{ for all } X \in W\} \tag{B.156}$$

the properties (B.151) through (B.154) are still valid.

B.16. NONHOMOGENEOUS LINEAR EQUATIONS.

The question of dependence of a vector K on a sequence $S = (A_1, \cdots, A_n)$ reduces to the question of solvability of the nonhomogeneous system,

$$AX = K \tag{B.157}$$

where A is the p by n matrix whose columns are $A_1, \cdots, A_n$. We consider in order the questions of Section B.13.

Question E. A solution of (B.157) exists if and only if K lies in the space spanned by the columns of A and hence if and only if

$$\text{rank } A = \text{rank } [A \quad K]. \tag{B.158}$$

We recall (compare Section B.10) that the matrix $[A \quad K]$ is called the augmented matrix of the system (B.157) and A is called the coefficient matrix. Thus, in words, *a nonhomogeneous system has a solution if and only if the rank of the coefficient matrix is equal to the rank of the augmented matrix.*

Question N. If $AY = 0$, $AX = K$ and $AX' = K$, then $AX + AY = K + 0$ and $A(X + Y) = K$; also $AX - AX' = K - K$ so that $A(X - X') = 0$; thus the difference of any solutions of (B.157) is a solution of (B.115) and the sum of any solution of (B.115) and a solution of (B.157) is again a solution of (B.157). Therefore, the set of all solutions of (B.157), if not empty, consists of all sums

$$X = X_0 + Y \tag{B.159}$$

where X_0 is a fixed solution of (B.157) and Y runs through the space $L^*(A)$ of all solutions of (B.115). We denote this solution set by $M^*(A, K)$ and call it a *linear manifold*. As a consequence of (B.159) we can write

$$M^*(A, K) = X_0 + L^*(A). \tag{B.160}$$

Geometrically interpreted a linear manifold is merely a vector space displaced by the addition of a constant vector to each of its elements. The dimension of a linear manifold is defined to be equal to the dimension of the corresponding vector space. For example, in three space there are four kinds of linear manifolds: (1) a single vector (dim $M^*(A, K) = 0$), (2) a line (dim $M^*(A, K) = 1$), (3) a plane (dim $M^*(A, K) = 2$), and (4) the entire space (dim $M^*(A, K) = 3$ or $A = 0$, $K = 0$).

For an example in V_2, the nonhomogeneous system $2x_1 - x_2 = 4$ and the related homogeneous system $2x_1 - x_2 = 0$ represent two parallel lines, and each vector X on the first line can be written as the sum of the fixed vector $X_0 = \begin{bmatrix} 2 \\ 0 \end{bmatrix}$ and a vector Y on the second line (see Figure B.5).

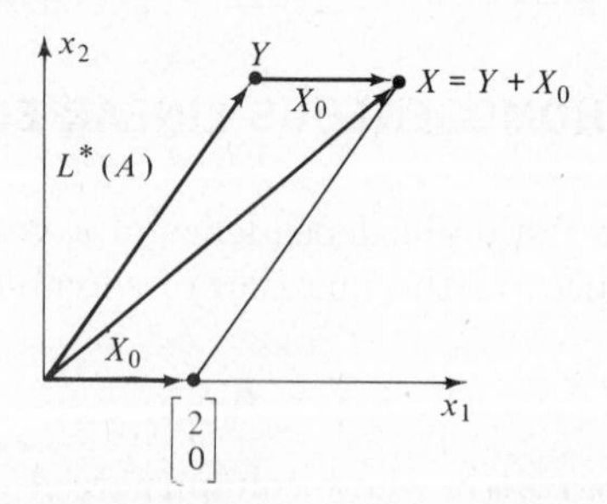

Figure B.5 $X_0 + L^*(A)$.

Question C. To solve a nonhomogeneous system we use the same operations as in the homogeneous case and reduce the augmented matrix $[A \quad K]$ to echelon form. The system is then consistent if and only if the largest echelon index does not exceed n. We can take for X_0 the solution then obtained by putting all of the non index variables equal to zero. For example, if instead of D in (B.130) we had

$$[D \quad L] = \begin{bmatrix} 0 & 1 & 0 & 3 & 0 & 4 & 2 \\ 0 & 0 & 1 & -2 & 0 & 7 & 4 \\ 0 & 0 & 0 & 0 & 1 & 6 & -3 \\ 0 & 0 & 0 & 0 & 0 & 0 & 0 \end{bmatrix}, \qquad (B.161)$$

then the echelon indices are still 2, 3, 5 and the particular solution

$$X_0 = \begin{bmatrix} 0 \\ 2 \\ 4 \\ 0 \\ -3 \\ 0 \end{bmatrix} \qquad (B.162)$$

is obtained by setting $x_1 = x_4 = x_6 = 0$. Note that in the particular solution each nonbasic variable is set equal to zero and each basic variable is equal to the constant term in the single equation in which it occurs.

If we insert zero rows so as to replace $[D \quad L]$ by the 7 by 8 matrix

$$[H \quad P] = \begin{bmatrix} 0 & 0 & 0 & 0 & 0 & 0 & 0 \\ 0 & 1 & 0 & 3 & 0 & 4 & 2 \\ 0 & 0 & 1 & -2 & 0 & 7 & 4 \\ 0 & 0 & 0 & 0 & 0 & 0 & 0 \\ 0 & 0 & 0 & 0 & 1 & 6 & -3 \\ 0 & 0 & 0 & 0 & 0 & 0 & 0 \end{bmatrix} \qquad (B.163)$$

whose first 7 columns constitute a Hermite matrix, then we note, first, that $P = X_0$ and, second, that $L^*(A)$ can be described by the columns of $I - H$ as before.

The process is valid in general. If $[D \quad L]$ is the echelon form of $[A \quad K]$, if H is the Hermite matrix related to D, and if P is obtained from L by insertion of zeros in the rows which are zero in H then $X_0 = P$ is a particular solution and $L^*(A) = L(I - H)$.

This answers the three questions, but some further discussion of the geometry and of double description of linear manifolds is desirable.

Equation (B.160) shows that a linear manifold can be regarded as a displaced vector space. It is easy to show that every displaced vector space is a linear manifold. Let S be any subspace and X_0 any vector; then the displaced vector space $X_0 + S$ can be described as a linear manifold in the following way. Let A be any matrix whose associated solution space $L^*(A)$

is S; such exists by the double-description theorem of Section B.15. Then let $AX_0 = K$ and it follows that $X_0 + S = M^*(A, K)$ because of the discussion leading to (B.160).

A linear combination

$$X = c_1X_1 + \cdots + c_tX_t \tag{B.164}$$

is sometimes called a *weighted sum.* If, in addition,

$$c_1 \geq 0, \cdots, c_t \geq 0 \tag{B.165}$$

we call X a *nonnegative weighted sum.* If

$$c_1 + \cdots + c_t = 1 \tag{B.166}$$

we call X a *weighted average*, and if both (B.165) and (B.166) hold we call X a *nonnegative weighted average,* or a *probability combination* or a *convex combination.* These various kinds of linear combinations arise in connection with studies of solutions of systems of equations and inequalities.

We consider one of these kinds of linear combinations, that of weighted averages. We are interested in this class because weighted averages of solutions of systems of linear equations are again solutions,

$$A(\Sigma c_iX_i) = \Sigma c_iAX_i = \Sigma c_iK = K, \qquad \text{since } \Sigma c_i = 1.$$

This suggests that we try to describe the linear manifold in terms of weighted averages. Let S be a subspace of V_n and let X_0 be a fixed vector in V_n. We noted earlier that a linear manifold M can be regarded as a vector space displaced by the addition of a fixed vector to each of the vectors in S,

$$M = X_0 + S.$$

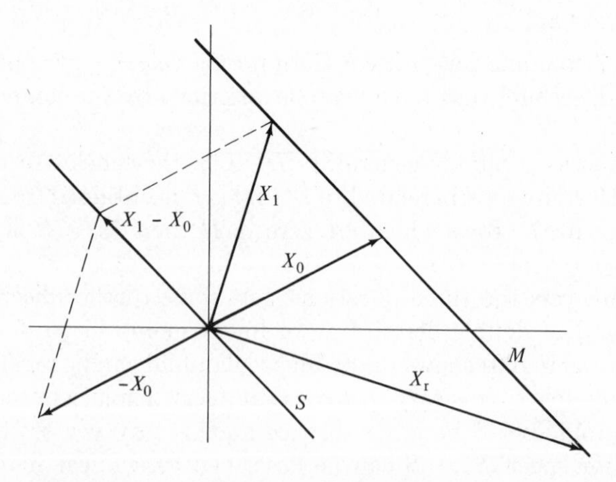

Figure B.6 Illustrating $X_i - X_0 \in S$ for $X_0, \ldots, X_r \in M$.

Suppose now that S has dimension r and that $(Y_1, \cdots, Y_r)$ is a sequence of basis vectors of S. Then any vector $X \in M$ can be written

$$X = X_0 + t_1 Y_1 + \cdots + t_r Y_r \tag{B.167}$$

where each t_i is a real number. From (B.167) we have

$$X - X_0 = t_1 Y_1 + \cdots + t_r Y_r,$$

so that $X - X_0 \in S$.

Now given any set of $r + 1$ vectors $X_0, X_1, \cdots, X_r$ in V_n, how can we determine the *smallest* linear manifold containing these vectors? We wish to know this because the set of vectors can be contained in many manifolds of different dimensions. If M is a manifold we know that

$$M = X + S$$

for some $X \in M$ and some subspace S. Let $X = X_0$; since $X_0, X_1, \cdots, X_r \in M$, then $X_i - X_0 \in S$ for $i = 1, \cdots, r$ (see Figure B.6). The set spanned by the vectors $X_1 - X_0, \cdots, X_r - X_0$, that is, the set $\langle X_1 - X_0, \cdots, X_r - X_0 \rangle$ is a subspace and it is clearly the smallest subspace containing these vectors. Then

$$M = X_0 + \langle X_1 - X_0, \cdots, X_r - X_0 \rangle$$

is the smallest linear manifold containing the $r + 1$ points $X_0, X_1, \cdots, X_r$. When we speak of the manifold containing a set of vectors we mean the smallest manifold containing the vectors.

Note that if $X_0, X_1, \cdots, X_r$ are any $r + 1$ points in V_n, then the sequence of vectors $(X_1 - X_0, \cdots, X_r - X_0)$ can be either dependent or independent. If the sequence is independent, then M has dimension r; if not, then M has dimension $k < r$.

An arbitrary vector X is in the linear manifold M if and only if

$$X = X_0 + t_1(X_1 - X_0) + \cdots + t_r(X_r - X_0)$$

for each t_i a real number and $X_0, X_1, \cdots, X_r \in M$. Thus

$$X = (1 - t_1 - \cdots - t_r)X_0 + t_1 X_1 + \cdots + t_r X_r. \tag{B.168}$$

Let

$$t_0 = 1 - \sum_{i=1}^{r} t_i;$$

then $t_0 + \sum_{i=1}^{r} t_i = \sum_{i=0}^{r} t_i = 1$ and (B.168) can be written

$$X = t_0 X_0 + t_1 X_1 + \cdots + t_r X_r. \tag{B.169}$$

$$\sum_{i=0}^{r} t_i = 1. \tag{B.170}$$

Conditions (B.169) and (B.170) are necessary and sufficient conditions for a vector X in V_n to be in the linear manifold M, so that any weighted average of vectors in M is in M.

Let B be any s by t matrix with column vectors $B_1, \cdots, B_t$. The set of all weighted averages of the columns of B constitute a linear manifold which we denote by

$$M(B). \tag{B.171}$$

The theorem of double description for linear manifolds states that *given any linear manifold M there exist matrices B, A and a vector K for which*

$$M = M(B) = M^*(A, K). \tag{B.172}$$

More precisely the theorem states that given B there exist A, K for which (B.172) holds, and given A, K for which $M^*(A, K)$ is not empty then there exists B for which (B.172) holds.

B.17. LENGTH, DISTANCE, AND ANGLE.

We now consider the problem of introducing the concepts of length, distance, and angle into the space $V_n(R)$ of all real column vectors of degree n. The key to this is the concept of inner product introduced in Section B.4.

In plane analytic geometry we encounter the formula $d = (x^2 + y^2)^{1/2}$ for the distance from the origin to the point (x, y). This can also be regarded as being the length of the column vector $X = [x \quad y]^T$. Note that using matrix multiplication this becomes

$$d = (x^2 + y^2)^{1/2} = (X^T X)^{1/2}. \tag{B.173}$$

A similar formula holds in three space.

If $X = [x_1\, x_2 \cdots x_n]^T$ then we introduce a function $|\ |$ by the formula

$$|X| = (X^T X)^{1/2} \tag{B.174}$$

and call $|X|$ the *length* of the vector X. If $Y = [y_1\, y_2 \cdots y_n]^T$ is a second vector we introduce a distance function $d(X,Y)$ by the formula

$$d(X, Y) = |X - Y|. \tag{B.175}$$

We define the angle θ, $0 \leq \theta \leq \pi$, between the vectors X and Y by the formula

$$\cos \theta = \frac{X^T Y}{|X|\,|Y|}. \tag{B.176}$$

We note a few obvious properties:

$$|kX| = |k|\,|X| \text{ for scalars } k \text{ and vectors } X, \tag{B.177}$$

(here $|k|$ is ordinary absolute value of k);

$$|X| \geq 0 \text{ for all } X \text{ and equality holds if and only if } X = 0; \quad (.B178)$$

$$d(X, Y) = d(Y, X); \qquad (B.179)$$

$$d(X, Y) \geq 0 \text{ for all vectors } X \text{ and } Y, \text{ and equality holds if and only if } X = Y; \qquad (B.180)$$

$$d(X + Z, Y + Z) = d(X, Y) \text{ for all vectors } X, Y, \text{ and } Z. \quad (B.181)$$

Our definitions would not be very satisfactory or useful unless the absolute value of the right-hand side of (B.176) never exceeds 1. That this is indeed the case is a consequence of the following property of inner products, known as the Schwartz Inequality:

$$(X^TY)^2 \leq (X^TX)(Y^TY) \text{ for all vectors } X \text{ and } Y. \qquad (B.182)$$

In component form this is

$$(x_1y_1 + \cdots + x_ny_n)^2 \leq (x_1^2 + \cdots + x_n^2)(y_1^2 + \cdots + y_n^2). \qquad (B.183)$$

In (B.182) the equality holds if and only if the vectors X and Y are proportional. This important inequality also guarantees that in any triangle no side has length exceeding the sum of the lengths of the remaining two sides. In view of (B.175) and (B.181) this statement is equivalent to the formula

$$|X + Y| \leq |X| + |Y|, \qquad (B.184)$$

which is a generalization to n dimensional-vector space of a familiar property of the ordinary absolute value of real scalars.

Two vectors X and Y are *perpendicular* or *orthogonal* if $\cos \theta = 0$, that is, if their inner product is zero:

$$X^TY = 0. \qquad (B.185)$$

A vector of length one is said to be a *unit vector*. For example,

$$\begin{bmatrix} 1 \\ 0 \\ 0 \end{bmatrix}, \quad \begin{bmatrix} 1/3 \\ 2/3 \\ 2/3 \end{bmatrix}, \quad \begin{bmatrix} 0 \\ 1 \\ 0 \end{bmatrix} \qquad (B.186)$$

are unit vectors. If X is any nonzero vector then $Y = (1/|X|)\, X$ is a unit vector.

A sequence $S = (X_1, \cdots, X_r)$ of vectors is said to be *orthonormal* if

(a) each element in S is a unit vector, (B.187)

(b) the elements in S are pairwise orthogonal, that is,

$$X_i^TX_j = 0 \text{ for } i \neq j, \ (i, j = 1, \cdots, n).$$

For example, the sequence

$$S = \left(\begin{bmatrix} 1 \\ 0 \\ 0 \end{bmatrix}, \begin{bmatrix} 0 \\ 1 \\ 0 \end{bmatrix}, \begin{bmatrix} 0 \\ 0 \\ 1 \end{bmatrix} \right) \tag{B.188}$$

is orthonormal, whereas the vectors in (B.186) are not pairwise orthogonal.

An orthonormal sequence S is said to be *maximal* if there is no orthonormal sequence T which has S as a proper subsequence. An orthonormal sequence in V_n is maximal if and only if it has n elements or vectors.

A sequence of nonzero vectors is said to be *orthogonal* if it satisfies (B.187b); an orthogonal sequence becomes orthonormal if each vector X in it is multiplied by the scalar $1/|X|$. For example, the sequence

$$S = \left(\begin{bmatrix} 1 \\ 2 \\ 2 \end{bmatrix}, \begin{bmatrix} 0 \\ 1 \\ -1 \end{bmatrix}, \begin{bmatrix} -4 \\ 1 \\ 1 \end{bmatrix} \right)$$

is orthogonal and the related sequence

$$T = \left(\begin{bmatrix} 1/3 \\ 2/3 \\ 2/3 \end{bmatrix}, \begin{bmatrix} 0 \\ 1/\sqrt{2} \\ 1/\sqrt{2} \end{bmatrix}, \begin{bmatrix} 4/\sqrt{18} \\ 1/\sqrt{18} \\ 1/\sqrt{18} \end{bmatrix} \right)$$

is orthonormal.

It is easy to prove that any orthogonal sequence of nonzero vectors is independent. For, if $S = (X_1, \cdots, X_s)$ is orthogonal, and if

$$c_1X_1 + \cdots + c_sX_s = 0, \tag{B.189}$$

then multiplying by X_j^T gives

$$c_j(X_j^TX_j) = c_j \mid X_j \mid^2 = 0$$

and hence $c_j = 0$, $(j = 1, \cdots, s)$, since the vectors X_j are nonzero. Therefore S satisfies only the trivial dependence relation in which each c_j is zero and so S is independent.

A vector Y is said to be orthogonal to a sequence of vectors $S = (X_1, \cdots, X_r)$ in $V_p(R)$ if $Y^TX_i = 0$ for $i = 1, \cdots, r$ and a vector Y is orthogonal to a subspace Q if Y is orthogonal to every vector in Q. It can be shown that if a vector Y is orthogonal to a sequence of vectors $S = (X_1, \cdots, X_r)$, then Y is orthogonal to the subspace spanned by S; that is, Y is orthogonal to $\langle S \rangle$.

A matrix A is said to be *orthogonal* if its columns form a maximal orthonormal sequence. We then have

$$A^TA = I; \tag{B.190}$$

and since A must be square (lest the sequence not be maximal) we also have

$$AA^T = I. \qquad \text{(B.191)}$$

Either condition (B.190) or (B.191) is sufficient to insure that a square matrix A be orthogonal. Note that either condition implies that $A^T = A^{-1}$.

PROBLEMS

B.1. Let $X = \begin{bmatrix} 2 \\ 5 \end{bmatrix}$, $Y = \begin{bmatrix} 1 \\ 7 \end{bmatrix}$. Graph the following vectors: $X + Y$, $2X$, $X - Y$, $-X$, $X + \frac{1}{2} Y$, $Y - X$.

B.2. Given X, Y in the preceding exercise and $Z = \begin{bmatrix} 3 \\ -2 \end{bmatrix}$, verify geometrically that $(X + Y) + Z = X + (Y + Z)$.

B.3. Let $X_1 = \begin{bmatrix} 5 \\ 1 \end{bmatrix}$, $X_2 = \begin{bmatrix} 1 \\ 2 \end{bmatrix}$; verify geometrically that $(2 + 3)X_1 = 2X_1 + 3X_1$, $2(X_1 + X_2) = 2X_1 + 2X_2$, $2(3X_2) = 6X_2$.

B.4. Consider the factory input-output problem whose matrix is shown in (B.15). Replace this matrix by the new input-output matrix,

$$\begin{bmatrix} 2 & 2 & 1 \\ 3 & 1 & 1 \\ 5 & 0 & 3 \\ 3 & 1 & 1 \end{bmatrix}.$$

Let the required output vector be

$$\begin{bmatrix} 1 \\ 3 \\ 18 \end{bmatrix}.$$

(a) What is the required input vector?
(b) Suppose the input cost vector D and the output price vector C are

$$D = \begin{bmatrix} 3 \\ 2 \\ 6 \\ 12 \end{bmatrix}, \qquad C = \begin{bmatrix} 61 \\ 78 \\ 36 \end{bmatrix};$$

calculate the total cost of the order, the total returns from the sale of the output, and the total profit (total returns less total cost). What are the unit profits of each of the outputs?

B.5. Consider the matrices:

$$A = \begin{bmatrix} 1 & 6 & 3 & -2 \\ 2 & 5 & -1 & 0 \\ 0 & 3 & 5 & 7 \end{bmatrix}, \quad B = \begin{bmatrix} -1 \\ -1 \\ 3 \\ 7 \end{bmatrix}, \quad C = \begin{bmatrix} -3 \\ -12 \\ 2 \\ 1 \end{bmatrix},$$

$$D = \begin{bmatrix} -2 \\ -2 \\ 4 \\ 6 \end{bmatrix}, \quad E = \begin{bmatrix} 3 \\ 8 \\ 11 \\ 9 \end{bmatrix}.$$

(a) What are the row and column degrees of A and of C?

(b) Is it true that $D < C$? Is $C < D$? Is $C \leq E$? Is $C < E$? Is $D \leq C$? Is $C < B$?

B.6. Give an example of two vectors X and Y for which $X \leq Y$ and $X \neq Y$.

B.7. Given the matrices

$$A = \begin{bmatrix} 2 & -1 & 3 \\ 1 & 0 & 2 \end{bmatrix}, \quad B = \begin{bmatrix} 1/5 & 2 & 4 \\ 7 & -3 & 1 \end{bmatrix}, \quad C = \begin{bmatrix} 1 & -1 \\ 2 & 3 \\ -2 & 5 \end{bmatrix},$$

$$D^T = [-1 \quad -2 \quad 6], \quad E = \begin{bmatrix} 2/7 \\ 1/5 \\ 1 \end{bmatrix}, \quad F = \begin{bmatrix} 1 & 0 & -1 & 2 \\ 1 & 5 & 3 & 1 \\ 2 & -1 & 0 & 1 \end{bmatrix},$$

$$G = \begin{bmatrix} 3 & 1 & 0 \\ 2 & 0 & 2 \\ -1 & 1 & 1 \\ 0 & 0 & 1 \end{bmatrix}, \quad H = \begin{bmatrix} -1 & 7 & 1 \\ 3 & 2 & 0 \end{bmatrix}, \quad I = \begin{bmatrix} 1 & 0 & 0 \\ 0 & 1 & 0 \\ 0 & 0 & 1 \end{bmatrix},$$

$$K = \begin{bmatrix} 1 & -1 \\ 3 & 5 \end{bmatrix}, \quad M = \begin{bmatrix} 1 & -1 & 0 \\ 1/3 & 2 & 5 \end{bmatrix}, \quad N = \begin{bmatrix} 0 & 0 \\ 1 & 1 \end{bmatrix}.$$

(a) Calculate $A + B$, $B + A$, $A + (B + M)$, $(A + B) + M$; verify that $A + B = B + A$, $A + (B + M) = (A + B) + M$.

(b) Compute $(A + B)E$. What are the row and column degrees of A, B, $A + B$, E and $(A + B)E$?

(c) Which of the following are defined: BE, BD^T, CE, FG, GF, FI, IF, $A + F$, $D^T + E$, GE, EG, HC, KM, KN, NK?

(d) Calculate AE, BC, FG, D^TF; is $A(3E) = 3AE$?

(e) Show that $IF = F$, $AI = A$; is $NK = KN$?

(f) Show that $(AC)K = A(CK)$.

(g) Show that $K(H + B) = KH + KB$. What are the row and column degrees of K, $H + B$, $K(H + B)$, KH, KB?

(h) Show that $(A + H)C = AC + HC$.

(i) Determine the matrices B^T, D^T, E^T, F^T, G^T; verify that $(F^T)^T = F$, that $(H + A)^T = H^T + A^T$, that $(A + B + H)^T = A^T + B^T + H^T$.

(j) Show that $(2F)^T = 2F^T$, that $(AC)^T = C^TA^T$, and that $(ACK)^T = K^TC^TA^T$.

(k) Show that $D^TE = (D^TE)^T = E^T(D^T)^T = E^TD$.

(l) Calculate GG^T, G^TG; is $GG^T = G^TG$?

B.8. Let

$$B = \begin{bmatrix} 1 & 2 & 3 \\ 1 & 2 & 3 \\ -1 & -2 & -3 \end{bmatrix};$$

show that $B^2 = 0$.

B.9. Suppose P and Q are matrices such that $PQ = 0$; show that it is not necessarily true that $P = 0$ or $Q = 0$.

B.10. Show that for any real numbers a, b, c, d the matrices

$$A = \begin{bmatrix} a & b \\ -b & a \end{bmatrix}, \quad B = \begin{bmatrix} c & d \\ -d & c \end{bmatrix}$$

commute with respect to matrix multiplication.

B.11. Let

$$R = \begin{bmatrix} 2 & -2 & -4 \\ -1 & 3 & 4 \\ 1 & -2 & -3 \end{bmatrix}, \quad T = \begin{bmatrix} -1 & 2 & 4 \\ 1 & -2 & -4 \\ -1 & 2 & 4 \end{bmatrix};$$

show that $R^2 = R$ and $T^2 = T$.

B.12. Let F and G be as given in Problem B.7. What are the following submatrices: $F(1 \mid 3, 4)$, $F(3 \mid 2, 3)$, $G(2 \mid 1)$, $F[1, 2 \mid 3, 4]$.

B.13. Let

$$A = \begin{bmatrix} -1 & 0 & 3 & 1 & 2 & 1 \\ 1 & 1 & 0 & 0 & -1 & 3 \\ 2 & -1 & 5 & 0 & 0 & 1 \\ 4 & 0 & 3 & 1 & -7 & 3 \end{bmatrix};$$

is $A(1 \mid 1, 2, 3) = A[2, 3, 4 \mid 4, 5, 6]$? Is $A(2, 4 \mid 2, 4, 6) = A[1, 3 \mid 1, 3, 5]$?

B.14. Let

$$A = \begin{bmatrix} 1 & 0 & a & b \\ 0 & 1 & c & d \\ 0 & 0 & -1 & 0 \\ 0 & 0 & 0 & -1 \end{bmatrix};$$

show that for any numbers a, b, c, d we have $A^2 = \begin{bmatrix} I_2 & 0 \\ 0 & I_2 \end{bmatrix} = I_4$, where I_n denotes the identity matrix of size n.

B.15. Let

$$B = \begin{bmatrix} 4 & 2 & -5 & 1 \\ 2 & -10 & 3 & 0 \\ -1 & 4 & 0 & 3 \end{bmatrix}, \quad C = \begin{bmatrix} 2 & -1 & 4 \\ 2 & 1 & 3 \\ 4 & 4 & -1 \\ 1 & 0 & -1 \end{bmatrix};$$

partition B so that $B = \begin{bmatrix} B_1 \\ B_2 \end{bmatrix}$, where $B_1 = B[1, 2 \mid 1, 2, 3, 4]$ and $B_2 = B[3 \mid 1, 2, 3, 4]$. How must C be partitioned so as to form the product BC? Compute this product by means of submatrices.

B.16. Let

$$A = \begin{bmatrix} 1 & 0 & 0 & 0 & 0 & 2 \\ 0 & 1 & 0 & 0 & 0 & 2 \\ 0 & 0 & 1 & 0 & 0 & 2 \\ 0 & 0 & 0 & 1 & 0 & 0 \\ 0 & 0 & 0 & 0 & 1 & 0 \\ 2 & 2 & 2 & 0 & 0 & 1 \end{bmatrix}$$

and partition A into submatrices A_i, where $A_1 = A[1, 2, 3 \mid 1, 2, 3]$, $A_2 = A[1, 2, 3 \mid 4, 5]$, $A_3 = A[1, 2, 3 \mid 6]$, $A_4 = A[4, 5 \mid 1, 2, 3]$, $A_5 = A[4, 5 \mid 4, 5]$, $A_6 = A[4, 5 \mid 6]$, $A_7 = A[6 \mid 1, 2, 3]$, $A_8 = A[6 \mid 4, 5]$, $A_9 = A[6 \mid 6]$. Compute A^2 using this partitioning.

B.17. E_m is the column vector of degree m each of whose entries is 1; let

$$X = \begin{bmatrix} 1 \\ -1 \\ 3 \\ 5 \end{bmatrix}, \quad Y = \begin{bmatrix} .1 \\ 0 \\ 0 \\ .2 \\ .7 \end{bmatrix}, \quad Q = \begin{bmatrix} 2 \\ 1 \\ 7 \\ -1 \\ 5 \\ 6 \end{bmatrix}.$$

Calculate $E_4^T X$, $E_5^T Y$, $E_6^T Q$. Is Y a probability vector? Is Q a probability vector? Find two vectors of degree 6 each of which is a probability vector.

B.18. Given the following matrices:

$$A = \begin{bmatrix} 2 & -1 & 3 \\ -1 & 1 & 0 \\ 3 & 0 & 0 \end{bmatrix}, \quad B = \begin{bmatrix} 1 & 0 & 1 & 1 \\ 0 & 0 & 1 & 1 \\ 0 & 0 & 0 & 0 \\ 0 & 0 & 0 & -1 \end{bmatrix},$$

$$C = \begin{bmatrix} 0 & 0 & 0 & 0 \\ 0 & -1 & 0 & 0 \\ 0 & 0 & 3 & 0 \\ 0 & 0 & 0 & 5 \end{bmatrix}, \quad D = \begin{bmatrix} -1 & 0 & 0 \\ 0 & -1 & 0 \\ 0 & 0 & -1 \end{bmatrix},$$

$$R = \begin{bmatrix} 1 & 0 & 0 \\ -1 & 4 & 0 \\ 1 & -1 & 9 \end{bmatrix}, \quad T = \begin{bmatrix} 0 & -1 & 3 & -2 \\ 1 & 0 & -7 & 1 \\ -3 & 7 & 0 & 6 \\ 2 & -1 & -6 & 0 \end{bmatrix}.$$

(a) Indicate a diagonal matrix.
(b) Indicate a skew-symmetric and two symmetric matrices.
(c) Which matrix is a scalar matrix?
(d) Indicate a lower and an upper triangular matrix.

B.19. In Section B.6 it was proved that any square matrix can be expressed as the sum of a symmetric and a skew-symmetric matrix. Prove that there is one and only one way of expressing a square matrix as such a sum.

B.20. Given the matrices

$$A = \begin{bmatrix} ai & b+ci & e+fi \\ -b+ci & di & h+ki \\ -e+fi & -h+ki & gi \end{bmatrix}, \quad B = \begin{bmatrix} a & b+ci & e+fi \\ b-ci & d & h+ki \\ e-fi & h-ki & g \end{bmatrix},$$

$$C = \begin{bmatrix} a & b & c \\ b & 0 & d \\ c & d & 1 \end{bmatrix},$$

where all symbols except i denote real numbers. (a) How many are Hermitian matrices? (b) How many are skew Hermitian?

B.21. Let

$$X_1 = \begin{bmatrix} 4 \\ 2 \end{bmatrix}, \quad X_2 = \begin{bmatrix} 1 \\ 4 \end{bmatrix}, \quad X_3 = \begin{bmatrix} 3 \\ -3 \end{bmatrix}, \quad Y = \begin{bmatrix} -5 \\ 3 \end{bmatrix}.$$

(a) Express Y as a linear combination of X_1 and X_2.
(b) Is Y dependent on the sequence (X_2, X_3)?
(c) Is Y dependent on the sequence $(X_1, -X_1)$?
(d) Let $S = (X_1, X_2, X_3)$; show two different linear combinations of Y in terms of the vectors in S.
(e) Let $S_1 = (X_1, X_2)$, $S_2 = (X_1, Y)$, $S = (X_1, X_2, X_3)$; what are the sets $\langle S_1 \rangle$, $\langle S_2 \rangle$, $\langle S \rangle$?
(f) What is the set spanned by the vector X_1?

B.22. Prove that if a sequence of vectors contains the zero vector then the sequence is (linearly) dependent.

B.23. Let $S = (X_1, \cdots, X_p)$ be a sequence of vectors, $p \geq 1$, and let S' be a subsequence of S; then if S' is dependent prove that S is dependent.

B.24. Solve the linear equation system

$$\begin{aligned} x_1 - 3x_2 + 4x_3 &= -1, \\ x_1 + \frac{1}{3}x_2 - \frac{1}{3}x_3 &= 1, \\ x_1 - \frac{3}{2}x_2 + x_3 &= 2. \end{aligned}$$

B.25. Form the augmented matrix of the linear equation system

$$\begin{aligned} x_1 - 3x_2 + 4x_3 &= -1, \\ 3x_1 + x_2 - 8x_3 &= 3, \\ 2x_1 - 3x_2 + 2x_3 &= 4. \end{aligned}$$

Does this system have a solution?

B.26. Reduce the augmented matrix of each of the following systems to echelon form. Which systems have solutions?

(a)
$$\begin{aligned} \frac{1}{2}x_1 - \frac{1}{4}x_2 + \frac{1}{4}x_3 &= \frac{1}{4}, \\ x_1 + 4x_2 - 3x_3 &= 19, \\ 2x_1 - 3x_2 + 4x_3 &= 3. \end{aligned}$$

(b)
$$\begin{aligned} 2x_1 + 2x_2 &= 15, \\ x_2 - x_3 &= 3, \\ x_1 - x_2 + x_3 &= 17. \end{aligned}$$

(c)
$$\begin{aligned} x_1 - 4x_2 &= 5, \\ 2x_1 + 3x_2 &= -1, \\ 4x_1 + x_2 &= 3. \end{aligned}$$

(d)
$$\begin{aligned} 7x_1 + 3x_2 - x_3 &= 0, \\ x_1 + 5x_2 + x_3 &= 0, \\ 2x_1 - 6x_2 + x_3 &= 0. \end{aligned}$$

B.27. Let $S = (X_1, X_2, X_3, X_4)$, where

$$X_1 = \begin{bmatrix} 1 \\ -1 \\ 1 \end{bmatrix}, \quad X_2 = \begin{bmatrix} 2 \\ 2 \\ 3 \end{bmatrix}, \quad X_3 = \begin{bmatrix} 1 \\ 1 \\ -1 \end{bmatrix}, \quad X_4 = \begin{bmatrix} 4 \\ 3 \\ 3 \end{bmatrix}.$$

Is S dependent or independent?

B.28. Show that the sequence $R = (R_1, R_2, R_3)$, where

$$R_1 = \begin{bmatrix} 1 \\ 1 \\ 2 \end{bmatrix}, \quad R_2 = \begin{bmatrix} -1 \\ 2 \\ 1 \end{bmatrix}, \quad R_3 = \begin{bmatrix} 1 \\ -1 \\ 3 \end{bmatrix},$$

is independent.

B.29. Given the vectors

$$Y_1 = \begin{bmatrix} 1 \\ 1 \\ 2 \end{bmatrix}, \quad Y_2 = \begin{bmatrix} -2 \\ -2 \\ -4 \end{bmatrix}, \quad Y_3 = \begin{bmatrix} 1 \\ -1 \\ -5 \end{bmatrix},$$

show that the sequence $Y = (Y_1, Y_2, Y_3)$ is dependent.

B.30. Given the sequence Y in the preceding exercise, let $X = \begin{bmatrix} 0 \\ 0 \\ 3 \end{bmatrix}$. Is X dependent on Y? If $X' = \begin{bmatrix} 1 \\ 1 \\ -1 \end{bmatrix}$, is X' dependent on Y?

B.31. Determine the rank of each of the following matrices,

$$\text{(a)} \begin{bmatrix} 1 & -1 & 2 & 4 \\ 1 & 0 & 1 & -2 \\ 2 & 1 & 1 & 3 \\ 4 & -1 & -2 & 7 \end{bmatrix}, \quad \text{(b)} \begin{bmatrix} -1 & -1 & 1 \\ 1 & 1 & 2 \\ 0 & 1 & -1 \end{bmatrix},$$

$$\text{(c)} \begin{bmatrix} 1 & -1 & 1 & 1 & 6 \\ 2 & 0 & 3 & -3 & 2 \\ 1 & -3 & 1 & 1 & 1 \end{bmatrix}.$$

B.32. Let

$$X_1 = \begin{bmatrix} 1 \\ 1 \\ 3 \end{bmatrix}, \quad X_2 = \begin{bmatrix} -1 \\ 1 \\ -1 \end{bmatrix}, \quad X_3 = \begin{bmatrix} 1 \\ -1 \\ 1 \end{bmatrix},$$

$$X_4 = \begin{bmatrix} 1 \\ 2 \\ 4 \end{bmatrix}, \quad B = \begin{bmatrix} 2 \\ -1 \\ 4 \end{bmatrix};$$

is B dependent on $S = (X_1, X_2, X_3, X_4)$?

B.33. Consider the homogeneous linear equation system

$$\begin{aligned} 2x_1 + 3x_2 - x_3 + x_4 &= 0, \\ 3x_1 + 2x_2 - 2x_3 + 2x_4 &= 0, \\ 5x_1 \qquad - 4x_3 + 4x_4 &= 0. \end{aligned}$$

whose coefficient matrix is A.
(a) What is the dimension of $L(A)$?
(b) What is the dimension of $L^*(A)$?
(c) Find a basis for $L^*(A)$.

B.34. Consider the system

$$\begin{aligned} 4x_1 + 3x_2 + 2x_3 + x_4 &= 0, \\ x_1 + x_2 + x_3 + x_4 &= 0, \\ x_1 + 2x_2 + 3x_3 + 4x_4 &= 0. \end{aligned}$$

(a) Let A denote the coefficient matrix of this system; find the echelon matrix D corresponding to A.
(b) Find a Hermite matrix H from D.
(c) Find a basis for the solution space of the system of equations.
(d) Verify that $H^2 = H$.

B.35. Consider the homogeneous linear equation system

$$\begin{aligned} 2x_1 + x_2 + x_3 + x_4 &= 0, \\ 3x_1 - x_2 + x_3 - x_4 &= 0, \\ x_1 + 2x_2 - x_3 + x_4 &= 0. \end{aligned}$$

(a) If A denotes the coefficient matrix of this system, find a matrix B apolar to A.
(b) Verify that B^T and A^T are an apolar pair.
(c) What is $L(B)$?
(d) Verify that $L^*(A) = L(B)$.
(e) Verify that $L^*(B^T) = L(A^T)$.

B.36. Let

$$\begin{aligned} 2x_1 - x_2 + 3x_3 + 2x_4 &= 0, \\ x_1 + 4x_2 \qquad - x_4 &= 0, \\ 2x_1 + 6x_2 - x_3 + 5x_4 &= 0; \end{aligned}$$

verify that the statements (B.119) through (B.119‴) of Section B.13 hold.

B.37. Determine which of the following matrices are nonsingular and calculate the inverse of each nonsingular matrix,

$$\text{(a)} \begin{bmatrix} 1 & 5 \\ -3 & 9 \end{bmatrix}, \quad \text{(b)} \begin{bmatrix} 1 & 2 & 3 \\ 4 & 5 & 6 \\ 7 & 8 & 9 \end{bmatrix},$$

$$\text{(c)} \begin{bmatrix} -1 & 2 & 4 \\ 3 & 2 & -1 \\ 4 & 3 & -1 \end{bmatrix}, \quad \text{(d)} \begin{bmatrix} 3 & 4 & 1 \\ 1 & 2 & 1 \\ 2 & 3 & 1 \end{bmatrix},$$

$$\text{(e)} \begin{bmatrix} 5 & 4 & 2 & 1 \\ 2 & 3 & 1 & -2 \\ 1 & -2 & -1 & 4 \\ -5 & -7 & -3 & 9 \end{bmatrix}, \quad \text{(f)} \begin{bmatrix} 1/2 & 1 & -1/2 \\ 2 & 4 & 2 \\ 1 & 2 & -1 \end{bmatrix}.$$

B.38. Which of the following matrices are idempotent?

$$\text{(a)} \begin{bmatrix} 4 & -2 \\ 6 & -3 \end{bmatrix}, \quad \text{(b)} \begin{bmatrix} 2 & -2 & -4 \\ -1 & 3 & 4 \\ 1 & -2 & -3 \end{bmatrix}, \quad \text{(c)} \begin{bmatrix} 1 & 1 & 0 \\ 2 & 1 & 3 \\ 1 & 1 & 1 \end{bmatrix}.$$

B.39. A square matrix A such that $A^p = 0$ for some positive integer p is called a nilpotent matrix, and the smallest positive integer p such that $A^p = 0$ is called the index of A. Which of the following are nilpotent? Determine the index of each nilpotent matrix.

$$\text{(a)} \begin{bmatrix} -4 & 4 & -4 \\ 1 & -1 & 1 \\ 5 & -5 & 5 \end{bmatrix}, \quad \text{(b)} \begin{bmatrix} 1 & 3 & -1 & 4 \\ 0 & 1 & 2 & 1 \\ 0 & 0 & 1 & 1 \\ 0 & 0 & 0 & 1 \end{bmatrix},$$

$$\text{(c)} \begin{bmatrix} 1 & 0 \\ 0 & -1 \end{bmatrix}, \quad \text{(d)} \begin{bmatrix} 1 & 1 \\ -1 & -1 \end{bmatrix}, \quad \text{(e)} \begin{bmatrix} 0 & 1 & 0 & 0 \\ 0 & 0 & 1 & 0 \\ 0 & 0 & 0 & 1 \\ 0 & 0 & 0 & 0 \end{bmatrix}.$$

B.40. Partition the matrix A as suggested by the dashed lines below and calculate A^{-1} by means of its submatrices,

$$A = \left[\begin{array}{cc:c} -2 & 1 & 1 \\ 5 & -2 & 4 \\ \hdashline 4 & 1 & 0 \end{array}\right].$$

B.41. Consider the system of nonhomogeneous linear equations $AX = K$, where

$$A = \begin{bmatrix} 1 & 0 & 0 & 1 & 4 \\ 1 & 1 & 0 & 3 & 9 \\ 2 & 0 & 1 & 5 & 14 \end{bmatrix}, \quad X = \begin{bmatrix} x_1 \\ x_2 \\ x_3 \\ x_4 \\ x_5 \end{bmatrix}, \quad K = \begin{bmatrix} 7 \\ 15 \\ 23 \end{bmatrix}.$$

(a) Express the linear manifold of solutions $M^*(A, K)$ of this system as the sum of a vector in the manifold and the span of some basis vectors for a subspace S. (b) Describe the subspace S in two ways.

B.42. Consider the linear manifold M, where

$$M = \begin{bmatrix} 3 \\ 1 \\ 1 \end{bmatrix} + \left\langle \begin{bmatrix} -2 \\ 2 \\ 1 \end{bmatrix}, \begin{bmatrix} 1 \\ -2 \\ 1 \end{bmatrix} \right\rangle;$$

find a system of linear equations whose solution set is this manifold.

B.43. Let

$$X_1 = \begin{bmatrix} -1 \\ 0 \\ 5 \\ 4 \end{bmatrix}, \quad X_2 = \begin{bmatrix} 2 \\ 1 \\ 1 \\ 3 \end{bmatrix}, \quad X_3 = \begin{bmatrix} 1 \\ 1 \\ -2 \\ -3 \end{bmatrix}, \quad X_4 = \begin{bmatrix} 1 \\ 4 \\ -1 \\ 0 \end{bmatrix}.$$

(a) Calculate $|X_1|$, $|X_3|$, $|X_4|$.
(b) Calculate $d(X_1, X_2)$, $d(X_1, X_3)$, $d(X_2, X_4)$.
(c) If θ is the angle between X_1 and X_3, what is $\cos\theta$?
(d) Verify the Schwartz inequality for vectors X_1 and X_3.
(e) Calculate the following vectors and verify that each is a unit vector:

$$\frac{1}{|X_1|}X_1, \frac{1}{|X_2|}X_2, \frac{1}{|X_3|}X_3.$$

(f) Calculate the following inner products: $X_1^TX_2$, $X_1^TX_3$, $X_1^TX_4$.
(g) Find a vector Y orthogonal to X_3.

B.44. Which of the following are orthogonal matrices?

(a) $\begin{bmatrix} 1 & 2 & 3 \\ 4 & 5 & 6 \\ 7 & 8 & 9 \end{bmatrix}$, (b) $\begin{bmatrix} 1/\sqrt{14} & 1/\sqrt{3} & 5/\sqrt{42} \\ 2/\sqrt{14} & 1/\sqrt{3} & -4/\sqrt{42} \\ 3/\sqrt{14} & -1/\sqrt{3} & 1/\sqrt{42} \end{bmatrix}$,

(c) $\begin{bmatrix} 1/\sqrt{6} & 1/\sqrt{3} & 1/\sqrt{2} \\ -2/\sqrt{6} & 1/\sqrt{3} & 0 \\ 1/\sqrt{6} & 1/\sqrt{3} & -1/\sqrt{2} \end{bmatrix}$, (d) $\begin{bmatrix} 3/5 & 4/5 \\ -4/5 & 3/5 \end{bmatrix}$,

(e) $\begin{bmatrix} -1/2 & 1/2 & 1/2 \\ 1/2 & -1/2 & 1/2 \\ 1/2 & 1/2 & -1/2 \end{bmatrix}$.

B.45. Prove that if A is an orthogonal matrix, then A^T and A^{-1} are orthogonal matrices.

B.46. Let X and Y be two vectors of degree n such that the Schwartz inequality is satisfied as an equality; prove that the sequence of vectors (X, Y) is dependent.

B.47. Prove that the product of two or more orthogonal matrices of the same size is an orthogonal matrix.

B.48. Let X and Y be any two vectors in $V_p(R)$ such that $|X| = |Y| = 1$; prove that $|(X^TY)| \leq 1$.

B.49. Prove the Schwartz inequality for any two vectors X, Y in $V_p(R)$ (see Section B.17).

B.50. Prove that if Y is orthogonal to the sequence $S = (X_1, \cdots, X_r)$ then Y is orthogonal to $\langle S \rangle$.

B.51. Suppose that by a sequence of elementary operations performed on the p by n matrix A we obtain a matrix R with the property that the last $p - r$ rows of R are zero and columns $s_1, \cdots, s_r$ of R are the first r unit vectors in V_p. Define T to be the matrix whose row s_i is row i of R, $i = 1, \cdots, r$ and row h of T is zero if h is not in the basic sequence $S = (s_1, \cdots, s_r)$. Prove that $T^2 = T$ and hence that the $n - r$ nonzero columns of $I - T$ are a basis of $L^*(A)$. (In the simplex algorithm A ordinarily has rank $r = p$ and the matrix R corresponds to a problem in canonical form with respect to a basic sequence S; the defining property for R is that $R_S = I_p$.)

Flow Charts for the Graves-Thrall Algorithm

Appendix C

C.1. GENERAL FLOW CHART FOR THE ASSIGNMENT ALGORITHM

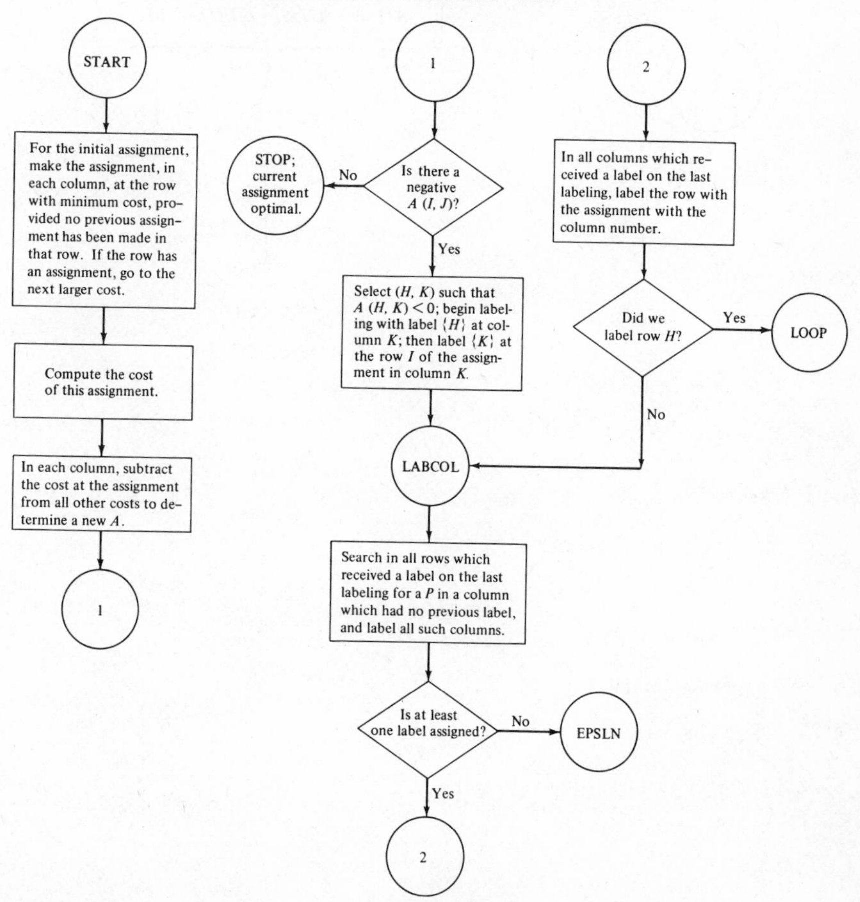

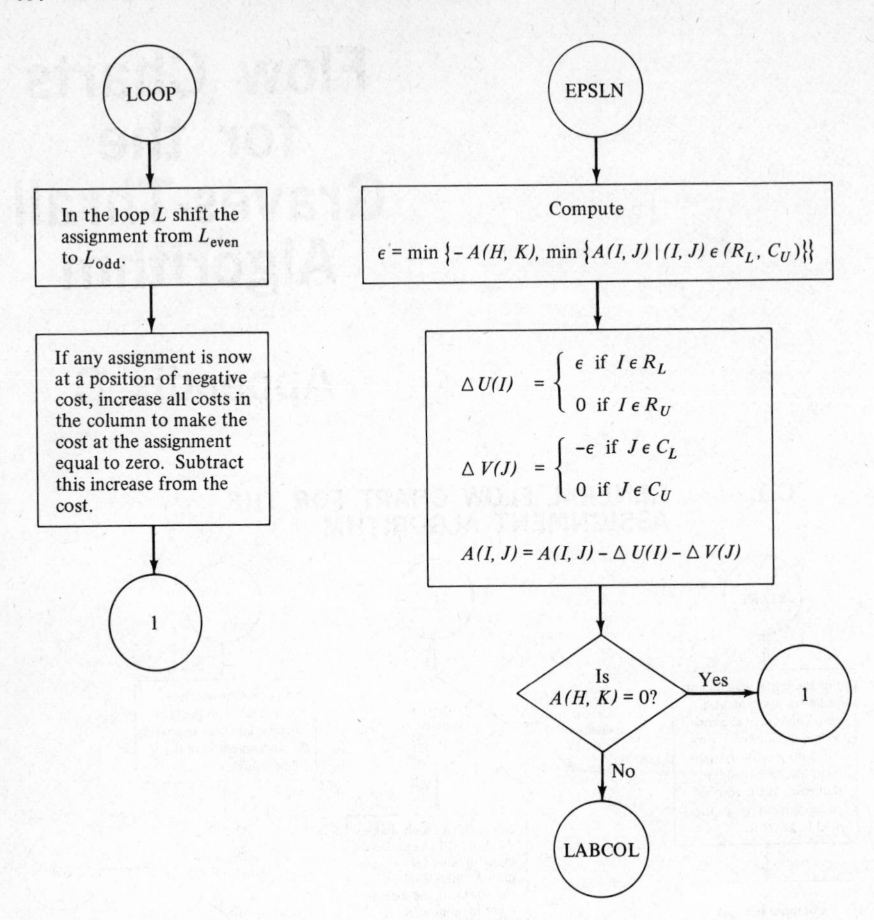
LOOP
In the loop L shift the assignment from L_{even} to L_{odd}.
If any assignment is now at a position of negative cost, increase all costs in the column to make the cost at the assignment equal to zero. Subtract this increase from the cost.
1
EPSLN
Compute
$\epsilon = \min \{-A(H, K), \min \{A(I, J) \mid (I, J) \in (R_L, C_U)\}\}$
$\Delta U(I) = \begin{cases} \epsilon & \text{if } I \in R_L \\ 0 & \text{if } I \in R_U \end{cases}$
$\Delta V(J) = \begin{cases} -\epsilon & \text{if } J \in C_L \\ 0 & \text{if } J \in C_U \end{cases}$
$A(I, J) = A(I, J) - \Delta U(I) - \Delta V(J)$
Is $A(H, K) = 0$?
Yes
1
No
LABCOL

C.2. DETAILED FLOW CHART FOR THE ASSIGNMENT ALGORITHM

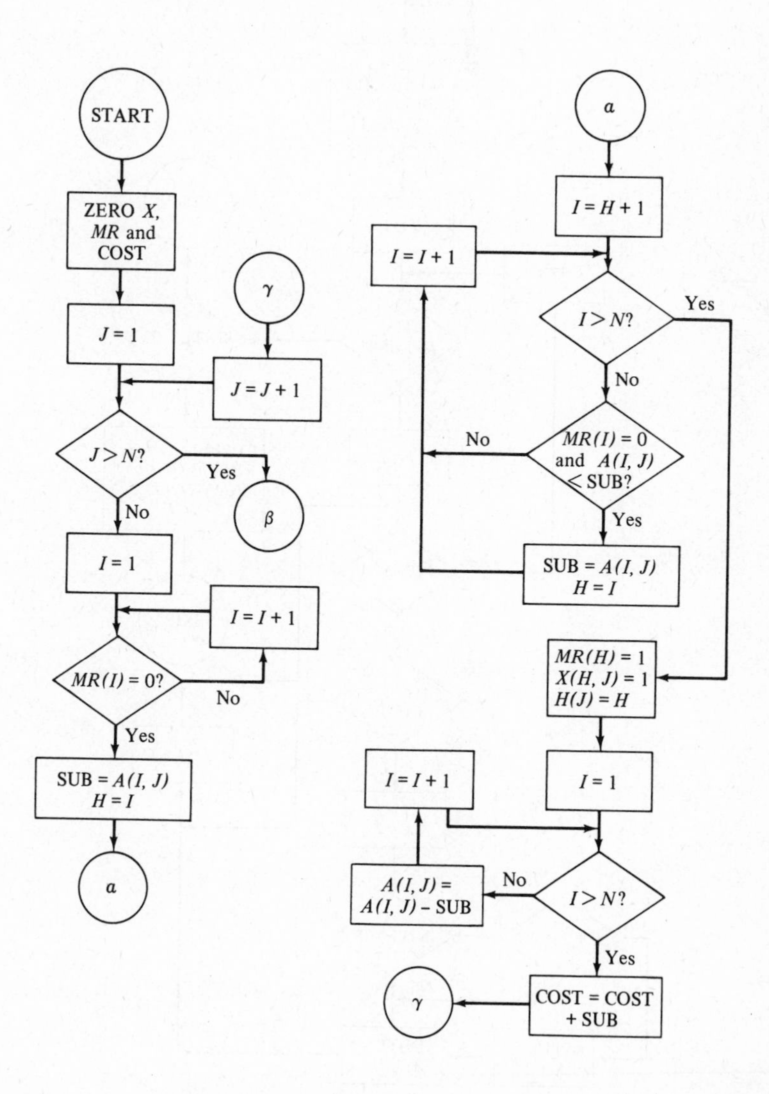

For each column J select a candidate (row H) for the min $A(I, J)$ of the column, then starting with this candidate search the rest of the column and find the smallest $A(I, J)$ in the column in a row without an assignment. Make the assignment, store its location in $H(J)$, and subtract $A(H, J)$ from each (AI, J) in this column. Add $A(H, J)$ to COST.

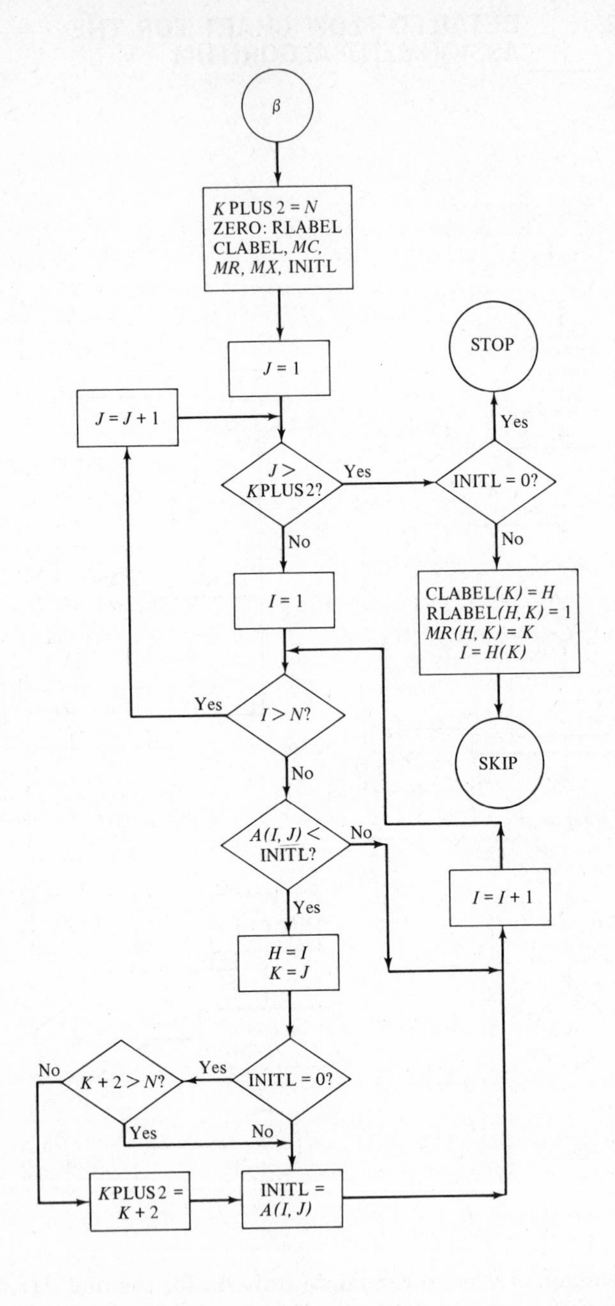

Search A for a negative element. Look in the next two columns for a smaller element. Call the starting position (H, K). If there is no negative element, we are at optimum. Otherwise label column K with H and row with assignment with K. Set MR to search next in row of assignment for a P position.

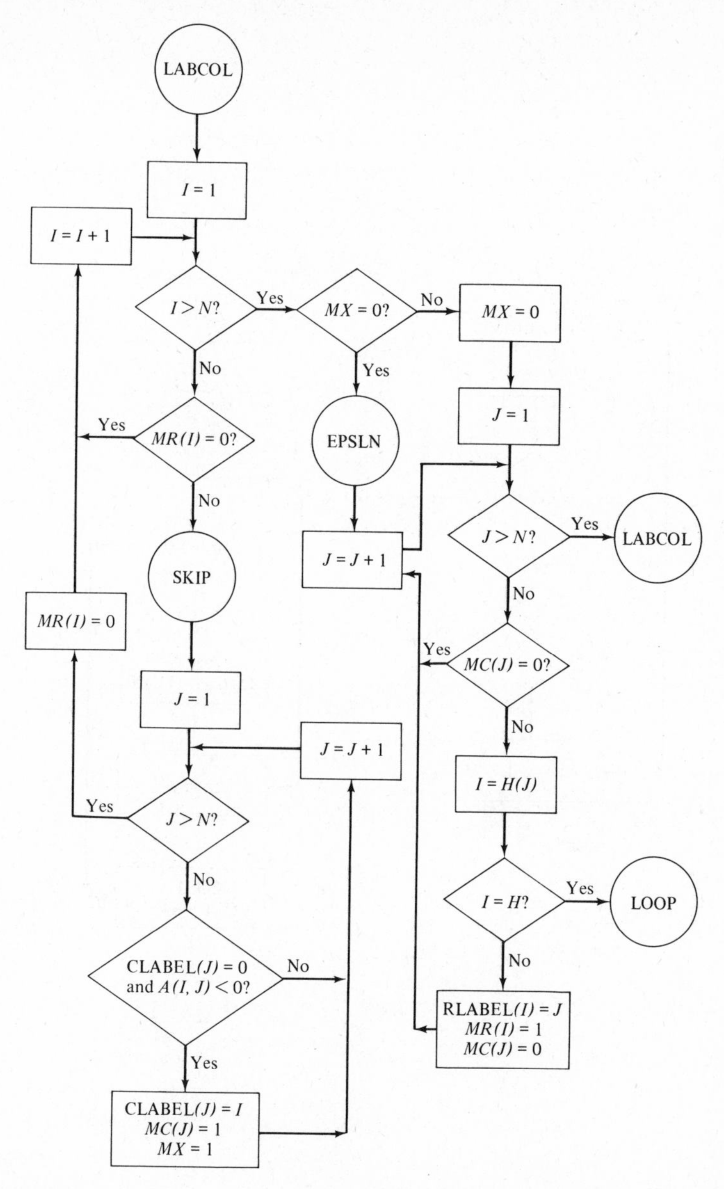

Check MR vector for a row to search in. If find a P in an unlabeled column, then label the column and set MC to locate the assignment in the column. After searching the row turn off row indicator. If no columns were labeled improve the dual vector (U, V). Reset indicator; scan MC for column or columns labeled on last pass. The row is labeled with the column number; the row to be labeled is stored in $H(J)$. We have a loop if the row is the row at which we started.

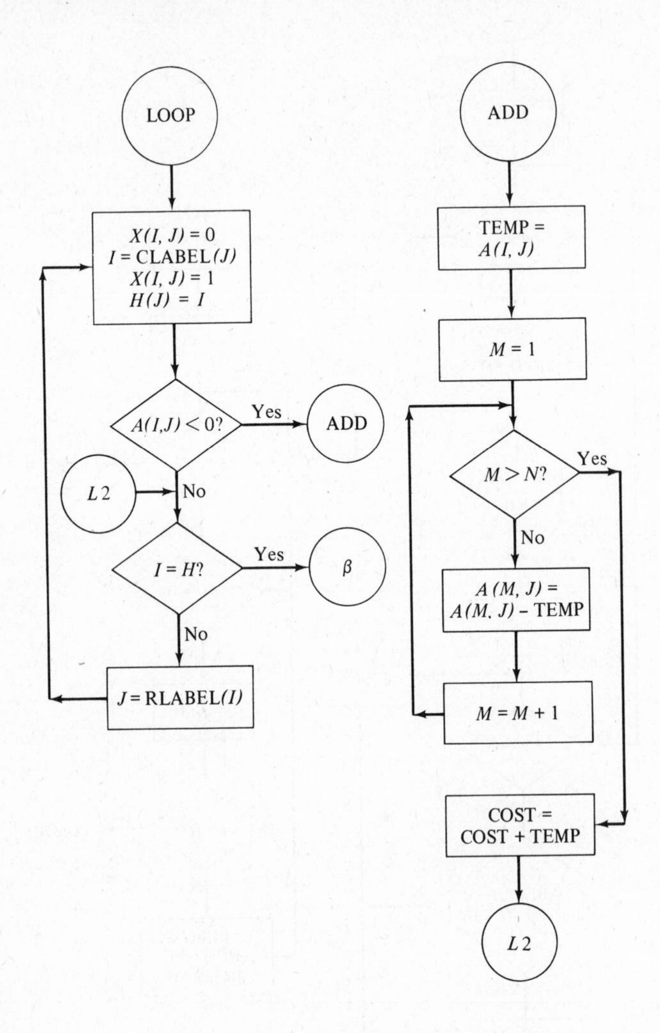

If a loop is found, follow the labels and shift the assignments from L_{even} to L_{odd}. Store row number of each assignment in $H(J)$. If a new assignment is made where $A(I, J) < 0$, go to ADD. When the loop is completed go back to β. If a new assignment is made where $A(I, J) < 0$, then subtract this $A(I, J)$ from each $A(M, J)$ in the column. COST is improved (reduced) by this assignment change.

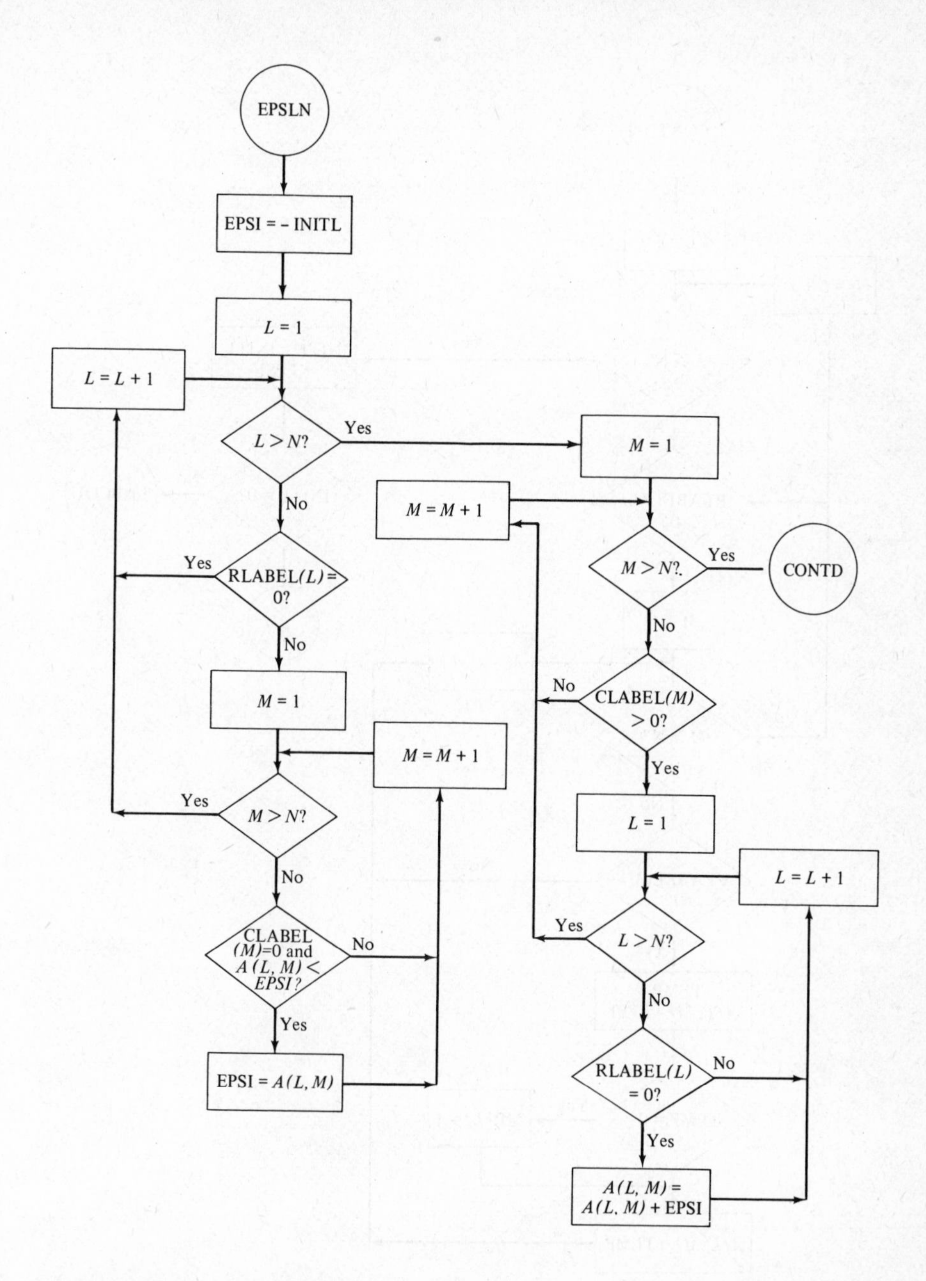

If no P's are found, then the dual must be improved to increase the number of P's. We have

$\epsilon = \min \{ -A(H, K), \min \{A(I, J) \mid (I, J) \in (R_L, C_U) \} \}$. Add epsilon to $A(L, M)$ at $\{ (L, M) \mid (L, M) \in (R_U, C_L) \}$.

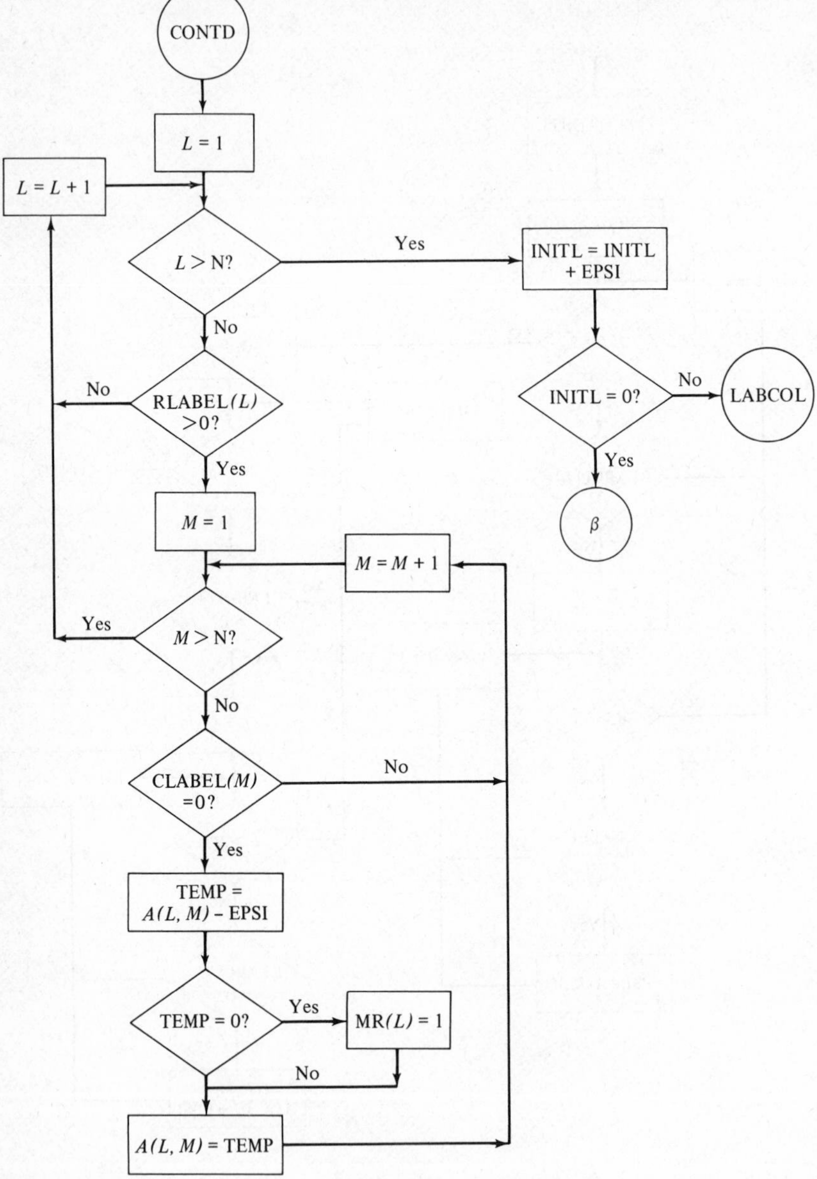

Subtract epsilon from $A(L, M)$ at $\{ (L, M) \mid (L, M) \in (R_L, C_U) \}$. If $A(L,M) = 0$, then indicate in $MR(L)$ to search the row on the next labeling. If $A(H, K) = 0$, go to β. Otherwise, continue the labeling.

C.3. FLOW CHART FOR THE GRAVES-THRALL CAPACITATED TRANSPORTATION ALGORITHM

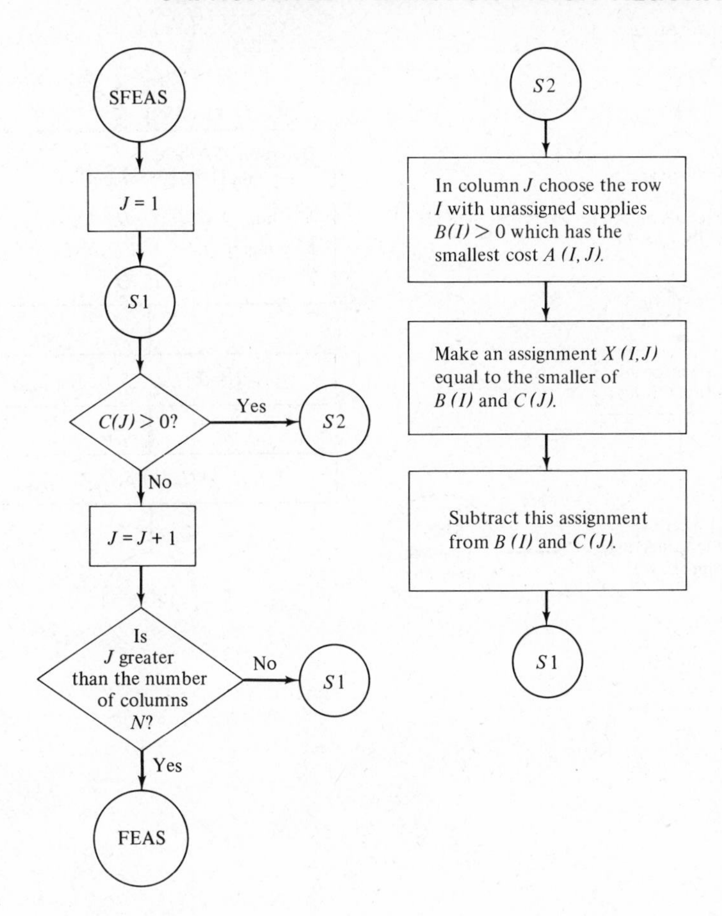

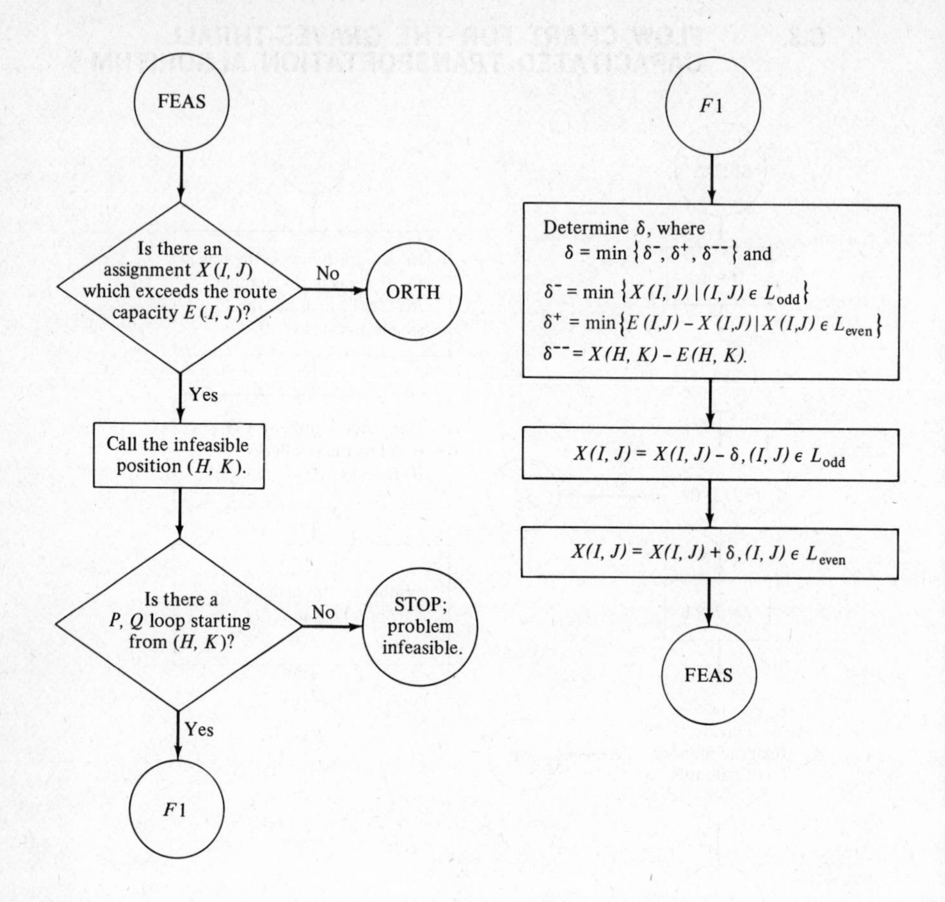
FEAS
Is there an assignment $X(I, J)$ which exceeds the route capacity $E(I, J)$?
No
ORTH
Yes
Call the infeasible position (H, K).
Is there a P, Q loop starting from (H, K)?
No
STOP; problem infeasible.
Yes
$F1$
$F1$
Determine δ, where
$\delta = \min\{\delta^-, \delta^+, \delta^{--}\}$ and
$\delta^- = \min\{X(I, J) \mid (I, J) \in L_{odd}\}$
$\delta^+ = \min\{E(I,J) - X(I,J) \mid X(I,J) \in L_{even}\}$
$\delta^{--} = X(H, K) - E(H, K)$.
$X(I, J) = X(I, J) - \delta, (I, J) \in L_{odd}$
$X(I, J) = X(I, J) + \delta, (I, J) \in L_{even}$
FEAS

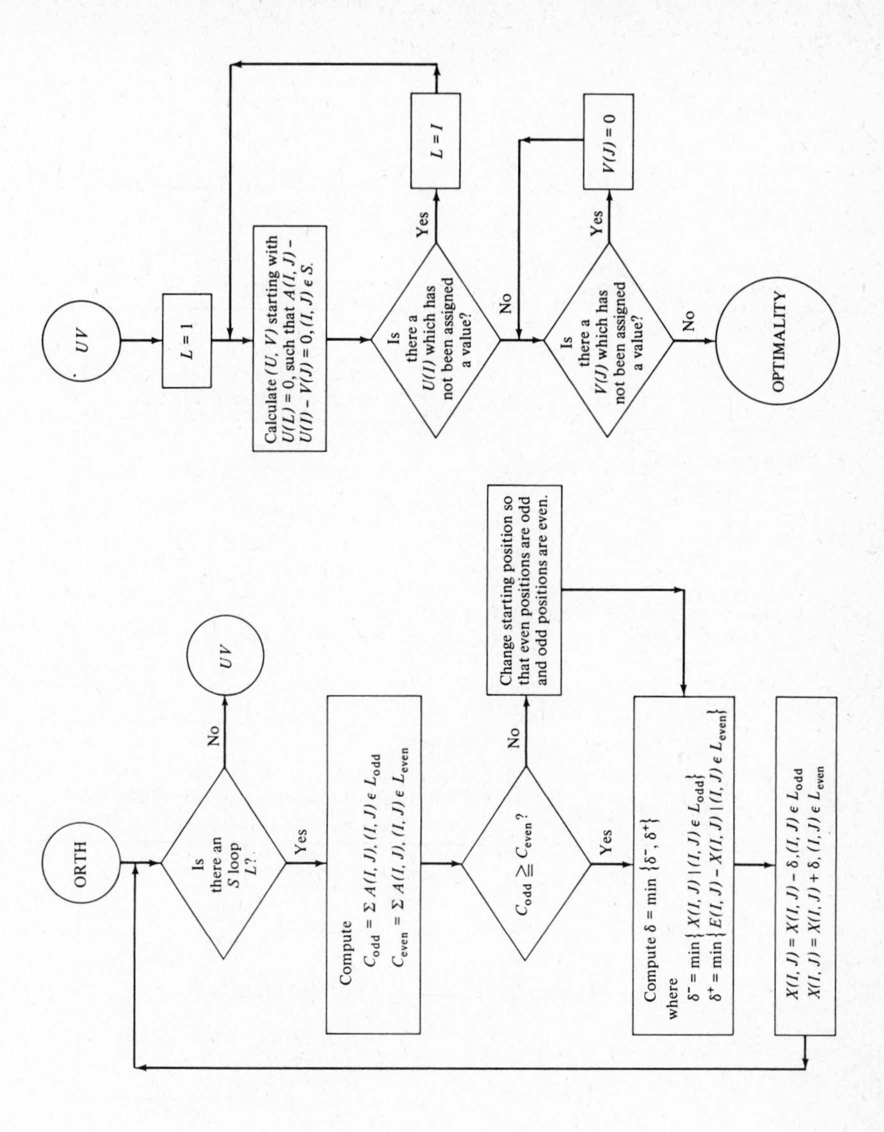

UV
L = 1
Calculate (U, V) starting with U(L) = 0, such that A(I, J) − U(I) − V(J) = 0, (I, J) ε S.
Is there a U(I) which has not been assigned a value?
Yes
L = I
No
Is there a V(J) which has not been assigned a value?
Yes
V(J) = 0
No
OPTIMALITY
ORTH
Is there an S loop L?
No
UV
Yes
Compute
C_odd = Σ A(I, J), (I, J) ε L_odd
C_even = Σ A(I, J), (I, J) ε L_even
C_odd ≧ C_even ?
No
Change starting position so that even positions are odd and odd positions are even.
Yes
Compute δ = min {δ⁻, δ⁺}
where
δ⁻ = min {X(I, J) | (I, J) ε L_odd}
δ⁺ = min {E(I, J) − X(I, J) | (I, J) ε L_even}
X(I, J) = X(I, J) − δ, (I, J) ε L_odd
X(I, J) = X(I, J) + δ, (I, J) ε L_even

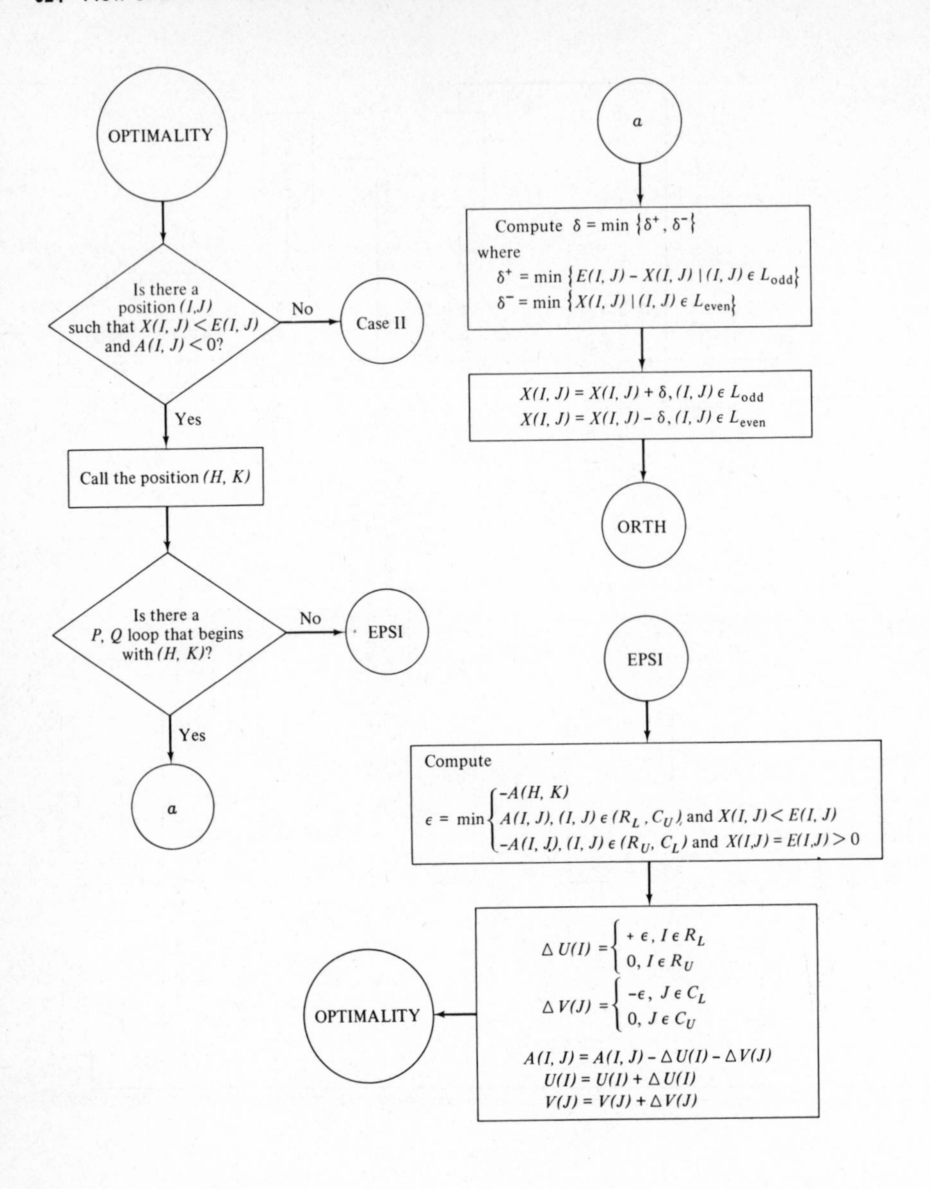

OPTIMALITY
Is there a position (I,J) such that $X(I, J) < E(I, J)$ and $A(I, J) < 0$?
No
Case II
Yes
Call the position (H, K)
Is there a P, Q loop that begins with (H, K)?
No
EPSI
Yes
a
a
Compute $\delta = \min \{\delta^+, \delta^-\}$
where
$\delta^+ = \min \{E(I, J) - X(I, J) \mid (I, J) \in L_{odd}\}$
$\delta^- = \min \{X(I, J) \mid (I, J) \in L_{even}\}$
$X(I, J) = X(I, J) + \delta, (I, J) \in L_{odd}$
$X(I, J) = X(I, J) - \delta, (I, J) \in L_{even}$
ORTH
EPSI
Compute
$\epsilon = \min \begin{cases} -A(H, K) \\ A(I, J), (I, J) \in (R_L, C_U) \text{ and } X(I, J) < E(I, J) \\ -A(I, J), (I, J) \in (R_U, C_L) \text{ and } X(I,J) = E(I,J) > 0 \end{cases}$
$\Delta U(I) = \begin{cases} +\epsilon, I \in R_L \\ 0, I \in R_U \end{cases}$
$\Delta V(J) = \begin{cases} -\epsilon, J \in C_L \\ 0, J \in C_U \end{cases}$
$A(I, J) = A(I, J) - \Delta U(I) - \Delta V(J)$
$U(I) = U(I) + \Delta U(I)$
$V(J) = V(J) + \Delta V(J)$
OPTIMALITY

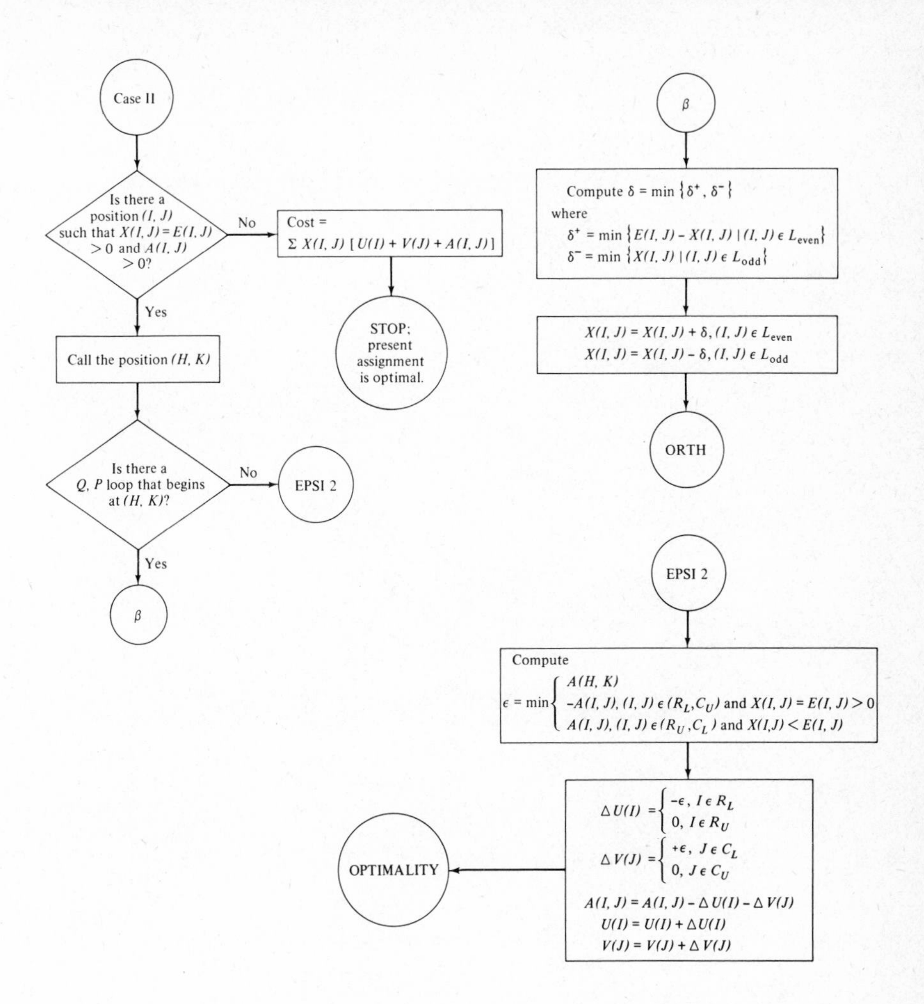
Case II
Is there a position (I, J) such that X(I, J) = E(I, J) > 0 and A(I, J) > 0?
No
Cost = Σ X(I, J) [U(I) + V(J) + A(I, J)]
STOP; present assignment is optimal.
Yes
Call the position (H, K)
Is there a Q, P loop that begins at (H, K)?
No
EPSI 2
Yes
β
β
Compute δ = min {δ⁺, δ⁻}
where
δ⁺ = min {E(I, J) − X(I, J) | (I, J) ε L_even}
δ⁻ = min {X(I, J) | (I, J) ε L_odd}
X(I, J) = X(I, J) + δ, (I, J) ε L_even
X(I, J) = X(I, J) − δ, (I, J) ε L_odd
ORTH
EPSI 2
Compute
ε = min { A(H, K); −A(I, J), (I, J) ε (R_L, C_U) and X(I, J) = E(I, J) > 0; A(I, J), (I, J) ε (R_U, C_L) and X(I,J) < E(I, J) }
ΔU(I) = { −ε, I ε R_L; 0, I ε R_U }
ΔV(J) = { +ε, J ε C_L; 0, J ε C_U }
A(I, J) = A(I, J) − ΔU(I) − ΔV(J)
U(I) = U(I) + ΔU(I)
V(J) = V(J) + ΔV(J)
OPTIMALITY

References

1. Balinski, M. L., Notes on a constructive approach to linear programming, *in* Dantzig, G. and A. F. Veinott, Jr., eds., "Mathematics of the Decision Sciences, Part I," American Mathematical Society, 1968.
2. Balinski, M. L. and R. E. Gomory, A mutual primal-dual simplex method, *in* Graves, R. L., and P. Wolfe, eds., "Recent Advances in Mathematical Programming," McGraw-Hill Book Co., 1963, 17–26.
3. ______, A primal method for the assignment and transportation problem, *Mgt. Science*, vol. 10, no. 3, April, 1964, 578–593.
4. Beale, E. M. L., Cycling in the dual simplex algorithm, *Naval Research Logistics Quarterly*, vol. 2, no. 4, December, 1955, 269–276.
5. ______, An algorithm for solving the transportation problem when the shipping cost over each route is convex, *Naval Research Logistics Quarterly*, vol. 6, no. 1, March, 1959, 43–56.
6. Beckenbach, E. F., ed., "Applied Combinatorial Mathematics," John Wiley and Sons, 1964.
7. Bellman, R., ed., "Mathematical Optimization Techniques," University of California Press, 1963.
8. Bellman, R. and M. Hall, Jr., "Proceedings of the Symposium on Applied Mathematics, Vol. X, Combinatorial Analysis," American Mathematical Society, 1960.
9. Charnes, A., W. W. Cooper, and A. Henderson, "An Introduction to Linear Programming," John Wiley and Sons, 1953.
10. Charnes, A. and W. W. Cooper, "Management Models and Industrial Applications of Linear Programming," Vol. I and Vol. II, John Wiley and Sons, 1961.
11. Dantzig, G., A proof of the equivalence of the programming problem and the game problem, *in* T. C. Koopmans, ed., "Activity Analysis of Production and Allocation," John Wiley and Sons, 1951, 330–335.
12. ______, "Notes on Linear Programming: Parts VIII, IX, X—Upper Bounds, Secondary Constraints, and Block Triangularity in Linear

Programming." Research Memorandum RM-1367, The RAND Corporation, 1954.

13. ———, "Linear Programming and Extensions," Princeton University Press, 1963.
14. Dantzig, G., D. R. Fulkerson, and S. M. Johnson, Solution of a large scale traveling salesman problem, *Jour. Opns. Res. Society of America*, vol. 2, no. 4, November, 1954, 393–410.
15. Dantzig, G. and A. F. Veinott, Jr., eds., "Mathematics of the Decision Sciences, Part I," American Mathematical Society, 1968.
16. Dantzig, G. and P. Wolfe, Decomposition principle for linear programs, *Operations Research*, vol. 8, no. 1, January–February, 1960, 101–111.
17. Engel, James F. and M. W. Warshaw, Allocating advertising dollars by linear programming, *Journal of Advertising Research*, vol. 4, no. 3, September, 1964, 42–48.
18. Ford, L. R. and D. R. Fulkerson, "Flows in Networks," Princeton University Press, 1963.
19. Fulkerson, D. R., An out-of-kilter method for minimal cost flow problems, *Journal of the Society for Industrial and Applied Mathematics*, vol. 9, no. 1, March, 1961, 18–27.
20. Godfrey, J. T., W. A. Spivey, and G. Stillwagon, Production and market planning with parametric programming, *Industrial Management Review*, vol. 10, no. 1, Fall, 1968, 61–76.
21. Graves, G. W., A complete constructive algorithm for the general mixed linear programming problem, *Naval Research Logistics Quarterly*, vol. 12, no. 1, March, 1965, 1–34.
22. Graves, G. W. and M. R. Thrall, "Capacitated Transportation Problem with Convex Polygonal Cost," RAND Research memorandum RM-4941-PR, The RAND Corporation, April, 1966.
23. Graves, R. L., and P. Wolfe, eds., "Recent Advances in Mathematical Programming," McGraw-Hill Book Co., 1963.
24. Hoffman, A. J., Some recent applications of the theory of linear inequalities to extremal combinatorial analysis, *in* Bellman, R. and M. Hall, Jr., "Proceedings of the Symposium on Applied Mathematics, Vol. X, Combinatorial Analysis," American Mathematical Society, 1960.
25. Jewell, W. S., "Optimal Flow Through Networks," Interim technical report No. 8 on fundamental investigations in methods of operations research, Massachusetts Institute of Technology, Cambridge, Mass., 1958.
26. Koopmans, T. C., ed., "Activity Analysis of Production and Allocation," John Wiley and Sons, 1951.
27. Kuhn, H. W., The Hungarian method for the assignment problem, *Naval Research Logistics Quarterly*, vol. 2, 1955, pp. 83–97.

28. Kuhn, H. W. and A. W. Tucker, eds., "Contributions to the Theory of Games," Vol. I, Annals of Mathematics Study No. 24, Princeton University Press, 1950.
29. ———, "Linear Inequalities and Related Systems," Annals of Mathematics Studies, 38, Princeton University Press, 1956.
30. Lee, Shao-ju, "An Experimental Study of Transportation Algorithms," M. S. thesis, Graduate School of Business Admin., Univ. of California at Los Angeles, 1968.
31. Lemke, C. E., The dual method of solving the linear programming problem, *Naval Research Logistics Quarterly*, vol. 1, 1954, 36–47.
32. Lucas, W. F., A game with no solution, *Bulletin of the American Mathematical Society*, vol. 74, no. 2, March, 1968, 237–239.
33. Luce, R. D. and H. Raiffa, "Games and Decisions: Introduction and Critical Survey," John Wiley and Sons, 1957.
34. Milnor, John, Games against nature, Chap. IV *in* Thrall, R. M., C. Coombs, and R. Davis, eds., "Decision Processes," John Wiley and Sons, 1954.
35. Reinfeld, N. V. and W. R. Vogel, "Mathematical Programming," Prentice-Hall, Inc., 1958, pp. 59–70.
36. Shapley, L. S. and R. N. Snow, Basic solutions of discrete games, *in* Kuhn, H. W. and A. W. Tucker, eds., "Contributions to the Theory of Games, Vol. I," Annals of Mathematics Study No. 24, Princeton University Press, 1950, pp. 27–35.
37. Stigler, G. J., The cost of subsistence, *Journal of Farm Economics*, vol. 27, no. 2, May, 1945, pp. 303–314.
38. Thrall, R. M., C. Coombs, and R. Davis, eds., "Decision Processes," John Wiley and Sons, 1954.
39. Tucker, A. W., Dual systems of homogeneous linear relations, *in* Kuhn, H. W. and A. W. Tucker, eds., "Linear Inequalities and Related Systems," Annals of Mathematics Studies, 38, Princeton University Press, 1956.
40. ———, Solving a matrix game by linear programming, *IBM Journal of Research and Development*, vol. 4, no. 5, November, 1960, pp. 507–517.
41. ———, A combinatorial equivalence of matrices, *in* Bellman, R. and M. Hall, Jr., "Proceedings of the Symposium on Applied Mathematics, Vol. X, Combinatorial Analysis," American Mathematical Society, 1960, 129–140.
42. Tucker, A. W., Combinatorial theory underlying linear programs, *in* Graves, R. L., and P. Wolfe, eds., "Recent Advances in Mathematical Programming," McGraw-Hill Book Co., 1963, pp. 1–16.
43. ———, Simplex method and theory, *in* Bellman, R., ed., "Mathematical Optimization Techniques," University of California Press, 1963, pp. 213–221.

44. ______, Combinatorial algebra of matrix games and linear programs, *in* Beckenbach, E. F., ed., "Applied Combinatorial Mathematics," John Wiley and Sons, 1964, pp. 320–347.
45. Vajda, S., "Mathematical Programming," Addison-Wesley Publishing Co., 1961.
46. von Neumann, J. and O. Morgenstern, "Theory of Games and Economic Behavior," Princeton University Press, 1944.
47. Williams, A. C., "Complementarity Theorems for Linear Programming," January, 1969, p. 5, unpublished.
48. Willner, L. B., On parametric linear programming, *SIAM Journal on Applied Mathematics*, vol. 15, no. 5, September, 1967, 1253–1257.
49. Zukhovitskiy, S. I., and L. I. Avdeyeva, "Linear and Convex Programming," W. B. Saunders Co., 1966.

Index